Mammalian Parenting

MAMMALIAN PARENTING

Biochemical, Neurobiological, and Behavioral Determinants

Edited by
NORMAN A. KRASNEGOR
ROBERT S. BRIDGES

New York   Oxford
OXFORD UNIVERSITY PRESS
1990

Oxford University Press

Oxford   New York   Toronto
Delhi   Bombay   Calcutta   Madras   Karachi
Petaling Jaya   Singapore   Hong Kong   Tokyo
Nairobi   Dar es Salaam   Cape Town
Melbourne   Auckland

and associated companies in
Berlin   Ibadan

Published by Oxford University Press, Inc.,
200 Madison Avenue, New York, New York 10016

Oxford is a registered trademark of Oxford University Press

Library of Congress Cataloging-in-Publication Data
Mammalian parenting : biochemical, neurobiological,
and behavioral determinants /
Norman A. Krasnegor, Robert S. Bridges, editors.
p.   cm.
Proceedings of the National Institute of Child Health
and Human Development Conference on
"Biological and Behavioral Determinants of Parental Behavior in Mammals,"
held in Leesburg, Va., Sept. 8–11, 1987.
Includes bibliographies and index.
ISBN 0-19-505600-0
1. Mammals—Behavior—Congresses.   2. Parental behavior in animals—Congresses.
3. Parenting—Congresses.
I. Krasnegor, Norman A.   II. Bridges, Robert S.
III. National Institute of Child Health
and Human Development Conference
on "Biological and Behavioral Determinants of
Parental Behavior in Mammals" (1987 : Leesburg, Va.)
IV. National Institute of Child Health
and Human Development (U.S.)
[DNLM: 1. Mammals—congresses.   2. Maternal Behavior—congresses.
3. Paternal Behavior—congresses.   WS 105.5.F2 M265 1987]
QL739.3.M34   1990 599'.056—dc19
DNLM/DLC   for Library of Congress
88-37152

1 2 3 4 5 6 7 8 9

Printed in the United States of America
on acid-free paper

To Our Parents

*Abraham J. Krasnegor and Julia Krasnegor
and Philip Barnet and Roberta Barnet
N.A.K.*

*Stafford E. Bridges and Marion W. Bridges
R.S.B.*

Acknowledgements

In addition to being subjected to the editors' scrutiny, each chapter was reviewed anonymously by the contributors and other readers. We gratefully acknowledge the assistance of the following individuals, who served as readers: Robert Barbieri, Mary Erskine, Michael Selmanoff, and Judith Stern.

Contents

Contributors

Jeffrey R. Alberts, Ph.D.
Department of Psychology
Indiana University
Bloomington, Indiana 47405

Laura J. Beasley, Ph.D.
Department of Psychobiology
University of California
Irvine, California 92717

Robert S. Bridges, Ph.D.
Department of Anatomy and Cellular Biology
Laboratory of Human Reproduction
    and Reproductive Biology
Harvard Medical School
Boston, Massachusetts 02115

Susan A. Brunelli, Ph.D.
Department of Developmental Psychobiology
New York State Psychiatric Institute
New York, New York 10032

Christopher L. Coe, Ph.D.
Wisconsin Regional Primate Research Center
Madison, Wisconsin 53715

Robert Coopersmith, Ph.D.
Department of Psychobiology
University of California
Irvine, California 92717

W. R. Crowley, Ph.D.
Department of Pharmacology
University of Tennessee Health
    Science Center
Memphis, Tennessee 38163

Martin Daly, Ph.D.
Department of Psychology
McMaster University
Hamilton L85 4K1
Ontario, Canada

Leon Eisenberg, M.D.
Department of Social Medicine
    and Health Policy
Harvard Medical School
Boston, Massachusetts 02115

Alison S. Fleming, Ph.D.
Department of Psychology
Erindale College
University of Toronto in Mississauga
Mississauga L5L 1C6
Ontario, Canada

Roger A. Gorski, Ph.D.
Chairman, Department of Anatomy
UCLA School of Medicine
Los Angeles, California 90024

C. E. Grosvenor, Ph.D.
Department of Molecular and Cell Biology
401 Althouse Labs
Pennsylvania State University
University Park, Pennsylvania 16802

David J. Gubernick, Ph.D.
Department of Psychology
Indiana University
Bloomington, Indiana 47405

Glenn I. Hatton, Ph.D.
Department of Psychology
Michigan State University
East Lansing, Michigan 48824

Myron Hofer, M.D.
New York State Psychiatric Institute
New York, New York 10032

Warren G. Holmes, Ph.D.
Psychology Department
University of Michigan
Ann Arbor, Michigan 48109

Thomas R. Insel, M.D.
Laboratory of Clinical Science
National Institute of Mental Health
Poolsville, Maryland 20837

K. M. Kendrick, Ph.D.
Department of Anatomy
Cambridge University-Downing Site
Cambridge CB2 3DY
England

E. B. Keverne, Ph.D.
Sub Department of Animal Behavior
Madingley
Cambridge CB3 8AA
England

Craig H. Kinsley, Ph.D.
Department of Psychology
Richmond University
Richmond, Virginia 23173

Norman A. Krasnegor, Ph.D.
National Institute of Child Health
  and Human Development
National Institutes of Health
Bethesda, Maryland 20892

Michael Leon, Ph.D.
Department of Psychobiology
University of California
Irvine, California 92717

Frederic Levy, Ph.D.
Laboratoire de Comportement
INRA
Nouzilly 37380
France

Barbara K. Modney
Department of Psychology
Michigan State University
East Lansing, Michigan 48824

Michael Numan, Ph.D.
Department of Psychology
Boston College
Chestnut Hill, Massachusetts 02167

Pascal Poindron, Ph.D.
Laboratoire de Comportement
INRA
Nouzilly 37380
France

Jay S. Rosenblatt, Ph.D.
Institute of Animal Behavior
Rutgers University
Newark, New Jersey 07102

G. V. Shah, Ph.D.
Department of Physiology and Biophysics
University of Tennessee
Memphis, Tennessee 38163

Barbara Shortle, M.D.
Department of Obstetrics and Gynecology
Lenox Hill Hospital
New York, New York 10021

Regina M. Sullivan, Ph.D.
Department of Psychobiology
University of California
Irvine, California 92717

Bruce B. Svare, Ph.D.
Department of Psychology
State University of New York
Albany, New York 12220

Michelle P. Warren, M.D.
St. Luke's Roosevelt Hospital Center
Antenucci Research Building
New York, New York 10019

Michael W. Yogman, M.D.
14 Wyman Road
Cambridge, Massachusetts 02138

Mammalian Parenting

Biochemical, Neurobiological, and Behavioral Determinants of Parenting in Mammals: An Introduction

NORMAN A. KRASNEGOR

This book deals with parenting in the class Mammalia: those behavioral patterns displayed characteristically, but not exclusively, by females during pregnancy, at parturition, and during lactation. More specifically, it focuses on research advances that shed light on biological and behavioral triggers and regulators of caregiving by parents to their newborn offspring. The book is divided into five sections: (1) perspectives, (2) biochemical correlates, (3) neurobiological correlates, (4) biobehavioral correlates, and (5) future directions.

Five main themes emerge from the chapters and are interwoven among them. They relate to evolutionary, hormonal, neurobiological, sensory, and experiential factors. These themes help to bind and unify the book into a conceptual whole and illustrate common mechanisms that underlie parenting among mammals.

EVOLUTIONARY THEMES

Students of behavior, psychologists and biologists alike, systematically study a wide variety of patterns among vertebrates that can be viewed as directed to the care and sustenance of their offspring. Among the Vertebrata, the Mammalia have evolved these patterns to such an extent that care and nourishment of the young have become two of the central inclusion criteria that taxonomists employ in defining the class as a whole. Echidnas, marsupials, and placental mammals, for example, though they each employ different strategies for reproduction and fetal development, all exhibit variants of care for their young, initiated at different times after parturition, that qualify them as members of the same class.

Modern biological theory focuses on caregiving in mammals as an evolutionary strategy that differs from reproductive strategy in which the metabolic cost of producing many young is pitted against survival. Caregiving trades off the lower metabolic cost of producing few young against the likelihood of greater reproductive fitness associated with the parents' investment of their time and energy in sustaining their offspring (see chapters 3, 4, 9, and 21). Of the mammals, the primates most clearly exhibit this caregiving strategy, which is commonly referred to as "parenting." Undoubtedly, parenting behavior has contributed to the reproductive success and other behavioral capacities that have helped humans to evolve, in a relatively short time span, to become the dominant mammalian form on earth.

The study of mechanisms, both biological and behavioral, that influence and regulate parenting is therefore essential to help gain a better understanding of ourselves. The knowledge gained from such research may materially aid in controlling some of the factors that have created population pressures that in some countries have resulted in public policies limiting family size (see chapter 2). In contrast, this era of high technology has witnessed the development of new approaches that enable persons once destined to be childless to become parents. Thus in vitro fertilization, sperm banking, and surrogate parenting, for example, have provided new hope for such couples but, in turn, have raised a number of moral, legal, ethical, and social questions.

HORMONAL THEMES

During the past decade, much new knowledge has been gained about the biochemical, neurobiological, and biobehavioral mechanisms that are considered necessary for the initiation and regulation of parenting in general and maternal behaviors in particular. Historically, progress was made in this area partly because of the establishment of effective animal models (see chapter 4). These have served as the frameworks within which empirical studies have been conducted by investigators with interdisciplinary expertise working in the fields of endocrinology, neuroanatomy, neurochemistry, sensory physiology, and behavior analysis. Through these efforts, they have begun to discover and unravel the mechanisms by which species-specific patterns of parental care are regulated.

Behavioral neuroscientists and developmental psychobiologists have revolutionized our understanding of parenting by making basic observations on specific hormones and their sites of action in the brain that are necessary for the initiation of caregiving behaviors in rodents and sheep. These investigators have described the roles of steroids (estradiol and progesterone), an anterior pituitary hormone (prolactin), and a neurohypophysial peptide (oxytocin) in triggering maternal patterns associated with pregnancy and parturition (see chapters 6, 7, 8, 13, and 14). In female rats, for example, these behaviors include building nests, retrieving pups, crouching over the litter, and anogenital licking. In sheep, the initial maternal repertoire includes avid licking of the lamb, emission of soft or low-pitched bleats, and acceptance of the lamb at the udder.

Research has also been carried out to characterize hormone–behavior relationships during lactation and at weaning (see chapters 6, 15, and 16). The results obtained demonstrate both the role of prolactin in these phases of parenting in rats and brain changes associated with this hormone—changes that disappear once the pups are weaned. In nonhuman primates and human females, there are a number of changes in the hormonal profile during these periods. However, the direct relationships between hormones and parental behavior observed in rats and sheep are not observed in these primates (see chapters 9, 10, and 11).

CENTRAL NERVOUS SYSTEM THEMES

Work on the central nervous system of rats has revealed that certain brain structures are sexually dimorphic (see chapter 5). The behavioral correlates of these structural dimorphisms raise the possibility that parenting may be similarly subsumed by unique neuroanatomical systems in the female that are configured differently from those in the males. Extensive investigations have already been carried out to characterize the neural substrate for parenting behaviors, as described above, for female rats (see chapter 12). Other, more recent data indicate that in rodents in which both male and female parents care for the young, brain structures of the male decrease in size to approximate those of the female (see chapter 20). Anatomical changes in brain areas associated with motherhood are elegantly detailed in female rats (see Chapter 15). These studies demonstrate the dynamic nature of neuronal connectivity in postpartum, lactating mothers.

SENSORY THEMES

Sensory factors have also been shown to be necessary for initiating parental behavior. Both rodents and sheep employ olfactory cues to stimulate their behavioral repertoire of caregiving (see chapters 8, 14, and 21). Neuroendocrine systems are also affected by odors and vomeronasal stimulation, as evidenced by changes in neural connections associated with the secretion of oxytocin (see chapter 15). Tactile stimulation has also been shown to be important in triggering maternal behavior in sheep. Thus data have been collected that support the conclusion that stimulation of the cervix in female sheep is a necessary cue for maternal behavior to appear (see chapters 8 and 14). Studies have also shown that thermal sensitivity of the mother rat and her litter can exert exquisite control over the contact between mother and offspring during lactation. The data suggest that the heating of the maternal brain is the mechanism at work (see chapter 19).

EXPERIENTIAL THEMES

Another uniquely important factor in the expression of parenting is experience. Research employing animal models has shown that environmental factors during the perinatal phase of development can affect parenting (see chapter 17). Moreover, studies reveal that juvenile rats acquire aspects of their adult repertoire of caregiving in association with play behaviors displayed when they are juveniles (see chapter 18).

Other findings demonstrate convincingly that parity (the number of offspring previously born) influences the latency and effectiveness of parental behaviors exhibited by the mother (see chapters 6 and 8). Prior experience with birth and delivery is of particular importance for nonhuman primates and human females (see chapters 9, 10, and 22).

The overall objective of the scheme outlined here is to make the book a structurally integrated whole and the individual chapters useful for investigators, clinicians, and students who are interested in this field of research.

I

PERSPECTIVES ON PARENTAL BEHAVIOR

The four chapters in this part provide viewpoints that are both theoretically broad and empirically rigorous, foster an appreciation for the field of mammalian parenting, and establish a context for the themes that appear throughout the remainder of the book. The first perspective is offered by Leon Eisenberg, who poses four questions:

1. Why study parenting at all?
2. Has anything been learned from clinical studies of abnormal parenting?
3. In what sense, if any, are animal studies relevant?
4. Does what the experts believe matter to parents?

These questions and their answers help define an array of clinical, social policy, and scientific issues related directly to investigations of mammalian parenting.

The second chapter, by Martin Daly, presents an overview of modern evolutionary theory as it relates to parental behavior. Daly focuses on the ideas of parent–offspring conflict, reproductive value, and discriminative parental solicitude as critical issues in the metatheory known as "evolutionary" or "Darwinian psychology." A central purpose is to show how this theoretical perspective can be a rich source of ideas in posing empirical questions concerning motivational systems that are deemed necessary for governing parenting behavior.

The next chapter, by Jay Rosenblatt, provides a historical perspective on research on the psychobiology of maternal behavior. Rosenblatt outlines the advances made in understanding the biological and behavioral factors that regulate the prepartum onset of maternal behavior and the motivational bases of maternal behavior in rats. He reviews experimental work on the hormonal bases of maternal behavior, as well as the sensory factors necessary for its expression and control. Rosenblatt's review of recent empirical findings on physiological aspects of maternal behavior demonstrates how far the field has come and the directions it is likely to take in the future.

Roger Gorski's chapter describes empirical findings that have revealed certain areas of the rat brain that are structurally differentiated in males and females of the species. The work of this author and others has demonstrated neuroanatomical differences in the sexually dimorphic nucleus of the preoptic area (SDN-POA) and the medial preoptic nucleus (MPN) of the medial preoptic area (MPOA). Gorski asserts that sexual differentiation, which in the rat occurs perinatally, has widespread effects on the developing brain. Brain sexual dimorphism, in turn, affects the expression of sexual behavior. This chapter provides an important background for the reader because it suggests the possibility that the discoveries related to structural dimorphisms, neural plasticity, and sexual behavior may also be true for parenting (see chapters 12, 15, and 20). This chapter therefore reveals the direction of future research, which, in this, the decade of the brain, may help to set a new agenda for the neurobiological study of parenting behavior.

2

The Biosocial Context
of Parenting in Human Families

LEON EISENBERG

Students of behavioral development start from the premise that research on parenting in humans and other animals is a legitimate and important field of inquiry. It may nonetheless be useful to take a step back and examine the grounds for that belief. To that end, four questions merit consideration:

1. Why study parenting at all?
2. Has anything been learned from clinical studies of abnormal parenting?
3. In what sense, if any, are animal studies relevant?
4. Does what the experts believe matter to parents?

WHY STUDY PARENTING AT ALL?

Parents have reared children without expert advice during the greatest part of human history. Indeed, it can be argued that parents have succeeded all too well; the human population now exceeds 5 billion and still counting, having tripled during this century (Bouvier, 1984)! This biblical injunction to "be fruitful and multiply," prudent advice on demographic policy from the Lord when our species numbered two, has become hazardous to our health. Perhaps a monograph on contraception might have been more appropriate than one on parenting. After all, it is self-evident that the prodigious growth in world population represents "success" in rearing as well as bearing children.

Then, in what meaningful sense do parents have to understand more clearly what they are doing? The obvious answer is that women and men the world over are concerned with optimizing the quality and not merely the quantity of human life. And it is quality that is at risk, in part because of that very increase in quantity (Blake, 1981a). Indeed, as clinical studies make evident, planning for parenthood— that is, the appropriate use of contraception to space childbirth and limit family size— has become a sign of good parenting. What gives research on parenting particular urgency today is the extent to which family structure is undergoing rapid and, in some cases, cataclysmic transformation.

For most of human history, men and women have lived in traditional societies characterized by glacial rates of social change (except in the face of local catastrophes). Under such circumstances, continuities in the structure of the community and its belief system ensured the transmission of culture-specific modes of child rearing from one generation to the next through precept, modeling, and social reinforcement. The mode of rearing in such stable societies produced adults adapted for the roles differentiated within each because those modes had been selected for in the course of social evolution. When a traditional way of life is interrupted and new adult behaviors are demanded for survival, not only is the taken-for-granted transmission of the old ways interrupted, but the old ways may no longer be adaptive for the new demands. Consider the most prominent features of change in the least developed, the rapidly developing, and the developed countries in recent decades.

In the less developed countries (LDCs) in Africa and Latin America, family life is undergoing major disruption. Whole villages have had to be abandoned in the course of civil wars; the populace has been translocated, whether by fleeing from a free-fire zone or by being forced to evacuate under military command (Clay & Holcomb, 1985; Rule, 1987), to areas where the old way of life can no longer be maintained. In South Africa and the countries adjacent to it, men have had to abandon their villages to find work in mines and industries far from home, with wives and children left behind because of apartheid. Mass migration to urban conurbations (Bouvier, 1984) in search of greater opportunity has generated enormous peripheral slums, lacking sanitation, a water supply, and transportation and characterized by staggering infant and child mortality rates. Life circumstances make it impossible for parents to rear their children as they had been reared. The wonder is not that so many children fail to survive but that so many do.

The People's Republic of China (PRC) exemplifies the rapid pace of development in what is still an LDC. The government's response to what had become an unsustainable rate of population growth is the "one-child-family" policy (Tien, 1983). Many urban families are now one-child families. Although the success of the policy has been far greater in the city than in the countryside, the total number of births in the PRC fell from 27 to 21 million between 1970 and 1981. By the second generation, to the extent that official policy is implementable, each only child will have no cousins, no uncles, and no aunts; only children who marry will carry the responsibility for four aging parents and one child. Those adults will not have had the opportunity to learn parenting by caring for sibs. So drastic a change will have an extraordinary impact on a culture whose traditions rest on a set of reciprocal intergenerational obligations and expectations within the large extended family (Kang, 1985).

Data from the United States highlight the rapid change in family life in developed countries. Over the past 30 years, because of a steadily declining birth rate, the total fertility rate per woman has fallen from 3.68 to 1.8. The proportion of the population under 15 has declined by 28%, whereas the proportion 65 and older has risen by 54% (Preston, 1984). Correspondingly, the political competition between the two groups of dependents for welfare transfer payments has shifted in favor of the elderly, to the detriment of the young. Half of all mothers with a child younger than 1 year are in the labor force, 50% more than the proportion employed outside the home a decade ago. The sexual "revolution" of the 1970s (a 60% increase between 1971 and 1979 in the percentage of teenagers engaging in premarital intercourse), coupled with failure to use contraception consistently (Zelnik & Kantner,

1980), has given the United States the dubious distinction of being first among the industrial countries in the teenage birth rate *despite* the fact that it also has the highest teenage abortion rate (Jones et al., 1985). As a consequence of out-of-wedlock births and a tripling in the divorce ratio over the past 15 years, almost one in four of the nation's 62 million children under 18 live with only one parent, almost always the mother. Of the children living in such female-headed households, more than one-third are reared in poverty; for black families, the proportion living in poverty is twice as high (U.S. Bureau of the Census, 1987).

These transformations in family life are having an enormous effect on the physical and mental health of populations. In the LDCs, infant mortality rates are staggering—more than an order of magnitude higher than those in industrial countries. The fundamental cause of this holocaust of the young is environmental: lack of a pure water supply and sanitary waste disposal; pervasive malnutrition; poorly ventilated, overcrowded housing; and lack of access to the rudiments of health care. However, it also reflects an absence of opportunity for parents to learn elementary hygienic practices. Comparative analysis of national infant mortality rates reveals that they are not a simple function of gross national product (GNP) per capita. For example, the Indian state of Kerala, Sri Lanka, and the PRC have *half* the rates of infant mortality of Oman, Saudi Arabia, and Iran despite a per capita GNP *one-twentieth* as great. The educational level of the population (particularly that of women) predicts infant mortality more powerfully than does income (Caldwell, 1986). However, the unreliability of data on mortality, education, and income in LDCs (Murray, 1987) limits the confidence to be placed on such generalizations.

In the PRC, where infant mortality has been sharply reduced, parents and professionals are reporting much higher frequencies of home and school behavior problems in children since the era of the one-child family began (Di, 1985; Tao, 1985). Chinese parents, having grown up in large extended families that placed a high value on producing many children, lack folkways to guide them in rearing an only child who has become the bearer of all their hopes. In Western countries, only children have, on average, done better than children with many sibs (Blake, 1981b), but they are the products of educated families and parental preferences for small family size. The consequences are far different when a historically determined preference for many children is submerged by adherence to a new doctrine, whether embraced or enforced (Eisenberg, 1985).

In the United States and other Western countries, there is a direct correlation between parental income and education and the academic and vocational attainments of children (Eisenberg & Earls, 1975). Parental rearing practices are one of the major pathways mediating between these macrosocial endpoints. Teenage motherhood, a correlate of low socioeconomic status, leads to prematurity and infant mortality, to worse developmental outcomes for the children (Institute of Medicine, 1985), and to poorer life chances for the mothers (Furstenberg, Lincoln, & Menken, 1981). Single mothers of whatever age, beset by poverty as well as by lack of emotional support from a husband, have difficulty meeting the needs of their children. Maternal employment per se need not have negative consequences for child development when developmentally appropriate child care is provided (Hayes & Kamerman, 1983), but child-care services in the United States remain lamentably inadequate (Gamble & Ziegler, 1986; Ziegler & Gordon, 1982).

Why study parenting? The answer to this question is that understanding the wellsprings of parenting behavior and the developmental needs of children is now more urgent than it has ever been. What we know is far short of what we have to know (Halpern, 1984), despite some encouraging successes in preventing developmental attrition by early intervention with children in at-risk families (Bryant & Ramey, 1987). Knowledge does not ensure effective social action; ignorance makes it impossible.

HAS ANYTHING BEEN LEARNED FROM CLINICAL STUDIES OF ABNORMAL PARENTING?

Among pediatricians and psychiatrists, it is taken as axiomatic that the quality of parenting is decisive for the outcome of the child. In this, they stand with the Anglican clergyman Robert Burton* (1621), who wrote in *The Anatomy of Melancholy:*

> . . . If a man escape a bad Nurse, he may be undone by evill bringing up. . . . Parents offend many times in that they are too sterne, always threatning, chiding, brawling, whipping, or striking; by means of which their poore children are so disheartned and cowed that they never after have . . . a merry houre in their lives.

If it is true that most clinicians overestimate the continuity of development from child to adult (Kagan, 1979), it is nonetheless striking how clearly recent research has documented what Burton discerned: that harsh discipline in childhood predicts depression (and alcoholism) in adult life (Holmes & Robins, 1987); that harsh and inconsistent discipline foreshadows juvenile delinquency (Rutter & Madge, 1976; West & Farrington, 1973); and that delinquency itself foretells adult psychiatric disorder and criminality (McCord, 1972, 1979; Robins, 1966).

Yet physicians have held so sentimental a view of the parent–child relationship that they have been slow to recognize gross physical abuse of children in Western society. The recency of its rediscovery merits emphasis. In the 1940s and 1950s, radiologists (Caffey, 1946; Silverman, 1953) identified a cluster of X-ray findings that they demonstrated to be diagnostic for repeated episodes of injury in infants and children after they had excluded all other càuses. The initial history provided by the parents did not include trauma; its occurrence was confirmed only by persistent questioning after the X-rays had been studied. Even then, the original investigators hesitated to conclude that the injuries had been deliberately inflicted. It was not until Kempe, Silverman, Steele, Droegemueller, and Silver (1962) coined the term "battered-child syndrome" that child abuse was brought to the consciousness of physicians and child-care workers.

The late Professor Kempe's tireless efforts led to legislation in all 50 states that mandates reporting of child abuse to official agencies. Between 1976, the first year the American Humane Association began collecting national data, and 1985, the last

---

*To give Burton his full due, he also recognized a role for heredity: "That other inward inbred cause of Melancholy, is our temperature [temperament] . . . which wee receive from our parents . . . such as is the temperature of the father, such is the sonnes' and looke what disease the father had when he begot him, such his son will have after him. . . ."

year for which complete data are available, the number of reports of child abuse *and* neglect almost tripled—from 669,000 to 1,928,000 (American Humane Association, 1987)! Of that number, about one-third are cases of physical abuse, with 40,000 classified as severe. The totals are indeed appalling, but the increase is primarily attributable to more complete case ascertainment. Two national probability samples of self-acknowledged family violence collected by Straus and Gelles (1986) during this time period show, if anything, a reduction rather than an increase in frequency.

Clinical research done in the past two decades has identified the characteristics of abusing parents, of the infants most vulnerable to abuse, and of the social context in which abuse is most likely to occur. The findings point to the impact of multiplicative social stresses: poor, ill-educated, and immature parents (Pelton, 1978), who themselves had often been victims of abuse (Kaufman & Zigler, 1987); infants who become targets for abuse (Altemeier, O'Connor, Vietz, Sandler, & Sherrod, 1982; Friedrich & Buriskin, 1976) because of prematurity (Hunter, Kilstrom, Kraybill, & Loda, 1978), physical handicap (White, Benedict, Wulff, & Kelley, 1987) or twinship (Groothuis et al., 1982), or short interbirth intervals (Light, 1973); and isolated families, bereft of support from friends or relatives, living in disorganized neighborhoods (Garbarino & Sherman, 1980).

Child abuse is customarily distinguished from infanticide on grounds of purposiveness. Although the death of a child is an all too frequent outcome of the former, it is not deliberately intended (and may indeed be lamented), whereas it is the deliberate goal of the latter. However, both child abuse and infanticide constitute the negation of parental care. Furthermore, what is to be said about "intent" when societies that profess solicitude for infants tolerate circumstances predictably associated with disproportionate infant mortality?

Infanticide is older than the recorded history of our species. It has been employed by many preliterate peoples (Whiting, Bogucki, & Kwong, 1977) as a culturally sanctioned "solution" to an unwanted birth, whether because it is female (Neel & Chagnon, 1968; Riches, 1974), defective (Lee, 1979), or otherwise inopportune (Whiting et al., 1977). It provides a telling example of the power of shared belief to transform a pattern of behavior into its opposite. The Netsilik Eskimo, whose habitat is northwest of Hudson Bay in Canada, are "extremely devoted to their children. . . . Children of all ages [are] an endless source of joy to their parents, despite the sacrifices that their rearing entails . . . . Relations between grandparents and grandchildren are marked by ceaseless fondling and joking" (Balikci, 1970). Yet if a female child is born into a community at a time when there is no suitable male child 2 to 5 years older for later betrothal, parents put their newborn daughter out in the snow to die before it is named; without a name, it is not human by definition. In Balikci's film on the Netsilik, one sees parents, otherwise gentle and loving, go about their chores with no visible response to the cries of the newborn dying in the snow. What internal distress they may experience is not revealed, but they adhere to an implacable demand placed on every member of the group. A female without prospect of marriage could not survive on her own as an adult; sharing resources in a marginal subsistence economy would endanger the community.

It is not feasible to review in this chapter the long history of infanticide and abandonment in "civilized" nations (Langer, 1974), except to note that both Plato in *The Republic* (Chapter V) and Aristotle in *Politics* (VII:16) justified infanticide as a necessary means to maintain quality and to control population size. Infanticide persisted long after it had been condemned by church authority, not by deliberate intent but by toleration of the conditions that gave rise to it. In Dublin in the last quarter of the eighteenth century, for example, only 45 of 10,000 infants abandoned at the foundling home survived; in England in the nineteenth century, child labor was so onerous that the Reform Law of 1838 constituted a considerable advance by providing that "children from 9 to 13 were not to work more than 48 hours in the week" (Kessen, 1965). In the United States today, a black infant is still twice as likely to die as a white one (U.S. Department of Health and Human Services, 1986), not, of course, by design, but through benign neglect (Eisenberg, 1984).

Culturally sanctioned infanticide may be said to represent an adaptation to the imbalance between local population size and available environmental resources, whatever the symbolic explanation given for the practice in a given society. Today all nation-states regard it as anathema. What made the shift in values possible was the marked improvement in living conditions and the availability of effective contraception and safe abortion. The persistence of infant and child abuse despite its official condemnation reflects the extent to which decent living conditions, methods of birth control, and family-planning education are still not available to large segments of our population.

Thus the question posed at the beginning of this section—What have we learned from clinical studies?—can be answered in this fashion: human parental behavior can be almost completely disrupted by severe and continuing environmental stress. Of course, individual psychopathology plays a role in child abuse, and abuse is found among socially advantaged parents, but social context is the overriding determinant of the quality of the relationship between parent and child.

IN WHAT SENSE, IF ANY, ARE ANIMAL STUDIES RELEVANT?

Given that the motor of evolution is the differential survival of those individuals best able to adapt to change as well as constancy in the environment, it follows that biological mechanisms must be built into each individual to ensure not only that it will produce offspring, but that it will act to enhance the likelihood that those offspring will survive to reproduce themselves. Parenting is the final phase in an integrated sequence of reproductive behaviors sculpted by selection to ensure the continuity of the genome. Darwin (1859) was explicit on this matter. In *On the Origin of Species*, he stated:

> I should premise that I use this term [struggle for existence] in a large and metaphorical sense, including dependence of one being on another, and including (which is more important) not only the life of the individual, but success in leaving progeny. . . . (p. 62)

> The real importance of a large number of eggs or seeds is to make up for . . . destruction at some period of life; and this period in the great majority of cases is an early one. If an animal can in any way protect its own eggs or young, a small number may be produced, and yet the average stock be fully kept up. (p. 66)

Darwin's emphasis on evolutionary continuities notwithstanding, some critics dismiss the relevance of animal studies on two grounds. First, they contend that the enormous variability in patterns of parenting among human communities "proves" that culture is the *sole* determinant of child rearing. Second, they argue that the differences in parenting behavior among species of a single genus, let alone among genera, call into question the legitimacy of any generalizations across species. The facts are correct; the inferences are unwarranted. Consider the arguments in turn.

Clearly, there is a remarkable variation among cultures in the ways children have been and are being reared (Levine, 1973; Whiting & Whiting, 1975). In one culture, the biological mother and all her sisters are denominated by the term "mother"; "father" applies equally to the mother's husband and his brothers. In another, child rearing is a task for the nuclear family. In one, the infant is kept in physical approximation to the mother's breast by a sling and feeds ad libitem in short bursts, with a marked effect on maternal hormones: high prolactin and suppressed estradiol and progesterone secretion (Konner & Worthman, 1980). In another, many infants are fed entirely with artificial foods on a predetermined schedule, and lactational amenorrhea in mothers is markedly foreshortened. In one, physical punishment is never used and in a second, regularly employed. Yet each child-rearing practice is compatible with the development of children into adults with enough social competence to ensure the continuity of the group within its own ecological niche.

As to the second point, the variability in parenting behavior among animal species of the same genus, the differences are indeed striking. Consider pigtail *(Macaca nemestrina)* and bonnet *(M. radiata)* macaque infant–mother pairs. A pigtail mother will not permit other adults to interact with her infant and limits its departures from her vicinity; in contrast, a bonnet mother allows conspecifics to explore and groom the infant and restrains its departures much less often. In consequence, the pigtail infant displays much greater distress than the bonnet following the removal of its mother from the colony (Rosenblum & Kaufman, 1968). Reared in social isolation, pigtail infants do not show the behavioral deficits that rhesus infants displayed in Harlow's original studies (Sackett, Holm, Ruppenthal, & Fahrenbruch, 1976). Furthermore, the response of the rhesus infant to separation differs, depending on whether it is the mother, the infant, or both that are removed from the group (McGinnis, 1980). In the case of interactions between adult males and infants, Japanese *(Macaca fuscata)* and Barbary *(M. sylvanus)* male macaques in the wild exhibit extensive paternal care toward infants, carrying, playing with, and defending them and sometimes "adopting" orphans, whereas rhesus, pigtail, and bonnet males do so far less often (Redican & Taub, 1981).

If culture has so great an influence on human parenting and if monkeys of a single genus behave so differently, as the critics contend, how can animal studies bear on the human condition? Indeed, skeptics add, what does biology have to do with the matter at all when human adoptive parents, in the absence of priming by pregnancy, parturition, or lactation, can do as well as and often better than biological parents, even when the adopted children are of a different race (Winick, Meyer, & Harris 1975)?

It is precisely the fact that only humans have the capacity for complex language and culture that emphasizes, rather than questions, the role of biology. It is the plasticity conferred by the human genome that permits culture to determine which of the many behaviors possible is displayed in a particular context. Moreover, commonalities underlie the manifest variations in modes of child rearing, commonalities dictated by the imperative needs of the infant. Adoption can be termed "nonbiological" only by ignoring the dependence of parental competence on the maturation of the child through developmentally appropriate interactions with peers and adults, including experiences in caring for younger children. If the display of parenting is ordinarily initiated during the hormonal and behavioral cascade accompanying courtship, pregnancy, and parturition, the origins of parenting are found in prior social experience; once initiated, the maintenance of parenting requires continuing stimulation of the parents by their offspring. Parental behavior does not emerge like Minerva, full grown from the head of Jove; it is the culmination of an epigenetic process reflecting past experience, level of maturation, and current context.

Indeed, studies with rodents demonstrate that juvenile animals exhibit caretaking behavior prior to hormonal priming, and that the persistence of parenting is a function of the stimulation provided by the young, rather than a programmed sequence determined by hormones (see chapters 4, 6, and 18). In jackals *(Canidae mesomelas* and *C. aureus)* and other canids (Moehlman, 1987), in elephants *(Loxodonta africana),* and in vervet monkeys *(Cercopithecus pygerythrus)* (Lee & Moss, 1987), auxiliary "mothering" by juveniles (allomothers) contributes substantially to the likelihood that the infants in a family, a troop, or a herd will survive. Among langurs *(Presbytis entellus),* infants initiate their own adoption by another adult female when the mother is removed from the group; following most experimental separations, the infant elects to remain with its adopted caretaker even after its biological mother is returned (Dolhinow, 1978).

The differences in parenting among primate troops are not a simple function of genetic instructions; patterns vary in response to the demands of the effective environment. In Sussman's (1977) field studies in Madagascar, *Lemur catta* and *L. fulvus* show clear habitat preferences (arboreal vs. terrestrial), even in areas where they overlap. The relatively more terrestrial *L. catta* infants are more precocious, move onto the mother's back and ultimately away from her at an earlier age, and are more often shared with other females. Laboratory experiments have shown that modifying the available manipulanda and the wall coverings of the cages in which squirrel monkeys *(Saimiri sciureus)* are housed during infant development uncovers sharp differences in the attachment behavior of male and female infants, differences that are not manifest in standard cages (Rosenblum, 1974).

Moreover, the occurrence of infanticide among many species (Hausfater & Hrdy, 1984) has become a topic of intense debate among behavioral biologists. Hrdy (1977) has observed among langurs that when a strange male takes over a female that is nursing an infant, the new consort is likely to kill it. The killing terminates maternal lactation, leads to the resumption of estrus, and permits the male to impregnate the female. Thus by the tenets of sociobiological theory, he ensures that his parental investment results in the perpetuation of his genes rather than those of the male he has displaced. But Curtin and Dolhinow (1978) report that infanticide is rarely seen among langurs living in the wild. Those studied by Hrdy congregated about Indian villages, with consequent changes in patterns of foraging and in colony behavior. Thus even within a single species, adult–infant relationships are influenced by the ecological context. The issue is not *whether* infanticide can occur under such circumstances, but how often it occurs, why it occurs, and how the variance in its occurrence is best accounted for.

Finally, just as commonalities underlie the manifest variation in parenting behavior, shared fundamental genetic structures have been conserved across the enormous evolutionary time span that separates flies, mice, and men. The homoeobox region of the genome, a DNA sequence of about 180 base pairs, first identified in *Drosophila*, determines segmentation along the vertical axis of the embryo (Snow, 1986). The homoeobox locus has now been found by cloned probes on chromosome 11 in the mouse and on chromosome 17 in the human (Joyner et al., 1985; Rabin et al., 1985). The genetic information specified in a DNA sequence present in mollusks, echinoderms, and flies (Holland & Hogan, 1986) has been conserved for use in regulating pattern formation in mammalian embryos. The way evolution has operated to construct the multiply branching limbs of the phylogenetic bush (Gould & Lewontin, 1979) is conveyed by an engaging metaphor proposed by the molecular biologist François Jacob (1977). He compares evolution to a tinkerer who makes do rather than an engineer with a forward plan:

> Natural selection . . . works like a tinkerer . . . who does not know exactly what he is going to produce but uses whatever he finds around him [and makes] a wing from a leg or a part of an ear from a piece of jaw. . . . (p. 1163)

> Evolution does not produce novelties from scratch. It works on what already exists. (p. 1164)

The "instructions" that reliably ensure the neurodevelopmental sequences typical for a species are to be found neither in the genome nor in the environmental surround, but in the degree of fit between the two. Organisms inherit not only their parents' genes but also their parents, peers, and places; that is, they inherit their ontogenetic niche (West & King, 1987) no less than Englishmen inherit England, as Francis Galton (1874/1970) recognized. That niche includes the prenatal environment in which exogenetic inheritance begins. Ducklings developing inside eggs receive acoustic stimulation from self-produced vocalizations and those of their unhatched sibs, stimulation that is essential to their postnatal ability to respond to maternal calls (Gottlieb, 1976). The sexual differentiation of fetal mice is influenced by the sex of their nearest intrauterine neighbors (Gandelman, vom Saal, & Reinisch, 1977). The naturally occurring exposure of the human fetus to maternal speech sounds in utero enhances the postnatal discrimination of and preference for its mother's voice (De-Casper & Spence, 1986).

Greenough and his colleagues (1987) distinguish two central nervous system (CNS) mechanisms for the storage of information derived from the environment: experience-*expectant* (niche related) and experience-*dependent* (niche unrelated). The first has been fashioned for experiences ubiquitous during the evolutionary history of the species; coded into the genes are instructions for the autonomous generation of surplus synaptic connections, from among which expectable stimulus input selects the survivors. It is sufficiently redundant that it goes awry only in the face of the gross deprivation produced in the experimental laboratory. The second, experience-dependent, records learning idiosyncratic to the individual organism; it requires the formation of new synaptic connections.

The challenge to students of behavior is to have the ingenuity to recognize the salience of those aspects of the environment that we fail to see precisely because they are ever present. When an adult organism displays a behavior without obvious isomorphic precursors, this does not establish that learning is not necessary. The relevant experience may not lie where we have looked for it (Eisenberg, 1971).

Thus the query Are animal studies relevant? merits the following response: because parenting is essential to survival of the species, marked deviations from species-specific patterns are uncommon. It is only by observing the "experiments of nature" seen in clinical practice, by studying animals in their econiche, by deliberately interfering with development in the animal laboratory, and by performing a comparative analysis of evidence from all three that we can illuminate the conditions both necessary and sufficient for the emergence of appropriate parental behavior.

DOES WHAT THE EXPERTS BELIEVE MATTER TO PARENTS?

Men and women, by their very essence as human beings, are theory-spinning animals. As parents, we do not just do "what comes naturally"; what comes "naturally" is what we have been taught to believe is natural (Eisenberg, 1972).

Preliterate peoples may lack child-development specialists, but they do not lack theories of development transmitted from elders to the young. For example, the Digo, the Masai, and the Kikuyu societies of East Africa hold different beliefs about infant socialization; correspondingly, they differ in child-rearing practices (DeVries & Sameroff, 1984). The Digo believe that infants are ready to learn at birth and expect a high degree of motor and social achievement from them; by the end of the first year, the Digo infant is relatively self-sufficient, having been toilet trained long before, and can be left in the care of others (DeVries & DeVries, 1977). The Masai do not believe that infants are ready to learn or need special teaching at all; little effort is put into elimination or motor training. According to the Kikuyu, serious education is possible beginning only in the second year of life; the infant is kept swaddled during its first year. Theory drives practice; practice "confirms" theory.

In Western societies, mothers (or at least educated mothers) do listen to the experts. From about the mid-sixteenth to the mid-eighteenth century, with the publication of William Cadogan's *Essay upon Nursing and the Management of Children* (1748), medical writings recommended against allowing the infant to nurse on colostrum because of its alleged harmful properties (Fildes, 1986); yet we now know that it contains valuable antibodies that protect against infection. As physicians after Cadogan began to emphasize the benefits of nursing, the use of wet nurses declined; women of the upper social classes began to breast-feed their own infants. In this century, the mass production of artificial foods and the ability to sterilize infant formulas have resulted in a decline of breast-feeding; more recently, as pediatricians have lauded the physiological and psychological benefits of breast-feeding, it is regaining popularity. Clearly, parents do listen to the experts, all too often when they shouldn't!

This excursion into history emphasizes the responsibility that "experts" bear in promulgating theories about parenting. The danger is that the public is all too ready to believe us; premature theorizing can be hazardous to the public health.

The concept of imprinting, based on methodologically flawed studies (Gottlieb, 1976) of a limited set of avian and ungulate species, entered the clinical literature as "bonding" theory: immediate postnatal skin-to-skin contact between infant and parent is held to be essential to the full development of parenting behavior. Klaus and Kennel (1976) have argued that "there is a sensitive period in the first minutes or hours of life during which it is *necessary* that the mother and father have close contact with their neonate for later development to be *optimal"* (italics added). That contention, coinciding with the thrust from the women's movement for the humanization of the birthing process, helped give impetus to changes in obstetrical practice that have been laudable. However, not only is there no compelling evidence for the validity of bonding theory (Lamb & Hwang, 1982), but the theory has aroused guilt and distress in parents who believe they have lost a crucial moment that can never be recaptured if early contact has been precluded by medical necessity or hospital rules. Further, the emphasis on bonding carries with it the inference that current patterns of infant care, consequent on new roles for women, may be harmful to both infants and mothers (Lozoff & Brittenham, 1979), a serious allegation to level without sufficient proof at a time when half of all mothers with a child younger than 1 year are in the labor force. Child abuse has been attributed to a failure of bonding; yet when the hypothesis was put to the test in a comparison of abused with well-cared-for infants born to primiparous mothers of low socioeconomic status, there were no discernible differences between the two groups in the amount of early contact (Egeland & Vaughn, 1981).

If bonding serves as an exemplar of flawed developmental theory, it is not for lack of other wrong-headed formulations. One could as readily have assailed the folly of "gentling" as a construct to explain the response of infant mice to handling, which in fact proved to be due to adrenal hypertrophy in response to stress, or of the introduction of Skinnerism into the automated nursery, or of Freud's (1933) dictum that "the discovery of her castration is a turning point in the life of the girl . . . (p. 173) [so that] women have but little sense of justice . . . their social interests are weaker . . . and their capacity for the sublimation of their instincts is less" (p. 183). Science may be self-correcting in the long run, but it is the short run that matters to parents looking for guidance. The media are avid for "one-armed" scientists and impatient with those who, because they are scrupulous about the evidence, formulate their public testimony in the form "on the one hand . . . but on the other hand. . . ."

Does what the experts say matter to parents? Indeed it does, and because it does, we bear a heavy responsibility when we go public. Silence is not a defensible stance; respect for the ambiguities and uncertainties of knowledge is, if we are to act responsibly.

SUMMARY

The provisional answers offered to the questions posed at the beginning of this chapter constitute a summary of the conclusions warranted by a critical review of the pertinent literature.

Understanding the sources of parental behavior as well as the developmental needs of children is today more urgent than it has ever been. If knowledge does not ensure that corrective social action will be taken, ignorance makes it impossible.

Clinical experience teaches us that human parenting can be completely disrupted by severe and continuing stress. Social context is the overriding determinant of the quality of the relationship between parent and children.

Only by observing the experiments of nature in clinical practice, by performing deliberate manipulations in the laboratory, and by studying animals in the wild can we dissect out the conditions both necessary and sufficient for effective parental behavior.

Because what experts say matters to parents, it is incumbent on developmentalists to respect and report the limits to certainty when going before the public.

Having earlier cited a seventeenth-century authority on melancholia, let this chapter conclude with an excerpt from an essay written 40 years before Burton's. Michel de Montaigne (1580/1958), advising Diane de Foix, Countess Gurson, on the education of children, cautioned:

> Just as in agriculture the operations that precede planting, and the planting itself, are certain and easy, but once the plant has taken life there are a variety of ways of cultivation and many difficulties; so with men, it requires little skill to plant them, but once they are born, training them and bringing them up demands care of a very different kind, involving much fear and tribulation. (p. 53)

REFERENCES

Altemeire, W. A., O'Connor, S., Vietz, P. M., Sandler, H. M., & Sherrod, K. B. (1982). Antecedents of child abuse. *Journal of Pediatrics, 100*, 823–829.

American Humane Association. (1987). *Highlights of official child neglect and abuse reporting*. Annual Report, 1985, Denver.

Balikci, A. (1970). *The Netsilik Eskimo* (pp. 104, 122). New York: Natural History Press.

Blake, J. (1981a). Family size and quality of children. *Demography, 18*, 421–442.

Blake, J. (1981b). The only child in America: Prejudice versus performance. *Population and Development Review, 7*, 43–54.

Bouvier, L. F. (1984). Planet earth 1984–2034. *Population Bulletin, 39*, 1–40.

Bryant, D. M., & Ramey, C. T. (1987). An analysis of the effectiveness of early intervention programs for environmentally-at-risk children. In M. Guralnick & F. C. Bennett (Eds.), *The effectiveness of early intervention for at-risk and handicapped children* (pp. 33–78), Orlando, FL: Academic Press.

Burton, R. (1621). *The anatomy of melancholy, what it is with all the kinds, causes, symptomes, prognostickes, and severall cures of it*. Quoted in R. Hunter & I. MacAlpine (Eds.), *Three hundred years of psychiatry 1535–1860* (pp. 96–97). London: Oxford University Press, 1963.

Cadogan, W. (1748). *An essay upon nursing and the management of children, from their birth to three years of age*. London: J. Roberts in Warwick Lane.

Caffey, J. (1946). Multiple fractures in long bones of infants suffering from chronic subdural hematoma. *American Journal of Roentgenology, 56*, 163–173.

Caldwell, J. C. (1986). Routes to low mortality in poor countries. *Population and Development Review, 12*, 171–220.

Clay, J. W., & Holcomb, B. K. (1985). *Politics and the Ethiopian family 1984–1985*. Cambridge: Cultural Survival.

Curtin, R., & Dolhinow, P. (1978). Primate social behavior in a changing world. *American Scientist, 66,* 468–475.

Darwin, C. (1859). *On the origin of species.* A 1964 facsimile of the first edition. Cambridge, MA: Harvard University Press.

DeCasper, A. J., & Spence, M. J. (1986). Prenatal maternal speech influences newborns' perception of speech sounds. *Infant Behavior and Development, 9,* 133–150.

DeVries, M., and DeVries, M. R. (1977). Cultural relativity of toilet training readiness. *Pediatrics, 77,* 170–179.

DeVries, M. W., and Sameroff, A. J. (1984). Culture and temperament: Influences on infant temperament in 3 East African societies. *American Journal of Orthopsychiatry, 54,* 83–96.

Di, G. (1985). Children's mental health care. In T-Y. Lin & L. Eisenberg (Eds.), *Mental health planning for one billion people: A Chinese perspective* (pp. 159–173). Vancouver: University of British Columbia Press.

Dolhinow, P. (1978). Commentary on infantile attachment theory. *Behavioral and Brain Sciences, 3,* 443–444.

Egeland, B., & Vaughn, B. (1981). Failure of "bond formation" as a cause of abuse, neglect and maltreatment. *American Journal of Orthopsychiatry, 51,* 78–84.

Eisenberg, L. (1965). Behavioral excellence. In A. C. Barnes (Ed.), *The social responsibility of gynecology and obstetrics* (pp. 53–73). Baltimore: Johns Hopkins University Press.

Eisenberg, L. (1971). Persistent problems in the study of the psychobiology of development. In E. Tobach, L. R. Aaronson & E. Shaw (Eds.), *The biopsychology of development* (pp. 515–529). New York: Academic Press.

Eisenberg, L. (1972). The *human* nature of human nature. *Science, 176,* 123–128.

Eisenberg, L. (1981). Cross-cultural and historical perspectives on child abuse and neglect. *Child Abuse and Neglect, 5,* 229–308.

Eisenberg, L. (1984). Rudolf Ludwig Karl Virchow, where are you now that we need you? *American Journal of Medicine, 77,* 524–532.

Eisenberg, L. (1985). Family policy and children's mental health in the People's Republic of China. In T-Y. Lin & L. Eisenberg (Eds.), *Mental health planning for one billion people: A Chinese perspective* (pp. 195–227). Vancouver: University of British Columbia Press.

Eisenberg, L., & Earls, F. J. (1975). Poverty, social depreciation and child development. In D. A. Hamburg (Ed.), *American handbook of psychiatry* (Vol. 6, pp. 275–291). New York: Basic Books.

Fildes, V. (1986). *Breasts, bottles and babies: A history of infant feeding.* Edinburgh: Edinburgh University Press.

Freud, S. (1933). *New introductory lectures on psycho-analysis* (pp. 172, 183). (J. W. H. Sprott, Trans.). New York: Norton.

Friedrich, W. N., & Boriskin, J. A. (1976). The role of the child in abuse. *American Journal of Orthopsychiatry, 46,* 580–590.

Furstenberg, F. F., Lincoln, R., & Menken, J. (Eds.). (1981). *Teenage sexuality, pregnancy and child bearing.* Philadelphia: University of Pennsylvania Press.

Galton, F. (1970). *English men of science: Their nature and nurture* (2nd ed.). London: Frank Cass. (Original work published 1874).

Gamble, T. J., Ziegler, E. (1986). Effects of infant day care: Another look at the evidence. *American Journal of Orthopsychiatry, 56,* 26–42.

Gandelman, R., vom Saal, F. S., & Reinisch, J. M. (1977). Contiguity to male fetuses affects morphology and behavior in female mice. *Nature, 226,* 722–723.

Garbarino, J., & Sherman, D. (1980). High-risk neighborhoods and high-risk families: The human ecology of child maltreatment. *Child Development, 51,* 188–198.

Gottlieb, G. (1976). Early development of species-specific perception in birds. In G. Gottlieb (Ed.), *Neural and behavioral specificity* (pp. 237–280). New York: Academic Press.

Gould, S. J., & Lewontin, R. C. (1979). The spandrels of Sn Marco and the Panglossian paradigm: A critique of the adaptionist programme. *Proceedings of the Royal Society of London, B 205*, 581–598.

Greenough, W. T., Black, J. E., & Wallace, C. S. (1987). Experience and brain development. *Child Development, 58*, 539–559.

Groothuis, J. R., Altemeier, W. A., Robarge, J. P., O'Connor, S., Sandler, H., Vietz, P., & Lustig, J. (1982). Increased child abuse in families with twins. *Pediatrics, 70*, 769–773.

Halpern, R. (1984). Lack of effects for home-based early intervention? Some possible explanations. *American Journal of Orthopsychiatry, 54*, 33–42.

Hausfater, G., & Hrdy, S. B. (Eds.). (1984). *Infanticide: Comparative and evolutionary perspectives*. New York: Aldine.

Hayes, C. D., & Kamerman, S. B. (1983). *Children of working parents: Experiences and outcomes*. Report of the Panel on Work, Family and Community. Washington, DC: National Academy Press.

Holland, P. W. H., & Hogan, B. L. M. (1986). Phylogenetic distribution of *Antennapedia*-like homoeo boxes. *Nature, 321*, 251–253.

Holmes, S. J., & Robins, L. N. (1987). The influence of childhood disciplinary experience on the development of alcoholism and depression. *Journal of Child Psycyology and Psychiatry, 28*, 399–416.

Hrdy, S. B. (1977). Infanticide as a primate reproductive strategy. *American Scientist, 65*, 40–49.

Hunter, R. S., Kilstrom, N., Kraybill, E. N., & Loda, F. (1978). Antecedents of child abuse and neglect in premature infants. *Pediatrics, 61*, 629–635.

Institute of Medicine. (1985). *Preventing low birth weight*. Washington, DC: National Academy Press.

Jacob, F. (1977). Evolution and tinkering. *Science, 196*, 1161–1166.

Jones, E. F., Forrest, J. D., Goldman, N., Henshaw, S. K., Lincoln, R., Rosoff, J., Westoff, C., & Wulf, D. (1985). Teenage pregnancy in developed countries. *Family Planning Perspectives, 17*, 53–63.

Joyner, A. L., Lebo, R. V., Kan, Y. W., Tijan, R., Cox, D. R., & Martin, G. R. (1985). Comparative chromosome mapping of a conserved homoeobox region in mouse and human. *Nature, 314*, 173–175.

Kagan, J. (1979). Perspectives on continuity. In O. G. Brim & J. Kagan (Eds.), *Constancy and change in human development* (pp. 26–74). Cambridge, MA: Harvard University Press.

Kang, W. (1985). Tradition and change: The structure and function of the Chinese family. In T-Y. Lin & L. Eisenberg (Eds.), *Mental health planning for one billion people: A Chinese perspective* (pp. 63–73). Vancouver: University of British Columbia Press.

Kaufman, J., & Zigler, E. (1987). Do abused children become abusive parents? *American Journal of Orthopsychiatry, 57*, 186–192.

Kempe, C. H., Silverman, F. N., Steele, B. F., Droegemueller, W., & Silver, H. K. (1962). The battered-child syndrome. *Journal of the American Medical Association, 181*, 17–24.

Kessen, W. (1965). *The child*. New York: Wiley.

Klaus, M. H., & Kennel, J. W. (1976). *Maternal–infant bonding*. St. Louis: Mosby.

Konner, M., & Worthman, C. (1980). Nursing frequency, gonadal function, and birth spacing among !Kung hunter-gatherers. *Science, 207*, 788–791.

Lamb, M. E., & Hwang, C. P. (1982). Maternal attachment and mother–neonate bonding: A critical review. In M. E. Lamb & A. L. Brown (Eds.), *Advances in developmental psychology* (Vol. 2, pp. 1–39). Hillsdale, NJ: Erlbaum.

Langer, W. A. (1974). Infanticide: A Historical survey. *History of Childhood Quarterly, 1,* 353–365.

Lee, P., & Moss, C. (1987). Cited in R. Lewin, Social life: A question of costs and benefits. *Science, 236,* 775–776.

Lee, R. B. (1979). *The !Kung-san: Men, women and work in the foraging society.* Cambridge: Cambridge University Press.

Levine, R. A. (1973). *Culture, behavior and personality.* Chicago: Aldine.

Light, R. (1973). Abused and neglected children in America. *Harvard Educational Review, 43,* 556–598.

Lin, T. Y., & Eisenberg, L. (Eds.). (1985). *Mental health planning for one billion people: A Chinese perspective.* Vancouver: University of British Columbia Press.

Lozoff, B., & Brittenham, G. (1979). Infant care: Cache or carry. *Journal of Pediatrics, 95,* 478–483.

McCord, J. (1972). Etiological factors in alcoholism: Family and personal characteristics. *Quarterly Journal of Studies on Alcohol, 33,* 1027–1033.

McCord, J. (1979). Some child-rearing antecedents of criminal behavior in adult men. *Journal of Personality and Social Psychology, 37,* 1477–1486.

McGinnis, L. M. (1980). Maternal separation studies in children and non-human primates. In R. W. Bell & W. P. Smotherman (Eds.), *Maternal influences and early behavior* (pp. 311–335). New York: Spectrum.

Moehlman, P. D. (1987). Social organization in jackals. *American Scientist, 75,* 366–375.

Montaigne, M. de (1958). *Essays.* (J. M. Cohen, Trans.). New York: Penguin Books. (Original work published 1580)

Murray, C. L. (1987). A critical review of international mortality data. *Social Science and Medicine, 25,* 773–781.

Neel, J. V., & Chagnon, N. A. (1968). The demography of two tribes of primitive, relatively unacculturated Indians. *Proceedings of the National Academy of Sciences USA, 59,* 680–689.

Pelton, L. H. (1978). Child abuse and neglect: The myth of classlessness. *American Journal of Orthopsychiatry, 48,* 608–617.

Preston, S. H. (1984). Children and the elderly: Divergent paths for America's dependents. *Demography, 21,* 435–457.

Rabin, R., Hart, C. P., Ferguson-Smith, A., McGinnis, W., Levine, M., & Ruddle, F. H. (1985). Two homoeobox loci mapped in evolutionarily related mouse and human chromosomes. *Nature, 314,* 175–178.

Redican, W. K., & Taub, D. M. (1981). Male parental care in monkeys and apes. In M. E. Lamb (Ed.), *The role of the father in child development* (pp. 203–258). New York: Wiley.

Riches, D. (1974). The Netsilik Eskimo: A special case of selective female infanticide. *Ethnology, 13,* 351–361.

Robins, L. N. (1966). *Deviant children grown up.* Baltimore: Williams & Wilkins.

Rosenblum, L. A. (1974). Sex differences, environmental complexity and mother–infant relationships. *Archives of Sexual Behavior, 3,* 117–128.

Rosenblum, L. A., & Kaufman, I. C. (1968). Variations in infant development and response to maternal loss in monkeys. *American Journal of Orthopsychiatry, 38,* 418–426.

Rule, S. (1987, July 13). Africans cite war's toll on children. *New York Times,* p. A-3.

Rutter, M., & Madge, N. (1976). *Cycles of disadvantage.* London: Heinemann.

Sackett, G. P., Holm, R. A., Ruppenthal, G. C., & Fahrenbruch, C. E. (1976). The effects of total isolation rearing on behavior of rhesus and pigtail macaques. In R. N. Walsh

& W. T. Greenough (Eds.), *Environment as therapy for brain dysfunction* (pp. 115–131). New York: Plenum Press.

Silverman, F. N. (1953). The roentgen manifestations of unrecognized skeletal trauma in infants. *American Journal of Roentgenology, 69,* 413–426.

Snow, M. H. L. (1986). New data from mammalian homoeobox containing genes. *Nature, 324,* 618–619.

Straus, M. A., & Gelles, R. J. (1986). Societal change and change in family violence from 1975 to 1985 as revealed by two national surveys. *Journal of Marriage and the Family, 48,* 465–479.

Sussman, R. W. (1977). Socialization, social structure and ecology of two sympatric species of lemur. In S. Chevalier-Skolnikoff & F. E. Poirer (Eds.), *Primate social development* (pp. 515–528). New York: Garland.

Tao, K-T. (1985). Children's mental health problems in China. In T-Y. Lin & L. Eisenberg (Eds.), *Mental health planning for one billion people: A Chinese perspective* (pp. 175–184). Vancouver: University of British Columbia Press.

Tien, H. Y. (1983). *China: Demographic billionaire.* Population Reference Bureau 38, No. 2.

U.S. Bureau of the Census. (1987). *Population profile of the United States: 1984–1985* (Current Population Reports, Series P. 23, No. 150). Washington, DC: Government Printing Office.

U.S. Department of Health and Human Services. (1986). *Health, United States, 1986* (DHHS Publication No. (PHS) 87–1232). Washington, DC: Government Printing Office.

West, D. J., & Farrington, D. P. (1973). *Who becomes delinquent?* London: Heinemann.

West, M. J., & King, A. P. (1987). Settling nature and nurture into an ontogenetic niche. *Developmental Psychology, 20,* 549–562.

White, R., Benedict, M. I., Wulff, L., & Kelley, M. (1987). Physical disabilities as risk factors for child maltreatment: A selective review. *American Journal of Orthopsychiatry, 57,* 93–101.

Whiting, B. B., & Whiting, J. W. M. (1975). *Children of six cultures: A psychocultural analysis.* Cambridge, MA: Harvard University Press.

Whiting, J. W. M., Bogucki, P., & Kwong, W. Y. (1977, February). *Infanticide.* Paper presented at the meeting of the Society for Cross-Cultural Research, East Lansing, MI.

Winick, M., Meyer, K., & Harris, R. (1975). Malnutrition and environmental enrichment by early adoption. *Science, 190,* 1173–1175.

Zelnik, M., & Kantner, J. F. (1980). Sexual activity, contraceptive use and pregnancy among metropolitan area teenagers: 1971–1979. *Family Planning Perspectives, 12,* 230–237.

Ziegler, E., & Gordon, E. (Eds.). 1982. *Day care: Scientific and social policy issues.* Boston: Auburn House.

3

Evolutionary Theory and Parental Motives

MARTIN DALY

This chapter is an exercise in "evolutionary" or "Darwinian psychology," by which is meant the use of contemporary evolutionary theory as a conceptual framework or metatheory for the study of psychological processes and mechanisms (Cosmides & Tooby, 1987; Daly & Wilson, 1988a, 1988b, 1988c; Symons, 1987). The focus of the discussion will not be "evolution" in the sense of phylogenetic reconstructions of the history of lactation or viviparity, topics that have been addressed by reproductive biologists (see, e.g., Blackburn, Evans, & Vitt, 1985; Calaby & Tyndale-Biscoe, 1977), but an exploration of the implications of recent evolutionary theorizing for the study of parental motivation. The aim is to illustrate how explicit consideration of the natural selective process can be a productive source of hypotheses about modulations in parental behavior and inclinations, and thus about the organization of the motivational systems responsible for variations in parental solicitude.

A BRIEF PRIMER OF RELEVANT EVOLUTIONARY THEORY

The processes of natural and sexual selection, as described by Darwin more than a century ago, imply that the adaptive functions of all evolved attributes are in a sense reproductive: attributes exist by virtue of their historical contributions to the relative reproductive success of their bearers. "Survival value" is a popular way of referring to a trait's adaptive significance, but the term is misleading in its implication that personal survival is the bottom line of adaptation. What selection favors is not mere longevity but whatever traits maximize the proportional representation of the trait bearer's genes in future gene pools. It follows that creatures have evolved to enter willingly into life-threatening contests in the pursuit of mating opportunities, to deplete their own bodily reserves to nourish dependent young, and in general to expend their very lives in the pursuit of posterity in the form of descendants.

In the early twentieth century, the developing science of genetics was widely misconstrued as invalidating Darwinism, because discrete Mendelian alternatives were seen as incompatible with Darwin's gradualistic notions of evolutionary change. By the 1930s, the theories of population genetics developed by Fisher, Wright, and Haldane had synthesized Mendel's and Darwin's contributions, establishing the (possibly unfortunate) tradition of invoking genes in discussions of the evolution and adaptive significance of traits (see especially Fisher, 1958). The reproductive posterity or "fitness" that Darwin's theory identified as the basic currency of adaptation was thereby given a more specific meaning in terms of the numbers of descendant gene copies. Arguably, however, Mendel's discovery that the structural basis of heredity consists of particulate alternatives had only minor relevance for Darwinian theory until 1964, when W. D. Hamilton published his theory of "inclusive fitness."

Hamilton explored the natural selective implications of an obvious but neglected fact: that collateral relatives as well as descendants are carriers of a focal individual's genes. For example, in a diploid sexual species like ourselves, one's outbred full sibling is just as closely related to oneself as one's outbred offspring, and is therefore just as effective a potential vehicle of one's genetic posterity. This is because sibling and offspring alike have a 50% chance of sharing any one of the focal individual's genes by recent descent from a common ancestor (or, if you like, because the sibling and the offspring have the same expected degree of genetic similarity to the focal individual), with the result that eventual reproduction by one's sibling has the same impact on the long-term survival and proliferation of the focal individual's alleles in future gene pools as an equivalent eventual reproduction by one's offspring. What this implies is that contributions to the survival and reproductive success of nondescendant relatives can be favored by selection in exactly the same way as can such contributions to descendants, and that fitness must be conceptualized not simply in terms of the number of offspring achieving maturity (classical Darwinian fitness), but also in terms of the focal individual's impact on the survival and proliferation, however achieved, of copies of its genes in competition with their alleles (Hamiltonian inclusive fitness). Hamilton thus replaced the classical Darwinian conception of organisms as "reproductive strategists" with a more complex conception of "nepotistic strategists," a development generally credited with launching the "sociobiological revolution," which has reinstated Darwinism as the theoretical centerpiece of animal behavior studies. ("Darwinism" is used here in its usual contemporary meaning to refer to the attribution of adaptation to evolution under the influence of "selection"—that is, nonrandomly differential reproductive success. "Darwinism" as used in this chapter thus encompasses such modern extensions as Hamilton's.)

Hamilton's analysis launched a whole new approach to the study of social organization, with its most dramatic successes to date concerning the elucidation of the diverse evolutionary bases for complex sociality in various invertebrate species with specialized, sterile phenotypes (see, e.g., Andersson, 1984). At first glance, Hamilton's elaboration of Darwinism would seem to have only slight relevance for the study of Mammalia, a group of animals with relatively simple social systems and infrequent nepotistic investment in nondescendant relatives, such as siblings. Unlike the situation in birds, for example, the care of mammalian young is usually the sole business of the mother, is occasionally biparental, and only rarely involves individuals other than the parents. However, a Hamiltonian understanding of sibling relations is essential to an evolutionary theoretical analysis of the parent–offspring relationship in any sexually reproducing creature, as will be shown in the next section.

The processes of natural and sexual selection imply not only that adaptive attributes exist to promote reproduction, but that they exist to promote success in reproductive *competition:* evolutionary change occurs because some phenotypes outreproduce others. Darwin clearly understood that selection is essentially a process of intraspecific competition, but until the 1960s, animal behaviorists were inclined to equate adaptation with social harmony: the common objective of conspecific animals was presumed to be the perpetuation of the species, a very different ultimate objective from the competitive ascendancy of one's genotypic elements over their alleles. This notion that organismic adaptations should have evolved to serve the interests of whole species rather than the separate and conflicting interests of individuals is the fallacy of "naive group selectionism" (Williams, 1966, 1971). For decades, such a position was almost universal among biologists, but not because of any compelling counterargument to the orthodox Darwinian view. (Until Wynne-Edwards, 1962, made group selectionism *non*naive, no such argument was offered, good *or* bad.) The contradiction was simply not recognized.

In retrospect, naive group selectionism provides a cautionary tale of the extent to which prevailing assumptions can influence not just our interpretations of behavioral phenomena, but even our ability to recognize their existence. Lorenz (1966), for example, was so thoroughly convinced that all evolved attributes function to "preserve the species" as to assert that the "aim of aggression" (p. 38) is never lethal, a claim that he attributed to "objective observation of animals" rather than to any theoretical predilections of his own. Lorenz simply dismissed the considerable evidence that animals are frequently killed by conspecifics as indicative of pathology or "mishap" and, by implication, unworthy of investigation by biologists interested in adaptation. Lethality was thus eliminated by fiat from Lorenz's analysis of intraspecific aggression and became invisible to his readers. His book was widely cited as having documented the sublethal restraint of animal aggression in natural environments and the unique murderousness of mankind, notions that were then and that remain wildly at odds with actual observations of animal conflict in the field (see, e.g., Wilson, 1975).

PARENT–OFFSPRING CONFLICT

In the world view of naive group selectionism, disputes between young animals and their (predominantly nurturant) parents provide a striking example of apparent maladaptation. Surely dependent young should have evolved parsimonious means of communicating their needs to parents, which should have evolved to respond with appropriate nurture. Why, then, do avian broods beg so loudly that they risk revealing their location to potential predators? Why do mammalian mothers fight off juvenile nursing attempts, and why do the rebuffed offspring sometimes respond with energetically wasteful tantrums? In a classic paper entitled "Parent–Offspring Conflict," Trivers (1974) provided an answer to such questions. His basic insight was that the fact that parent and offspring are not genetically identical implies that the course of action that maximizes the fitness of one party will not necessarily match the other's optimum. A conflict of interests between parent and offspring is thus endemic to sexually reproducing species.

Consider the case of a pair of littermate kittens (Ego and Sib), still dependent on maternal hunting. Pickings have been slim, but today the mother brings home a small carcass. Let us suppose that consuming half the carcass would increase either kitten's *reproductive value* (the animal's age-, sex-, phenotype-, and condition-specific expected future fitness, hereafter abbreviated RV) by 4 arbitrary units, and that consuming the entire carcass would increase it by 7. (This assumption of a diminishing marginal utility of successive meal-size increments probably applies in a wide range of natural situations, such as whenever averting starvation is the principal short-term consideration affecting variation in RV, or whenever digestive or assimilative efficiency declines with meal size.) How, then, should the carcass be divided? If the mother's interests prevail, each kitten should get half the food, thus augmenting the total RV of the mother's brood by 8 units. Ego, however, should have evolved to value increments to her own RV more highly than equal increments to Sib's: 4 units of RV gained by Sib make a contribution to Ego's inclusive fitness that is equivalent to only a 2-unit gain in her own RV if the kittens are full siblings or a 1-unit gain if their sires are different and unrelated to each other; in either case, monopolizing the carcass for a 7-unit gain is the optimal strategy for Ego. Of course, Sib's view of the matter is expected to be no less selfish than Ego's, and so the combined effects of a balance of power between the littermate rivals plus the mother's efforts to impose her own optimum will often dictate an equitable outcome. Each offspring is expected, however, to exploit opportunities to sabotage that equity.

This argument is artificial and specific, but the point is extremely general: the optimal allocation of *parental investment,* or PI (Trivers, 1972), from a particular offspring's point of view seldom or never matches the parental optimum. This is true whether one is considering the allocation of PI among littermates or among successive offspring (Macnair & Parker, 1979), and it extends to nepotistic investment in collateral kin (Trivers, 1985). From the mother's point of view, for example, a daughter is twice as closely related to herself as a niece, and hence is twice as valuable as a potential vehicle of maternal fitness, but from the daughter's perspective, the mother's niece is a mere cousin and only one-eighth as valuable as herself. It follows that parents are likely to perceive their offspring as unduly self-centered, and to attempt to induce those offspring to take a greater benevolent interest in collateral relatives than the offspring would be spontaneously inclined to exhibit in the absence of parental pressure. This last sentence was written, of course, with the case of *Homo sapiens* in mind; it would be interesting to investigate just how general such conflict really is in our own species and whether it is demonstrable in other creatures.

Selection may be expected to favor any tactics in the offspring that will successfully elicit PI surpassing the parent's optimum and more closely approaching the offspring's. Such tactics include persistent begging, which may be effective either by exaggerating one's needs or simply by raising the parent's costs of noncompliance, and "regression," which entails feigning the dependency of earlier life stages and is, in the case of human children, especially a response to competition from younger siblings (Dunn & Kendrick, 1982). Likewise, selection may be expected to favor tactics by the parents that discount offspring propaganda to arrive at accurate assessments of offspring need and quality (e.g., Stamps, Clark, Arrowood, & Kus, 1985) and that manipulate offspring to comply with parental agendas, such as punishing manifestations of sibling rivalry. Alexander (1974) suggested that parent–offspring conflict must be resolved in favor of parents because any allele that conferred advantage in such conflict in the offspring stage would only damage its carriers' fitness when they became "losing" parents of "winning" offspring. This seemingly plausible argument is false: there is nothing in the genetics of parent–offspring conflict to make the contest asymmetrical (see, e.g., Blick, 1977). Parents do, of course, often have greater power than their young and may thus be in a position to impose their will, but there is no reason to suppose that the evolutionary "arms race" (Dawkins, 1982) is any likelier to have a parental victory as its equilibrium than an offspring victory (Parker & Macnair 1979).

Because offspring have their own self-interested agendas, a knee-jerk parental responsiveness to their demands would represent a maladaptive distribution of valuable parental efforts and resources. It follows that the evolved mechanisms of parental motivation must be subtle. Parents may be expected to have evolved to modulate their investment of efforts in the development and welfare of offspring as though they (the parents) had assessed the expected impacts of those efforts on their own fitness.

DISCRIMINATIVE PARENTAL SOLICITUDE

Theoretical evolutionary considerations suggest that parents should have evolved to allocate their efforts with sensitivity to any cues that are predictive of the fitness returns of alternative allocations. The problem is subtle. In determining the optimal allocation of parental resources among offspring differing in intrinsic quality or RV, for example, we cannot predict on a priori grounds which ones (if any) the parent should favor. The answer depends not on the offspring's present RVs, but on their ability to translate PIs into RV increments; PI might be wasted, for example, if bestowed on offspring that are capable of thriving without it or if bestowed on offspring that are incapable of thriving in any case.

Trivers (1972) defined PI as any parental action that contributes to offspring RV while diminishing parental capacity to invest elsewhere. The definition is theoretically appropriate because it allows us to combine in a single conceptual framework such disparate sorts of "expenditure" as foraging effort, depleting one's own bodily reserves to build up those of the young, and risking death to defend them. All these efforts entail diminution (whether certain or probabilistic) of the parent's ability to promote her fitness through other avenues, in the interest of enhancing or maintaining the RV of a present offspring.

The *strategic commonality* of the various forms of investments subsumed by Trivers's definition provides reason to expect that there will be some commonality of *causation* as well. From the strategic perspective, for example, any grounds for doubt that an offspring is indeed the parent's own should lead to a discounting of the perceived benefit to the parent of the offspring's RV. If the motivational mechanisms modulating parental behavior have been shaped by selection to effectively promote fitness (and, of course, we have no alternative to supposing that they have), then a modicum of doubt that a child is indeed one's own should ideally diminish parental willingness to expend PI of *any* sort (and by a constant factor). The same sort of argument applies to any circumstantial or phenotypic cue that is predictive of offspring fitness: a decrement in the offspring's expected fitness for any reason should cause an adaptive parent to expend less PI of any sort on that offspring. These considerations suggest that the sort of system of parental motivation that selection would favor would incorporate some instantiation of a unitary "intervening variable" (Miller, 1959), effectively representing the expected fitness value to the parent of the individual offspring or present brood and having a causal impact on the whole gamut of parental activities. Such an intervening variable could be labeled "offspring-specific (or brood-specific) parental solicitude" (Daly & Wilson, 1988a). Notwithstanding the diverse causal factors that affect the various parental actions, such as nest building, feeding, and protecting the young, then, these considerations point to the likelihood of a common causal core such as the maternal "state" long studied by Rosenblatt and his colleagues (see chapter 4). It is not generally to be expected, however, that this state would motivate parental action without regard to the identity and parenthood of the recipients, as seems to be the case in laboratory rats. PI that is discriminative with respect to parental condition and offspring demands, and yet is indiscriminate with respect to offspring identity, is apparently a peculiarity of only a minority of mammals with a special ecology, as discussed below ("Own versus Alien Offspring").

Animal behaviorists have found the concept of PI compelling in the abstract, as witness the many theoretical papers that have elaborated on Trivers's models and the frequency with which his papers are cited by field workers. However, rigorous operationalization of the concept has been rare or nonexistent. It is extremely difficult to assess the cost in expected future parental fitness of a tolerated risk or a small depletion of parental reserves, and no easier to measure the impact of individual acts of parental care on the RV of the offspring (see, e.g., Clutton-Brock, 1984). There is, however, one paradigm within which it has proved possible to assess parental sensitivity to variables predicted to affect the costs and benefits of PI allocation alternatives: the measurement of parental response to a standard experimental presentation of a real or simulated predator threatening to reduce the present brood's RV to zero.

Reproductive Value of the Offspring

The first obvious prediction that can be tested in such a paradigm is that the parent should be willing to incur greater risk to own life in defense of a brood of greater RV. One determinant of RV is brood size, each additional nestling representing an additional projection of 50% of a parent's genotype into the offspring generation. Another is offspring age, since the RV of immature animals increases over time, at least until maturity, simply by virtue of having survived a period of potential mortality. Many studies of parental defense in birds and fishes have documented the predicted effects of these determinants of brood RV (e.g., Andersson, Wiklund, & Rundgren, 1980; Barash, 1975; Carlisle, 1985; Curio, Regelmann, & Zimmermann, 1984; Greig-Smith, 1980; Patterson, Petrinovich, & James, 1980; Pressley, 1981).

Surprisingly, however, it is much less clear that the expected effect of offspring age is exhibited by mammals. More often, the mother's willingness to defend her offspring seems, paradoxically, to *decline* as their RV increases. Svare (1981), for example, in reviewing laboratory studies of several species of myomorph rodents, maintains that "most authors consistently find that maternal aggression is most intense early in the postpartum period and declines as lactation advances" (p. 184); see also Ostermeyer's (1983) review and chapter 7. In precocial terrestrial mammals, such as many ungulates, it may often be the case that the developing self-defensive or escape capabilities of the young mean that the offspring's need for parental defense (or more precisely, their expected profit in RV therefrom) declines so rapidly with age as to obscure the opposing influence of the offspring's changing value to the parent. However, it is not in cursorial ungulates that the paradoxical time course of maternal defensiveness has been documented, but in precisely those mammals in which we might have especially expected results to parallel those in songbirds and nest-guarding fishes: altricial, burrow-dwelling rodents whose vulnerability to predators seems likely to be more or less constant over the course of the early developmental period during which "maternal aggression" declines.

Several possible resolutions of this paradox could be suggested. One hypothesis might be that psychological mechanisms functioning to generate adaptive variation in maternal protectiveness in relation to offspring age have simply never evolved. Adaptation is, after all, never perfect, but is instead jerry-built by selection from preexisting mechanisms. Perhaps protectiveness shares a causal core of "maternal motivation" with other activities having other optimal time courses and is poorly adapted to the time course of the demands on it. This is not a particularly appealing or heuristic alternative for two reasons. The first is the rather abstract philosophical consideration that nonadaptation is not so much a testable hypothesis in its own right as a refuge from the necessity to formulate one. The second, more concrete, consideration is that rodent mothers do *not* simply exhibit a correlated waxing and waning of all parental activities in concert, but instead manifest a functional succession of stage-appropriate behaviors (e.g., Daly, 1972). Certainly there are things that are positively correlated with brood RV that *could* be used to modulate maternal protective inclinations, such as the thermal and gaseous by-products of the brood's gross metabolism. Thus the reason why small mammals exhibit a seemingly maladaptive time course of variable maternal defensiveness remains a puzzle.

One interesting possibility is that changing maternal protectiveness is in fact perfectly well tuned to changes in the demand for it, because pups are especially in need of protection *from conspecifics* when very young (Ostermeyer, 1983). This sort of reasoning would be interesting to explore because the utility of pup killing as a function of the pups' age might constitute a quite different function for conspecific female versus male killers or for juvenile versus adult males, according to whether pup killing is adaptive in the context of energy accrual, of male–male reproductive competition, or of clearing some local *Lebensraum*. (For discussions of the evidence for adaptive variation in infanticidal tendencies of conspecifics, see Brooks, 1984; Hrdy, 1979; Sherman, 1981; vom Saal, 1985.) One might then predict differential postpartum time courses of maternal aggression according to the age–sex class of conspecific intruders, as well as differential time courses of reaction to conspecific intrusions versus encounters with predators of other species. Evidence of such differentials would not only support the hypothesis that maternal aggression has been fine-tuned by selection, but also suggest where further causal analysis of maternal motivation might prove revealing.

Reproductive Value of the Parent

A much more controversial prediction than that of increasing parental solicitude with increasing offspring age and RV is that parental solicitude toward offspring of a given RV should be an increasing function of parental age. The prediction follows from the consideration that the parent's own RV declines with age, so that the parent should have evolved to modulate her PI decisions as though weighing her own future less and less heavily against the fitness potential of her present brood. This prediction is controversial for at least two reasons. The first is that according to present theory (Alexander, 1987; Hamilton, 1966), the senescent decline in RV with age is a consequence of the expenditure of "reproductive effort" (including PI), and it therefore seems suspiciously circular to suggest that escalated expenditure is a response to senescence. The second reason for doubting the prediction is that as long as senescent processes impose significant fitness costs, selection will push such processes back in the life span until most mortality is presenescent and RV is a constant or even increasing function of age over the adult life span of all but a few long-lived individuals. This circumstance seems especially applicable to species such as migratory songbirds, which experience high overwinter mortality that is relatively age independent; similarly, many rodents may rarely manifest senescence in natural populations and may therefore have evolved no particular strategy of escalated parental effort with old age.

Notwithstanding these difficulties, there is some good evidence that parents of at least a few species escalate their effort and in effect devalue their own survival as they age. Still the cleanest demonstration of such an effect is that of Pugesek (1981), who found that California gulls left their nests unattended less with increasing parental age, and yet managed to feed their young more frequently. While this pattern of results might be attributed to improving parental skills, additional facts force the conclusion that older birds really expend greater effort: despite feeding their young of a given age at a higher rate and thus enabling them to grow faster, older parents persisted in feeding them for a longer period as well. Unfortunately, Pugesek's study did not incorporate the method of assessing tolerated risk to self in the face of a standardized nest predator.

Field studies of large mammals have generally found that older females are likelier to wean their offspring successfully than younger ones, but it is difficult to distinguish life-span changes in the expenditure of PI from life-span changes in parental skills. (Most field workers seem to have assumed the latter, especially in the case of primate studies; the alternative hypothesis, that what looks like a practice effect is in fact an adaptive shift in effort, is seldom considered.) Ozoga and Verme (1986) provide evidence that the greater survival of white-tailed deer fawns born to older mothers is at least partly due to greater maternal effort to protect the young from predators. The most convincing demonstration of increased expenditure of PI over the life-span of a mammal was provided by Clutton-Brock (1984), who showed that the condition of red deer calves at weaning relative to that of their mothers is an increasing function of maternal age, "condition" being assessed by a kidney fat measure of bodily stores that is predictive of overwinter survival. Furthermore, Clutton-Brock was able to demonstrate an increase with maternal age in the survival cost of lactation, as indicated by elevated winter mortality of hinds of a given age that had weaned calves as compared with same-age hinds that had not.

Experimental students of the physiology of maternal motivation in mammals have apparently paid rather little attention to life-span developmental changes in maternal inclinations and behavior. The subject warrants some comparative research.

Own versus Alien Offspring

If selection favors the discriminative allocation of PI where it is likely to enhance parental fitness, a consideration of obvious relevance to allocation decision should be the reality or probability that the offspring is indeed the parent's own. Seals that deliver and suckle their young in close proximity generally attack unrelated pups that try to nurse, even while nursing their own (LeBoeuf & Briggs, 1977). Similar discrimination with respect to individual identity is exhibited by ungulates that raise precocious young in herds in which nursing mothers and same-stage youngsters all mingle; postpartum bonding guarantees that maternal nurture is directed to appropriate vehicles of fitness (see, e.g., Gubernick, 1981).

Things are quite different in the muroid rodents that have been the subjects of most laboratory work on maternal motivation. They seem oblivious to the own-versus-alien distinction, blithely giving suck to whatever young they happen to encounter in the nest, including even those of other species. But in effect, maternal discrimination is not so much absent in these burrow-dwelling rodents as it is dependent on other cues than those presented by the young themselves. The mother could be said to use her nest site as the cue by which she "recognizes" her offspring. The relative immobility and sequestered situation of these altricial pups mean that mix-ups, with concomitant risk of misdirected maternal investments, are probably rare or nonexistent in nature. In comparative studies among gulls, swallows, and other groups of birds, it has been clearly shown that parents recognize their own young in those species in which precocious mobility or close proximity of nests demands such discrimination, but behave parentally toward whatever young they encounter in the nest in related species in which young do not intermingle (see, e.g., Daly & Wilson, 1988b; Medvin & Beecher, 1986). The point has not been explored as systematically or elegantly in the mammals as in the birds, but it appears to apply to mammals as well.

Possibly the most perplexing allegation of indiscriminate mammalian mothering in nature is to be found in Davis, Herreid, and Short's (1962) account of the behavior of Mexican free-tailed bats. These bats roost in dark caves, sometimes in aggregations of millions of individuals, and within hours of giving birth to her single pup, the female leaves it hanging in a crèche while she goes foraging. The crèche may contain several thousand infants per square meter, and they are not immobile, sometimes crawling a meter or more between nursing bouts. How could their mothers hope to find them? Davis et al. concluded that they cannot, reasoning both from the considerations above and from experimental demonstrations that pups would latch on to any female held near them and that females would not reject them once attached. Davis et al.'s conclusion that the females act as an anonymous "dairy herd" has been widely accepted and cited as a remarkable fact (e.g., Riedman, 1982), but the conclusion is astonishing if true. How could lactation be evolutionarily stable in such a case? The nursing bat incurs both energetic depletion and predation risk in order to deliver 16% of her body weight per day in milk. Surely, selection must favor the female that deposits her baby in the care of the dairy herd, dries up, and opts out. It is therefore satisfying to learn that the bats do not nurse indiscriminately after all. McCracken (1984) genotyped mothers and the infants nursing from them in the field and showed that while some mismatches indeed occurred, female–pup pairs exhibited a hugely greater genetic concordance than random mixing would have dictated. McCracken estimated that about 83% of nursing mothers were actually feeding their own offspring, and although a 17% loss of milk to nonrelatives represents a truly remarkable failure of parental discrimination, it is not such an egregious failure as to make the very existence of parental care evolutionarily unstable, as truly indiscriminate feeding would.

The problem of identifying one's own offspring correctly is, of course, much more severe for male mammals than for females, and this provides at least a partial explanation for the overwhelming predominance of female care in the Mammalia. If one partitions reproductive efforts into "mating effort" (expended in the quest to produce offspring) and "parental effort" (expended to enhance the RV of offspring already produced), following Low (1978), the expected returns from parental effort are devalued for males by some statistical probability that one's putative offspring were sired by a rival. It follows that the optimal allocation for females will be more toward parental effort than the optimal allocation for males. (Maynard Smith, 1978, has argued that a male's returns from mating effort must be devalued by the same risk-of-cuckoldry factor as his returns from parental effort, so that variable "confidence of paternity" has no effect on the optimal allocation of male reproductive efforts, but this argument is fallacious; see Daly & Wilson, 1983, pp. 166–167.) When females can rear young alone, their parental efforts become the "resource" limiting male fitness, a circumstance that leads to escalated male mating competition and selection for still greater male specialization in mating effort (Trivers, 1972). Little wonder that male parental care is rare in animals in which fertilization occurs within the female's body. In externally fertilizing vertebrates (i.e., most fishes and amphibia), by contrast, male care of the young actually predominates (see, e.g., Ridley, 1978), while in birds, in which both sexes are susceptible to mistaken identity of the young thanks to egg dumping (see, e.g., Brown, 1984), care is predominantly biparental. In those few mammals in which males do in fact invest in their young, paternal behavior has clearly evolved secondarily, consisting of some protective or provisioning behavior complementary to the investments provided by the mother, rather than entailing an absence of sex-role differentiation (Daly, 1979).

For a more thorough discussion of preferential parental treatment of own offspring and of the recognition mechanisms that permit such behavior, see chapter 21.

DISCRIMINATIVE PARENTAL SOLICITUDE AND HOMICIDE

The distal objective of all organismic adaptations is the promotion of fitness. Proximal perceptions of self-interest—a full belly, delivery from fear, sexual satisfaction, and so forth; in short, pleasure and pain in a broad sense—must therefore have evolved as predictive tokens of fitness in historical environments. This perspective suggests that creatures are likely to have evolved to *experience* conflict in their social relations when their fundamental fitness interests conflict in the sense that events that could be expected to promote one party's fitness would be likely to damage the other's.

Violence is a response to experienced conflict and can therefore be used as a sort of assay thereof, regardless of whether the violence itself is adaptive for its perpetrator. If one wishes to test theoretically derived propositions about the determinants of interpersonal conflict in our own species, lethal violence (homicide) provides an assay with obvious face validity and with less reporting bias than can be achieved using measures of any lesser manifestation of conflict (Daly & Wilson, 1988a). Homicides by parents (filicides) provide a case in point. The risk that a human parent will kill her or his own child is generally low, but it exhibits systematic variation that can be examined in the light of evolutionary models of parental motivation. Again, it must be stressed that this entails no assumption that filicide is adaptive, but only that the likelihood of this rare lethal event is (at least in part) an inverse function of child-specific parental solicitude.

Consider first the reproductive value of the child. One may predict that parents will value their children increasingly until at least the age of maturity, because age-specific reproductive value increases over that period. Supportive evidence comes from cross-cultural studies showing that hard-pressed parents who cannot raise children born too close together universally sacrifice the younger one (Daly & Wilson, 1984). A more subtle observation is that the probability of filicide declines monotonically with the child's age, despite the adolescent's rapidly increasing conflictual inclinations and likelihood of entering into lethal conflict with *non*relatives. Moreover, evolutionary considerations can be used to predict further details of this age-related decline in filicides (Daly & Wilson, 1988c). The decline with child's age may be predicted to be negatively accelerated and concentrated in the first year of life because of ancestral patterns of juvenile changes in RV, and the decline in filicides by the mother may be predicted to have a steeper slope than those by the father, primarily because the utility of alternative reproductive efforts declines more steeply with own age for women than for men (and secondarily because of the time course of maturation of paternity markers). These and other hypotheses derived from an evolutionary model of parental motives are supported by data on Canadian filicides (Daly & Wilson, 1988c); they have not yet been tested elsewhere.

What about parental RV? *Homo sapiens* would seem an especially likely creature in whom to seek effects of declining maternal RV in view of the phenomenon of menopause, but several authors have argued that our forebears had such short life spans that women could not have evolved adaptive life schedules for the modulated expenditure of PI, since few would have faced the selection pressure of impending menopause. This argument is not convincing. It is generally based on archaeological or other estimates of ancestral *mean* life spans, which are short because infantile and juvenile mortality is averaged in; it remains perfectly plausible that a significant proportion of those early human and protohuman females who lived long enough to bear children also lived long enough to experience a plummeting RV. A more interesting quarrel with the prediction that women should have evolved an age-dependent maternal psychology is the fact that whereas *reproductive* value, as usually defined, indeed drops to zero at menopause, women's capacity to promote their fitness obviously does not. The concept of RV should therefore be replaced by a concept of "nepotistic value"—that is, a quantity representing the organism's capacity to promote its inclusive fitness (Daly & Wilson 1988a). But with these reservations expressed, there still *is* evidence that older human mothers cherish their infants more than do younger mothers: infanticide rates decline with maternal age (Bugos & McCarthy, 1984; Daly & Wilson, 1984). This decline cannot be attributed to age differences in the proportion of mothers who are unwed, for it is manifested within both married and unmarried new mothers considered separately (Daly & Wilson, 1988a).

Finally, as regards the dimension of own versus alien offspring, there is only anecdotal evidence that suspicions or revelations of nonpaternity are stimuli to filicide by fathers (Daly & Wilson, 1988a), but it is also worth remarking that stepparents (of both sexes) are far more likely to kill children in their care than are natural parents (Daly & Wilson, 1988a). The elevated risks to children living with stepparents cannot be attributed to a confound with poverty, family size, personality type, or any other correlate yet suggested (Wilson & Daly, 1987). The explanation seems simply to be that stepparents do not love their wards as natural parents do, and are therefore likelier not only to resent and exploit them, but also to lose their tempers even when striving to treat the children fairly and well.

CONCLUDING COMMENTS

This discussion has highlighted three considerations that ought to be relevant to the adaptive modulation of parental efforts: offspring age, parental age, and the reality of the putative parent–offspring link. Much more could be said about these topics; the implications of internal fertilization and concomitant paternity doubt in mammals, for example, have barely been touched on. Further discussion and references can be found in Daly and Wilson (1988b). A complex issue that is avoided completely in this chapter is differential treatment of offspring according to sex; see Charnov (1982) and Hrdy (1987).

Little has been done in linking evolutionary theoretical models of parental behavior and the field studies they have inspired to the analyzed mechanisms of motivation, as studied by laboratory psychobiologists. The gap seems especially large in the case of Mammalia. Such a synthesis would be useful not only by elucidating the mechanistic underpinnings of reproductive strategies, but also by opening the way to a newly principled *comparative* analysis of the diversity in motivational mechanisms among the mammals.

REFERENCES

Alexander, R. D. (1974). The evolution of social behavior. *Annual Review of Ecology and Systematics, 5,* 325–383.
Alexander, R. D. (1987). *The biology of moral systems.* Hawthorne, NY: Aldine de Gruyter.

Andersson, M. (1984). The evolution of eusociality. *Annual Review of Ecology and Systematics, 15,* 165–189.

Andersson, M., Wiklund, C. G., & Rundgren, H. (1980). Parental defence of offspring: A model and an example. *Animal Behaviour, 28,* 536–542.

Barash, D. P. (1975). Evolutionary aspects of parental behavior: The distraction display of the Alpine accentor, *Prunella collaris. Wilson Bulletin, 87,* 367–373.

Blackburn, D. G., Evans, H. E., & Vitt, L. J. (1985). The evolution of fetal nutritional adaptations. In H.-R. Duncker & G. Fleischer, eds., *Functional morphology of vertebrates.* Stuttgart: Fischer.

Blick, J. (1977). Selection for traits which lower individual reproduction. *Journal of Theoretical Biology, 67,* 597–601.

Brooks, R. J. (1984). Causes and consequences of infanticide in populations of rodents. In G. Hausfater & S. B. Hrdy (Eds.), *Infanticide* (pp. 331–348). New York: Aldine.

Brown, C. R. (1984). Laying eggs in a neighbor's nest: Benefit and cost of colonial nesting in swallows. *Science, 224,* 518–519.

Bugos, P. E., & McCarthy, L. M. (1984). Ayoreo infanticide: A case study. In G. Hausfater & S. B. Hrdy (Eds.) *Infanticide* (pp. 503–520). New York: Aldine.

Calaby, J. H., & Tyndale-Biscoe, C. H. (Eds.). (1977). *Reproduction and evolution.* Canberra: Australian Academy of Science.

Carlisle, T. R. (1985). Parental response to brood size in a cichlid fish. *Animal Behaviour, 33,* 234–238.

Charnov, E. L. (1982). *The theory of sex allocation.* Princeton, NJ: Princeton University Press.

Clutton-Brock, T. H. (1984). Reproductive effort and terminal investment in iteroparous animals. *American Naturalist, 123,* 212–229.

Cosmides, L., & Tooby, J. (1987). From evolution to behavior: Evolutionary psychology as the missing link. In J. Dupré (Ed.), *The latest on the best: Essays on evolution and optimality* (pp. 277–306). Cambridge, MA: MIT Press.

Curio, E., Regelmann, K., & Zimmermann, U. (1984). The defence of first and second broods by great tit *(Parus major)* parents: A test of a predictive sociobiology. *Zeitschrift für Tierpsychologie, 66,* 101–127.

Daly, M. (1972). The maternal behaviour cycle in golden hamsters *(Mesocricetus auratus). Zeitschrift für Tierpsychologie, 31,* 289–299.

Daly, M. (1979). Why don't male mammals lactate? *Journal of Theoretical Biology, 78,* 325–345.

Daly, M., & Wilson, M. (1983). *Sex, evolution and behavior* (2nd ed.). Belmont, CA: Wadsworth.

Daly, M., & Wilson, M. (1984). A sociobiological analysis of human infanticide. In G. Hausfater & S. B. Hrdy (Eds.), *Infanticide* (pp. 487–502). New York: Aldine.

Daly, M., & Wilson, M. (1988a). *Homicide.* Hawthorne, NY: Aldine de Gruyter.

Daly, M., & Wilson, M. (1988b). The Darwinian psychology of discriminative parental solicitude. *Nebraska Symposium on Motivation, 35,* 91–144.

Daly, M., & Wilson, M. (1988c). Evolutionary social psychology and family violence. *Science, 242,* 519–524.

Davis, R. B., Herreid, C. F., & Short, H. L. (1962). Mexican free-tailed bats in Texas. *Ecological Monographs, 32,* 311–346.

Dawkins, R. (1982). *The extended phenotype.* Oxford: Freeman.

Dunn, J., & Kendrick, C. (1982). *Siblings.* Cambridge, MA: Harvard University Press.

Fisher, R. A. (1958). *The genetical theory of natural selection* (2nd ed.). New York: Dover.

Greig-Smith, P. W. (1980). Parental investment in nest defense by stonechats *(Saxicola torquata). Animal Behaviour, 28,* 604–619.

Gubernick, D. J. (1981). Parent and infant attachment in mammals. In D. J. Gubernick & P. H. Klopfer (Eds.), *Parental care in mammals.* (pp. 243–305). New York: Plenum Press.

Hamilton, W. D. (1964). The genetical evolution of social behaviour. I & II. *Journal of Theoretical Biology, 7,* 1–52.

Hamilton, W. D. (1966). The moulding of senescence by natural selection. *Journal of Theoretical Biology, 12,* 12–45.

Hrdy, S. B. (1979). Infanticide among animals: A review, classification and examination of the implications for the reproductive strategies of females. *Ethology and Sociobiology, 1,* 13–40.

Hrdy, S. B. (1987). Sex-biased parental investment among primates and other mammals: A critical evaluation of the Trivers-Willard hypothesis. In R. J. Gelles & J. B. Lancaster (Eds.), *Child abuse and neglect: Biosocial dimensions* (pp. 97–147). Hawthorne, NY: Aldine de Gruyter.

LeBoeuf, B. J., & Briggs, K. T. (1977). The cost of living in a seal harem. *Mammalia, 41,* 167–195.

Lorenz, K. Z. (1966). *On aggression.* London: Methuen.

Low, B. S. (1978). Environmental uncertainty and the parental strategies of marsupials and placentals. *American Naturalist, 112,* 197–213.

Macnair, M. R., & Parker, G. A. (1979). Models of parent–offspring conflict. III. Intra-brood conflict. *Animal Behaviour, 27,* 1202–1209.

Maynard Smith, J. S. (1978). *The evolution of sex.* Cambridge: Cambridge University Press.

McCracken, G. F. (1984). Communal nursing in Mexican free-tailed bat maternity colonies. *Science, 223,* 1090–1091.

Medvin, M. B., & Beecher, M. D. (1986). Parent–offspring recognition in the barn swallow *(Hirundo rustica). Animal Behaviour, 34,* 1627–1639.

Miller, N. E. (1959). Liberalization of basic S–R concepts. In S. Koch (Ed.), *Psychology: A study of a science* (Vols. I-2, pp. 196–292). New York: McGraw-Hill.

Ostermeyer, M. C. (1983). Maternal aggression. In R. W. Elwood (Ed.), *Parental behaviour of rodents* (pp. 151–179). Chichester: Wiley.

Ozoga, J. J., & Verme, L. J. (1986). Relation of maternal age to fawn-rearing success in white-tailed deer. *Journal of Wildlife Management, 50,* 480–486.

Parker, G. A., & Macnair, M. R. (1979). Models of parent–offspring conflict. IV. Suppression: Evolutionary retaliation by the parent. *Animal Behaviour, 27,* 1210–1235.

Patterson, T. L., Petrinovich, L., & James, D. K. (1980). Reproductive value and appropriateness of response to predators by white-crowned sparrows. *Behavioral Ecology and Sociobiology, 7,* 227–231.

Pressley, P. H. (1981). Parental effort and the evolution of nest-guarding tactics in the three-spine stickleback, *Gasterosteus aculeatus* L. *Evolution, 35,* 282–295.

Pugesek, B. H. (1981). Increased reproductive effort with age in the California gull *(Larus californicus). Science, 212,* 822–823.

Ridley, M. (1978). Paternal care. *Animal Behaviour, 26,* 904–932.

Riedman, M. L. (1982). The evolution of alloparental care and adoption in mammals and birds. *Quarterly Review of Biology, 57,* 405–435.

Sherman, P. W. (1981). Reproductive competition and infanticide in Belding's ground squirrels and other animals. In R. D. Alexander & D. W. Tinkle (Eds.), *Natural selection and social behavior* (pp. 311–331). New York: Chiron.

Stamps, J., Clark, A., Arrowood, P., & Kus, B. (1985). Parent–offspring conflict in budgerigars. *Behaviour, 94,* 1–40.

Svare, B. B. (1981). Maternal aggression in mammals. In D. J. Gubernick & P. H. Klopfer (Eds.), *Parental care in mammals* (pp. 179–210). New York: Plenum Press.

Symons, D. (1987). If we're all Darwinians, what's the fuss about? In C. Crawford, M. Smith, & D. Krebs (Eds.), *Sociobiology and psychology* (pp. 121–146). Hillsdale, NJ: Erlbaum.

Trivers, R. L. (1972). Parental investment and sexual selection. In B. Campbell (Ed.), *Sexual selection and the descent of man 1871–1971* (pp. 136–179). Chicago: Aldine.

Trivers, R. L. (1974). Parent–offspring conflict. *American Zoologist, 14,* 249–264.

Trivers, R. L. (1985). *Social evolution.* Menlo Park, CA: Benjamin/Cummings.

vom Saal, F. S. (1985). Time-contingent change in infanticide and parental behavior induced by ejaculation in male mice. *Physiology and Behavior, 34,* 7–16.

Williams, G. C. (1966). *Adaptation and natural selection.* Princeton, NJ: Princeton University Press.

Williams, G. C. (Ed.). (1971). *Group selection.* Chicago: Aldine Atherton.

Wilson, E. O. (1975). *Sociobiology.* Cambridge, MA: Belknap.

Wilson, M., & Daly, M. (1987). Risk of maltreatment of children living with stepparents. In R. J. Gelles & J. B. Lancaster (Eds.), *Child abuse and neglect: Biosocial dimensions* (pp. 215–232). Hawthorne, NY: Aldine de Gruyter.

Wynne-Edwards, V. C. (1962). *Animal dispersion in relation to social behaviour.* Edinburgh: Oliver & Boyd.

4

Landmarks in the Physiological Study of Maternal Behavior with Special Reference to the Rat

JAY S. ROSENBLATT

The purpose of this chapter is to present landmarks in the study of the physiological basis of maternal behavior among the mammals, with special reference to the study of the rat. Landmarks in research point the way to further understanding and mark a road that is followed by later investigators. Of greatest value are those that combine empirical findings with conceptual formulations, opening up new ways of seeing a phenomenon and of studying it. The tracing of landmarks is necessarily a historical process, but the aim is to understand the contemporary situation and to project into the future by following the path indicated by the landmarks. Therefore, the ultimate purpose of this chapter is to provide a review of the current status of research on maternal behavior.

EVOLUTION OF MATERNAL BEHAVIOR

Maternal behavior among the vertebrates evolved in gradual steps to replace the high metabolic/energetic cost to the female—among fishes, for example—of producing and casting a large number of eggs, fertilized externally, into the marine environment, only a few of which resulted in viable young. By fish producing fewer eggs and either placing them in sheltered nests, previously selected or constructed, or carrying them in the mouth or in special structures on the body, a greater proportion were hatched and survived (Amoroso, Heap, & Renfree, 1979; Breder & Rosen, 1966). Even further protection could be given the young by gestating them in special structures within the mother to which the eggs, fertilized internally, could become attached and in which a small number of embryos could grow until they hatched and were released into the environment (Amoroso et al., 1979; Browning, 1973). Parental care followed the release of the newly hatched young, but generally it was of short duration and only rarely did it involve providing nutrition to the young by either parent (Amoroso et al., 1979; Browning, 1973).

Mammals have specialized in maternal care, but it evolved slowly even in this class. Among the echidna (spiny anteater, platypus), egg laying and incubation are retained, and maternal care is initiated only after the young hatch, when lactation is also initiated. The marsupial mammals (kangaroos, wallabys) also do not initiate maternal behavior at parturition: their embryonic young crawl into the pouch, grasp a nipple, and suckle, a sign that lactation has been initiated. Maternal behavior appears only months later, when the young leave the pouch for the first time, returning to it to suckle, until they leave it permanently (Russell, 1982; Sharman, Calaby, & Poole, 1966).

Among the marsupials, there is no true gestation. The period of gestation is contained within the estrous cycle, and there are no hormonal changes during gestation that distinguish it from a normal estrous cycle. Nor is there evidence that the initiation of maternal behavior months after parturition is accompanied by specific hormonal changes (Harder & Fleming, 1981; Sharman et al., 1966).

Among the placental mammals, maternal behavior accompanies parturition and lactation and shares with them a physiological (i.e., hormonal) basis. There is a true gestation period (i.e., gestation extends beyond the normal estrous cycle), which is based on the formation of functional corpora lutea or the retention of those that are otherwise active for a shorter period during nonfertile reproductive cycles (Heap, Galil, Harrison, Jenkin, & Perry, 1977). The number of young that the female produces during each cycle ranges from one to several, and synchrony of delivery with lactation and of both with the onset of maternal behavior ensures that the needs of the young will be satisfied. The reduced metabolic cost of producing a few young compared with the high metabolic cost of producing many has been made possible among the mammals by substituting for this high cost the less costly, but much more effective, parental care of the young.

MATERNAL RESPONSIVENESS DURING PREGNANCY

Although there is a sudden initiation of parturition and lactation at the end of gestation, in reality they are the result of gradual changes during pregnancy in uterine responsiveness to hormones and in hormonally stimulated mammary gland development and milk synthesis. Maternal behavior appears suddenly at parturition, but studies have shown that maternal responsiveness also develops gradually during pregnancy (Rosenblatt & Siegel, 1983; Rosenblatt, Siegel, & Mayer, 1979).

Early research on maternal behavior in the rat (Kinder, 1927; Sturman-Hulbe & Stone, 1929) was not physiologically oriented: retrieving and nest building were studied as adaptations that ensured that the young would be cared for in a nest in the face of environmental hazards. Perhaps the first physiologically oriented study of maternal behavior was that of Stone (1925), who investigated the causal connection between maternal behavior and the hormones of pregnancy. In his early attempt (which failed) to stimulate maternal behavior in a nonpregnant cycling female, he joined her in parabiosis with a pregnant female. This biobehavioral relationship was made explicit by Wiesner and Sheard (1933), who noted that the onset of maternal behavior is synchronized with the onset of parturition and the initiation of lactation. Since parturition and lactation were known to depend on the hormones of pregnancy, noting this synchrony implied that perhaps maternal behavior is also based on these hormones. This idea guided these investigators in their subsequent studies of the hormonal basis of maternal behavior.

Among mice and hamsters, nest building, an important component of maternal behavior, is initiated soon after conception at a time when the estrogen in circulation declines and the progesterone increases (McCormack & Greenwald, 1974). This is the earliest time during pregnancy that maternal behavior has been observed, and in mice its appearance has been causally linked with the synergistic action of progesterone with estrogen (Koller, 1955; Lisk, 1971; Lisk, Prelow, & Friedman, 1969). In the rabbit, nest building normally appears shortly before parturition and is associated with hair loosening and pulling, the hair being incorporated into the nest to form a soft inner lining (Ross, Sawin, Zarrow, & Denenberg, 1963). The readiness to build a nest in response to the hormonal changes at the end of pregnancy develops gradually during pregnancy. Zarrow and his colleagues terminated pregnancies prematurely by ovariectomizing females or treating them with estrogen. They observed that only when pregnancies were terminated after 13 days in the Dutch-belted strain and after 22 days in the Grey Chinchilla strain did females initiate nest building. Estrogen and progesterone treatment of ovariectomized female rabbits was also effective, when administered for 13 or 22 days, in stimulating maternal nest building (Zarrow, Farooq, & Denenberg, 1962; Zarrow, Sawin, Ross, & Denenberg, 1962).

Pregnancy termination by hysterectomy in rats also stimulates maternal behavior. The full complement of nest building, nursing behavior (without lactation), retrieving and anogenital licking has been observed; more recently, nest defense has been added to this list (Mayer, Reisbick, Siegel, & Rosenblatt, 1987; Rosenblatt & Siegel, 1975). Pregnant females also show increasing maternal responsiveness to foster young as pregnancy advances (Bridges, Feder, & Rosenblatt, 1977; Mayer & Rosenblatt, 1984; Siegel & Rosenblatt, 1975a). Hysterectomy of pregnant mice also stimulates maternal responsiveness, as measured by the operant response of lever pressing to obtain a pup (Hauser & Gandelman, 1985). Several strains show an increase in nest defense during pregnancy. In these strains, virgin females readily exhibit maternal behavior toward young, but during early pregnancy they lose this responsiveness and then regain it later in pregnancy (Noirot & Goyens, 1971).

The discovery, in nearly all species that have been studied, that maternal responsiveness develops during pregnancy challenged the earlier belief that pregnancy is a behaviorally silent period. It raised questions about the similarly held idea that short-latency maternal behavior could be elicited only when females had begun to deliver their young. How is the onset of maternal behavior so closely correlated with the birth of the young? Either the onset is endogenously regulated to coincide with the birth of the young, or it is dependent on stimulation received before parturition or from the young during parturition. It could be a combination of endogenous regulation and dependence on stimulation associated with parturition that enables the onset of maternal behavior to be coordinated with the delivery of the young.

Investigators were led by their observations of maternal behavior at parturition to conclude that maternal responsiveness begins at parturition. However, this was not borne out when they began to present newborns to females at varying intervals before parturition to observe whether they exhibited maternal behavior toward the young and, in those species that exhibit defense of the young, whether they defended them against an intruder. In many rodent species (hamster, rat, gerbil), cannibalism of young is prevalent during pregnancy; females have been tested during pregnancy to see when it subsides, as it does at parturition.

In all rodent species that have been studied, including the rabbit, as was noted, and in sheep (only multiparous ewes, though), females exhibit maternal behavior more readily later in pregnancy than earlier; certain responses often appear immediately when females are presented with newborns (Poindron & Le Neindre, 1980; Rosenblatt & Siegel, 1983). Different components come to the fore in the different species: in the gerbil, nest building appears and cannibalism declines prepartum (Rosenblatt & Siegel, 1981); hamster females exhibit all components of maternal behavior, including retrieving, adopting a nursing position, and terminating pup killing (Siegel & Greenwald, 1975); in sheep, the prospective mother stands and accepts the nursing approaches of the newborn lamb and emits low bleats, evidence of her maternal responsiveness toward it (Poindron & Le Neindre, 1980).

In the rat, all components of maternal behavior can be elicited starting 24 to 34 hr before parturition (Mayer & Rosenblatt, 1984; Rosenblatt & Siegel, 1975). These include the behaviors noted above for the hamster and, in addition, nest defense and approach responses to nest odors and to the calls of hidden pups (Dunbar, Ranson, & Buehler, 1981; Koranyi, Lissak, Tamasy, & Kamaras, 1976). The former is also seen in dogs during parturition and the latter in cats before parturition (Dunbar et al., 1981). Female rats that have not exhibited maternal behavior or nest defense earlier rapidly change their behavior and become maternal around 3.5 hr before parturition begins. Such maternal behavior has been correlated with the onset of vigorous and regular uterine contractions that are a prelude to parturition (Mayer & Rosenblatt, 1984). This will be discussed later.

MOTIVATIONAL BASIS OF MATERNAL BEHAVIOR

There are two problems concerning the motivational basis of maternal behavior in the rat (and other mammals) that should be discussed at this point. The first is whether motivation to exhibit maternal behavior is conceived of as a threshold phenomenon or as the resolution of conflicting tendencies in the female's response to her young.

On the basis of early studies of postpartum, lactating mothers, it was believed that pup odors were attractive to mothers and formed a component of the stimuli to which the mother responded in exhibiting maternal behavior (Beach & Jaynes, 1956). While this may be true of lactating females (Smotherman, Bell, Starzec, Elias, & Zachman, 1974), the situation is quite different for nonpregnant females; the change from nonmaternal to maternal behavior during pregnancy and parturition may involve an important change in the female's olfactory response to the young.

Nonpregnant females can be induced to exhibit maternal behavior by continuous exposure to pups, a process that is called either "sensitization," "concaveation," or "pup induction" (Cosnier & Couturier, 1966; Rosenblatt, 1967; Wiesner & Sheard, 1933). Latencies to sensitization range from 2 days in females confined to small cages to 4–6 days in females living in larger cages, but strains differ in this respect: some strains may require longer periods for sensitization to occur, and one strain has proved completely unresponsive in this respect (Jakubowski & Terkel, 1985; Terkel & Rosenblatt, 1971).

When the phenomenon of sensitization was discovered (Cosnier & Couturier, 1966; Rosenblatt, 1967; Wiesner & Sheard, 1933), it was proposed that exposing females to pups day by day increased their responsiveness to the pups until a threshold level of responsiveness was reached and they exhibited maternal behavior (Rosenblatt, 1967). The motivational process was believed to be similar to sexual stimulation, which is still viewed as increasing under the influence of sexual stimulation or hormones until a threshold level is reached and sexual behavior is exhibited.

Studying the role of olfaction in the sensitization process made it necessary to change this conception of maternal motivation. The most important finding in this respect was that females that were made anosmic with a nasal lavage of zinc sulfate or in which the lateral olfactory nerve was cut bilaterally became highly responsive to pups almost immediately, and nearly all exhibited maternal behavior during the first day of pup exposure (Fleming & Rosenblatt, 1974a, 1974b). Instead of delaying the appearance of maternal behavior, as would be expected if olfactory stimuli from the pups enhanced maternal responsiveness, making females anosmic accelerated its onset.

From these findings, it was concluded that the delay in the onset of maternal behavior in nonpregnant females undergoing sensitization is due to their need to overcome their aversive response to pup odors. Aversive responses in nonpregnant females presented with pups can be observed: they approach the pups out of curiosity, sniff them and draw back, closing their eyes, and often cover the pups with nest material and remain at a distance from them for a long period (i.e., 2–3 days; Terkel & Rosenblatt, 1971). When maternal behavior appears during sensitization, it is not only because a level of maternal responsiveness has been reached, but also because an aversive response to newborn pups has gradually waned; that is, conflicting tendencies in the female have been resolved. The aversive response may be based on other stimuli in addition to olfaction and may consist of fear and timidity related to newborn young (Fleming & Luebke, 1981). As females become sensitized and overcome their timidity, they spend more time in proximity to the pups, often in contact with them; they resemble more closely postparturient, lactating mothers in their emotional responses to pups and in an open-field test. The reduced fearfulness of lactating females, measured by decreased freezing in response to a loud auditory stimulus, can be increased by the use of benzodiazepine antagonists, implying that the neurotransmitter gamma aminobutyric acid (GABA) plays a role in reducing the fearfulness of mothers (Hansen, Ferreira, & Selart, 1985; Hard & Hansen, 1985).

The concept that maternal behavior arises in female rats after withdrawal responses to the pups have waned, following which approach responses increase, has implications for the development of maternal behavior and for its neural basis. Components of maternal behavior can be readily elicited in prepubertal pups, and the pattern gradually becomes distinct from play behaviors, better organized and more complete than earlier, during puberty (Brunelli, Shindledecker, & Hofer, 1987). However, latencies abruptly increase around 23–24 days of age, and this change in behavior is correlated with growing fearfulness, evident in their responses to newborns and in open-field behavior (Bridges, Zarrow, Goldman, & Denenberg, 1974; Mayer, 1983; Mayer & Rosenblatt, 1979). After puberty, the aversive response to pup odors appears, and the sensitization latencies of adults remain above those of the less fearful prepubertal adolescents (Mayer, Freeman, & Rosenblatt, 1979).

Males and females differ in their maternal responsiveness, and this difference may be related to gender differences in the morphology of the medial preoptic area (MPOA), a critical area for the performance of maternal behavior in the rat (see chapter 5). After puberty (45 days), differences in maternal responsiveness between males and females arise: males that were as responsive as or even more responsive than females from day 18 to day 30 gradually become less responsive as females become more responsive, and by day 90, many males are unresponsive to pups despite prolonged exposure to them (Bridges et al., 1974; Gray & Chesley, 1984; Mayer et al., 1979; Mayer & Rosenblatt, 1979). Postpubertal males appear to become more timid than females toward pups, and it has also been reported that males at this age also become more timid in the open-field situation (Slob, Huizer, & Ten Bosch, 1986). In part, the difference that arises between males and females is based on concurrent hormonal factors and can be altered by ovariectomizing females; another method is by handling males to reduce their timidity, but there remains a residual gender difference in responsiveness that appears to be based on the pre- and early postnatal effects of gonadal hormones (Mayer, 1983).

The balance of approach and withdrawal responses to young in the rat determines whether females will display maternal behavior. However, once females have exhibited this behavior, having overcome their aversion to pup odors, the balance remains in favor of approach for some time, perhaps for life. Even the small amount of contact with the pups during parturition, admittedly when the female is highly excited and very responsive to them, is sufficient to maintain the balance in favor of approach for at least 25 days (Bridges, 1975, 1977; Mayer & Rosenblatt, 1975). Although females do not immediately exhibit maternal behavior when presented with pups once again, indicating that withdrawal responses have increased somewhat from the earlier period, they require only a short time (e.g., 24 hr) to reestablish their earlier behavior.

The second problem concerning the motivational basis of maternal behavior is the degree to which all the components of maternal behavior are included within a single motivational system. Is there a broadly based maternal motivation, or are the different components motivated individually or in groups? Different ways of studying maternal behavior in the rat give different answers to these questions. During observations of a lactating female performing maternal behavior, the different components—nursing, retrieving the pups to the nest, anogenital licking, nest defense, and nest building—occur in succession over a short period of time if the stimulus situations that elicit these behaviors are present. Moreover, these behaviors are maintained over the course of the 3-week lactation period, and they decline together or within a short time of one another at the end of this period (Rosenblatt & Lehrman, 1963; Slotnick, 1967). In addition, when hormones that stimulate maternal behavior in nonpregnant or in pregnant hypophysectomized-ovariectomized females are administered, all the components of maternal behavior are stimulated (Moltz, Rowland, Steele, & Halaris, 1970; Siegel & Rosenblatt, 1975a, 1975b; Zarrow, Gandelman, & Denenberg, 1971). These observations suggest that the various components are tied to one another by an underlying motivational organization.

Yet there are situations that suggest that the various components are at least partially independent of one another. Nursing may persist in females that have declined in all other components of maternal behavior (Rosenblatt, unpublished observations), nest building may decline earlier than other components (Rosenblatt & Lehrman, 1963), and nest defense cannot be elicited except with great difficulty in nonpregnant females that are exhibiting all other components of maternal behavior (Erskine, Barfield, & Goldman, 1980; Mayer & Rosenblatt, 1987). Retrieving may be retained in females that have had experience with maternal behavior and may be elicited almost immediately, although other components (nursing and taking care of pups in the nest) require an additional 2 days to reappear (Mayer & Rosenblatt, 1975). The prepartum onset of maternal behavior is often led by nest building and is then followed by the other components (Rosenblatt & Siegel, 1975). Finally, somewhat different stimuli elicit the different components: nest material elicits nest building; pup odors and calls elicit retrieving; pups in the nest elicit anogenital licking and nursing; and intruding animals elicit nest defense. An analysis of the neural basis of maternal behavior provides evidence for both a common site of neural organization of maternal behavior and pathways more or less specific for particular components (Numan, 1988; chapter 12). Lesions of the MPOA in lactating females prevent the performance of maternal behavior, while estrogen implants in this region stimulate maternal behavior. Yet it has been possible to block retrieving behavior in lactating females by systemic injections of haloperidol without blocking the other components of maternal behavior (Giordano, Johnson, & Rosenblatt, unpublished observations). In addition, nest defense is almost absent in lactating females with peripeduncular nuclei lesions, with no interference with other components of maternal behavior (Hansen & Ferreira, 1986).

As noted above, a developmental analysis of maternal behavior provides still another approach to understand the motivational basis of maternal behavior. Maternal responses performed by adolescent young are components of a generalized social behavior that includes play behavior, which itself is not well understood (Mayer & Rosenblatt, 1979; see chapter 18). However, even in 18- to 23-day-old pups, there are specific responses to younger pups that differ from those to older siblings and to nonliving objects: adolescents prefer to crouch over young pups (as in nursing) than over siblings and prefer pups to similarly warm, soft objects; in addition, they bite pup-size toys, while they lick and carry pups and never bite them (Mayer & Rosenblatt, 1979; Rosenblatt, 1975). The components of maternal behavior begin to form a more coherent pattern resembling that of the adult around the time of puberty (see chapter 18).

We can conclude tentatively that the motivational basis of maternal behavior in the adult female is general sociability, with the approach and withdrawal aspects common to all social responses. It is a specialized motivational system, however, that arises from the action of hormones on specific regions of the central nervous system and on peripheral structures (e.g., mammary glands, snout sensitivity, auditory and olfactory responsiveness), on the one hand, and, on the other hand, from the nature of the stimuli to which the female responds, each of which has specific eliciting properties.

HORMONAL BASIS OF MATERNAL BEHAVIOR

The hormones of pregnancy—estrogen, progesterone, and prolactin—are the principal candidates for stimulation of maternal behavior. The pattern of circulating levels of these hormones is quite similar in nearly all mammals that have been studied, with hamsters and monkeys as notable exceptions (Rosenblatt & Siegel, 1981). Too few species have been studied, however, to allow us to determine whether the uniformity in hormonal secretions during pregnancy implies a similar uniformity in the hormones that stimulate maternal behavior.

Two important concepts have emerged in this area of study: hormonal "priming" and "triggering" of maternal behavior. Hormonal priming refers to hormonal effects that do not in themselves stimulate maternal behavior but that enable a subsequent hormonal stimulus to elicit the behavior. Pregnant females are primed by the hormones of pregnancy to respond to the terminal rise in estrogen (and perhaps other hormones; see below) by initiating maternal behavior (see below for further discussion of this example). The two effects are also illustrated when 16-day pregnant females are hysterectomized: the termination of pregnancy results in a rise in estrogen, which, together with the previous priming, stimulates maternal behavior (Rosenblatt & Siegel, 1975). If the female's ovaries are removed at the time of hysterectomy, she is only primed and therefore does not display as rapid a short-latency maternal behavior; a single triggering dose of estrogen causes the rapid onset of maternal behavior (Siegel & Rosenblatt, 1975a).

Estrogen injected into *nonpregnant* females that have been ovariectomized and hysterectomized also stimulates maternal behavior (Siegel & Rosenblatt, 1975b). However, higher doses of estrogen are required (at least five times higher), and latencies are longer than in hysterectomized and ovariectomized pregnant females. The previous period of pregnancy, therefore, primes the females to respond to estrogen by exhibiting maternal behavior.

Prolonged treatment with estrogen and progesterone, combined with endogenous prolactin, which simulates the hormones of pregnancy, can also prime females to respond to the triggering stimulus of estrogen; maternal behavior is blocked, however, if progesterone levels remain high (Bridges, Rosenblatt, & Feder, 1978; Siegel & Rosenblatt, 1975b).

The behavioral effects of priming females with estrogen and progesterone, or of 16 days of pregnancy followed by estrogen triggering, are dependent on the uptake of estrogen in the preoptic area (POA) of the brain (Giordano, 1987; Giordano, Siegel, & Rosenblatt, in press). Uptake of estrogen is measured by the concentration of cytosol and nuclear estrogen receptors over the 48 hr following injection of estrogen in hormonally primed females. Primed females have a high initial concentration of nuclear estrogen receptors and a low concentration of cytosol estrogen receptors. Nonprimed females have the opposite receptor pattern. In primed females, triggering is correlated with a rise in nuclear estrogen receptors over 12–24 hr following estrogen stimulation.

The gradual increase in maternal responsiveness in the rat during pregnancy may be an expression of the continual priming by ovarian estrogen and progesterone and possibly by placental lactogen during pregnancy (Bridges, 1984; Bridges, Loundes, DiBiase, & Tate-Ostroff, 1985; Giordano et al., in press). The terminal surge of estrogen secretion by the ovaries may function as the triggering stimulus, causing a prepartum onset of maternal behavior when pups are presented; this may be the basis for the normal postpartum appearance of maternal behavior. Blocking the triggering action of estrogen at the MPOA with brain implants of an antiestrogen (4-hydroxy tamoxifen) starting on day 20 of pregnancy prevents the prepartum onset of maternal behavior. In addition, if females are delivered by caesarean section, thereby preventing them from having contact with pups during parturition, it blocks the appearance of maternal behavior "postpartum" in a significant proportion of females (Ahdieh, Mayer, & Rosenblatt, 1987).

The concept of two phases in the hormonal stimulation of maternal behavior in the rat was present in one earlier study but was absent in another. Moltz, Lubin, Leon, and Numan (1970) used prolonged treatment with estrogen and progesterone followed by prolactin to stimulate maternal behavior in ovariectomized females and proposed that the ovarian hormones, in particular the withdrawal of progesterone, primed the female to respond to the injections of prolactin. Zarrow et al. (1971) used a similar hormonal regimen but did not propose a two-phase theory of hormonal action.

Studies have identified the hormone or hormones involved in stimulating maternal behavior in species other than the rat, but they have not been as extensive as the studies with the rat. In the rabbit, estrogen and progesterone also appear to prime females, but the triggering stimulation may be due to prolactin. In the mouse, as noted earlier, progesterone, primed by estrogen, stimulates nest building and maternal aggression during pregnancy (Lisk, 1971; see chapter 7). In the hamster, blocking the release of prolactin either prepartum or early postpartum results in an absence of maternal behavior and maternal aggression, but prolactin has not yet been shown to restore this behavior. In the ewe, estrogen is effective in inducing maternal behavior in experienced females, and prolactin is not needed (Poindron & Le Neindre, 1980; Siegel & Greenwald, 1975). More recent studies have implicated oxytocin (see chapters 8, 13, and 14).

SENSORY CONTROL OF MATERNAL BEHAVIOR

The release of hormones that stimulate maternal behavior is believed to be an endogenous process involving several of the hormones that stimulate parturition and lactation. It has now been shown, however, that the fine-tuning of the onset of maternal behavior in the rat may be dependent on sensory stimuli that arise in preparation for parturition. Earlier, it was proposed (Birch, 1956) that female licking of newborn pups, bathed in uterine fluids during parturition, stimulates the onset of maternal behavior in the rat. The female is attracted to this salty fluid because of a salt deficiency developed during pregnancy, and through her licking of the newborn, she transfers this attraction to them (Steinberg & Bindra, 1962). This hypothesis has not been substantiated by several other investigators, despite the original report (Rosenblatt, 1984).

The stimulation that contributes to the onset of maternal behavior may, however, be uterine contractions in the immediate prepartum period and during parturition itself. The correlation between the onset of maternal behavior about 3.5 hr before parturition and the start of vigorous uterine contractions has already been noted; we can add the observation that severing the pelvic nerve that carries afferent stimulation from the lower uterus to the central nervous system delays the onset of maternal behavior (Mayer & Rosenblatt, unpublished observations). In addition, nonpregnant females, ovariectomized and treated with estrogen, show a rapid onset of maternal behavior (less than 3 hr) when given vaginal-cervical stimulation with a rod, provided, however, that they have previously exhibited maternal behavior while rearing a litter (Yeo & Keverne, 1986). Moreover, uterine distention (i.e., by injection of saline into the uterus) of pseudopregnant females with or without deciduoma (i.e., progesterone-secreting cells that prolong pseudopregnancy from 12 to 22 days) shortened latencies for maternal behavior to less than 2 days compared with 5–6 days without uterine distention (Graber & Kristal, 1977).

Among sheep, as reported in this volume and elsewhere (Keverne, 1988; Poindron, Lévy, & Krehbel, 1988), vaginal-cervical stimulation appears to be important in stimulating the onset of maternal behavior in first-time parturient ewes. Blocking this stimulation interferes with the onset of maternal responsiveness to the newborn lamb, and applying this stimulation to estrogen-primed, experienced females stimulates a rapid onset of maternal behavior. Ewes that have already formed a selective bond to their own lambs and rebuff alien lambs can be made to accept alien lambs by a short period of vaginal-cervical stimulation about 2 hr after parturition (Keverne, Lévy, Poindron, & Lindsay, 1983).

Another source of stimulation contributing to the initiation of maternal behavior in the rat may be tactile stimulation of the snout and mouth area during parturition (Stern, Johnson, & McDonald, 1987). Nonpregnant females presented with a single pup invariably sniff it and make snout contact, but they rarely pick it up in the mouth (Rosenblatt, 1975). The parturient female sniffs and contacts the pup, and most often picks it up in her mouth and carries it to her nest. The difference in her response when making snout contact with pups may be based on increased tactile sensitivity of the facial region and enlargement of the sensory receptive fields induced in females during pregnancy by prolonged estrogen exposure (Berieter & Barker, 1975, 1980). Eliminating snout sensitivity by lesioning the infraorbital nerve severely interferes with all aspects of parturitive behavior (Stern et al., 1987). Snout contact with the young during parturition is perhaps the most universal behavior among all mammals; its importance in stimulating the onset of maternal behavior has only recently been recognized and investigated.

ADDITIONAL FACTORS INVOLVED IN THE ONSET
OF MATERNAL BEHAVIOR

The belief that the physiology of the onset of maternal behavior could be understood with reference chiefly to the ovarian steroids has given way to a more complex view in which many additional factors are seen as contributing to the onset. Nevertheless, research on the ovarian steroids has established the paradigms that have served as the springboards for the discovery of new factors involved in initiating maternal behavior.

Prolactin, as noted earlier, has emerged as one of the hormonal components of the priming process during pregnancy and as a possible triggering stimulus in the rat, as it is in the rabbit. It has been proposed that estrogen stimulates maternal behavior in the rat through its effect on prolactin synthesis and release (Bridges & Dunckel,

1987; Bridges et al., 1985). One study has reported that prolactin may effectively stimulate maternal behavior in the absence of estrogen when there is priming by progesterone alone. During pregnancy, it is possible that placental lactogen acts as prolactin to prime the female to respond to the terminal rise in estrogen, initiating maternal behavior (Bridges et al., 1985). This raises the interesting possibility that the fetus, through its secretion of lactogen from the placenta, acts on the mother to initiate the care that it will require at parturition and thereafter.

Prolactin has been implicated in the maternal behavior of the hamster and the mouse (Rosenblatt et al., 1979), but in the ewe the high level of prolactin present at parturition and afterward can be blocked (while estrogen levels remain high) without affecting maternal behavior (Poindron & Le Neindre, 1980). Whether prolactin has a priming action in the ewe has not been investigated.

In the rat, mouse, and ewe, oxytocin, a peptide normally associated with uterine contractions during pregnancy and milk letdown during lactation, has been shown to be involved in the stimulation of maternal behavior. It has been proposed that in the rat and ewe, oxytocin is released centrally, possibly from the paraventricular nucleus into the cerebral ventricles, but is principally transmitted to various brain sites that mediate maternal behavior (at least in the rat; see chapter 12). The release of oxytocin centrally may be in response to vaginal-cervical stimulation during parturition, but stimulation of maternal behavior in females requires that they receive prior estrogen stimulation (Baldwin, Kendrick, & Keverne, 1986; Fahrbach, Morrell, & Pfaff, 1984, 1985; McCarthy, Bare, & vom Saal, 1986; Pedersen & Prange, 1979; Van Leengoed, Kerker, & Swanson, 1987). The presence of estrogen-stimulated oxytocin receptors in one of these brain regions is reported by Insel in chapter 13. The studies that have provided the evidence for the ewe are presented in detail in chapters 8 and 14. The evidence for the rat raises questions about strain specificity of the effect of oxytocin on maternal behavior (Bolwerk & Swanson, 1984; Rubin, Menniti, & Bridges, 1983) and, where oxytocin has been shown to be effective, the possible mechanism. A critique has been published elsewhere (Rosenblatt, Mayer, Aldieh, & Siegel, 1988): the main question is whether oxytocin, when it is effective, is acting on the background of reduced olfactory function, itself a condition that favors the rapid onset of maternal behavior in nonpregnant females (Wamboldt & Insel, 1987).

Although the results are too few to warrant a detailed review, it is worth mentioning a number of additional factors that are currently being studied for their effects on maternal behavior. Norepinephrine in the MPOA and in the olfactory bulbs has been investigated. In the rat, norepinephrine involvement at the MPOA, reported earlier has not been confirmed (Bridges, Clifton, & Sawyer, 1982; Moltz et al., 1975; Steele, Rowland, & Moltz, 1979); but in the ewe, in which olfactory recognition of the mother's own young plays a crucial role in her maternal behavior, interrupting norepinephrine input to the bulb eliminates this selectivity, and maternal behavior is shown to alien lambs as well (Pissonnier, Thierry, Fabre-Nys, Poindron, & Keverne, 1985; see chapter 14).

Several studies have suggested that the endogenous opioids β-endorphin and met-enkephalin, by decreasing in the brain just before parturition, may play a permissive role in the onset of maternal behavior and postpartum aggression (Bridges & Grimm, 1982; Bridges & Ronsheim, 1983; Grimm & Bridges, 1983; Kinsley & Bridges, 1986) but a mixed role in parturition and parturitive behavior (Leng et al.,

1985, 1987; Mayer, Faris, Komisaruk, & Rosenblatt, 1985). Opiate treatment inhibits maternal behavior, and reduced levels of opiate activity promote maternal care. Ovarian steroids modulate opiate receptor and endogenous opiate concentrations in the MPOA, with lowered concentrations associated with the onset of maternal behavior around the time of parturition and an increase in maternal aggression during lactation (Hammer & Bridges, 1987). Antagonizing the action of endogenous opiates during parturition severely reduces pup cleaning and placenta eating (Mayer et al., 1985), while inducing the release of endogenous opiates through stress interferes with delivery by inhibiting the pituitary release of oxytocin, which is needed to stimulate vigorous uterine contractions (Leng et al., 1985, 1987).

PARENT–YOUNG CONFLICT THEORY AND MATERNAL BEHAVIOR

Maternal behavior and mother–young relationships have been shaped by natural selection. An important objective of this selection has been to speed up reproduction to its maximum, and within this larger objective is the specific one of weaning young at an age when they are capable of surviving the postweaning environment they enter. Focus on this period has given rise to the parent–young conflict theory proposed by Trivers (1974), and documented and extended by Daly (see chapter 3). The general objective of speeding reproduction will be discussed first, and then the specific one of parent–young conflict.

In many species, exemplified by the rat, reproduction is accelerated by the occurrence of a postpartum estrous period when females permit the male to mate with them (otherwise they attack males that approach them and their nest) during the first 24 hr after parturition (Gilbert, Pelchat, & Adler, 1980). Conception occurs, but implantation of the blastocyts is delayed up to 9 days (above the 6-day delay that occurs during regular estrous cycling). The delay occurs because suckling by the young inhibits the pituitary release of gonadotropins that stimulate estrogen release, needed for implantation (Rothchild, 1962). By the time the present litter is weaned, the female has initiated her second pregnancy, and a new litter is born soon after the weaning (Gilbert, 1984). By becoming pregnant during postpartum estrus, females accelerate reproduction by 25% over those that do not become pregnant and mate again only after the third week of lactation (Rosenblatt & Lehrman, 1963).

The acceleration of reproduction is made possible because maternal behavior is no longer dependent on hormonal stimulation once it is initiated at parturition (Rosenblatt & Lehrman, 1963). There is residual stimulation from the hormones that initiated the behavior, but the female becomes increasingly dependent on stimulation from the young to maintain and alter her maternal responsiveness (Orpen, Furman, Wong, & Fleming, 1987). Of course, lactation depends on pituitary hormones during this period. Thus the endocrine system can initiate and maintain the second pregnancy while the first litter is receiving care from the mother.

A corollary of this is that the mother's behavior is closely attuned to the young, just as their behavior is adjusted to hers and to changes in her responsiveness. It has been suggested that in the early phase, the two exist in a symbiotic relationship in which the mother provides resources to the young and, in turn, receives from the young the water and salt she has given up and relief from her unusually high body temperature (Rosenblum & Moltz, 1983).

The synchrony in the behavioral interaction between the mother and her off-spring (Rosenblatt, 1965) has been termed a "complementary relationship" (Hinde, 1979) because the contribution of each to the relationship complements that of the other. When the young are helpless and unable to satisfy their needs independently, the mother's responsiveness is high and she plays the major role in providing for the needs of the young; they, in turn, evoke caregiving behavior from her. As the young develop their capacities for taking the initiative in satisfying their own needs, the mother's caregiving behavior begins to decline. This initiates the process of weaning, which eventually terminates the relationship (although other relationships may develop between mother and young) as the young attain complete independence of the mother, and her responsiveness to this litter wanes. Weaning, therefore, is a phase of the relationship that grows out of earlier phases.

Weaning may be viewed in another light: parent–infant conflict theory proposes that weaning is the occasion of a conflict between the mother and her young over different interests having to do with the inclusive fitness of each. The mother's interest is in weaning her young as early as is compatible with their postweaning survival, and this is in conflict with the interest of the young in prolonging their feeding and dependent relationship with the mother in order to increase their chances of postweaning survival to reproductive age (Trivers, 1974). This is a conflict of interests and, in fact, exists from the moment the young are conceived, since at that time genotypic differences between mother and young come into being that are the basis for their conflict of interests. This theory has significant predictive value, as shown by Daly in chapter 3 of this volume: it can predict the times during the relationship when the young are more vulnerable to attack by the parents and when they are less vulnerable, and when parents are themselves most likely to disrupt the relationship. There is little doubt that the outcomes of behavior shape the mechanisms underlying behavior, just as these mechanisms provide the basis for achieving these outcomes.

Little contact has been made between the kind of analysis of underlying mechanisms of maternal behavior and, ultimately, the mother–young relationship that we have pursued in this chapter and parent–young conflict theory. One thing is clear, however, and Daly has acknowledged this in chapter 3: "parent–young conflict" is not a behavioral term but a term describing a conflict of interests or outcomes, and it cannot be used as an explanatory term with reference to the process of weaning in the rat and other animals. The conflict referred to in the theory of parent–young conflict is not the behavior often seen in litters in which the young try to nurse from the mother while she avoids and evades them, sometimes exhibiting hostile behavior toward them (Rosenblatt, 1965). This is a behavioral conflict; the conflict referred to is a conflict of interest between the mother and her offspring arising from their somewhat different genotypes and the effort of each to improve its own inclusive fitness.

CONCLUDING REMARKS

The physiological basis of maternal behavior is now a well-established fact, as well as a thriving field of study that has attracted an increasing number of investigators from a wide variety of fields having a broad range of methodologies to apply to its study. It is a complex field of study because the behavior pattern itself is composed of many different interrelated behavioral items. It has a developmental aspect in two senses of the term: the pattern itself develops during the reproductive cycle and undergoes its own cycle of growth, maintenance, and decline. In addition, during ontogeny, maternal behavior can be elicited from very young animals of both sexes in the rat and perhaps other species (e.g., hamsters, humans), and from then on it undergoes developmental changes into adulthood.

Maternal behavior is subject to many levels of analysis from different viewpoints. The endocrine basis of maternal behavior has received attention over the longest period and continues to do so. Endocrine studies point in two directions: toward the behavior patterns typifying maternal care and toward the nervous system mediating these patterns. The neural basis of maternal behavior has been studied at the levels of regional, cellular, and intracellular analysis. It has now become necessary to study the interactions of hormones with a wide variety of neuroactive substances (e.g., neuropeptides, catecholamines) that mediate the effects of the hormones or act as modulators of hormone action. A new entry into the field is the role of sensory stimuli from the birth process itself that appear to mobilize and trigger endocrine and related responses necessary for the timing of the onset of maternal behavior. Stimuli from the pups for initiating and maintaining maternal behavior have been known for some time, but the endocrinological basis for the female's responsiveness to these stimuli is a new field of investigation.

While this survey has been limited to the physiological analysis of maternal behavior, functional aspects of maternal care are equally important. Chapter 3 deals with one aspect of this subject—the outcomes of evolutionary pressures shaping the mother's relationship to her young in different species. There is a very large literature on the study of the interaction of mother and young from the point of view of the behavioral and physiological mechanisms underlying mother–young interaction in different species. As yet, except for broad statements concerning evolutionary processes and maternal behavior, it is not yet possible to relate these in detail to behavioral mechanisms.

For the first time, efforts have begun to study maternal behavior in humans along the lines indicated by studies of animals. The previous direct application of findings from lower animals to humans has stimulated only a limited range of studies of parturitional contact between mother and newborn paralleling similar studies in rats, goats, and sheep (Klaus & Kennell, 1976). The newer studies, exemplified in chapter 10 in this volume, have adopted a broader view and are based on a thorough knowledge of the animal studies and a more subtle and sophisticated psychological approach to human maternal behavior. One can look forward to an even more exciting era of research on maternal behavior than the one through which we have just passed.

ACKNOWLEDGEMENTS

This chapter was written while the author was supported by Biomedical Research Support and Busch Memorial grants. The research reported in this chapter from the author's laboratory was supported by grant MH-08604 and was done by a large number of former students, particularly my associates Anne D. Mayer, Harold Siegel, and Harry Ahdieh. Winona Cunningham provided secretarial services. Institute of Animal Behavior Publication No. 477.

REFERENCES

Ahdieh, H. B., Mayer, A. D., & Rosenblatt, J. S. (1987). Effects of brain antiestrogen implants on maternal behavior and on postpartum estrus in pregnant rats. *Neuroendocrinology, 46,* 522–531.

Amoroso, E. C., Heap, R. B., & Renfree, M. B. (1979). Hormones and evolution of viviparity. In E. J. W. Barrington (Ed.), *Hormones and evolution* (Vol 1, pp. 925–989). New York: Academic Press.

Baldwin, B. A., Kendrick, K. M., & Keverne, E. B. (1986). Intraventricular oxytocin stimulates maternal behaviour in the sheep. *Journal of Physiology, 381,* 80P.

Bauer, J. H. (1983). Effects of maternal state on the responsiveness to nest odors of hooded rats. *Physiology and Behavior, 30,* 229–232.

Beach, F. A., & Jaynes, J. (1956). Studies of maternal retrieving in rats: III. Sensory cues involved in lactating female's response to her young. *Behaviour, 10,* 104–125.

Berieter, D. A., & Barker, D. J. (1975). Facial receptive fields of trigeminal neurons: Increased size following estrogen treatment in female rats. *Neuroendocrinology, 18,* 115–124.

Berieter, D. A., & Barker, D. J. (1980). Hormone-induced enlargement of receptive fields in trigeminal mechanoreceptive neurons: I. Time course hormone sex and modality specificity. *Brain Research, 184,* 395–410.

Birch, H. G. (1956). Sources of order in the maternal behavior of animals. *American Journal of Orthopsychiatry, 26,* 279–284.

Bolwerk, E. L. M., & Swanson, H. H. (1984). Does oxytocin play a role in the onset of maternal behaviour in the rat? *Journal of Endocrinology, 101,* 353–357.

Breder, C. M., Jr., & Rosen, D. E. (1966). *Modes of reproduction in fishes.* New York: Natural History Press.

Bridges, R. S. (1975). Long-term effects of pregnancy and parturition upon maternal responsiveness in the rat. *Physiology and Behavior, 14,* 245–249.

Bridges, R. S. (1977). Parturition: Its role in the long-term retention of maternal behavior in the rat. *Physiology and Behavior, 18,* 487–490.

Bridges, R. S. (1984). A quantitative analysis of the roles of dosage, sequence, and duration of estradiol and progesterone exposure in the regulation of maternal behavior in the rat. *Endocrinology, 114,* 930–940.

Bridges, R. S., Clifton, D. K., & Sawyer, C. H. (1982). Postpartum luteinizing hormone release and maternal behavior in the rat after late-gestational depletion of hypothalamic norepinephrine. *Neuroendocrinology, 34,* 286–291.

Bridges, R. S., & Dunckel, P. T. (1987). Hormonal regulation of maternal behavior in rats: Stimulation following treatment with ectopic pituitary grafts plus progesterone. *Biology of Reproduction, 37,* 518–526.

Bridges, R. S., Feder, H. H., & Rosenblatt, J. S. (1977). Induction of maternal behaviors in primigravid rats by ovariectomy, hysterectomy, or ovariectomy, plus hysterectomy: Effect of length of gestation. *Hormones and Behavior, 9,* 156–169.

Bridges, R. S., & Grimm, C. T. (1982). Reversal of morphine disruption of maternal behavior by concurrent treatment with the opiate antagonist naloxone. *Science, 218,* 166–168.

Bridges, R. S., Loundes, D. D., DiBiase, R., & Tate-Ostroff, B. A. (1985). Prolactin and pituitary involvement in maternal behavior in the rat. In R. M. MacLeod, M. O. Thorner, and U. Scapagnini (Eds.), *Prolactin. Basic and clinical correlates,* Fidia Research Series, Vol. 1 (pp. 591–599). Padua: Liviana.

Bridges, R. S., & Ronsheim, P. M. (1983). Changes in beta-endorphin concentrations in the medial preoptic area during pregnancy in the rat. *Society for Neuroscience,* Abstract, 9, 799.

Bridges, R. S., & Ronsheim, P. M. (1986). Prolactin regulation of maternal behavior in female rats: Bromocriptine treatment delays and prolactin treatment reinstates the rapid onset of maternal behavior. *Society for Neuroscience,* Abstract, 12, 1160.

Bridges, R. S., Rosenblatt, J. S., & Feder, H. H. (1978). Serum progesterone concentrations and maternal behavior in rats after pregnancy termination: Behavioral stimulation after progesterone withdrawal and inhibition by progesterone maintenance. *Endocrinology, 102,* 258–267.

Bridges, R. S., Zarrow, M. X., Goldman, B. D., & Denenberg, V. H. (1974). A developmental study of maternal responsiveness in the rat. *Physiology and Behavior, 12,* 149–151.

Browning, H. C. (1973). The evolutionary history of the corpus luteum. *Biology of Reproduction, 8,* 128–157.

Brunelli, S. A., Shindledecker, R. D., & Hofer, M. A. (1987). Behavioral responses of juvenile rats *(Rattus norvegicus)* to neonates after infusion of maternal blood plasma. *Journal of Comparative Psychology, 1,* 47–59.

Cosnier, J., & Couturier, C. (1966). Comportement maternel provoqué chez les rattes adultes castrées. *Comptes Rendus des Séances de la Société de Biologie et de Ses Filiales, 160,* 789–791.

Dunbar, I. Ranson, E., & Buehler, M. (1981). Pup retrieval and maternal attraction to canine amniotic fluids. *Behavior Proceedings, 6,* 249–260.

Erskine, M. S., Barfield, R. J., & Goldman, B. D. (1980). Postpartum aggression in rats: II. Dependence upon maternal sensitivity to young and effects of experience with pregnancy and parturition. *Journal of Comparative and Physiological Psychology, 94,* 495–505.

Fahrbach, S. E., Morrell, J. I., & Pfaff, D. W. (1984). Oxytocin induction of short-latency maternal behavior in nulliparous, estrogen-primed female rats. *Hormones and Behavior, 18,* 267–286.

Fahrbach, S. E., Morrell, J. I., & Pfaff, D. W. (1985). Possible role of endogenous oxytocin in estrogen-facilitated maternal behavior in rats. *Neuroendocrinology, 40,* 526–532.

Fleming, A. S., & Luebke, C. (1981). Timidity prevents the virgin female rat from being a good mother: Emotionality differences between nulliparous and parturient females. *Physiology and Behavior, 27,* 863–868.

Fleming, A. S., & Rosenblatt, J. S. (1974a). Olfactory regulation of maternal behavior in rats: I. Effects of olfactory bulb removal in experienced and inexperienced lactating and cycling females. *Journal of Comparative and Physiological Psychology, 86,* 221–232.

Fleming, A., & Rosenblatt, J. S. (1974b). Olfactory regulation of maternal behavior in rats: II. Effects of peripherally induced anosmia and lesions of the lateral olfactory tract in pup-induced virgins. *Journal of Comparative and Physiological Psychology, 86,* 233–246.

Gilbert, A. N. (1984). Postpartum and lactational estrus: A comparative analysis in rodentia. *Journal of Comparative Psychology, 98,* 232–245.

Gilbert, A. N., Pelchat, R. J., & Adler, N. T. (1980). Postpartum copulatory behaviour in Norway rats under seminatural conditions. *Animal Behaviour, 28,* 989–995.

Giordano, A. L. (1987). *Relationship between nuclear estrogen receptor binding in the preop-*

tic area and hypothalamus and the onset of maternal behavior in female rats. Unpublished doctoral dissertation, Rutgers University, New Brunswick, N.J.

Giordano, A. L., Johnson, A. E., & Rosenblatt, J. S. Haloperidol-induced disruption of retrieving in lactating rats and reinstatement with apomorphine. Unpublished observations.

Giordano, A. L., Siegel, H. I., & Rosenblatt, J. S. (in press). Nuclear estrogen receptor binding in the preoptic area and hypothalamus of pregnancy-terminated rats: Correlation with the onset of maternal behavior. *Neuroendocrinology.*

Graber, G. C., & Kristal, M. B. (1977) Uterine distention facilitates the onset of maternal behavior in pseudopregnant but not in cycling rats. *Physiology and Behavior, 19,* 133–137.

Gray, P., & Chesley, S. (1984). Development of maternal behavior in nulliparous rats *(Rattus norvegicus):* Effects of sex and early maternal experience. *Journal of Comparative Psychology, 98,* 91–99.

Grimm, G. T., & Bridges, R. S. (1983). Opiate regulation of maternal behavior in the rat. *Pharmacology, Biochemistry and Behavior, 19,* 609–616.

Hammer, R. P., Jr., & Bridges, R. S. (1987). Preoptic area opioids and opiate receptors increase during pregnancy and decrease during lactation. *Brain Research, 420,* 48–56.

Hansen, S., & Ferreira, A. (1986). Food intake, aggression, and fear behavior in the mother rat: Control by neural systems concerned with milk ejection and maternal behavior. *Behavioral Neuroscience, 100,* 64–70.

Hansen, S., Ferreira, A., & Selart, M. F. (1985). Behavioural similarities between mother rats and benzodiazepine-treated non-maternal animals. *Psychopharmacology, 86,* 344–347.

Hard, E., & Hansen, S. (1985). Reduced fearfulness in the lactating rat. *Physiology and Behavior, 35,* 641–643.

Harder, J. D., & Fleming, M. A. (1981). Estradiol and progesterone profiles indicate a lack of endocrine recognition of pregnancy in the Opossum. *Science, 212,* 1400–1402.

Hauser, H., & Gandelman, R. (1985). Lever pressing for pups: Evidence for hormonal influence upon maternal behavior of mice. *Hormones and Behavior, 19,* 454–468.

Heap, R. B., Galil, A. K. A., Harrison, F. A., Jenkin, G., & Perry, J. S. (1977). Progesterone and oestrogen in pregnancy and parturition: comparative aspects of hierarchical control. In Ciba Foundation Symposium *47* (new series), *The fetus and birth* (pp. 127–150). Amsterdam: Elsevier.

Hinde, R. A. (1979). *Towards understanding relationships.* European Monographs in Social Psychology, Vol. 18. New York: Academic Press.

Jakubowski, M., & Terkel, J. (1985). Transition from pup killing to parental behavior in male and virgin female albino rats. *Physiology and Behavior, 34,* 683–686.

Keverne, E. B. (1988). Central mechanisms underlying the neural and neuroendocrine determinants of maternal behaviour. *Psychoneuroendocrinology, 13,* 127–141.

Keverne, E. B., Lévy, F., Poindron, P., & Lindsay, D. (1983). Vaginal stimulation: An important determinant of maternal bonding in sheep. *Science, 219,* 81–83.

Kinder, E. F. (1927). A study of the nestbuilding activity of the albino rat. *Journal of Experimental Zoology, 47,* 117–161.

Kinsley, C. H., & Bridges, R. S. (1986). Opiate involvement in postpartum aggression in rats. *Pharmacology, Biochemistry and Behavior, 15,* 1007–1011.

Klaus, M. H., & Kennell, J. H. (1976). *Maternal–infant bonding.* St. Louis: Mosby.

Koller, G. (1955). Hormonale unde psychische Steuerung beim Nestbau weiser Mause. *Vehrhandlungen der Deutschen Zoologie Gesellschaft, 20,* 123–132.

Koranyi, L., Lissak, K., Tamasy, V., & Kamaras, L. (1976). Behavioral and electrophysiological attempts to elucidate central nervous system mechanisms responsible for maternal behavior. *Archives of Sexual Behavior, 5,* 503–510.

Leng, G., Mansfield, S., Bicknell, R. J., Brown, D., Ingram, C. D., Marsh, M. I. C., Yates, J. O., & Dyer, R. G. (1987). Stress-induced disruption of parturition in the rat may be mediated by endogenous opioids. *Journal of Endocrinology, 114,* 247–252.

Leng, G., Mansfield, S., Bicknell, R. J., Dean, A. D. P., Ingram, C. D., Marsh, M. I. C., Yates, J. O., & Dyer, R. G. (1985). Central opioids: A possible role in parturition? *Journal of Endocrinology, 106,* 219–224.

Lisk, R. D. (1971). Oestrogen and progesterone synergism and elicitation of maternal nest-building in the mouse *(Mus musculus). Animal Behaviour, 19,* 606–610.

Lisk, R. D., Prelow, R. A., & Friedman, S. A. (1969). Hormonal stimulation necessary for eliciting of maternal nest building in the mouse. *Animal Behaviour, 17,* 730–738.

Mayer, A. D. (1983). The ontogeny of maternal behaviour in rodents. In R. W. Elwood (Ed.), *Parental behaviour of rodents* (pp. 1–21). Chichester: Wiley.

Mayer, A. D., Faris, P. L., Komisaruk, B. R., & Rosenblatt, J. S. (1985). Opiate antagonism reduces placentophagia and pup cleaning by parturient rats. *Pharmacology, Biochemistry, and Behavior, 22,* 1035–1044.

Mayer, A. D., Freeman, N. G., & Rosenblatt, J. S. (1979). Ontogeny of maternal behavior in the laboratory rat: Factors underlying changes in responsiveness from 30 to 90 days. *Developmental Psychobiology, 12,* 425–439.

Mayer, A. D., Reisbick, S., Siegel, H. I., & Rosenblatt, J. S. (1987). Maternal aggression in rats: Changes over pregnancy and lactation in a Sprague-Dawley strain. *Aggressive Behavior, 13,* 29–43.

Mayer, A. D., & Rosenblatt, J. S. (1975). Olfactory basis for the delayed onset of maternal behavior in virgin female rats: Experiential basis. *Journal of Comparative and Physiological Psychology, 89,* 701–710.

Mayer, A. D., & Rosenblatt, J. S. (1979). Ontogeny of maternal behavior in the laboratory rat: Early origins in 18–27-day old young. *Developmental Psychobiology, 12,* 407–424.

Mayer, A. D., & Rosenblatt, J. S. (1984). Prepartum changes in maternal responsiveness and nest defense in *Rattus norvegicus. Journal of Comparative Psychology, 98,* 177–188.

Mayer, A. D., & Rosenblatt, J. S. (1987). Hormonal factors influence the onset of maternal aggression in laboratory rats. *Hormones and Behavior, 21,* 253–267.

McCarthy, M. M., Bare, J. E., & vom Saal, F. S. (1986). Infanticide and parental behavior in wild female house mice: Effects of ovariectomy adrenalectomy and administration of oxytocin and prostaglandin F2 alpha. *Physiology and Behavior, 36,* 17–23.

McCormack, J. T., & Greenwald, G. S. (1974). Progesterone and oestradiol-17 beta concentrations in the peripheral plasma during pregnancy in the mouse. *Journal of Endocrinology, 62,* 101–107.

Moltz, H., Lubin, M., Leon, M., & Numan, M. (1970). Hormonal induction of maternal behavior in the ovariectomized nulliparous rat. *Physiology and Behavior, 5,* 1373–1377.

Moltz, H., Rowland, D., Steele, M., & Halaris, A. (1975). Hypothalamic norepinephrine concentrations and metabolism during pregnancy and lactation in the rat. *Neuroendocrinology, 19,* 252–258.

Noirot, E., & Goyens, J. (1971). Changes in maternal behavior during gestation in the mouse. *Hormones and Behavior, 2,* 207–215.

Numan, M. (1988). Maternal behavior. In E. Knobil and J. Neill (Eds.), *The physiology of reproduction* (pp. 1569–1645). New York: Raven Press.

Orpen, B. G., Furman, N., Wong, P. Y., & Fleming, A. S. (1987). Hormonal influences on the duration of postpartum maternal responsiveness in the rat. *Physiology and Behavior, 40,* 307–315.

Pedersen, C. A., & Prange, Jr., A. J. (1979). Induction of maternal behavior in virgin rats after intracerebroventricular administration of oxytocin. *Proceedings of the National Academy of Sciences USA, 76,* 6661–6665.

Pissonnier, D., Thiery, J. C., Fabre-Nys, C., Poindron, P., & Keverne, E. B. (1985). The importance of olfactory bulb noradrenalin for maternal recognition in sheep. *Physiology and Behavior, 35,* 361–363.

Poindron, P., & Le Neindre, P. (1980). Endocrine and sensory regulation of maternal behavior in the ewe. In J. S. Rosenblatt, R. A. Hinde, C. Beer, & M.-C. Busnel (Eds.), *Advances in the study of behavior* (Vol. 11, pp. 75–119). New York: Academic Press.

Poindron, P., Lévy, F., & Krehbiel, D. (1988). Genital, olfactory and endocrine factors in the development of maternal behaviour in the parturient ewe. *Psychoneuroendocrinology, 13,* 99–125.

Rosenblatt, J. S. (1965). The basis of synchrony in the behavioral interaction between the mother and her offspring in the laboratory rat. In B. M. Foss (Ed.), *Determinants of infant behavior* (Vol. 3, pp. 3–41). London: Methuen.

Rosenblatt, J. S. (1967). Nonhormonal basis of maternal behavior in the rat. *Science, 156,* 1512–1514.

Rosenblatt, J. S. (1975). Selective retrieving by maternal and nonmaternal female rats. *Journal of Comparative and Physiological Psychology, 88,* 678–686.

Rosenblatt, J. S. (1984). The Schneirla/Birch trophallaxis hypothesis: Reformulation in relation to maternal behavior in the rat. In G. Greenberg & E. Tobach (Eds.), *Behavioral evolution and integrative levels* (pp. 259–276). Hillsdale, NJ: Erlbaum.

Rosenblatt, J. S., & Lehrman, D. S. (1963). Maternal behavior in the laboratory rat. In H. L. Rheingold (Ed.), *Maternal behavior in mammals* (pp. 8–57). New York: Wiley.

Rosenblatt, J. S., Mayer, A. D., Ahdieh, H., & Siegel, H. I. (1988). Hormonal basis during pregnancy for the onset of maternal behavior in the rat. *Psychoneuroendocrinology, 13,* 29–46.

Rosenblatt, J. S., & Siegel, H. I. (1975). Hysterectomy-induced maternal behavior during pregnancy in the rat. *Journal of Comparative and Physiological Psychology, 89,* 685–700.

Rosenblatt, J. S., & Siegel, H. I. (1981). Factors governing the onset and maintenance of maternal behavior among nonprimate mammals. The role of hormonal and nonhormonal factors. In D. J. Gubernick & P. H. Klopfer (Eds.), *Parental care in mammals* (pp. 13–76). New York: Plenum Press.

Rosenblatt, J. S., & Siegel, H. I. (1983). Physiological and behavioural changes during pregnancy and parturition underlying the onset of maternal behaviour in rodents. In R. W. Elwood (Ed.), *Parental behaviour of rodents* (pp. 23–66). London: Wiley.

Rosenblatt, J. S., & Siegel, H. I., & Mayer, A. D. (1979). Progress in the study of maternal behavior in the rat: Hormonal, nonhormonal, sensory, and developmental aspects. In J. S. Rosenblatt, R. A. Hinde, G. C. Beer, & M.-C. Busnel (Eds.), *Advances in the study of behavior* (Vol. 10, pp. 225–311). New York: Academic Press.

Rosenblum, L. A., & Moltz, H. (1983). *Symbiosis in parent–offspring interactions.* New York: Plenum Press.

Ross, S., Sawin, P. B., Zarrow, M. X., & Denenberg, V. H. (1963). Maternal behavior in the rabbit. In H. L. Rheingold (Ed.), *Maternal behavior in mammals* (pp. 94–121). New York, Wiley.

Rothchild, I. (1962). Corpus luteum–pituitary relationship: The effect of progesterone on the folliculotropic potency of the pituitary in the rat. *Endocrinology, 70,* 303–313.

Rubin, B. S., Menniti, F. S., & Bridges, R. S. (1983). Intracerebroventricular administration of oxytocin and maternal behavior in rats after prolonged and acute steroid pretreatment. *Hormones and Behavior, 17,* 45–53.

Russell, E. M. (1982). Patterns of parental care and parental investment in marsupials. *Biological Reviews of the Cambridge Philosophical Society, 57,* 423–486.

Sharman, G. B., Calaby, J. H., & Poole, W. E. (1966). Patterns of reproduction in female diprotodont marsupials. In I. W. Rolands (Ed.), *Comparative biology of reproduction in mammals* (pp. 205–232). New York: Academic Press.

Siegel, H. I., & Greenwald, G. S. (1975). Prepartum onset of maternal behavior in hamsters and the effects of estrogen and progesterone. *Hormones and Behavior, 6,* 237–245.

Siegel, H. I., & Rosenblatt, J. S. (1975a). Hormonal basis of hysterectomy-induced maternal behavior during pregnancy in the rat. *Hormones and Behavior, 6,* 211–222.

Siegel, H. I., & Rosenblatt, J. S. (1975b). Estrogen-induced maternal behavior in hysterectomized-ovariectomized virgin rats. *Physiology and Behavior, 14,* 465–471.

Slob, A. K., Huizer, T., & Ten Bosch, J. J. V. D. W. (1986). Ontogeny of sex differences in open-field ambulation in the rat. *Physiology and Behavior, 37,* 313–315.

Slotnick, B. M. (1967). Intercorrelations of maternal activity in the rat. *Animal Behaviour, 15,* 267–269.

Smotherman, W. P., Bell, R. W., Starzec, Elias, J., & Zachman, T. A. (1974). Maternal responses to infant vocalizations and olfactory cues in rats and mice. *Behavioral Biology, 12,* 55–66.

Steele, M. K., Rowland, D., & Moltz, H. (1979). Initiation of maternal behavior in the rat. Possible involvement of limbic norepinephrine. *Pharmacology, Biochemistry and Behavior, 11,* 123–130.

Steinberg, J., & Bindra, D. (1962). Effects of pregnancy and salt intake on genital licking. *Journal of Comparative and Physiological Psychology, 55,* 103–106.

Stern, J. M., Johnson, S., & McDonald, C. (1987). Prepartum infraorbital denervation severely disrupts parturition behavior in rats. *International Society for Developmental Psychobiology Abstracts.*

Stone, C. P. (1925). Preliminary note on maternal behavior of rats living in parabioses. *Endocrinology, 9,* 505–512.

Sturman-Hulbe, M., & Stone, C. P. (1929). Maternal behavior in the albino rat. *Journal of Comparative Psychology, 9,* 203–237.

Terkel, J., & Rosenblatt, J. S. (1971). Aspects of nonhormonal maternal behavior in the rat. *Hormones and Behavior, 2,* 161–171.

Trivers, R. L. (1974). Parent-offspring conflict. *American Zoology, 14,* 249–264.

Van Leengoed, E., Kerker, E., & Swanson, H. H. (1987). Inhibition of post-partum maternal behaviour in the rat by injecting an oxytocin antagonist into the cerebral ventricles. *Journal of Endocrinology, 112,* 275–282.

Wamboldt, M. Z., & Insel, T. R. (1987). The ability of oxytocin to induce short latency maternal behavior is dependent on peripheral anosmia. *Behavioral Neuroscience, 101,* 439–441.

Wiesner, B. P., & Sheard, N. M. (1933). *Maternal behavior in the rat.* London: Oliver & Boyd.

Yeo, J. A. G., & Keverne, E. B. (1986). The importance of vaginal-cervical stimulation for maternal behaviour in the rat. *Physiology and Behavior, 37,* 23–26.

Zarrow, M. X., Farooq, A., & Denenberg, V. H. (1962). Maternal behavior in the rabbit:

Critical period for nestbuilding following castration during pregnancy. *Proceedings of the Society for Experimental Medicine and Biology, 111,* 537–538.

Zarrow, M. X., Gandelman, R., & Denenberg, V. H. (1971). Prolactin: Is it an essential hormone for maternal behavior in mammals? *Hormones and Behavior, 2,* 343–354.

Zarrow, M. X., Sawin, P. B., Ross, S., & Denenberg, V. H. (1962). Maternal behavior and its endocrine basis in the rabbit. In E. L. Bliss (Ed.), *Roots of behavior* (pp. 187–197). New York: Harper.

5

Structural Sexual Dimorphisms in the Brain

ROGER A. GORSKI

Although a major theme of this book is the biological determination of a specific set of behaviors collectively called "parenting," this chapter deals with the sexual differentiation of the brain, which represents one of the most remarkable demonstrations of the biological determination of behavior. Since the statement that the mammalian brain is inherently female, both functionally and structurally, or at least is undifferentiated in these respects, has profound biological and behavioral implications, some general comments are in order before a detailed discussion of the process of sexual differentiation is presented.

The concept of the sexual differentiation of peripheral components of the reproductive system is widely accepted (but see Wai-Sum, Short, Renfree, & Shaw, 1988). Thus although the gonad is determined at fertilization by the sex chromosome of the sperm cell, the process of masculine differentiation of the other internal reproductive organs and the genitalia depends on two factors: the production of hormones by the testes and the ability of the tissue of the individual to respond to these hormones (Jost & Magre, 1984; Wilson, Griffin, Leskin, & George, 1981). Because of its critical role in endocrine function (Kalra & Kalra, 1983) and sexual behavior (Gorski, 1974), the brain can be considered an integral part of the reproductive system. In that sense, the existence of functional and sexual differences in the brain and their dependence during early development on testicular hormones (DeVries, DeBruin, Uylings, & Corner, 1984; Gorski, 1985a; Goy & McEwen, 1980; Serio, Motta, Zanisi, & Martini, 1984) are actually predictable from what is known about the peripheral components of the reproductive system.

For sexual reproduction to succeed, sex differences in gamete production, internal sex organs, the genitalia, and *behavior* must exist. Why in mammals nature has chosen to require the male, through his testicular activity, to masculinize his apparently fundamentally female body and brain is unknown. One can argue, however, that since appropriate testicular activity is necessary for the masculinization of the reproductive system, including the brain, and for successful reproduction, the process of sexual differentiation helps to ensure that genetic males will be reproductively competent. If masculine differentiaton were programmed genomically and independent of the gonad, the incidence of infertile males might well be much greater.

Although sex differences are obviously required for reproduction per se, are they necessary for parenting behaviors? This complex question can be answered positively, but perhaps simplistically. Leaving aside the potential need for the male to forage for food or defend the "nest" while the female is lactating, a male that exhibited complete maternal behavior would only serve to compete with the lactating female, to the possible detriment of the offspring, which require the nourishment that only the female can provide.

THE RAT AS AN EXPERIMENTAL ANIMAL MODEL

The discussion that follows concentrates on studies of the laboratory rat because the process of sexual differentiation in the rat brain has been most thoroughly studied. Moreover, there are species differences in this process, and by stressing such differences, one can obfuscate fundamental principles that may have general applicability across species. With specific reference to parenting behavior, as will become clear, sex differences in this parameter do exist in rats, and studies of sexual differentiation in rats now under way in the author's laboratory are directed to an understanding of the hormone-dependent development of the medial preoptic area (MPOA), one region strongly implicated in the control of maternal behavior (Jacobson, Terkel, Gorski, & Sawyer, 1980; Numan, 1974; Numan, Rosenblatt, & Komisaruk, 1977).

Nevertheless, the applicability of the results of studies of the rat to the human being is very difficult to establish at the present time. Interestingly, just as social factors have been implicated in human parenting behavior (chapter 2), the same concern has characterized attempts to understand human psychosexual differentiation, in spite of the clear hormonal effects demonstrated in experimental animals. Although a detailed consideration of human psychosexual differentiation is beyond the scope of this discussion, one observation is particularly germane. Ehrhardt and Baker (1974) have analyzed a number of behavioral traits in women exposed prenatally to abnormally high levels of adrenal androgens. Such women have been characterized as tomboys, but included among Ehrhardt and Baker's statistically significant findings is the observation that these androgen-exposed women play significantly less with dolls and show an aversion to infants. Whether this is a masculine trend is certainly debatable, but it is an observation consistent with the concept of the sexual differentiation of the brain as derived from studies of the rat and other laboratory animals. Only future research will determine the validity of the rat as a model of human behavioral development. As a minimum, however, current understanding of the process of sexual differentiation of the brain—or, for that matter, of the regulation of parenting behavior—provides a theoretical construct to allow us to interpret data based on human studies that, appropriately, are subject to often severe ethical restrains.

THE FUNCTIONAL SEXUAL DIFFERENTIATION OF THE RAT BRAIN

For any consideration of the sexual differentiation of the brain, it is very helpful to adopt a classical concept first proposed in 1959—that is, the conceptual difference between the "activational" and the "organizational" actions of gonadal steroids

(Phoenix, Goy, Gerall, & Young, 1959). It is generally thought that in the adult animal, gonadal steroids enter neurons and bind to stereospecific receptors. This steroid–receptor complex then binds to the nuclear chromatin and alters genomic function, resulting ultimately in altered protein synthesis and altered function. This classical concept can be called "activational," with the key being that steroids induce a *transient* change in neuronal function. Since this change may be either excitatory or inhibitory, the term "activational" reflects only the temporal characteristics of the modification of neuronal function. In contrast, certainly during early development, gonadal hormones have the ability to induce *permanent* changes in brain structure and function. These actions of steroids can be labeled "organizational."

Although it may ultimately be shown that at the biochemical level the organizational and activational effects of steroids are actually comparable, this concept is very useful, as will become evident. Although a relatively large number of functional systems have been shown to undergo sexual differentiation, this concept can be well illustrated with the two functional systems that have been most thoroughly studied. In the female rat, ovulation appears to be induced by an action of ovarian estrogen on a cyclic neural system within the brain. This cyclic system has a 24-hr periodicity and, when activated by estrogen, triggers a release of luteinizing hormone releasing hormone (LHRH) (Fink & Jamieson, 1976; Porter et al., 1977) and, thus, the surge of luteinizing hormone (LH) and ovulation. The interesting observation is that the genetic male rat does not appear to have a functioning cyclic neural system. Thus when an adult male rat is gonadectomized and primed with ovarian hormones in a regime that in females induces positive feedback, an expression of the cyclic neural system, positive feedback is totally absent (Harlen & Gorski, 1977; Mennin & Gorski, 1975; Taleisnik, Caligaras, & Astrada, 1971). In earlier studies using ovarian transplantation, it was also demonstrated that the male brain is incapable of providing the hormonal stimulus for ovulation (Gorski & Wagner, 1965; Harris, 1964).

In the rat, sexual behavior is dependent on the activational effects of gonadal hormones. This dependence is absolute in the case of female sex behavior, and although male rats will exhibit masculine sexual behavior for a period of time after castration, reproductive behavior eventually disappears from their repertoire. In rats, female sexual behavior is highly dimorphic (Beach, 1971; Gorski, 1974). When mounted, the sexually receptive female exhibits a classical posture called "lordosis." Thus it is relatively simple to quantify female sexual behavior in terms of a lordosis quotient (LQ, number of lordoses/number of mounts $\times$ 100). Genetic male rats, when castrated as adults and given ovarian hormones that in the ovariectomized female stimulate lordosis behavior, only rarely exhibit lordosis; their LQ is very low (Clemens, Shryne, & Gorski, 1970), but not necessarily zero.

Table 5.1 illustrates the classical experimental approach that has led to the concept of sexual differentiation of the brain. As already indicated, male and female rats differ in their control of the release of LH and in the ease with which hormones can activate lordosis behavior. Table 5.1 indicates the consequences in the adult of the manipulation of the hormone environment perinatally. Exposing the rat to an injection of testosterone propionate (TP) in the first week of life has little, if any, effect on the genetic male: he remains neuroendocrinologically masculine. Note, however, that a single exposure to TP permanently affects the female rat, so that she no longer exhibits lordosis behavior or ovulates (Barraclough & Gorski, 1961; Clemens et al.,

Table 5.1 Classical experimental approach that established the concept of the functional sexual differentiation of the rat brain

		Functional consequences in adults	
Genetic sex	Functional activity in the normal adult	Exposure to gonadal steroids perinatally	Gonadectomy perinatally
Female	Cyclic release of LH High lordosis response rate	*Tonic release of LH Lordosis rarely displayed*	Cyclic release of LH High lordosis response rate
Male	Tonic release of LH Lordosis rarely displayed	Tonic release of LH Lordosis rarely displayed	*Cyclic release of LH High lordosis response rate*

1970). Thus exposure to testosterone during the first week of postnatal life "masculinizes" (or "defeminizes" in terms of ovulation and lordosis behavior) the female rat.

Of even greater significance are the functional consequences of removing the gonads during early development. If female rats are gonadectomized shortly after birth, as adults they are still neuroendocrinologically feminine. They will ovulate if given ovarian grafts or show positive feedback when injected with ovarian hormones, and it is easy to activate lordosis behavior by exogenous hormone treatment. Note that the same is true of the male rat that is gonadectomized within the first 3 days of life. As an adult, he is neuroendocrinologically female, and will support ovulation or positive feedback and show female levels of lordosis response (Feder & Whalen, 1964; Gorski & Wagner, 1965; Grady, Phoenix, & Young, 1965; Harris, 1964).

That the process of functional sexual differentiation of the brain is not limited to the control of the cyclic release of LH, to the control of lordosis behavior, or to the rat is illustrated in Table 5.2. From this table it is apparent that there are many functional sexual dimorphisms, and several points should be emphasized. In terms of the cyclic release of LH, the sex difference in rats appears to be absolute, while in terms of lordosis behavior, it is only relative. In terms of masculine copulatory behavior, the sex difference is even less marked. Female rats will exhibit mounting behavior (Whalen, 1968); nevertheless, this behavior, (i.e., masculinization) is increased after exposure to TP perinatally (Christensen & Gorski, 1978). The sex difference in maternal behavior in rats appears to be more similar to masculine copulatory behavior. Male rats exhibit good maternal behavior (Noirot, 1972; Rosenblatt, 1967) but do show differences in the quality of that behavior (Bridges, 1983; Bridges, Zarrow, & Denenberg, 1973) and in the prolactin response to pups (Bridges, 1983). Given the less robust sex difference, not surprisingly, the effect of the hormonal environment perinatally on adult maternal behavior in rats is still controversial (Bridges, 1983; Bridges et al., 1973; Quadagno & Rockwell, 1972).

It is likely that Table 5.2 is incomplete, in terms of both the number of functional sex differences and the animal species listed, but certain generalizations can be made. It is important to note that animals of various species are born at different stages of maturity. Thus in the rat, sexual differentiation of the brain occurs perinatally; in other species, such as the guinea pig (Hines, Davis, Coquelin, Goy, & Gorski, 1985; Phoenix, Goy, Gerall & Young, 1959), and the rhesus monkey (Goy

& Phoenix, 1971; Phoenix, Goy, & Resko, 1968), it occurs during gestation. Moreover, sexual dimorphisms are obviously linked to the behavioral repertoire of a given species—for example, urination posturing in the dog. Moreover, not all components of reproductive control are sexually dimorphic in all species. The most notable example is the control of ovulation. Exposing the embryonic rhesus monkey to TP

Table 5.2 Examples of brain functions that undergo sexual differentiation in various mammalian species

---

Gonadotropin Secretion
Rat Barraclough & Gorski (1961), Harlan & Gorski (1977), Mennin & Gorski (1975), Taleisnik et al. (1971)
Mouse Barraclough & Leathem (1954)
Guinea pig Brown-Grant & Sherwood (1971)
Hamster Alleva, Alleva, & Umberger (1969), Swanson (1970)
Pig Elsaesser & Parvizi (1979)
Sheep Karsch & Foster (1975), Short (1974)

Reproductive Behavior
Rat Beach (1971), Clemens et al. (1970), Feder & Whalen (1964), Grady et al. (1965), Whalen (1968)
Mouse Edwards & Burge (1971)
Guinea pig Phoenix et al. (1959)
Hamster Carter, Clemens, & Hoekema (1972), Paup, Coniglio, & Clemens (1972)
Pig Ford (1982)
Sheep Clarke, Scaramuzzi, & Short (1976), Short (1974)
Dog Beach & Kuehn (1970), Beach, Kuehn, Sprague, & Anisko (1972)
Ferret Baum (1976)
Rhesus monkey Goy & Phoenix (1971), Phoenix et al. (1968)

Play Behavior
Rat Beatty, Dodge, Taylor, & Meany (1981), Meany & Stewart (1981)
Rhesus monkey Goy & Resko (1972)

Open Field Behavior
Rat Gray, Levine, & Broadhurst (1965), Pfaff & Zigmond (1971)
Hamster Swanson (1967)

Territorial Marking Behavior
Gerbil Commins & Yahr (1984), Lumia, Raskin, & Eckhert (1977), Turner (1975)

Urination Posturing
Dog Beach (1974), Martins & Valle (1967)

Aggressive Behavior
Rat Barr, Gibbons, & Moyer (1976), Conner, Levine, Wertheim, & Cummer (1969), Powell, Francis, & Schneiderman (1971)
Mouse Bronson & Desjardins (1968), Edwards (1969), Vale, Ray, & Valle (1972), vom Saal, Svare, & Gandelman (1976)
Hamster Payne & Swanson (1972)

Taste Preference
Rat Valenstein, Kakolewski, & Cox (1967), Wade & Zucker (1969)

Feeding and Body Weight
Rat Beatty, Powley, & Keesey (1970), Bell & Zucker (1971), Tarttelin, Shryne, & Gorski (1975)

Learning Behavior
Rat Beatty & Beatty, (1970), Dawson, Cheung, & Lau (1975), Denti & Negroni (1975), Scouten, Groteleuschen, & Beatty (1975), Stewart, Skvarenina, & Pottier (1975)

---

masculinizes behaviors but does not prevent ovulation (Goy & Resko, 1972), and it has been reported that male monkeys show positive feedback of estrogens on gonadotropins (Hodges & Hearn, 1978; Karsch, Dierschke, & Knobil, 1973) and can even support ovulation when given ovarian grafts (Norman, Spies, Brenner, & Malley, 1983). Species differences are readily apparent; nevertheless, it has to be considered very likely that sexual differentiation of the brain is characteristic of the mammalian species. In addition, this process is not restricted to mammals (Goy & McEwen, 1980). Thus sexual differentiation of the brain appears to be a fundamental biological principle that has widespread effects on the developing brain. Yet until recently, very little was known about the mechanisms of steroid action in this process.

The results of early studies suggested that the MPOA is at least one site of testosterone action (Christensen & Gorski, 1978; Hayashi & Gorski, 1974; Wagner, Erwin, & Critchlow, 1966), and the temporal pattern of exposure to hormones was characterized to a degree. An exposure of 6–12 hr to agents that interfere with testosterone action appeared necessary to masculinize gonadotrophin control (Arai & Gorski, 1968a, 1968b). Similarly, the results of several studies suggested that protein synthesis is involved (Gorski & Shryne, 1972; Salaman, 1974; Salaman & Birkett, 1974), which would be consistent with what is known about steroid action and the permanency of the organizational effects of steroids. Other than these general statements, however, little else was known.

This situation changed dramatically about a decade ago when the significance of structural sex differences was recognized. Although some structural sex differences in the rat brain had been known for many years (Dörner & Staudt, 1968; Pfaff, 1966; Smith, 1934; Sugita, 1917), perhaps the fact that many functional sex differences are not all-or-none discouraged experimental and conceptual pursuit of this approach. Remember that the normal male rat can exhibit lordosis, although much less frequently than the female; the neural circuitry for this behavior must be present in his brain. Why were structural sex differences thought to exist? The basis of behavioral sex differences was thought to lie in the realm of the functional accessibility of such neuronal circuits. Thus it was thought that sex differences might be explained solely in terms of the sensitivity of existing neuronal circuits to hormones or to environmental stimuli, rather than to structural sex differences.

The publication, in relatively rapid succession, of three papers changed the thinking in this area very dramatically. First, Raisman and Field (1973) demonstrated that in a specific area of the MPOA of the rat, there are statistically significant sex differences in neuronal connectivity at the ultrastructural level. Importantly, these investigators established that these structural sex differences are indeed dependent on the perinatal hormonal environment. Three years later, Nottebohm and Arnold (1976) published their report of the existence of dramatic sex differences in terms of volume in a system of neuronal nuclei involved in the control of song production in the zebra finch and canary. This observation was soon followed by the identification of a marked sex difference in the MPOA of the rat brain.

THE SEXUALLY DIMORPHIC NUCLEUS OF THE PREOPTIC AREA

As shown in Figure 5.1 and Figure 5.2, bar graphs A and B, an intensely staining component of the MPOA is approximately five times greater in volume in male rats

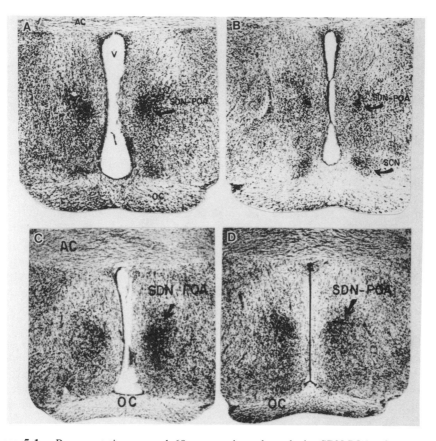

Figure 5.1. Representative coronal 60-μm sections through the SDN-POA of two control (A,B) and two experimental (C,D) groups of rats. The difference in overall SDN-POA volume between male (A) and female (B) rats is suggested by these illustrations. The animals in (C) and (D) are genetic females that were exposed from embryonic day 16 through postnatal day 10 to daily injections of testosterone propionate (C) or diethylstilbestrol (D). All at the same magnification. Abbreviations: AC, anterior commissure; OC, optic chiasm; V, third ventricle. (From Gorski, 1987a)

than in female rats (Gorski, Gordon, Shryne, & Southam, 1978). Because neuronal density in this region is significantly greater than in the surround, this structure has been labeled a "nucleus"—specifically, the sexually dimorphic nucleus of the preoptic area (SDN-POA) (Gorski, Harlan, Jacobson, Shryne, & Southam, 1980). Importantly, neuronal density is similar in the SDN-POA of both males and females (Gorski et al., 1980; Jacobson & Gorski, 1980). Since the SDN-POA of the male is much larger in volume, it could be inferred that the SDN-POA of the male is composed of more neurons than is that of the female. If this sex difference could be shown to be hormone dependent, it would be possible to make a significant conceptual advance in terms of possible mechanisms: testosterone exposure could determine the number of neurons that form the SDN-POA in males.

The results of the study of the ontogeny of the sex difference in the volume of the SDN-POA showed that this nucleus was first significantly different on the day of

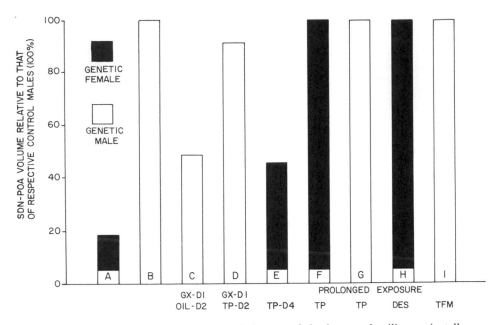

Figure 5.2. Schematic illustration of the influence of the hormonal milieu perinatally on SDN-POA volume in young adult rats. SDN-POA volume is expressed as a percentage of the volume of this nucleus in control males from several independent studies: (A,B) the normal sex difference in nuclear volume; (C) males were castrated on postnatal day 1 (GX-D1; Gorski et al., 1978); (D) similarly castrated males were then injected with 100 μg TP on postnatal day 2 (TP-D2; Jacobson et al., 1981); (E) females were injected with 1 mg TP on postnatal day 4 (TP-D4; Gorski et al., 1978); (G) males were exposed to daily injections of TP from embryonic day 16 to postnatal day 10 (Döhler, Coquelin, et al., 1982); (F) results of similar exposure to TP perinatally in females (Döhler, Coquelin, et al., 1982); (H) results of the treatment of female rats with diethylstilbestrol perinatally (Döhler, Hines et al., 1982). Group I indicates that genetic male rats with the testicular feminizing mutation (TFM) have a totally masculine SDN-POA (Gorski et al., 1981). See the individual studies for statistical analyses. (From Gorski, 1985a)

birth, and that during the course of about the first 10 days of postnatal life the nucleus grew in the male, while there was no significant effect of age in the female (Jacobson, Shryne, Shapiro, & Gorski, 1980). Thus the development of the SDN-POA parallels the so-called critical period for functional sexual differentiation of the brain.

The initial studies on the hormonal sensitivity of the developing SDN-POA were positive but inconclusive. An injection of 1 mg TP on day 5 significantly increased the SDN-POA volume attained in the adult female (Figure 5.2E), but a dose of 90μg TP, which is very effective in permanently changing gonadotropin control mechanisms, failed to alter SDN-POA volume (Gorski et al., 1978). Castration of the newborn male resulted in a 50% reduction in SDN-POA volume in the adult (Figure 5.2C), but in spite of the fact that such a male is feminine in terms of gonadotropin control and lordosis, his SDN-POA is still significantly larger than that of the normal female. Interestingly, as shown in Figure 5.2D, SDN-POA volume could be restored to normal male levels if castration on day 1 was followed by TP injection 24 hr later

(Jacobson, Csernus, Shryne, & Gorski, 1981). Thus about 50% of the SDN-POA, at least in terms of its volume, is sensitive to testosterone postnatally.

The fact that androgenization of the postnatal female or castration of the newborn male fails to sex-reverse the SDN-POA volume suggests several possibilities. Is there a genomic component to this sex difference? Is a single exposure to exogenous TP inadequate? Is there a prenatal component, since a male-specific surge of testosterone has been observed on embryonic day 18 (Weisz & Ward, 1980)? To address this issue, developing rats were exposed to TP for a prolonged period. Pregnant rats were injected with 2 mg TP daily beginning on day 16 of gestation; when the pups were born, both males and females received 100 μg TP daily through postnatal day 10 (Döhler, Coquelin, et al., 1982). As seen in Figure 5.2G, such prolonged exposure to exogenous TP had no effect on the male, but completely sex-reversed SDN-POA volume in the genetic female (Figure 5.2F). Although these results do not rule out a possible contribution of genomic factors to the sex difference in SDN-POA volume under physiological conditions, they clearly indicate that hormones *alone* can determine the structure of this nucleus. The SDN-POA can indeed be viewed as a "morphological signature" of the developmental action of gonadal steroids.

Before possible mechanisms are considered in more detail, there is an important point to emphasize. As shown in Figure 5.2H, exposure to the nonsteroidal estrogen diethylstilbestrol for the same prolonged period perinatally also sex-reverses SDN-POA volume in the female rat (Döhler, Hines, et al., 1982). In fact, there is compelling evidence that it is actually estradiol that affects the masculine differentiation of the brain, including that of the SDN-POA. First, it was shown many years ago that estradiol benzoate is actually more potent than TP in masculinizing brain function (Gorski, 1971). Dihydrotestosterone, which cannot be aromatized, is very ineffective (Korenbrot, Paup, & Gorski, 1975; Whalen & Rezek, 1974). The enzyme aromatase, which converts testosterone to estradiol, is present in the rat hypothalamus (Naftolin et al., 1975); moreover, the inhibition of aromatase activity prevents masculine differentiation in the male (McEwen, Lieberburg, Chaptal, & Krey, 1977; McEwen, Lieberburg, Maclusky, & Plapinger, 1977), as does treatment with antiestrogen (Booth, 1977; Doughty & McDonald, 1974). Although antiandrogen treatment also blocks masculine differentiation (Arai & Gorski, 1968b), this may be mediated by an inhibition of aromatase activity (Naftolin et al., 1975). Finally, in the male rat with the testicular feminizing mutation (Bardin, Bullock, Schneider, Allison, & Stanley, 1970)—that is, a marked deficit in androgen receptors (Naess et al., 1976)—the SDN-POA is indistinguishable from that of a normal male (Gorski, Csernus, & Jacobson, 1981). This observation further suggests that estrogen receptors mediate brain masculinization.

Although it may be surprising that for the rat hypothalamus estrogen appears to be the masculinizing hormone, there was an even greater surprise in store for researchers in this field. When radioimmunoassays became available, it was possible to measure estradiol levels in newborn rats. These levels in females were found to be very high during the first week of postnatal life (Ojeda, Kalra, & McCann, 1975; Weisz & Gunsalus, 1973), so high that if the concept of estrogen-induced masculine sexual differentiation of the brain was correct, there could not be normal females! The solution to this dilemma appears to be α-fetoprotein, which in the rat binds

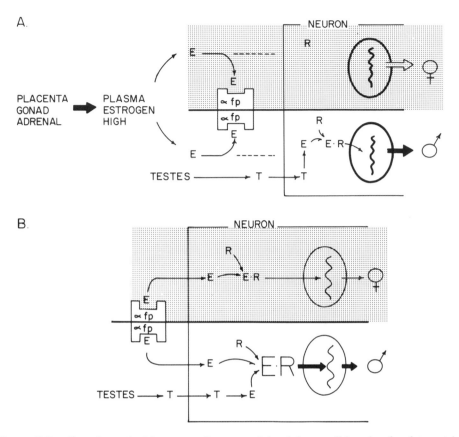

Figure 5.3. Two theoretical but contradictory models of the possible role of α-fetoprotein (α fp) during the perinatal period of sexual differentiation of the rat brain (A) α fp serves to protect the brain of both the female (shaded) and the male from exposure to high plasma titers of estradiol (E). (B) α fp delivers E to the female brain for its normal development. In both models, the E derived from the intraneuronal aromatization of testosterone (T) is postulated to be responsible for the masculine differentiation of the male brain. Other abbreviations: E–R estrogen–receptor complex; R, estrogen receptor, which actually is probably nuclear. (From Gorski, 1983)

estradiol but not testosterone (Nunez et al., 1971; Plapinger, McEwen, & Clemens, 1973; Raynaud, Mercier-Bodard, & Baulieu, 1971) and exists in high concentration for the first several postnatal weeks (Plapinger et al., 1973). As illustrated in Figure 5.3, α-fetoprotein is thought to bind estradiol, thus protecting the brain. Since testosterone is not bound peripherally, this testicular hormone can enter neurons and be converted to estradiol, the masculinizing hormone.

Figure 5.3, however, illustrates another hypothesis. Döhler and Hancke (1978) have reported that injecting newborn rats with the antiestrogen tamoxifen blocks ovulation and lordosis behavior without enhancing male behavior. Moreover, tamoxifen treatment postnatally reduces SDN-POA volume in male rats, as expected, but does so in female rats as well (Döhler, Srivastasa et al., 1984). Thus Döhler has postulated that exposure to low titers of estrogen is *required* for the normal development of the female brain; in species such as the rat, which are born before this process is complete, α-fetoprotein serves as a reservoir of estrogen to prolong the perinatal hormonal environment (Döhler, Hancke et al., 1984). In this regard, the results of the studies of Toran-Allerand (1976, 1980) have shown that estradiol is a neurite growth-promoting factor in vitro, which would also support the possibility that estradiol is required for normal development in the female. The story is more complex, however, since Toran-Allerand (1984) has also reported that α-fetoprotein is found in, but not made in, neurons. Thus α-fetoprotein might also serve as a carrier to deliver estradiol to specific neurons. In any case, it is clear that in the male rat, masculine differentiation of the brain is due to the aromatization of testicular testosterone. The controversial role of α-fetoprotein has an implication for the true nature of the brain—is it inherently female, undifferentiated, or bipotential?—but not for the general concept of masculine differentiation, which is clearly hormone dependent.

POSSIBLE MECHANISMS OF STEROID ACTION

Since neuronal density within the SDN-POA is not sexually dimorphic (Gorski et al., 1980; Jacobson & Gorski, 1980), the larger volume of the male nucleus implies that more neurons are included within the SDN-POA of the male, and this has been confirmed with actual cell counts (Gorski, unpublished observations). Also, in the experimentally sex-reversed female, the number of neurons in the posterior component (see below) of the SDN-POA is comparable with the number in the nucleus of the normal male (Dodson, Shryne, & Gorski, 1988). Clearly, gonadal hormones determine the number of neurons that form the SDN-POA. Table 5.3 presents six possible levels at which hormones can act to influence SDN-POA development: neurogenesis per se; migration from the subependymal lining of the ventricular system; survival during the migration process; aggregation or coalescence into a nucleus; survival after migration is complete, which may be dependent on the growth of neuronal processes; and morphological or chemical specification of neurons.

A major goal of this laboratory is to obtain evidence to establish or refute the existence of each of these possible mechanisms. Data from the literature suggest that the neurons that form the MPOA become postmitotic around embryonic day 16 (Altman & Bayer, 1978; Anderson, 1978; Ifft, 1972). If true, this finding would support the argument that an effect of steroids on neurogenesis is an unlikely mechanism,

Table 5.3 Possible developmental processes through which gonadal hormones may influence the development of the SDN-POA

Neurogenesis	Are steroids mitogenic?
Migration	Do steroids guide neurons?
Survival during migration	Are steroids neuronotrophic?
Aggregation	Do steroids modify cell–cell communication?
Survival	Are steroids neurite growth-promoting?
Specification	Do steroids determine morphological and/or chemical differentiation?

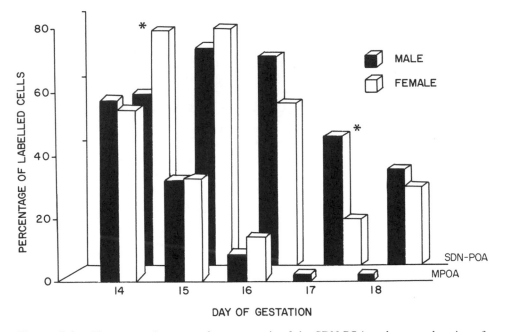

Figure 5.4. The temporal pattern of neurogenesis of the SDN-POA and a central region of the medial preoptic area (MPOA) just lateral to the former. Tritiated thymidine was injected into pregnant rats on the indicated days, and the offspring were sacrificed on postnatal day 30. Asterisks indicate a statistically significant sex difference in the percentage of labeled cells within the SDN-POA. (Data from Jacobson & Gorski, 1980; reprinted from Gorski, 1987a)

since the volume of the SDN-POA can be altered by the hormonal environment postnatally (Gorski et al., 1978; Jacobson et al., 1981). However, as shown in Figure 5.4, the formation period of neurons of the SDN-POA is rather specifically prolonged (Jacobson & Gorski, 1980). In fact, the exposure to tritiated thymidine on embryonic day 18 specifically labels a subcomponent of SDN-POA neurons a "late-arising" subcomponent. By exposing rats to tritiated thymidine on embryonic day 18 and sacrificing them at different perinatal ages, the probable migratory pathway followed by at least the late-arising SDN-POA neurons has been identified (Jacobson, Davis, & Gorski, 1985). Although a sex difference in this parameter has not been identified, this possibility is still under active investigation. However, these late-arising neurons are clearly sensitive to steroid hormones. In the female rat sex-reversed by prolonged TP exposure beginning on either embryonic day 16 or day 20, the number of tritiated, thymidine-labeled neurons is increased to levels equivalent to those seen in the male (Dodson et al., 1988).

The development of the nervous system is characterized by a period of pronounced cell death (Cowan, 1978; Hamburger & Oppenheim, 1982; Varon & Adler, 1980). In many areas of the brain, many more neurons are produced mitotically than ultimately survive development. Although the precise mechanism of neuronal survival is not known, one hypothesis is that postmigratory neurons become dependent for survival on a neuronotrophic substance(s) produced by appropriate target neurons (Hamburger & Oppenheim, 1982). In this regard, the demonstration by Toran-

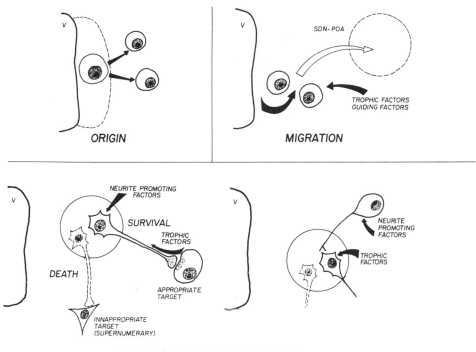

DEVELOPMENT OF CONNECTIONS

Figure 5.5. Highly schematic illustration of the sexually dimorphic development of the SDN-POA and several points at which gonadal steroids might act to produce or induce the observed sex differences in the SDN-POA. These include the actual mitotic formation of neurons, their migration, and/or their survival due to a direct neuronotrophic action of steroids or to the stimulated development of critical afferent or efferent connections. (From Gorski, 1987b)

Allerand (1984) that in vitro, estradiol is a neurite growth-promoting factor may be of considerable conceptual importance.

Figure 5.5 presents a current model. A possible mitogenic action of estradiol on neuroblasts cannot be ruled out, but this appears to be verifiable only through in vitro studies. It is possible that during migration, estradiol is directly neuronotrophic. Finally, after neurons reach the SDN-POA, the local hormonal environment may promote the outgrowth of efferent or afferent neural processes and thus the procurement of critical neuronotrophic factors necessary for the survival of the SDN-POA neurons. Further study of the SDN-POA is necessary to elucidate the precise mechanism of steroid action. At present, the key action appears to be the promotion of neuronal survival or the prevention of neuronal death.

THE ANTEROVENTRAL PERIVENTRICULAR NUCLEUS:
ANOTHER MODEL, ANOTHER CONCEPT

The hypothesis that gonadal steroids prevent neuronal death is supported by circumstantial evidence for the SDN-POA, but by more direct experimental findings for the

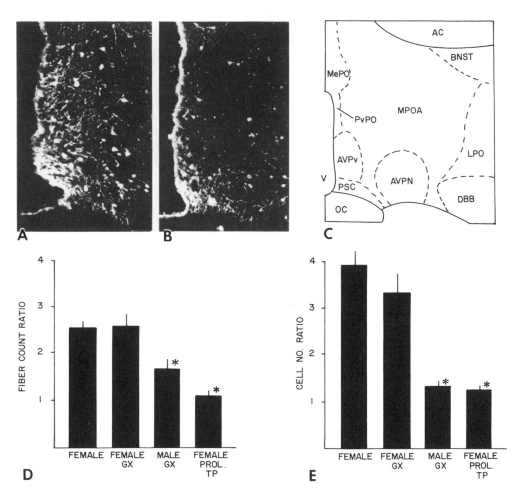

Figure 5.6. The sex difference in the distribution of putative dopaminergic cells and fibers in the anteroventral periventricular nucleus (AVPV) and the influence of prolonged exposure perinatally to TP. (A,B) Fluorescence photomicrographs of tyrosine hydroxylase immunoreactive cells and fibers in a colchicine-treated female (A) and male (B) rat. (From Simerly et al., 1985a) (C) General location of the AVPV. The effect of gonadectomy (GX) and of prolonged exposure to TP (PROL.TP) is shown as an index of fiber count density (D) and the number of immunoreactive cells (E). (Data from Simerly et al., 1985; reprinted from Gorski, 1985a)

sexually dimorphic spinal nucleus of the bulbocavernosus (SNB), (Nordeen, Nordeen, Sengelaub, & Arnold, 1985). However, before a general statement can be made, it will be necessary to evaluate other regions. The anteroventral periventricular nucleus (AVPV), for example, has been reported to be sexually dimorphic, but in the opposite direction; the AVPV of the female is larger than that of the male (Bleier, Byne, & Siggelkow, 1982; Bloch and Gorski, 1988a).

In immunohistochemical studies of this nucleus, females were found to have about four times as many putative dopaminergic somata as males (Figure 5.6) (Simerly, Swanson, & Gorski, 1985a). Most interestingly, treatment with TP, which

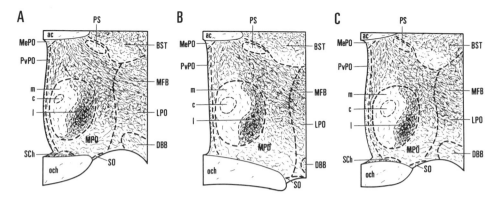

Figure 5.7. Line drawings of the medial preoptic area (MPO) at the level of the medial preoptic nucleus (MPN) to illustrate the distribution of serotonin immunoreactive fibers in the MPN and adjacent regions in the oil-treated female (A), oil-treated male (B), and perinatally TP-treated female (C) rat. The divisions of the MPN are labeled m, medial; c, central; l, lateral; other abbreviations in Simerly et al. (1984). (From Simerly et al., 1985b)

sex-reverses SDN-POA volume, also sex-reverses the number of dopamine immunoreactive somata (Simerly, Swanson, Handa, & Gorski, 1985), but in this case, sex reversal means a great *reduction* in dopamine-positive neurons. Perhaps in the AVPV, hormones specify the chemical nature of neurons rather than affect their survival or death. Nevertheless, until cell counts are made, the possibility that in certain areas brain exposure to TP actually *increases* neuronal death must be entertained. Thus the SDN-POA is an excellent model system, but it may not be representative of all possible actions of gonadal steroids.

THE MEDIAL PREOPTIC NUCLEUS

This discussion has focused on the SDN-POA as a relatively dramatic structural sex difference. However, more detailed anatomical immunohistochemical studies have produced results that suggest that the SDN-POA is a component of a much larger nuclear system that also exhibits sex differences (Simerly, Swanson, & Gorski, 1984). The major nucleus in the MPOA is the medial preoptic nucleus (MPN), and by applying cytoarchitectonic and antiserotonin immunohistochemical analyses to this larger nucleus, three subdivisions of the MPN were identified (Figure 5.7). Each of these subdivisions is sexually dimorphic in volume and can be sex-reversed by the prolonged administration of TP perinatally to the female (Simerly, Swanson, & Gorski, 1985b).

Because of the importance of the MPOA, and presumably the MPN, to the regulation of maternal behavior, it is useful to summarize what is currently known of the neurochemical distribution in the MPN. It must be emphasized that the relationship between the SDN-POA and the various subcomponents of the MPN is still being evaluated. Although it was initially concluded that the cell-dense central com-

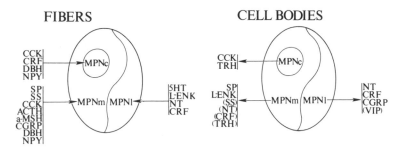

Figure 5.8. Schematic representation of the topographic organization of specific neurochemical inputs (fibers) and outputs (cell bodies) of the three components of the medial preoptic nucleus (MPN), medial, MPNm; central, MPNc; lateral, MPNl. Abbreviations: ACTH, adrenocorticotrophic hormone; a-MSH, melanocyte-stimulating hormone; CCK, cholecystokinin; CGRP, calcitonin gene-related protein; CRF, corticotropin-releasing factor; DBH, dopamine β-hydroxylase; L-ENK, leucine-enkephalin; NPY, neuropeptide Y; NT, neurotensin; SP, substance P; SS, somatostatin; TRH, thyrotropin-releasing hormone; VIP, vasoactive intestinal protein; 5HT, serotonin. (From Simerly et al., 1986)

ponent of the MPN, the MPNc, was equivalent to the SDN-POA (Simerly et al., 1984), the results of further research have modified this view (Bloch & Gorski, 1988b). The SDN-POA clearly includes the MPNc in its posterior aspect, but the SDN-POA also includes, anteriorly, part of the medial division of the MPN.

The distribution of immunohistochemically identified neuronal somata and fibers very closely parallels the compartmentalization of the MPN, as studied by cytoarchitectonic and antiserotonin immunohistochemical analyses (Simerly, Gorski, & Swanson, 1986). Figure 5.8 summarizes the neurochemical nature of neuronal cell bodies and neuronal afferents to the various subdivision of the MPN. At present, cholecystokinin (CCK) and thyrotropin releasing hormone (TRH) appear to be localized rather specifically to the MPNc, the posterior aspect of the SDN-POA. Other results suggest that the MPN and/or the SDN-POA may be particularly rich in opiate receptors (Hammer, 1985). Much further study is required to elucidate the functional significance of this neurochemical distribution, but present findings emphasize that great caution must be applied to the interpretation of the effects of MPOA lesions on maternal behavior or on any functional system; this region is very complex neurochemically.

FUNCTION(S) OF THE SDN-POA

Any study of the putative function or functions of the SDN-POA is complicated by the small size of this nucleus, its relation to the MPN itself, and the plasticity of the nervous system following surgical intervention. Nevertheless, several attempts have been made to identify the possible functions of the SDN-POA.

An initial approach involved the technique of brain-tissue transplantation. The region of the SDN-POA (which included a significant part of the MPN) was punched out from 1-day-old male rats and transplanted stereotaxically into the region of the

SDN-POA of littermate female rats. This tissue survived and markedly altered the behavior of the hosts in adulthood. (Arendash & Gorski, 1982).

Actually, two groups of animals were tested. In one, the transplantation of male brain tissue into the female brain was followed by oil injections. It was considered possible that by transplanting hormone-sensitive male brain tissue into the female, which does not have the appropriate hormonal environment, there would be no graft development. Therefore, some grafted female hosts were injected with 8 μg TP, a dose that would not be expected to masculinize the hosts' behavior.

As illustrated in Figure 5.9, transplantation of the male SDN-POA/MPN into females markedly facilitated masculine copulatory behavior in adulthood, whether or not TP was administered after grafting. However, such transplants also facilitated female copulatory behavior. In the latter case, TP treatment was found to be quite important when amygdala tissue was transplanted (Arendash & Gorski, 1982). Further studies of this type are required before such results can be interpreted clearly in terms of the functional activity of the SDN-POA.

The SDN-POA has also been lesioned in adult animals. In one study, masculine copulatory behavior was evaluated: although lesions restricted to the SDN-POA were without effect, small lesions placed dorsal to the SDN-POA significantly inhibited copulatory behavior in males (Arendash & Gorski, 1983). The influence of MPOA lesions on maternal behavior in female rats was also investigated. The lesions that disrupted maternal behavior were not related to SDN-POA destruction, but again involved regions just dorsal to the SDN-POA (Jacobson et al., 1980).

Two caveats to the lesion approach to the functions of the SDN-POA must be emphasized: (1) Since it is not known if the function of the SDN-POA is stimulatory of inhibitory, it is possible that animals of the wrong sex were given lesions in the study of maternal behavior. Would lesions of the SDN-POA, which is larger in males than in females, actually facilitate the male's expression of maternal behavior? (2) As mentioned above, the SDN-POA is part of a more complex morphological and neurochemical organization, the MPN. The specific loci of lesions must be carefully verified. This is emphasized by the recent report that lesions of the SDN-POA block the postcastrational rise in LH, follicle-stimulating hormone (FSH), and prolactin (Preslock & McCann, 1987). As judged from the lesion localization discussed in this publication, the SDN-POA was actually undamaged, and the lesions appear to have destroyed the periventricular nucleus.

In a final study, which had tantalizing results, advantage was taken of the fact that the in utero position of rats and mice affects the process of sexual differentiation of the brain and the reproductive system (Clemens, Gladue, & Coniglio, 1978; vom Saal, 1983). In this study, male rats resident in utero between two females ejaculated significantly more often before reaching sexual saturation, and their SDN-POA was significantly larger in volume (vom Saal, Coquelin, Schoonmaker, Shryne, & Gorski, 1984).

Thus although the SDN-POA represents a unique model system to elucidate the mechanisms by which testicular hormones bring about sexual differentiation of the brain, the function or functions of this nucleus is unknown. The elucidation of SDN-POA function represents a major and important challenge for future research.

Although lesions of the SDN-POA did not disrupt maternal behavior in female rats (Jacobson et al., 1980), several observations in the literature suggest a possible

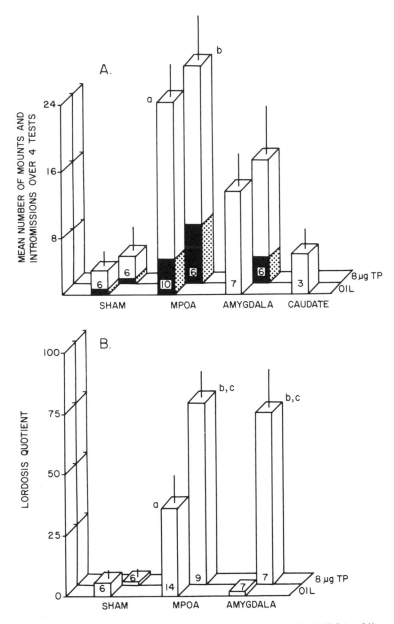

Figure 5.9. The influence of male brain tissue transplants into the MPOA of littermate new-born female rats on the sexual behavior in adulthood. (A) Masculine behavior (mounts, open bars; intromissions, solid bars) over four tests conducted during exposure to TP in adult rats given sham implants, or implants of the MPOA (including the SDN-POA), amygdala, or caudate and treated after transplantation with oil or 8 μg TP (B) Female behavior displayed by these same animals (minus the caudate group) following treatment with 2 μg estradiol benzoate daily for 3 days; progesterone was not administered, so lordosis behavior in controls is low. (a) Significantly different from all groups given oil after surgery. (b) Significantly different from the comparable sham group. (c) Significantly different from the same group given oil after surgery. (Data from Arendash & Gorski, 1982; reprinted from Gorski, 1986b)

Table 5.4   Structural sex differences in the rat brain

Nuclear Volume ("neuronal number")
Components of the medial preoptic nucleus (MPN)   Simerly et al. (1984, 1985b)
Sexually dimorphic nucleus of the preoptic area (SDN-POA)   Gorski et al. (1978, 1980)
Anteroventral periventricular nucleus (AVPV)   Bleier et al. (1982), Bloch & Gorski (1988a), Simerly et al. (1985a), Swanson & Gorski (1985a), Simerly et al. (1985)
Bed nucleus of the stria terminalis (BNST)   Hines & Gorski (unpublished observations)
Ventromedial nucleus (VMN)   Matsumoto & Arai (1983)
Spinal nucleus of the bulbocavernosus (SNB)   Breedlove & Arnold (1980), Nordeen et al. (1985)

Connectivity
Medial preoptic area   Raisman & Field (1973)
Arcuate nucleus   Matsumoto & Arai (1980)
Medial amygdaloid nucleus   Nishizuka & Arai (1981)
Lateral septum   De Vries, Fuijs, & van Leeuwen (1984)
Suprachiasmatic nucleus   Guldner (1982)

---

role of the SDN-POA or another subcomponent of the MPN in the regulation of this behavior. For example, Rubin and Bridges (1984) found that the direct administration of morphine, apparently in the region that Jacobson et al. (1980) lesioned, suppressed maternal behavior. Moreover, as indicated above, the distribution of opiate receptors in the region of the SDN-POA is sexually dimorphic (Hammer, 1985). The serotoninergic input to the MPN is sexually dimorphic (Simerly et al., 1984), and serotoninergic elements have been shown to play an obligatory role in maternal behavior (Barofsky, Taylor, Tizabi, Kumar, & Jones-Quartey, 1983). Finally, lesions of the amygdala or its major efferent pathway, the stria terminalis, facilitate the temporal development of maternal behavior (Fleming, Vaccarino, & Luebke, 1980), and the stria terminalis has a pronounced terminal field in the region of the SDN-POA (Simerly & Swanson, 1986). It is clear that the negative results of a single study with electrolytic lesions (Jacobsen et al., 1980) do not prove that the SDN-POA does not play a role in influencing maternal behavior.

OTHER SEX DIFFERENCES

This discussion has focused on the SDN-POA because it represents the most dramatic structural sex difference observed in the mammalian brain. Three other structural sex dimorphisms have already been mentioned: the AVPV (Bleier et al., 1982; Bloch & Gorski, 1988a; Simerly et al., 1985a), the SNB (Breedlove & Arnold, 1980; Nordeen et al., 1985), and the sex differences within the ultrastructural connectivity of the MPOA (Raisman & Field, 1973). Actually, a growing number of structural sex differences have been discovered in the rat brain (Table 5.4). Moreover, structural sex differences have also identified in the brains of the guinea pig (Hines et al., 1985), gerbil (Commins & Yahr, 1984), ferret (Tobet, Gallagher, Zahniser, Cohen, & Baum, 1983), and human being.

Swaab and Fliers (1985) have reported the existence in the human brain of an SDN-POA in the hypothalamus that is larger in volume in the male than in the female. Although this finding has not been confirmed, two small "interstitial nuclei of the anterior hypothalamus" were significantly larger in volume in males in the sample of human brains analyzed in this laboratory (Allen, Hines, Shryne, & Gorski, 1989).

As indicated earlier, three papers (Gorski et al., 1978; Nottebohm & Arnold, 1976; Raisman & Field, 1973), published in rapid succession, launched a new era in neurobiology. Structural sex differences in the brain clearly exist and presumably play an important causal role in functional sex differences.

ARE THE MORPHOLOGICAL ACTIONS OF GONADAL STEROIDS RESTRICTED TO DEVELOPMENT?

The developing brain clearly is sensitive to the organizational actions of gonadal steroids, which are manifested in permanent changes in both the functional and the structural characteristics of the brain. Although the developmental period when sexual differentiation occurs is a period of great, perhaps greatest, sensitivity to steroids, growing evidence suggests that the morphological effects of gonadal steroid action can be manifested throughout life. Estradiol appears to be synaptogenic near puberty (Clough & Rodriguez-Sierra, 1982) and following surgical intervention in adults (Arai, Matsumoto, & Nishizuka, 1978). The synaptogenic effect of estradiol is even demonstrable in grafted neural tissue (Nishizuka & Arai, 1982). It has also been shown that steroid administration promotes cranial nerve regeneration (Yu & Yu, 1983) and alters regional nuclear volume in the gerbil (Commins & Yahr, 1984) and the rat (Bloch & Gorski, 1988a). But perhaps the most exciting observation relates to the songbird, in which morphological changes occur in the brain as a result of seasonal fluctuations in testicular activity (DeVoogd, Nixdorf, & Nottebohm, 1982; DeVoogd & Nottebohm, 1981). Thus there is a distinct possibility that the gonadal hormones can cause morphological changes in the brain throughout life. Certainly, this action is more pronounced during development, but currently, one cannot ignore the possibility that hormonal actions in the adult include a modification of neuronal circuitry and, through this mechanism, alter behavior.

CONCLUDING REMARKS

The data that have been reviewed in this chapter clearly establish as fact the principle that the brain of the laboratory rat undergoes sexual differentiation. The functional activities of the rat brain considered typical of the male are established during perinatal life by the hormonal environment. More recently, structural sex differences in the rat brain have been identified and have been shown to be determined by the hormonal environment perinatally. The most marked structural sex dimorphism in the rat brain, the SDN-POA, represents a very useful model system to elucidate the fundamental mechanisms of the permanent or organizational effects of testicular hormones. Although a direct relationship between the SDN-POA and parenting behavior has not been established, it is probable that at least some components of the regulation of parenting behavior also undergo the process of sexual differentiation.

ACKNOWLEDGEMENT

Original research of the author was supported by USPHS grants HD01182 and NS 21060.

REFERENCES

Allen, L. S., Hines, M., Shryne, J. E., & Gorski, R. A. (1989). Two sexually dimorphic cell groups in the human brain. *Journal of Neuroscience, 9,* 497–506.

Alleva, F. R., Alleva, J. J., & Umberger, E. J. (1969). Effect of a single prepubertal injection of testosterone propionate on later reproductive function of the female golden hamster. *Endocrinology, 84,* 312–318.

Altman, J., & Bayer, S. A. (1978). Development of the diencephalon in the rat: I. Autoradiographic study of the time of origin and settling patterns of neurons of the hypothalamus. *Journal of Comparative Neurology, 182,* 945–972.

Anderson, C. H. (1978). Time of neuron origin in the anterior hypothalamus of the rat. *Brain Research, 154,* 119–122.

Arai, Y., & Gorski, R. A. (1968a). The critical exposure time for androgenization of the developing hypothalamus in the female rat. *Endocrinology, 82,* 1010–1014.

Arai, Y., & Gorski, R. A. (1968b). Critical exposure time for androgenization of the rat hypothalamus determined by antiandrogen injection. *Proceedings of the Society for Experimental Biology and Medicine, 127,* 590–593.

Arai, Y., Matsumoto, A., & Nishizuka, M. (1978). Synaptogenic action of estrogen on the hypothalamic arcuate nucleus (ARCN) of the developing brain and of the deafferented adult brain in female rats. In G. Dorner & M. Kawakami (Eds.), *Hormones and brain development* (pp. 43–48). Amsterdam: Elsevier/North Holland.

Arendash, G. W., & Gorski, R. A. (1982). Enhancement of sexual behavior in female rats by neonatal transplantation of brain tissue from males. *Science, 217,* 1276–1278.

Arendash, G. W., & Gorski, R. A. (1983). Effects of discrete lesions of the sexually dimorphic nucleus of the preoptic area or other medial preoptic regions on the sexual behavior of male rats. *Brain Research Bulletin, 10,* 147–154.

Bardin, C. W., Bullock, L., Schneider, G., Allison, J. E., & Stanley, A. J. (1970). Pseudohermaphrodite rat: End organ insensitivity to testosterone. *Science, 167,* 1136–1137.

Barofsky, A., Taylor, J., Tizabi, Y., Kumar, R., & Jones-Quartey, K. (1983). Specific neurotoxin lesions of median raphe serotonergic neurons disrupt maternal behavior in the lactating rat. *Endocrinology, 113,* 1184–1893.

Barr, G. A., Gibbons, J. L., & Moyer, K. E. (1976). Male–female differences and the influence of neonatal and adult testosterone on intraspecies aggression in rats. *Journal of Comparative and Physiological Psychology, 90,* 1169–1183.

Barraclough, C. A., & Gorski, R. A. (1961). Evidence that the hypothalamus is responsible for androgen-induced sterility in the female rat. *Endocrinology, 68,* 68–79.

Barraclough, C. A., & Leathem, J. H. (1954). Infertility induced in mice by a single injection of testosterone propionate. *Proceedings of the Society for Experimental Biology and Medicine, 85,* 673–674.

Baum, M. J. (1976). Effects of testosterone propionate administered perinatally on sexual behavior of female ferrets. *Journal of Comparative and Physiological Psychology, 90,* 399–410.

Beach, F. A. (1971). Hormonal factors controlling the differentiation, development, and display of copulatory behavior in the ramstergig and related species. In E. Tobach, K. R. Aronson, & E. Shaw (Eds.), *The biopsychology of development* (pp. 249–296). New York: Academic Press.

Beach, F. A. (1974). Effects of gonadal hormones on urinary behavior in dogs. *Physiology and Behavior, 12,* 1005–1013.

Beach, F. A., & Kuehn, R. E. (1970). Coital behavior in dogs: X. Effects of androgenic stimulation during development on feminine mating responses in females and males. *Hormones and Behavior, 1,* 347–367.

Beach, F. A., Kuehn, R. E., Sprague, R. H., & Anisko, J. J. (1972). Coital behavior in dogs: XI. Effects of androgenic stimulation during development on masculine mating responses in females. *Hormones and Behavior, 3,* 143–168.

Beatty, W. W., & Beatty, P. A. (1970). Hormonal determinants of sex differences in avoidance behavior and reactivity to electric shock in the rat. *Journal of Comparative and Physiological Psychology, 73,* 446–455.

Beatty, W. W., Dodge, A. M., Taylor, K. L., & Meaney, M. J. (1981). Temporal boundary of the sensitive period for hormonal organization of social play in juvenile rats. *Physiology and Behavior, 26,* 241–243.

Beatty, W. W., Powley, T. L., & Keesey, R. E. (1970). Effects of neonatal testosterone injection and hormone replacement in adulthood on body weight and body fat in female rats. *Physiology and Behavior, 5,* 1093–1098.

Bell, D. D., & Zucker, I. (1971). Sex differences in body weight and eating: Organization and activation by gonadal hormones in the rat. *Physiology and Behavior, 7,* 27–34.

Bleier, R., Byne, W., & Siggelkow, I. (1982). Cytoarchitectonic sexual dimorphisms of the medial preoptic and anterior hypothalamic areas in guinea pig, rat, hamster and mouse. *Journal of Comparative Neurology, 212,* 118–130.

Bloch, G. J., & Gorski, R. A. (1988a). Estrogen/progesterone treatment in adulthood affects the size of several components of the medial preoptic area in the male rat. *Journal of Comparative Neurology, 275,* 613–622.

Bloch, G. J., & Gorski, R. A. (1988b). Cytoarchitectonic analysis of the SDN-POA of the intact and gonadectomized rat. *Journal of Comparative Neurology, 275,* 604–612.

Booth, J. E. (1977). Sexual behavior of male rats injected with the antioestrogen MER-25 during infancy. *Physiology and Behavior, 19,* 35–39.

Breedlove, S. M., & Arnold, A. P. (1980). Hormone accumulation in a sexually dimorphic motor nucleus of the rat spinal cord. *Science, 210,* 564–566.

Bridges, R. S. (1983). Sex differences in prolactin secretion in parental male and female rats. *Psychoneuroendocrinology, 8,* 109–116.

Bridges, R. S., Zarrow, M. S., & Denenberg, V. H. (1973). The role of neonatal androgen in the expression of hormonally induced maternal responsiveness in the adult rat. *Hormones and Behavior, 4,* 315–322.

Bronson, F. H., & Desjardins, C. H. (1968). Aggression in adult mice: Modification by neonatal injections of gonadal hormones. *Science, 161,* 705–706.

Brown-Grant, K., & Sherwood, M. R. (1971). The "early androgen syndrome" in the guinea pig. *Journal of Endocrinology, 49,* 277–291.

Carter, C. S., Clemens, L. G., & Hoekema, D. J. (1972). Neonatal androgen and adult sexual behavior in the golden hamster. *Physiology and Behavior, 9,* 89–95.

Christensen, L. W., & Gorski, R. A. (1978). Independent masculinization of neuroendocrine systems by intracerebral implants of testosterone or estradiol in the neonatal female rat. *Brain Research, 146,* 325–340.

Clarke, I. J., Scaramuzzi, R. J., & Short, R. V. (1976). Sexual differentiation of the brain: Endocrine and behavioural responses of androgenized ewes to oestrogen. *Journal of Endocrinology, 71,* 175–176.

Clemens, L. B., Gladue, B. A., & Coniglio, L. P. (1978). Prenatal endogenous androgenic influences on masculine sexual behavior and genital morphology in male and female rats. *Hormones and Behavior, 10,* 40–53.

Clemens, L. G., Shryne, J., & Gorski, R. A. (1970). Androgen and development of progesterone responsiveness in male and female rats. *Physiology and Behavior, 5,* 673–678.

Clough, R. W., & Rodriguez-Sierra, J. F. (1982). Puberty associated neural synaptic changes in female rats administered estrogen. *Society for Neuroscience Abstracts, 8,* 196.

Commins, D., & Yahr, P. (1984). Adult testosterone levels influence the morphology of a sexually dimorphic area in the Mongolian gerbil brain. *Journal of Comparative Neurology, 224,* 132–140.

Conner, R. L., Levine, S., Wertheim, G. A., & Cummer, J. F. (1969). Hormonal determinants of aggressive behavior. *Annals of the New York Academy of Sciences, 159,* 760–776.

Cowan, W. N. (1978). Aspects of neural development. In R. Porter (Ed.), *International review of physiology and neurophysiology* (Vol. 3, pp. 149–191). Baltimore: University Park Press.

Dawson, J. L. M., Cheung, Y. M., & Lau, R. T. S. (1975). Developmental effects of neonatal sex hormones on spatial and activity skills in the white rat. *Biological Psychology, 3,* 213–229.

Denti, A., & Negroni, J. A. (1975). Activity and learning in neonatally hormone treated rats. *Acta Physiological Latinoamerica, 25,* 99–106.

DeVoogd, T. J., Nixdorf, B., & Nottebohm, F. (1982). Recruitment of additional synapses into a brain network takes extra brain space. *Society for Neuroscience Abstracts, 8,* 140.

DeVoogd, T., & Nottebohm, F. (1981). Gonadal hormones induce dendritic growth in the adult avian brain. *Science, 214,* 202–204.

DeVries, G. J., DeBruin, J. P. C., Uylings, H. B. M., & Corner, M. A. (Eds.). (1984). *Progress in brain research: Vol. 61. Sex differences in the brain. The relation between structure and function.* Amsterdam: Elsevier.

DeVries, G. J., Fuijs, R. M., & van Leeuwen, F. W. (1984). Sex differences in vasopressin and other neurotransmitter systems in the brain. *Progress in Brain Research, 71,* 185–203.

Dodson, R. E., Shryne, J. E., & Gorski, R. A. (1988). Hormonal modification of the number of total and late-arising neurons in the central part of the medial preoptic nucleus of the rat. *Journal of Comparative Neurology, 275,* 623–629.

Döhler, K. D., Coquelin, A., Davis, F., Hines, M., Shryne, J. E., & Gorski, R. A. (1982). Differentiation of the sexually dimorphic nucleus in the preoptic area of the rat brain is determined by the perinatal hormone environment. *Neuroscience Letters, 33,* 295–298.

Döhler, K. D., & Hancke, J. L. (1978). Thoughts on the mechanism of sexual brain differ-
    entiation. In G. Dörner & M. Kawakami (Eds.), *Hormones and brain development*
    (pp. 153–157). Amsterdam: Elsevier/North Holland.
Döhler, K. D., Hancke, J. L., Srivastava, S. S., Hofmann, C., Shryne, J. E., & Gorski,
    R. A. (1984). Participation of estrogens in female sexual differentiation of the brain:
    Neuroanatomical, neuroendocrine and behavioral evidence. In G. J. DeVries,
    J. P. C. DeBruin, H. B. M. Uylings, & M. A. Corner (Eds.), *Progress in brain
    research: Volume 61. Sex differences in the brain* (pp. 99–117). Amsterdam: Elsev-
    ier.
Döhler, M. D., Hines, M., Coquelin, A., Davis, F., Shryne, J. E., & Gorski, R. A. (1982).
    Differentiation of the sexually dimorphic nucleus in the preoptic area of the female
    rat brain. *Neuroendocrinology Letters, 4*, 361–365.
Döhler, K. D., Srivastava, S. S., Shryne, J. E., Jarzab, B., Sipos, A., & Gorski, R. A.
    (1984). Differentiation of the sexually dimorphic nucleus in the preoptic area of the
    rat brain is inhibited by postnatal treatment with an estrogen antagonist. *Neuroen-
    docrinology, 38*, 297–301.
Dörner, G., & Staudt, J. (1968). Structural changes in the preoptic anterior hypothalamic area
    of the male rat following neonatal castration and androgen substitution. *Neuroendo-
    crinology, 3*, 136–140.
Doughty, C., & McDonald, P. G. (1974). Hormonal control of sexual differentiation of the
    hypothalamus in the neonatal female rat. *Differentiation, 2*, 275–285.
Edwards, D. A. (1969). Early androgen stimulation and aggressive behavior in male and
    female mice. *Physiology and Behavior, 4*, 333–338.
Edwards, D. A., & Burge, K. G. (1971). Early androgen treatment and male and female
    sexual behavior in mice. *Hormones and Behavior, 2*, 49–58.
Ehrhardt, A. A., & Baker, S. W. (1974). Fetal androgen, human CNS differentiation and
    behavior sex differences. In R. C. Friedman, R. M. Richart, & R. L. Vandewiele
    (Eds.), *Sex differences in behavior* (pp. 53–76). New York: Wiley.
Elsaesser, F., & Parvizi, N. (1979). Estrogen feedback in the pig: Sexual differentiation and
    the effect of prenatal testosterone treatment. *Biology of Reproduction, 20*, 1187–
    1193.
Feder, H. H., & Whalen, R. E. (1964). Feminine behavior in neonatally castrated and estro-
    gen-treated male rats. *Science, 147*, 306–307.
Fink, G., & Jamieson, M. G. (1976). Immunoreactive luteinizing hormone releasing factor in
    rat pituitary stalk blood: Effects of electrical stimulation of the medial preoptic area.
    Journal of Endocrinology, 68, 71–87.
Fleming, A. S., Vaccarino, R., & Luebke, C. (1980). Amygdaloid inhibition of maternal
    behavior in the nulliparous female rat. *Physiology and Behavior, 25*, 731–743.
Ford, J. J. (1982). Testicular control of defeminization in male pigs. *Biology of Reproduction,
    27*, 425–430.
Gorski, R. A. (1971). Gonadal hormones and the perinatal development of neuroendocrine
    function. In L. Martini & W. G. Ganong (Eds.), *Frontiers in neuroendocrinology,
    1971* (pp. 237–290). New York: Oxford University Press.
Gorski, R. A. (1974). The neuroendocrine regulation of sexual behavior. In G. Newton &
    A. H. Riesen (Eds.), *Advances in psychobiology* (Vol. 2, pp. 11–58). New York:
    Wiley.
Gorski, R. A. (1983). Steroid-induced sexual characteristics in the brain. In E. E. Muller &
    R. M. MacLeod (Eds.), *Neuroendocrine perspectives* (Vol. 2, pp. 1–35). Amster-
    dam: Elsevier/North Holland.
Gorski, R. A. (1985a). Sexual differentiation of the brain: Possible mechanisms and implica-
    tions. *Canadian Journal of Physiology and Pharmacology, 63*, 577–594.

Gorski, R. A. (1985b). Gonadal hormones as putative neurotropic substances. In C. W. Cotman (Ed.), *Synaptic plasticity* (pp. 287–310). New York: Guilford Press.

Gorski, R. A. (1987a). Sex differences in the rodent brain: Their nature and origin. In J. M. Reinisch, L. A. Rosenblum, & S. A. Sanders (Eds.), *Masculinity/femininity: Basic perspectives* (pp. 37–67). New York: Oxford University Press.

Gorski, R. A. (1987b). Sexual differentiation of the brain: Mechanisms and implications for neuroscience. In S. S. Easter, K. F. Barald, & B. M. Carlson (Eds.), *From message to mind: Directions in developmental neurobiology* (pp. 256–271). Sunderland, MA: Sinauer.

Gorski, R. A., Csernus, V. J., & Jacobson, C. D. (1981). Sexual dimorphism in the preoptic area. In B. Flerko, G. Setalo, & L. Tima (Eds.), *Advances in physiological sciences: Vol. 15. Reproduction and development* (pp. 121–130). Budapest: Akadamiai Kiado.

Gorski, R. A., Gordon, J. H., Shryne, J. E., & Southam, A. M. (1978). Evidence for a morphological sex difference within the medial preoptic area of the rat brain. *Brain Research, 148*, 333–346.

Gorski, R. A., Harlan, R. E., Jacobson, C. D., Shryne, J. E., & Southam, A. M. (1980). Evidence for the existence of a sexually dimorphic nucleus in the preoptic area of the rat. *Journal of Comparative Neurology, 193*, 529–539.

Gorski, R. A., & Shryne, J. (1972) Intracerebral antibiotics and androgenization of the neonatal female rat. *Neuroendocrinology, 10*, 109–120.

Gorski, R. A., & Wagner, J. W. (1965). Gonadal activity and sexual differentiation of the hypothalamus. *Endocrinology, 76*, 226–239.

Goy, R. W., & McEwen, B. S. (1980). *Sexual differentiation of the brain*. Cambridge, MA: MIT Press.

Goy, R. W., & Phoenix, C. H. (1971). The effects of testosterone propionate administered before birth on the development of behavior in genetic female rhesus monkeys. In C. Sawyer & R. Gorski (Eds.), *Steroid hormones and brain function* (pp. 193–201). Berkeley: University of California Press.

Goy, R. W., & Resko, J. A. (1972). Gonadal hormones and behavior of normal and pseudohermaphroditic nonhuman female primates. *Recent Progress in Hormone Research, 28*, 707–733.

Grady, K. L., Phoenix, C. H., & Young, W. C. (1965). Role of the developing rat testis in differentiation of the neural tissues mediating mating behavior. *Journal of Comparative Physiological Psychology, 59*, 176–182.

Gray, J. A., Levine, S., & Broadhurst, P. L. (1965). Gonadal hormone injections in infancy and adult emotional behavior. *Animal Behaviour, 13*, 33–45.

Guldner, F. H. (1982). Sexual dimorphisms of axo-spine synapses and postsynaptic density material in the suprachiasmatic nucleus of the rat. *Neuroscience Letters, 28*, 145–150.

Hamburger, B., & Oppenheim, R. S. (1982). Naturally occurring neuronal death in vertebrates. *Neuroscience Comment, 1*, 39–55.

Hammer, R. P., Jr. (1985). The sex hormone dependent development of opiate receptors in the rat medial preoptic area. *Brain Resarch, 360*, 65–74.

Harlan, R. E., & Gorski, R. A. (1977). Steroid regulation of luteinizing hormone secretion in normal and androgenized rats at different ages. *Endocrinology, 101*, 741–749.

Harris, G. W. (1964). Sex hormones, brain development and brain function. *Endocrinology, 75*, 627–648.

Hayashi, R., & Gorski, R. A. (1974). Critical exposure time for androgenization by intracranial crystals of testosterone propionate in neonatal female rats. *Endocrinology, 94*, 1161–1167.

Hines, M., Davis, F. C., Coquelin, A., Goy, R. W., & Gorski, R. A. (1985). Sexually dimorphic regions in the medial preoptic area and the bed nucleus of the stria terminalis of the guinea pig brain: A description and an investigation of their relationship to gonadal steroids in adulthood. *Journal of Neuroscience, 5,* 40–47.

Hodges, J. K., & Hearn, J. P. (1978). A positive feedback effect of oestradiol on LH release in the male marmoset monkey, *Callithrix jaccbus. Journal of Reproduction and Fertility, 52,* 83–86.

Ifft, J. D. (1972). An autoradiographic study of the time of final division of neurons in rat hypothalamic nuclei. *Journal of Comparative Neurology, 144,* 192–204.

Jacobson, C. D., Csernus, V. J., Shryne, J. E., & Gorski, R. A. (1981). The influence of gonadectomy, androgen exposure, or a gonadal graft in the neonatal rat on the volume of the sexually dimorphic nucleus of the preoptic area. *Journal of Neuroscience, 1,* 1142–1147.

Jacobson, C. D., Davis, F. C. & Gorski, R. A. (1985). Formation of the sexually dimorphic nucleus of the preoptic area: Neuronal growth, migration and changes in cell number. *Developmental Brain Research, 21,* 7–18.

Jacobson, C. D., & Gorski, R. A. (1980). Neurogenesis of the sexually dimorphic nucleus of the preoptic area of the rat. *Journal of Comparative Neurology, 196,* 519–529.

Jacobson, C. D., Shryne, J. E., Shapiro, F., & Gorski, R. A. (1980). Ontogeny of the sexually dimorphic nucleus of the preoptic area. *Journal of Comparative Neurology, 193,* 541–549.

Jacobson, C. D., Terkel, J., Gorski, R. A., & Sawyer, C. H. (1980). Effects of small medial preoptic area lesions on maternal behavior: Retrieving and nest building in the rat. *Brain Research, 194,* 471–478.

Jost, A., & Magre, S. (1984). Testicular development phases and dual hormonal control of sexual organogenesis. In M. Serio, M. Motta, M. Zanisi, & L. Martin (Eds.), *Sexual differentiation: Basic and clinical aspects* (pp. 1–15). New York: Raven Press.

Kalra, S. P., & Kalra, P. S. (1983). Neural regulation of luteinizing hormone secretion in the rat. *Endocrine Review, 4,* 311–351.

Karsch, F. J., Dierschke, D. J., & Knobil, E. (1973). Sexual differentiation of pituitary function: Apparent difference between primates and rodents. *Science, 179,* 484–486.

Karsch, F. J., & Foster, D. L. (1975). Sexual differentiation of the mechanisms controlling the preovulatory discharge of luteinizing hormone in sheep. *Endocrinology, 97,* 373–379.

Korenbrot, C. C., Paup, D., & Gorski, R. A. (1975). Effects of testosterone or dihydrotestosterone propionate on plasma FSH and LH levels in neonatal rats and on sexual differentiation of the brain. *Endocrinology, 97,* 709–717.

Lumia, A. R., Raskin, L. A., & Eckhert, S. (1977). Effects of androgen on marking and aggressive behavior of neonatally and prepubertally bulbectomized and castrated male gerbils. *Journal of Comparative and Physiological Psychology, 91,* 1377–1389.

Martins, T., & Valle, S. R. (1967). Micturition behaviour of neonatally testosterone treated female dogs. *Experientia, 23,* 921–922.

Matsumoto, A., & Arai, Y. (1980). Sexual dimorphism in "wiring pattern" in the hypothalamic arcuate nucleus and its modification by neonatal hormonal environment. *Brain Research, 190,* 422–426.

Matsumoto, A., & Arai, Y. (1983). Sex difference in volume of the ventromedial nucleus of the hypothalamus in the rat. *Endocrinologia Japonica, 30,* 277–280.

McEwen, B. S., Lieberburg, I., Chaptal, C., & Krey, L. C. (1977). Aromatization: Important for sexual differentiation of the neonatal rat brain. *Hormones and Behavior, 9,* 249–263.

McEwen, B. S., Lieberburg, I., Maclusky, N., & Plapinger, L. (1977). Do estrogen receptors play a role in the sexual differentiation of the rat brain? *Journal of Steroid Biochemistry, 8,* 593–598.

Meany, M. J., & Stewart, J. (1981). Neonatal androgens influence the social play of prepubescent rats. *Hormones and Behavior, 15,* 197–213.

Mennin, S. P., & Gorski, R. A. (1975). Effects of ovarian steroids on plasma LH in normal and persistent estrous adult female rats. *Endocrinology, 96,* 486–491.

Naess, O., Haug, E., Attramadal, A., Aakvaag, A., Hansson, V., & French, F. (1976). Androgen receptors in the anterior pituitary and central nervous system of the androgen "insensitive" (Tfm) rat: Correlation between receptor binding and effects of androgens on gonadotropic secretion. *Endocrinology, 99,* 1295–1303.

Naftolin, F., Ryan, K. J., Davies, I. J., Reddy, V. V., Flores, F., Petro, Z., Kihn, M., White, R. J., Takaoka, Y., & Wolin, L. (1975). The formation of estrogens by central neuroendocrine tissues. *Recent Progress in Hormone Research 31,* 295–319.

Nishizuka, M., & Arai, Y. (1981). Organizational action of estrogen on synaptic pattern in the amygdala: Implications for sexual differentiation of the brain. *Brain Research, 213,* 422–426.

Nishizuka, M., & Arai, Y. (1982). Synapse formation in response to estrogen in the medial amygdala developing in the eye. *Proceedings of the National Academy of Sciences USA, 79,* 7024–7026.

Noirot, E. (1972). The onset of maternal behavior in rats, hamsters, and mice: A selective review. In D. S. Lehrman, R. A. Hinde, & E. Shaw (Eds.), *Advances in the study of behavior* (Vol. 4, pp. 106–145). New York: Academic Press.

Nordeen, E. J., Nordeen, K. W., Sengelaub, D. R., & Arnold, A. P. (1985). Androgens prevent normally occurring cell death in a sexually dimorphic spinal nucleus. *Science, 229,* 671–673.

Norman, R. L., Spies, R. M., Brenner, R. M., & Malley, A. (1983). Subcutaneous ovarian transplants show cyclic steroid function in male rhesus macaques. *Endocrinology, 112*(Suppl., abstracts, 94).

Nottebohm, F., & Arnold, N. P. (1976). Sexual dimorphism in vocal control areas of the songbird brain. *Science, 194,* 211–213.

Numan, M. (1974). Medial preoptic area and maternal behavior in the female rat. *Journal of Comparative and Physiological Psychology, 87,* 746–759.

Numan, M., Rosenblatt, J. S., & Komisaruk, B. R. (1977). Medial preoptic area and onset of maternal behavior of female rats. *Hormones and Behavior, 3,* 29–38.

Nunez, E., Savu, L., Engelmann, F., Benassayag, C., Crepy, O., & Jayle, N. F. (1971). Origine embryonnaire de la protéine sérique fixant l'oestrone et l'oestrone et l'estradiol chez la ratte impubère. *Comptes Rendus Hebdomadaires des Séances de l'Academie des Sciences, D, 273,* 242–245.

Ojeda, S. R., Kalra, P. S., & McCann, S. M. (1975). Further studies on the maturation of the estrogen negative feedback on gonadotropin release in the female rat. *Neuroendocrinology, 18,* 242–255.

Paup, D. C., Coniglio, L. P., & Clemens, L. G. (1972). Masculinization of the female golden hamster by neonatal treatment with androgen or estrogen. *Hormones and Behavior, 3,* 123–131.

Payne, A. P. & Swanson, H. H. (1972). Neonatal androgenization and aggression in the male golden hamster. *Nature, 239,* 282–283.

Pfaff, D. W. (1966). Morphological changes in the brains of adult male rats after neonatal castration. *Journal of Endocrinology, 36,* 415–416.

Pfaff, D. W., Zigmond, R. E. (1971). Neonatal androgen effects on sexual and non-sexual behavior of adult rats tested under various hormone regimes. *Neuroendocrinology, 7,* 129–145.

Phoenix, C. H., Goy, R. W., Gerall, A. A., & Young, W. C. (1959). Organizing action of prenatally administered testosterone propionate on the tissues mediating mating behavior in the female guinea pig. *Endocrinology, 65,* 369–382.

Phoenix, C. H., Goy, R. W., & Resko, J. A. (1968). Psychosexual differentiation as a function of androgenic stimulation. In M. Diamond (Ed.), *Perspectives in reproduction and sexual behavior* (pp. 33–49). Bloomington: Indiana University Press.

Plapinger, L., McEwen, B. S. & Clemens, L. W. (1973). Ontogeny of estradiol-binding sites in rat brain: II. Characteristics of a neonatal binding macromolecule *Endocrinology, 93,* 1129–1139.

Porter, J. C., Eskay, R. L., Oliver, C., Ben-Jonathan, N., Warberg, J., Parker, C. R., Jr., & Barnea, A. (1977). Release of hypothalamic hormones under *in vivo* and *in vitro* conditions. In J. C. Porter (Ed.), *Hypothalamic peptide hormones and pituitary regulation* (pp. 181–202). New York: Plenum Press.

Powell, D. A., Francis, J., & Schneiderman, N. (1971). The effect of castration, neonatal testosterone, and previous experience with fighting on shock-elicited aggression. *Communications in Behavioral Biology, 5,* 371–377.

Preslock, J. P., & McCann, S. M. (1987). Lesions of the sexually dimorphic nucleus of the preoptic area: Effects upon LH, FSH and prolactin in rats. *Brain Research Bulletin, 18,* 127–134.

Quadagno, D. M., & Rockwell, J. (1972). The effect of gonadal hormones in infancy on maternal behavior in the adult rat. *Hormones and Behavior, 3,* 55–62.

Raisman, G. & Field, P. M. (1973). Sexual dimorphism in the neuropil of the preoptic area of the rat and its dependence on neonatal androgen. *Brain Research, 54,* 1–29.

Raynaud, J. P., Mercier-Bodard, C., & Baulieu, E. E. (1971). Rat estradiol binding plasma protein (EBP). *Steroids, 18,* 767–788.

Rosenblatt, J. D. (1967). Nonhormonal basis of maternal behavior in the rat. *Science, 156,* 1512–1514.

Rubin, B. S., & Bridges, R. S. (1984). Disruption of ongoing maternal responsiveness in rats by central administration of morphine sulfate. *Brain Research, 307,* 91–97.

Salaman, D. F. (1974). The role of DNA, RNA and protein synthesis in sexual differentiation of the brain. *Progress in Brain Research, 41,* 349–362.

Salaman, D. F., & Birkett, S. (1974). Androgen-induced sexual differentiation of the brain is blocked by inhibitors of DNA and RNA synthesis. *Nature, 247,* 109–112.

Scouten, C. W., Groteleuschen, L. K., & Beatty, W. W. (1975). Androgens and the organization of sex differences in active avoidance behavior in the rat. *Journal of Comparative Physiological Psychology, 88,* 264–270.

Serio, M., Motta, M., Zanisi, M., & Martini, L. (Eds.). (1984). *Sexual differentiation: Basic and clinical aspects.* New York: Raven Press.

Short, R. V. (1974). Sexual differentiation in the brain of the sheep. In M. G. Forest & J. Bertrand (Eds.), *Endocrinologie sexuelle de la periode perinatale* (pp. 121–142). Paris: INSERM 32.

Simerly, R. B., Gorski, R. A., & Swanson, L. W. (1986). Neurotransmitter specificity of cells and fibers in the medial preoptic nucleus: An immunohistochemical study in the rat. *Journal of Comparative Neurology, 246,* 343–363.

Simerly, R. B., & Swanson, L. W. (1986). The organization of neural inputs to the medial preoptic nucleus of the rat. *Journal of Comparative Neurology, 246,* 312–342.

Simerly, R. B., Swanson, L. W., & Gorski, R. A. (1984). Demonstration of a sexual dimorphism in the distribution of serotonin immunoreactive fibers in the medial preoptic nucleus of the rat. *Journal of Comparative Neurology, 225,* 151–166.

Simerly, R. B., Swanson, L. W., & Gorski, R. A. (1985a). The distribution of monoaminergic cells and fibers in a periventricular preoptic nucleus involved in the control of gonadotropin release: Immunohistochemical evidence for a dopaminergic sexual dimorphism. *Brain Research, 330,* 55–64.

Simerly, R. B., Swanson, L. W., & Gorski, R. A. (1985b). Reversal of the sexually dimorphic distribution of serotonin-immunoreactive fibers in the medial preoptic nucleus by treatment with perinatal androgen. *Brain Research, 340,* 91–98.

Simerly, R. B., Swanson, L. W., Handa, R. J., & Gorski, R. A. (1985). The influence of perinatal androgen on the sexually dimorphic distribution of tyrosine-hydroxylase immunoreactive cells and fibers in the anteroventral periventricular nucleus of the rat. *Neuroendocrinology, 40,* 501–510.

Smith, C. G. (1934). The volume of the neocortex of the albino rat and the changes it undergoes with age after birth. *Journal of Comparative Neurology, 60,* 319–347.

Stewart, J., Skvarenina, A., & Pottier, J. (1975). Effects of neonatal androgens on open-field and maze learning in the prepubescent and adult rat. *Physiology and Behavior, 14,* 291–295.

Sugita, N. (1917). Comparative studies on the growth of the cerebral cortex: I. On the change in the size and shape of the cerebrum during the postnatal growth of the brain. Albino rat. *Journal of Comparative Neurology 28,* 495–510.

Swaab, D. F., & Fliers, E. (1985). A sexually dimorphic nucleus in the human brain. *Science, 228,* 1112–1115.

Swanson, H. H. (1967). Alteration of sex-typical behaviour of hamsters in open field and emergence tests by neo-natal administration of androgen or estrogen. *Animal Behaviour, 15,* 209–216.

Swanson, H. H. (1970). Effects of castration at birth in hamsters of both sexes on luteinization of ovarian implants, oestrous cycles and sexual behavior. *Journal of Reproduction and Fertility, 21,* 183–186.

Taleisnik, S., Caligaris, L., & Astrada, J. J. (1971). Sex difference in hypothalamo-hypophyseal function. In C. H. Sawyer, & R. A. Gorski (Eds.), *Steroid hormones and brain function* (pp. 171–184). Berkeley: University of California Press.

Tarttelin, M. F., Shryne, J. E., & Gorski, R. A. (1975). Patterns of body weight change in rats following neonatal hormone manipulation: A "critical period" for androgen-induced growth increases. *Acta Endocrinologica, 79,* 177–191.

Tobet, S. A., Gallagher, C. A., Zahniser, D. J., Cohen, M. H., & Baum, M. J. (1983). Sexual dimorphism in the preoptic/anterior hypothalamic area of adult ferrets. *Endocrinology, 112*(Suppl.), 240.

Toran-Allerand, C. D. (1976). Sex steroids and the development of the newborn mouse hypothalamus and preoptic area *in vitro:* Implications for sexual differentiation. *Brain Research, 106,* 407–412.

Toran-Allerand, C. D. (1980). Sex steroids and the development of the newborn mouse hypothalamus and preoptic area *in vitro:* II. Morphological correlates and hormonal specificity. *Brain Research, 189,* 413–427.

Toran-Allerand, C. D. (1984). On the genesis of sexual differentiation of the central nervous system: Morphogenetic consequences of steroidal exposure and possible role of α-fetoprotein. In C. J. DeVries, J. P. C. DeBruin, H. B. M. Uylings, & M. A. Corner (Eds.), *Progress in brain research* (Vol. 61, pp. 63–98). Amsterdam: Elsevier.

Turner, J. W. (1975). Influence of neonatal androgen on the display of territorial marking in the gerbil. *Physiology and Behavior, 15,* 265–270.

Vale, J. R., Ray, D., & Vale, C. A. (1972). The interaction of genotype and exogenous neonatal androgen: Agonistic behavior in female mice. *Behavioral Biology, 7,* 321–334.

Valenstein, E. S., Kakolewski, J. W., & Cox, V. C. (1967). Sex differences in taste preference for glucose and saccharin solutions. *Science, 156,* 942–943.

Varon, S., & Adler, R. (1980). Nerve growth factors and control of nerve growth. In A. A. Moscona & A. Monroy (Eds.), *Current topics in developmental biology: Vol. 16.*

Neural development. II. Neural development in model systems (R. K. Hunt, Ed.), (pp. 207–252). New York: Academic Press.

vom Saal, F. S. (1983). The interaction of circulating oestrogens and androgens in regulating mammalian sexual differentiation. In J. Balthazart, E. Prove, & R. Gilles (Eds.), *Hormones and behaviour in higher vertebrates* (pp. 159–177). Berlin: Springer--Verlag.

vom Saal, F., Coquelin, A., Schoonmaker, J., Shryne, J., & Gorski, R. A. (1984). Sexual activity and sexually dimorphic nucleus volume in male rats are correlated with prior intra-uterine position. *Society for Neuroscience Abstracts, 10,* 927.

vom Saal, F. S., Svare, B., & Gandelman, R. (1976). Time of neonatal androgen exposure influences length of testosterone treatment required to induce aggression in adult male and female mice. *Behavioral Biology, 17,* 391–397.

Wade, G. N., & Zucker, I. (1969). Taste preferences of female rats: Modification by neonatal hormones, food deprivation, and prior experience. *Physiology and Behavior, 4,* 935–943.

Wagner, J. W., Erwin, W., & Critchlow, V. (1966). Androgen sterilization produced by intracerebral implants of testosterone in neonatal female rats. *Endocrinology, 79,* 1135–1142.

Wai-Sum, O., Short, R. V., Renfree, M. B., & Shaw, G. (1988). Primary genetic control of somatic sexual differentiation in a mammal. *Nature, 331,* 716–717.

Weisz, J., & Gunsalus, P. (1973). Estrogen levels in immature female rats: True or spurious—ovarian or adrenal? *Endocrinology, 93,* 1057–1065.

Weisz, J., & Ward, I. L. (1980). Plasma testosterone and progesterone titers of pregnant rats, their male and female fetuses, and neonatal offspring. *Endocrinology, 106,* 306–316.

Whalen, R. E. (1968). Differentiation of the neural mechanisms which control gonadotropin secretion and sexual behavior. In M. Diamond (Ed.), *Perspectives in reproduction and sexual behavior* (pp. 303–340). Bloomington: Indiana University Press.

Whalen, R. E., & Rezek, D. L. (1974). Inhibition of lordosis in female rats by subcutaneous implants of testosterone, androstendione or dihydrotestosterone in infancy. *Hormones and Behavior, 5,* 125–138.

Wilson, J. D., Griffin, J. E., Leshin, M., & George, F. W. (1981). Role of gonadal hormones in development of the sexual phenotypes. *Human Genetics, 58,* 78–84.

Yu, A., & Yu, M. C. (1983). Acceleration of the regeneration of the crushed hypoglossal nerve by testosterone. *Experimental Neurology, 80,* 349–360.

II

BIOCHEMICAL CORRELATES OF PARENTING

This part contains six chapters that focus on hormonal, sensory, and experiential factors involved in governing parental behavior. It is comparative in approach in that the subjects of interest are, respectively, rodents, sheep, nonhuman primates, and human females.

In the first chapter, Robert Bridges reviews the endocrinological and neurochemical factors that have been shown to be necessary for regulating parenting in laboratory rats. His overview provides an assessment of the endocrine profile of rats during pregnancy and lactation, the role of endocrines in the initiation and maintenance of maternal behavior, and the way in which endogenous opiates contribute to the regulation of maternal behavior. Bridges also discusses how studies of endocrine factors that regulate parenting in rodents can shed light on the psychobiology of parenting in nonhuman primates and humans.

Next, Bruce Svare describes empirical studies on an aspect of maternal behavior that has not been widely investigated: aggression. During pregnancy and lactation, this behavior is manifested by female mice toward conspecific intruders. Observations of behaviors performed by such mice during pregnancy and lactation are summarized. Also discussed are the endocrinological factors thought to be necessary for the expression of aggression and its observed variations. Finally, Svare suggests future directions for research on this topic.

In the following chapter, Pascal Poindron and Frédéric Lévy outline research that implicates the roles of estrogens, genital stimulation, sensation, and experience in the initiation of maternal behavior in female sheep. They point out how olfactory and genital stimulation cues act together to powerfully regulate maternal responsivity. Poindron and Lévy summarize their own research, which demonstrates the role of prior parenting (experience) as a unique factor and its interaction with hormonal, sensory, and genital stimulation in regulating the initiation of maternal behavior in female sheep. Their findings provide new data that may help to elucidate how learning processes are influenced by sensory and physiological factors that regulate the onset and facilitation of maternal behavior.

Christopher Coe next provides a comprehensive overview of parental behavior in nonhuman primates. He describes the evolutionary factors that have influenced and helped to shape the development of parental behaviors. Following is a discussion of the characteristic behaviors observed in female primates during pregnancy, at parturition, and in caring for their offspring. Coe reviews the literature on the role of hormones in the regulation of maternal behavior in nonhuman primates and summarizes the evidence that the primate offspring is a stimulus that elicits such behavior. The final part of the chapter focuses on a unique aspect of the mother–offspring dyad: the placental transfer of maternal immunity to the fetus.

The next chapter, by Alison S. Fleming, provides an overview of the literature on maternal responsivity from a comparative perspective. The author outlines the findings on maternal responsivity in women and discusses how to define this concept. Fleming then evaluates the evidence on hormonal correlates of maternal responsivity and on the behavioral mechanisms of hormonal action just after delivery. The studies on postpartum experience and retention of maternal responsivity are also reviewed. Finally, the author assesses the effects of parity on maternal behavior and discusses the interactions between affect (mood) and maternal behavior.

In the last chapter of this section, Michelle Warren and Barbara Shortle review the endocrinological profile of the human female during pregnancy. They outline the endocrine status of the mother, placenta, and fetus. Next, they discuss what is known about the role of hormones in the initiation of labor and describe the hormonal profile that ensues postpartum and during lactation. Warren and Shortle summarize the data on a condition known as "pseudocyesis," commonly called "false pregnancy." During this condition, women experience aspects of pregnancy, but the observed hormonal profile is quite different from that of a genuine pregnancy. The authors advocate that more research be carried out on teenage mothers, which could provide new information on the relationship(s) between hormones and parental behaviors.

6

Endocrine Regulation of Parental Behavior in Rodents

ROBERT S. BRIDGES

The behavioral responses toward young shown by parturient and lactating mammals function to foster the growth and maturation of their offspring. Rodent mothers engage in a number of quantifiable behavioral responses that nurture and protect their young. The behaviors exhibited by the females during pregnancy and lactation are distinct and are tied to each species's reproductive and behavioral repertoires. For example, rats, hamsters, and mice display high levels of pup retrieval, increased nest building, and increased levels of aggression toward intruders, while rabbits primarily engage in nest-building and nursing behaviors. The apparent absence of direct pup retrieval by the lactating rabbit appears to be tied to the fact that lactating rabbits spend little time near the nest and pups. Rather, they visit the nest about twice daily to nurse the developing offspring, which are kept in a maternal nest lined and covered with the mother's hair (Sawin & Crary, 1953). In contrast, rat, hamster, and mouse mothers spend a high percentage of their time in close proximity to the young. Moreover, their young are quite mobile and capable of leaving the nest site during early development. Retrieval behavior by these mothers is, therefore, much more likely to be a normal component of the responses shown during the first few weeks postpartum.

The behaviors exhibited by biological mothers are often behaviorally distinct from those displayed by inexperienced females. In rodents, the female's behavior is noticeably altered as a consequence of pregnancy and parturition. Whereas nulliparous animals display fairly low levels of interest toward young, often ignoring them, parturient animals exhibit high levels of maternal responsiveness, such as retrieval, nest building, defense of the young, nursing behavior, and pup-oriented homeostatic responses like anogenital licking.

That changes in behavioral responses to the young are temporally associated with changes in reproductive physiology has led researchers to investigate the possible causal relationships between reproductive physiology and parental behavior. Behavioral and physiological profiles themselves have provided a strong empirical basis for subsequent studies of behavioral–physiological interactions and relationships in rodents. Considerable attention has focused on one set of physiological events that is significantly altered during gestation and lactation—changes in endocrine function.

The objective of this chapter is to characterize the behavioral, endocrine, and neurochemical profiles present in one rodent, the rat, during pregnancy and lactation and to identify those hormonal and neurochemical factors that help to regulate maternal behavior in this species. Such an evaluation should help one gain an appreciation of the specific behavioral–physiological relationships associated with parental behavior, as well as the extent and nature of hormonal regulation of parental care in this mammal.

DEVELOPMENTAL ASPECTS OF MATERNAL BEHAVIOR

Newly parturient rats display a rapid onset of caregiving behaviors toward their young. Among these behaviors are pup retrieval, crouching over the young in order to provide warmth and access to the dam's nipples, pup cleaning, nest building, and defending the young from intruders. Whereas mothers giving birth for the first time or at subsequent times display these responses spontaneously and with high frequency, maternal or maternal-like responses are first shown by adult nulliparous famales only after a number of days of constant exposure to rat young. Specifically, when a behaviorally inexperienced female is housed daily with three 3- to 8-day-old foster young and her behavioral responses are recorded during daily 1-hr test sessions (Bridges, 1984), maternal behaviors—retrieval, grouping, crouching, and anogenital licking responses—are displayed after about 5–7 test days (Cosnier & Couturier, 1966; Rosenblatt, 1967). The basic capacity to respond maternally, therefore, is present in inexperienced female rats. The events associated with mating, pregnancy, parturition, and lactation thus help to transform the female from a slow responder or nonresponder to a fully maternal animal at the time of parturition.

Whereas the most profound behavioral changes occur as a result of pregnancy and parturition, other events during the course of development can also modify the female's (and male's) responses to young. The neural substrate that underlies the regulation of maternal behavior (see chapter 12) is present and can be utilized very early in development, well before puberty. Twenty-four-day-old female (and male) rats will retrieve, group, and crouch over 3- to 8-day-old rat pups after as little as 1–2 days of constant pup exposure (Bridges, Zarrow, Goldman, & Denenberg, 1974; Mayer & Rosenblatt, 1979) (Figure 6.1). Then, between 24 and 30 days of age, the latencies to respond initially to foster young lengthen. This waning of behavioral responsiveness may indicate that some developmental event, possibly the maturation of a behavioral inhibitory neuroregulatory system, becomes functional between 24 and 30 days of age. From 30 days of life to the time well past puberty (days 40–45) and into adulthood, nulliparous female rats require an average of 5–7 days of constant pup exposure before they display pup-induced maternal behavior. When the female mates and becomes pregnant, her behavioral responses toward foster young again change. These changes are most evident during late pregnancy, when females display spontaneous maternal care toward foster young during the immediate prepartum period (Mayer & Rosenblatt, 1984; Slotnick, Carpenter, & Fusco, 1973), as well as heightened levels of aggression toward intruders (Mayer & Rosenblatt, 1984). At parturition, females display high levels of maternal responsivity. Parturient dams actively lick newborn pups and consume the attached placentas (Dollinger, Holloway, & Denenberg, 1980). Nursing and licking, as well as retrieval behaviors, are

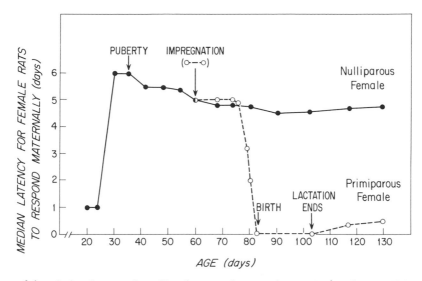

Figure 6.1. A developmental profile of maternal responsiveness in female rats. Values represent the median number of days of pup exposure necessary to induce maternal behavior—retrieval, crouching, and grouping responses—in a 1-hour test session. Developmental landmarks such as puberty, pregnancy, parturition, and weaning are designated on this schematic figure.

evident within the first hour postpartum. As lactation progresses, mothers continue to engage in a full complement of pup-oriented responses and protective behaviors, displaying the highest levels of postpartum aggression (see chapter 7) at the end of the first week of lactation (Erskine, Barfield, & Goldman, 1978). Mother-initiated behavior begins to wane during the third week of lactation as the female begins to wean the developing young (Rosenblatt, 1969). Although the experienced mother no longer actively retrieves her pups during the third week postpartum, once initiated, the more active maternal behaviors, including retrieval and crouching, are retained by the female. Maternal behavior reappears spontaneously in primparous rats presented with 3- to 8-day-old foster young 1–2 months after the weaning of their own litters (Bridges, 1975). This retention of maternal responsiveness results from the state of pregnancy combined with a brief period (1–2 hr) of mother–young interactions postpartum (Bridges, 1975, 1977). The behavioral consequence of this immediate postpartum maternal experience is that maternal care is rapidly displayed by experienced parous females throughout adult life, whether or not the females are lactating.

In the field of maternal behavior, it has been conceptually useful to divide maternal behavior into specific phases. In the rat, three major phases of maternal care have been described: (1) the initiation phase, brought about by pregnancy and evident postpartum; (2) the maintenance phase, which is present throughout lactation until weaning of the young; and (3) the retention phase, which reflects the female's past maternal experience. In the last phase, maternal behavior is retained or remembered independently of the state of lactation. The sections that follow discuss the roles of the endocrine system and of one neurochemical system, opioids, in regulating the initiation and maintenance of maternal behavior in the rat.

HORMONAL REGULATION OF THE INDUCTION
OF MATERNAL BEHAVIOR

That behavioral changes occur in association with significant endocrine changes during pregnancy and lactation makes it important to highlight both the endocrine and the behavioral changes that occur in these animals during these reproductive states. Therefore, in the subsections that follow, the normative endocrine profiles found during pregnancy and lactation in the rat will be presented first. Then the behavioral work that serves as the background and the more recent research on the involvement of prolactin or prolactin-like molecules in maternal behavior will be developed.

Endocrine Profiles During Pregnancy and Lactation

During the past 20 years, research on the initiation or induction of maternal behavior in rats has focused on the role of two classes of hormones: steroids and proteins. Specifically, the involvement of estradiol and progesterone, two steroid hormones, and prolactin, an anterior pituitary protein hormone, has been the subject of numerous investigations. Blood concentrations of these hormone, as well as of another prolactin-like molecule, placental lactogen, are shown in Figure 6.2. Pregnancy in the rat lasts for an average of 22 days; the morning after mating is typically considered day 1 of pregnancy. In the pregnant rat, the ovaries are the primary source of serum progesterone (P) and estradiol (E_2). P levels are significantly elevated by day 3 of pregnancy in the rat. They reach an initial plateau on days 7 to 10 and a second plateau between days 15 and 20 of pregnancy. In the rat (Morishige, Pepe, & Rothchild, 1973; Wiest, 1970), as in mice (McCormick & Greenwald, 1974), hamsters (Baranzcuk & Greenwald, 1974) and rabbits (Challis, Davies, & Ryan, 1973), blood P concentrations decline during late pregnancy prior to parturition. This decline in P levels helps to induce parturition and stimulates the onset of maternal behavior in the rat (Bridges, Rosenblatt, & Feder, 1978). Circulating concentrations of E_2 also increase during pregnancy in the rat, peaking during the latter stages of gestation (Bridges, 1984; Soloff, Alexandrova, & Fernstrom, 1979). An examination of the pattern of prolactin levels in blood during gestation in the rat reveals that prolactin is released phasically from the pituitary gland during the first half of pregnancy (Butcher, Fugo, & Collins, 1972; Gunnet & Freeman, 1983; Neill, 1974). As the conceptus mass becomes endocrinologically active at mid-pregnancy, large amounts of placental lactogens (rPL-I and rPL-II) are secreted (Robertson & Friesen, 1981). Blood concentrations of rPLs increase significantly around day 12 of pregnancy and remain elevated until the end of gestation in the rat (Robertson & Friesen, 1981; Robertson, Owen, McCoshen, & Friesen, 1984). Evolutionarily, the placental lactogen molecules are closely related to prolactin and growth hormone, sharing a high degree of homology in primary structure (Nicoll, 1982; Nicoll, Mayer, & Russell, 1986) and, like prolactin, help to stimulate mammary gland development and milk

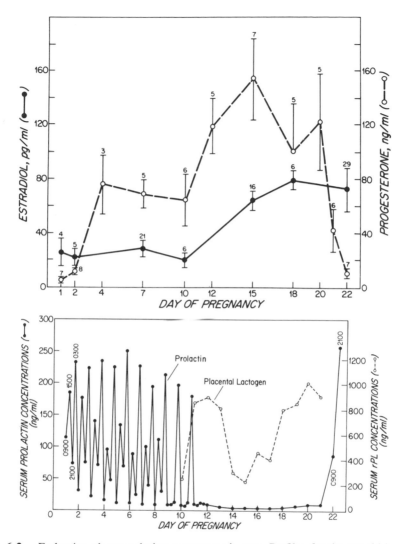

Figure 6.2. Endocrine changes during pregnancy in rats. Profiles for the steroid hormones estradiol and progesterone are shown in the upper panel (from Bridges, 1984), while the levels and patterns of prolactin (adapted from Bridges, unpublished data; Butcher et al., 1972; Gunnet & Freeman, 1983; Linkie et al., 1972; NIADDK rat reference preparation RP-1) and rat placental lactogen (rPL; adapted from Robertson & Friesen, 1981) are depicted in the lower panel.

production. During the 24–48 hr period prior to parturition, which occurs on days 22–23 in the rat, serum prolactin levels increase dramatically, as the result of the shift in balance in steroid exposure, from P to E_2 dominance (Bridges & Goldman, 1975; Linkie & Niswender, 1972). Although the patterns of secretion of other hormones (e.g., growth hormone, corticoids, thyroid stimulating hormone [TSH], androgens) also change as a function of pregnancy, less is known about their behavioral actions.

During lactation, females release large quantities of hormones into the blood-stream. These hormones stimulate lactogenesis as well as milk letdown. Specifically, in response to suckling stimuli, female rats release prolactin, growth hormone, oxytocin, corticosterone (via adrenocorticotrophic hormone [ACTH]), and insulin (see chapter 16). Present thinking in the field of maternal behavior is that changes in hormone levels during lactation regulate milk synthesis and release but do not affect the expression of ongoing maternal behavior (Erskine, Barfield, & Goldman, 1980; Rosenblatt, 1967). However, since thorough studies to examine the possible endocrine regulation of the maintenance of maternal behavior have not been conducted, it may be premature to conclude that the intensity or tone of maternal behavior is unaffected by the endocrine state of the lactating mother. This issue is discussed in greater detail later in this chapter.

Behavioral–Endocrine Studies and the Induction of Maternal Behavior

The early work of Oscar Riddle and his colleagues was the first to indicate that maternal behavior could be affected by hormonal manipulation. Riddle, Lahr, and Bates (1935) reported that injections of crude prolactin preparations into virgin female rats facilitated retrieval. This early work, however, was not well controlled, and the conclusions were seriously questioned over the next three to four decades because a number of researchers were unable to replicate Riddle's work or demonstrate a role for hormones in the induction of maternal behavior in mammals (Beach & Wilson, 1963; Lott & Fuchs, 1962). Then, in the later 1960s and early 1970s, hormonal effects on maternal behavior were demonstrated in a number of species. In rabbits, Zarrow, Farooq, Denenberg, Sawin, and Ross (1963) showed that maternal nest building could be induced in virgin females by the long-term (8–9 weeks) administration of E_2, P, and prolactin. These workers further demonstrated that nest building failed to occur in rabbits that were hypophysectomized during late gestation (Anderson, Zarrow, Fuller, & Denenberg, 1971), but could be induced when pregnant, ergocornine-treated (to reduce prolactin secretion) rabbits were injected with prolactin (Zarrow et al., 1971). In mice, nest building occurs throughout pregnancy and can be induced by P treatment (Koller, 1955). The first studies that convincingly demonstrated hormonal stimulation of pup-oriented maternal behavior were conducted in rats by Moltz, Lubin, Leon, and Numan (1970) and Zarrow et al. (1971). In these reports the latencies of ovariectomized, virgin female rats to retrieve, group, and crouch over young were significantly reduced after 2- to 3-week treatments with pregnancy-like hormone regimens that consisted of E_2, P, prolactin, and, in the study by Zarrow et al. (1971), cortisol.

Shortly thereafter, some technically elegant work by Terkel and Rosenblatt (1972) provided strong evidence that a humoral factor present in the periparturitional rat could stimulate maternal behavior in nulliparous females. Using a parabiotic apparatus that permitted blood exchange between pregnant donor and nulliparous recipient rats, Terkel and Rosenblatt found that some factor(s) in the blood of the female prior to and after delivery stimulated retrieval and grouping responses in the nulliparous recipients. Blood transferred from pup-induced maternal virgins or donor blood exchanged after 12 hr postpartum failed to affect the behavioral responses of the nulliparous recipients. More recent studies have shown that the injection of maternal blood potentiates the expression of maternal or maternal-like behavior in juvenile rats, which are predisposed to respond to young (Brunelli, Shindledecker, & Hofer, 1987; Koranyi, Lissak, Tamasy, & Kamaras, 1976; see also chapter 18). To date, the factor or combination of factors responsible for facilitating maternal behavior in these model systems has not been identified, although E_2 and prolactin have been suggested as potential effective factors.

The work of Rosenblatt and his colleagues in recent years suggests that ovarian steroids, especially estrogens, are responsible for stimulating the rapid onset of maternal behavior (Rosenblatt & Siegel, 1975; Siegel & Rosenblatt, 1975a, 1975b). Nulliparous rats whose ovaries and uteri were removed and that were then injected with estradiol benzoate (EB) at a dose of 100μ/kg 48 hr prior to behavioral testing became maternal faster than did vehicle-injected controls (Siegel & Rosenblatt, 1975a). Likewise, removal of the pregnant uterus on day 16 of gestation stimulated a high incidence of maternal behavior 2 days later (Rosenblatt & Siegel, 1975). At this time of maximal behavioral responsiveness, serum P levels decline (Bridges et al., 1978) and estrogen secretion is increased, as indicated by the vaginal cytology of the rat (Rosenblatt & Siegel, 1975). Using this same day 16 pregnant rat preparation, Numan, Rosenblatt, and Komisaruk (1977) identified the medial preoptic area (MPOA) as a site of estrogen stimulation of maternal behavior (see chapter 12), a finding later confirmed by Fahrbach and Pfaff (1986). It is noteworthy that in virtually all reports of endocrine regulation of the induction of maternal behavior, estrogens are a common and necessary endocrine component. The precise mechanism of estrogenic stimulation of maternal behavior, however, remains to be elucidated.

The role of P in the induction of maternal behavior has proved to be somewhat enigmatic. That is, while a number of studies have implicated P in the induction of maternal behavior in the rat, sites of P action (Numan, 1978; Rubin & Bridges, unpublished data), as well as its mode of action, have not been identified. Nevertheless, P appears to have two important physiological functions that affect the expression of maternal behavior. First, prolonged P exposure in the presence of E_2 (and/or prolactin) primes the female to respond maternally to young. The longer inexperienced females are exposed to the combination of P and E_2 prior to behavior testing, the shorter are their behavioral latencies (Bridges, 1984) (Figure 6.3). Second, the decline in circulating P concentrations prepartum signals the female to engage in maternal behavior; short-latency maternal behavior appears some 24–42 hr after P levels drop from more than 100 ng/ml to less than 10 ng/ml (Bridges et al., 1978). The establishment of maternal behavior is delayed when high circulating titers of P are maintained in rats ovariectomized and hysterectomized during late pregnancy (Bridges, et al., 1978; Numan, 1978; Siegel & Rosenblatt, 1975b). The prepartum decline in P concentrations also appears to be an important signal in stimulating maternal nest building in rabbits (Zarrow et al., 1963) and may help induce maternal behavior in parturient hamsters (Siegel & Rosenblatt, 1980)

Whereas many investigators have examined the roles of steroids in the induction of maternal care, less attention has been paid to the possible involvement of protein and peptide hormones. Recently, however, attention has been given to the involvement of the protein prolactin, the neuropeptide oxytocin (Pedersen & Prange, 1979;

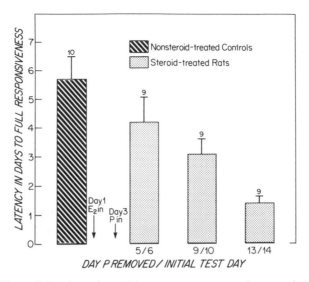

Figure 6.3.   Effect of duration of steroid exposure on maternal responsiveness in ovariectomized, nulliparous rats. Estradiol (E_2)-filled (2-mm) and progesterone (P)-filled (3×30-mm)
Silastic capsules were implanted on treatment days 1 and 3, respectively. P capsules were
removed on days 5, 9, and 13, and testing began 1 day after P removal. The latency to exhibit
maternal behavior toward foster young decreased in proportion to the length of steroid treatment. (From Bridges, 1984)

see chapter 13), and endogenous opioid peptides in the regulation of maternal behavior. Evidence for the role of prolactin in maternal behavior is presented below, while
that of the endogenous opioids is discussed later in this chapter.

Evidence of a role for prolactin in maternal care in the rat reemerged (Bridges,
DiBiase, Loundes, Doherty, 1985; Bridges & Dunckel, 1987; Loundes & Bridges,
1986) after a series of reports indicated that prolactin did not stimulate the induction
of maternal behavior (Baum, 1978; Beach & Wilson, 1963; Lott & Fuchs, 1962;
Numan et al., 1977). In recent work, it was found that treating female rats with a
combination of E_2 and P for 11 days prior to the start of behavioral testing stimulated
maternal behavior. However, when hypophysectomized females (their pituitary glands
had been surgically removed, thereby depriving them of prolactin and other pituitary
hormones) were given this same sequential steroid treatment, maternal behavior was
not stimulated (Bridges et al., 1985). These findings indicated that the pituitary gland
and its secretions were involved in stimulating the onset of maternal behavior. When
another set of nulliparous female rats were hypophysectomized, treated with a combination of steroid hormones (E_2 and P), and given donor pituitary glands that were
surgically grafted beneath the kidney capsule and that secrete prolactin, maternal
behavior rapidly appeared. In fact, the combination of graft treatment and steroid
priming stimulated full maternal behavior in 40% of the nulliparous recipients within
10 min and in 50% of the animals within 1 hr. As predicted, these ectopic pituitary
grafts secreted large amounts of prolactin. A positive correlation between the rate of
onset of maternal behavior and circulating titers of prolactin was found in the graft
recipients (Figure 6.4), a finding that gave greater support to a role for prolactin in

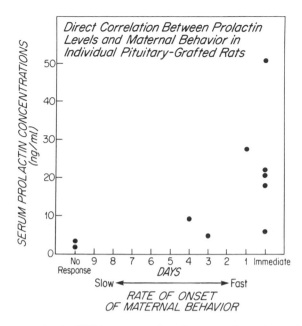

Figure 6.4. Serum prolactin (PRL) concentrations in maternal ovariectomized, hypophysectomized rats treated with steroid implants (3×30 mm P, days 1–11; 2 mm E_2, days 12–22) and bearing two ectopic anterior pituitary gland drafts (days 1–22). PRL concentrations are shown as a function of the rate of onset of maternal responsiveness for each animal. PRL levels were highest in the graft recipients that responded rapidly to foster young. The mean latencies of the graft recipients ($N = 10$) and controls ($N = 11$) to exhibit full maternal behavior were 2 and 8 days, respectively. (Adapted from Bridges et al., 1985)

the induction of maternal behavior. In a subsequent experiment, prolactin injections also stimulated maternal behavior in steroid-treated, hypophysectomized females (Bridges et al., 1985).

In another set of experiments, it was found that suppressing endogenous prolactin secretion in nonhypophysectomized, ovariectomized, nulliparous rats with bromocriptine (CB-154), a dopamine agonist, also blocked the rapid onset of maternal behavior (Bridges & Ronsheim, 1986). As described below, this inhibition of maternal behavior by CB-154 can be prevented by treating the females with exogenous prolactin. In these studies, blood concentrations of prolactin were first quantified in nonhypophysectomized females (all rats were ovariectomized) treated with a steroid regimen that consisted of P (days 1–11) and E_2 (days 11–22) (Bridges, 1984). Plasma prolactin levels in these females remained low throughout the period of P treatment but increased significantly after the introduction of the E_2-filled Silastic capsules. When similar groups of steroid-treated females were administered 0.5 or 2.0 mg CB-154/kg twice daily from day 11 on, circulating prolactin was reduced to nondetectable levels (Figure 6.5). The behavioral effects of this higher dose of CB-154 were then measured in steroid-primed females. In CB-154-treated females, latencies to display aspects of and full maternal behavior increased from about 2–5 days. Thus suppression of endogenous prolactin secretion prevented the rapid onset of maternal

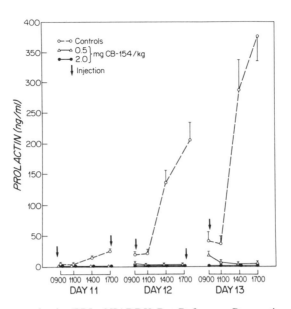

Figure 6.5. Plasma prolactin (PRL; NIADDK Rat Reference Preparation RP-3) concentrations in steroid-treated, ovariectomized rats after bromocriptine (CB-154) or saline administration. All rats were exposed sc to P-filled implants (3×30mm) from days 1 to 11. Prior to P implant removal and E_2 (2 mm) implant insertion on day 11, the animals were injected sc with saline (controls) or one of two doses of CB-154. Both doses of CB-154 suppressed the diurnal surge of PRL induced by estradiol exposure. (From Bridges & Ronsheim, 1986)

behavior. Concurrent administration of exogenous prolactin to steroid-primed, CB-154-injected rats interfered with the disruptive actions of CB-154 (Figure 6.6). Together, these data provide more support for the hypothesis that prolactin is involved in the rapid induction of maternal behavior at parturition.

Under what conditions does prolactin stimulate maternal behavior? First, it appears that the actions of prolactin occur over the course of pregnancy, not just near parturition. In both rats and rabbits, the first effects of pregnancy on maternal behavior are detected after surgical pregnancy termination—on day 16 in the rabbit, prior to mid-pregnancy (Zarrow, Farooq, & Denenberg, 1962), and as early as day 11 in the rat (Rosenblatt, 1969). These findings indicate that the biochemical events that promote the induction of maternal responsiveness are occurring well before birth and that some gradual cumulative priming develops over the course of gestation. This cumulative behavioral priming appears to be hormonally mediated, since the length of time that a female is exposed to hormones affects her maternal response latencies toward young. Specifically, the latencies of female rats to respond maternally to foster young are inversely related to the duration of steroid (E_2 and P) priming prior to behavioral testing; as the length of prior steroid priming increases, the latency to display maternal behavior decreases (see Figure 6.3) (Bridges, 1984). The longer term actions of hormones on the induction of maternal behavior are further supported by the finding that hypophysectomized, steroid-primed rats respond differently when they are treated with exogenous prolactin for either 11 days or 1 day prior to the

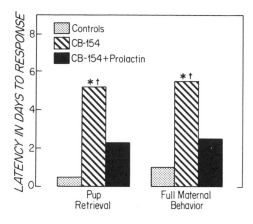

Figure 6.6. The effects of prolactin (PRL-ovine) injections on maternal behavior in steroid-primed, bromocriptine (CB-154)-treated female rats. Ovariectomized rats were given Silastic capsules that contained P (days 1–11) and E_2 (days 11–22), and were then injected sc twice daily from day 11 to day 22 with vehicle, CB-154 (2 mg/kg) or CB-154 plus PRL (0.5 mg). Exogenous prolactin administration prevented the behavioral inhibitory actions of CB-154 on maternal behavior. Behavioral testing was conducted daily from day 12 to day 22. (From Bridges & Ronsheim, 1986)

onset of behavioral testing. Females treated chronically with prolactin for 11 days prior to testing display a rapid onset of all components of maternal behavior, whereas those treated acutely (starting 1 day prior to testing) fail to exhibit a facilitation of full maternal behavior (Loundes & Bridges, 1986). Interestingly, females treated acutely with prolactin show a slight enhancement of pup retrieval when compared with controls.

Evidence that priming may occur over the course of gestation does not preclude a role for further endocrine modulation of maternal behavior just prior to or at parturition. In fact, it seems likely that there is redundancy in the endocrine regulation of the induction of maternal behavior; that is, both long- and short-term mechanisms operate in the rat and other rodents. Whether the neural mechanisms that regulate the expression of maternal behavior are similar for the longer and more acute actions of hormones in the stimulation of maternal behavior has not been elucidated. However, it seems reasonable to assume that these hormones act through a common mechanism.

That prolactin stimulates maternal behavior in steroid-primed rats raises the question of the relationship between prolactin and the steroids E_2 and P. Moltz and his colleagues (1970) previously proposed that a triad of hormones—PRL, P, and E_2—acted together to stimulate the onset of maternal behavior in the rat. What is the nature of this hormonal interaction? Is one hormone primary, while the others are supplemental or secondary? E_2 is a potent stimulator of prolactin synthesis and secretion in mammals (Amenomori, Chen, & Meites, 1970). In a recent study, the possibility that E_2's behavioral action was through its stimulation of prolactin release was examined by eliminating E_2 from the endocrine regimen (Bridges & Dunckel, 1987). Both hypophysectomized and nonhypophysectomized, nulliparous female rats were gonadectomized and given a combination of subcutaneous P implants and ectopic pituitary grafts that secrete prolactin. The P implants were removed after 10 days, and the females were tested for maternal behavior daily, starting 1 day after P removal. Maternal behavior was facilitated in both the hypophysectomized and the nonhypophysectomized females given ectopic grafts and P, whereas grafting or P alone did not facilitate maternal behavior. Again, these effects were evident in the absence of estrogen treatment. This finding, along with the finding that high doses of EB fail to stimulate maternal behavior in hypophysectomized, nulliparous rats (Dunckel & Bridges, unpublished data), indicates that a major, but not necessarily the only, role of E_2 in stimulating maternal behavior is through its stimulation of prolactin release.

While the evidence for prolactin involvement in the induction of maternal behavior in the rat is strong, the specific biobehavioral information present in the prolactin molecule has not been identified. In an attempt to determine the specificity of the message in the prolactin hormone, the ability of a second pituitary hormone, growth hormone, to stimulate the onset of maternal behavior was evaluated. Evolutionarily, growth hormone is closely related to prolactin, the two having a high degree of homology in their primary structures (Wallis, 1978). Like prolactin, growth hormone (250 μg ovine growth hormone twice daily on days 1–22) stimulated a rapid onset of maternal behavior in hypophysectomized, ovariectomized, nulliparous rats treated with a sequential steroid regimen of P (days 1–11) and E_2 (days 11–22; Bridges & Millard, 1988). The average latency for the growth hormone–injected rats to behave maternally was 2 days, while that of steroid-treated controls was >10 days. These findings indicate that growth hormone, like prolactin and possibly rat placental lactogen (rPL), has the capacity to stimulate maternal behavior.

Although unproved, it is intriguing to postulate that during the normal course of pregnancy, all these members of the prolactin family (prolactin, growth hormone, rPL) may contribute to priming the female to respond maternally to the newborn rat pups. As schematically illustrated in Figure 6.7, the pregnant rat is exposed to elevated titers of these proteins throughout gestation. Females are exposed to high levels of prolactin during the first half of gestation and again prepartum, to high levels of circulating rPLs from mid-gestation to just prior to parturition, and to variable, yet apparently elevated, blood concentrations of growth hormone during the early (Blake et al., 1983) and latter (Klindt, Robertson, & Friesen, 1981; Terry et al., 1977) phases of pregnancy. Thus on any single day of gestation, the gravid female is exposed to and can receive hormonal priming from at least one prolactin-like set of molecules. Additional studies are needed to delineate the specific physiological contributions of growth hormone and placental lactogens to the hormonal priming process. Finally, it should be noted that, to date, the full actions of prolactin and growth hormone on maternal responsiveness occur only in the presence of P and E_2. How these steroids interact with prolactin or growth hormone to stimulate maternal behavior is not known.

Given an important role for prolactin and/or prolactin-like molecules in the induction of maternal behavior in the rat, the question arises as to where prolactin is acting to stimulate these behaviors. Although central sites of prolactin stimulation of maternal behavior in rats have not been elucidated, it seems likely that prolactin acts centrally to promote maternal care. Prolactin has access to regions both outside and

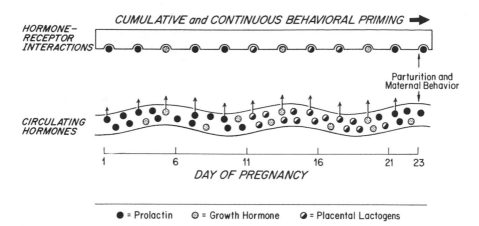

Figure 6.7. A schematic representation of the behavioral priming actions of prolactin and prolactin-like molecules during pregnancy in rats. The patterns of PRL-like hormones in the circulation during pregnancy indicate that rats are exposed to high titers of one or more of these hormones throughout the course of gestation. It is proposed that during pregnancy (in the presence of progesterone and possibly estradiol), prolactin, growth hormone, and placental lactogens contribute to the continuous and cumulative priming of maternal behavior shown by the female at parturition.

inside the blood–brain barrier. That is, prolactin is transported by blood to a number of areas where there is no blood–brain barrier (i.e., circumventricular organs). Some of these sites—for example, the subfornical organ—have neurons that project directly to the vicinity of the MPOA, where they may influence maternal behavior (Numan, 1974, 1985; see also chapter 12). Prolactin can also bypass the blood–brain barrier, being transported into the cerebrospinal fluid (CSF), where prolactin then has access to a host of neural structures previously protected by the blood–brain barrier. Evidence indicating that circulating prolactin reaches the CSF is as follows: prolactin is found in CSF of rabbits and rats (Barbanel, Ixar, Arancibia, & Assenmacher, 1986; Clemens & Sawyer, 1974; Perlow, 1982; Rubin, Ronsheim, & Bridges, 1988); prolactin receptors have been localized on the choroid plexus (Walsh, Posner, Kopriwa, & Brawer, 1978); and a prolactin-receptor-mediated transport system has been identified within the choroid plexus (Walsh, Slaby, & Posner, 1987). To date, the sole demonstration of a central site of prolactin action on maternal behavior has been in maternal mice, in which crystalline prolactin placed into the anterior hypothalamus adjacent to the preoptic region enhanced pup retrieval and nest building (Voci & Carlson, 1973). Future work directed at identifying the sites and modes of action of large proteins like prolactin in the regulation of maternal behavior is needed.

Why should maternal behavior be under the regulation of pituitary secretions? It seems rather inefficient for the organism to require pituitary gland participation in the stimulation of maternal behavior. It would be much easier just to have a direct endocrine action of steroids on the brain and not to require activation of a longer behavioral feedback loop—that is, the subsequent prolactin action on the central nervous system. The answer to this question may relate to the normal physiological state of pregnancy. During gestation, large amounts of prolactin-like molecules, placental lactogens, are secreted by the placenta. These protein hormones may, like prolactin, be transported across the blood–brain barrier to prime maternal behavior. That the developing conceptus could directly prime maternal behavior through its secretions is an attractive hypothesis. If, indeed, the placental lactogens play an important role in stimulating maternal behavior, the effects of pituitary prolactin and growth hormone may be less important physiologically. Thus prolactin-like molecules, and not necessarily prolactin or growth hormone, may be the key factors in stimulating the onset of maternal behavior.

ENDOCRINE MODULATION OF ONGOING MATERNAL BEHAVIOR

The vast majority of research on endocrine involvement in parental behavior has focused on the role of hormones in the induction of maternal behavior; much less attention has focused on the role of hormones in the maintenance or modulation of maternal behavior in mammals. Most species spend large amounts of time parenting after the postpartum period—that is, after parental care is initiated. The results of a number of studies indicate that during this maintenance phase, maternal behavior is not under endocrine control. Maternal female rats and mice exhibit relatively normal pup-directed maternal care after hypophysectomy (Leblond & Nelson, 1937; Obias, 1957; Rosenblatt, 1967), and hypophysectomized mice and rats exhibit apparently normal patterns of maternal aggression (Erskine et al., 1980; Svare, Mann, Broida, & Michael, 1982). Yet, while these findings demonstrate that maternal behavior is not disrupted after removal of the pituitary gland, it is conceivable that the level of intensity of maternal behavior may be altered by endocrine conditions during lactation. Evidence that hormones may modulate the level of parental care is present, although limited. In mice, Voci and Carlson (1973) found that intracerebral implantation of P and prolactin into the hypothalamus potentiated pup retrieval and nest building. Maternal mice given hormonal treatment retrieved pups faster and built better nests than did maternal controls.

The possible endocrine modulation of ongoing maternal behavior was recently reevaluated in female rats (Bridges & Ronsheim, unpublished data). In this study, ovariectomized, nulliparous rats were housed daily with three foster young (sensitization procedure) until the females displayed 2 consecutive days of full maternal care. Maternal rats were then assigned to one of two groups. Experimental females had two ectopic anterior pituitary glands surgically grafted beneath their right renal capsule. Controls were sham operated. All females were then tested once daily between 0900 and 1100 h for 5 days for maternal behavior with three recently fed, healthy foster young. Pup retrieval, grouping, and crouching responses were scored. From the time of onset of maternal behavior to the end of testing, all maternal animals were given three rat pups to maintain maternal behavior. At the end of testing on day 7 of maternal behavior, the pups were removed from each animal, and at 1600 h that day, blood samples were collected. These samples were subsequently assayed for prolactin content. As shown in Figure 6.8, the retrieval responses of the two groups differed over the course of testing. The median latencies (based on the median value for each animal over the 5 test days) to retrieve one, two, or three pups were

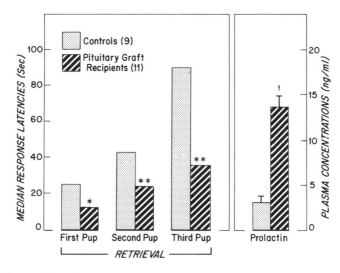

Figure 6.8.   Pup retrieval latencies and prolactin levels in maternal, nulliparous, ovariectomized rats. Females were gonadectomized and induced to display maternal behavior by constant exposure to 3- to 10-day-old rat young. Rats were then given two ectopic anterior pituitary gland grafts or sham-operated (controls) on the second day of full maternal responsiveness. From days 3 to 7, the latencies to retrieve three foster young were recorded for each female. Trunk blood was collected 4–6 hr after testing on day 7 in the absence of rat young. The median latencies to retrieve the three test pups over five test sessions are shown. Graft recipients, which retrieved the test pups significantly faster than did the controls over the 5-day test period, had higher circulating prolactin levels (RP-3) on day 7 of maternal behavior. $*p < .05$; $**p < .01$; $+p < .001$.

significantly faster in the pituitary graft recipients. Serum prolactin levels were about four times higher in experimental, graft-recipient rats, although considerably below those normally found in lactating rats. Thus the intensity of maternal care, as assessed by measuring retrieval latencies, was enhanced or potentiated in maternal rats exposed to higher circulating levels of prolactin (and possibly growth hormone). Additional studies of the possible endocrine modulation of maternal care are needed to elucidate the extent of hormonal influence on ongoing parenting in lactating mothers. These findings in the rat are of particular importance, since maternal as well as paternal animals are normally exposed to varying endocrine profiles throughout parenthood, and changes in endocrine states might therefore affect or modulate a constellation of parental responses, some in more subtle and others in more obvious manners. Whether the endocrine system is involved in affecting parental behavior in humans remains to be determined. Such studies in humans appear to be warranted.

OPIOID-ENDOCRINE REGULATION OF MATERNAL BEHAVIOR

Opioid systems, some of which are under endocrine control, appear to modulate the onset and maintenance of maternal behavior in female rats. *When opiate activity is high, maternal behavior is inhibited or delayed* (Bridges & Grimm, 1982; Hammer & Bridges, 1987). Experiments that support a role for endogenous opiates as one neurochemical system that regulates maternal behavior fall into two categories: (1) physiological studies that quantify endogenous opiate and opiate-receptor concentrations during pregnancy, during lactation, and after treatment with pregnancy-like steroid hormone regimens, and (2) studies that examine behavioral responses of females to morphine and endogenous opiates. Findings from these studies are presented below.

In an initial report, Gintzler (1980) found that opiate activity, measured by changes in reactivity to shock, increased during pregnancy and was diminished during the lactational period. Measurements of endogenous opioids in specific brain regions (Petraglia et al., 1985; Wardlaw & Frantz, 1983; Wardlaw, Thoron, & Frantz, 1982), including the behaviorally important preoptic area (POA), have demonstrated that β-endorphin concentrations are elevated throughout pregnancy and decline prepartum in the POA (Bridges & Ronsheim, 1987). Furthermore, the concentrations of mu-type opiate receptors are higher in the MPOA during mid-pregnancy and are reduced during lactation when β-endorphin concentrations in the POA are also reduced (Hammer & Bridges, 1987).

The higher levels in POA β-endorphin concentrations during pregnancy can be mimicked by treating ovariectomized, nulliparous rats with pregnancy levels of E_2 and P (Bridges & Ronsheim, 1987). This increase in POA opioid concentrations probably results from exposure to high circulating levels of P, since treatment with E_2 alone causes a decrease in the hypothalamic content of β-endorphin (Wardlaw et al., 1982). Opiate-receptor concentrations in the MPOA are also elevated in females treated with pregnancy levels of P and E_2 (Hammer & Bridges, 1987).

Like the regulation of opioid ligand and receptor concentrations during pregnancy, the reductions in opioid concentrations and mu-type opiate receptors during lactation appear to be under hormonal regulation. Hyperprolactinemic female rats bearing prolactin-secreting tumors, like lactating rats, have low neural concentration of both β-endorphin and met-enkephalin (Panerai, Sawynok, LaBella, & Friesen, 1980). Under physiological conditions, the increased secretion of prolactin and possibly suckling-related sensory input during lactation may reduce brain concentrations of these opioids.

These physiological studies characterizing opioid states during pregnancy, during lactation, and after treatment with behaviorally effective steroid regimens help us to understand more about the status of opioid systems during different reproductive conditions. Specifically, the reduced concentrations of opioid ligands in the behaviorally important POA prepartum and of ligands and receptors during lactation appear to reflect a reduced opiate tone at these times. This interpretation is supported by the behavioral finding that lactating rats that have less endogenous ligand in the POA and fewer opiate receptors in the MPOA (Hammer & Bridges, 1987) are less sensitive to the disruptive effects of morphine on maternal behavior than are maternal virgins (Mann, Kinsley, & Bridges, 1988). Such reductions in opiate tone during lactation are probably adaptive in a behavioral sense, since stressors that would stimulate the release of endogenous opioids would be less apt to disrupt maternal behavior under conditions when opiate action is attenuated.

Behavioral studies have provided the strongest empirical support for an inhibitory role for opiates in the regulation of maternal behavior (Bridges & Grimm, 1982; Grimm & Bridges, 1983; Kinsley & Bridges, 1986, 1988b; Rubin & Bridges, 1984). Specifically, systemic administration of morphine sulfate to pregnancy-terminated or lactating rats disrupts both the initiation and the maintenance of maternal behavior (Bridges & Grimm, 1982; Kinsley & Bridges, 1988b). The disruptive effects of morphine can be prevented by administering the opiate antagonist naloxone at the time of morphine treatment. Furthermore, when morphine sulfate is infused bilaterally into the MPOA of maternal lactating rats, both active (e.g., retrieval) and passive (e.g., crouching) components of maternal behavior are significantly impaired, an effect that can be prevented by concurrent central administration of the opiate antagonist naloxone (Rubin & Bridges, 1984). More recently, we have found that infusion of β-endorphin into the region of the MPOA of lactating rats also disrupts pup retrieval, crouching, and grouping behaviors (Apter & Bridges, unpublished data). Thus administration of exogenous and endogenous opiates that bind to the mu-type opiate receptor interferes with the normal expression of parenting behaviors in rats.

Although these findings indicate that increased exposure to opiates disrupts ongoing maternal behavior, the role of opioids in the physiological context of pregnancy and lactation in regulating maternal behavior has not been established. One likely mode of opiate regulation of maternal behavior is through the effects of opiates on the sensory processing of olfactory cues. Morphine and possibly endogenous opioids appear to interfere with maternal behavior by making the smell of the pups adverse to the lactating female (Kinsley & Bridges, 1988a). When lactating primparous rats are injected systemically with morphine at a dose of 5 mg/kg on days 5–6 postpartum, the mothers spend significantly less time in proximity to pup-scented wooden shavings and more time close to unscented shavings. Control females, in contrast, show a strong preference for the pup-scented shavings. Although unproved, the reduction in β-endorphin concentrations found in the POA prepartum (Bridges & Ronsheim, 1987) may alter the female's olfactory acuity or preference so that the smell of pups is no longer adverse to her. In sheep, acute changes in gustatory preference occur prepartum and are thought to be linked to the establishment of maternal behavior (see chapter 8).

In addition to altering sensory processing, opiates may affect maternal behavior through their central actions on motor and thermoregulatory functions. Specifically, during lactation, endogenous opioids may control the nursing behavior of the mother by first keeping her with the young and then driving her away from them. To begin with, in lactating rats, suckling causes the release of β-endorphin, an endogenous opioid, which, in turn, stimulates the secretion of pituitary prolactin (Selmanoff & Gregerson, 1986). It is hypothesized that when this endogenous β-endorphin binds to opiate receptors in the brain (e.g., MPOA), it may initially quiet the female, so that she remains in contact with her young and provides them with easy access to her nipples and milk. As suckling continues, however, the increased exposure to endogenous opiates in the MPOA may elevate the female's core temperature (Martin & Morrison, 1978). This functional hyperthermia, in turn, might induce the mother to leave the young and her nest, thereby terminating the suckling bout (Leon, Crosberry, & Smith, 1978; see chapter 19). Since the actions of morphine on maternal behavior are short-lived (Grimm & Bridges, 1983; Kinsley & Bridges, 1988b), the actions of endogenous opiates in the context of the nursing bout should dissipate within minutes. The female would then become interested in reestablishing contact with her litter and would initiate another nursing bout. This proposed series of actions of endogenous opiates within the physiological and behavioral conditions of lactation is consonant with the available data; however, studies of endogenous-opiate release using techniques such as push-pull cannulas and measurement of opioid synthesis using modern biochemical techniques—for example, proopiomela-nocortin (POMC) transcriptase activity—are needed to evaluate critically this purported role for endogenous opioids in the regulation of ongoing maternal behavior.

The biochemical mode of action of opioids on maternal behavior is not well understood. Whether opioids act, for example, by suppressing the release of endogenous oxytocin or other neurotransmitters, such as norepinephrine, has not been determined. In rats, treatment with morphine interferes with the actions of oxytocin on parturition (Cutting et al., 1985) and decreases the in vitro release of norepinephrine from POA tissues (Diez-Guerra, Augood, Emson, & Dyer, 1987). Likewise, how hormones such as prolactin regulate opioid function is not well known. Additional work is needed to establish both what regulates opioid activity and how changes in opioid activity mediate changes in maternal behavior.

PARITY-ASSOCIATED CHANGES IN THE BIOCHEMICAL REGULATION OF PARENTAL BEHAVIOR

While all biological mothers are primiparous caregivers during some phase of their lives, most adult females become pregnant and give birth many times. Thus the so-called normative adult female mammal spends most of her adult life as a multiparous parent. Although we tend to think that mothers remain fairly uniform in their biological and behavioral states during adulthood, recent findings suggest that this may not be a valid assumption. That is, neurochemical as well as behavioral changes occur between the female's first and second pregnancies and lactations, changes that probably continue to develop as females raise additional young and age. In mother rats that are multiparous and lactating, morphine administration is less effective in disrupting ongoing maternal behavior and producing analgesia than it is in age-matched, primiparous lactating animals (Kinsley & Bridges, 1988b) (Figure 6.9). The reasons for this change in opiate sensitivity are unclear, but the implications are quite interesting. For example, if endogenous-opiate activity is functionally reduced as a female undergoes repeated pregnancies and lactations, one would predict that the potential inhibitory actions of opiates on parenting would be diminished in multiparous females. Such a female would be expected to and does exhibit more consistent and intense maternal behavior. In fact, multiple birthing experiences can override earlier deficits in maternal responsiveness. In rhesus monkeys, for example, Ruppenthal, Arling, Harlow, Sackett, and Suomi (1976) found that mothers that were deprived of stimulation during early development frequently failed to care for their young at the end of their first pregnancy. However, these rhesus mothers displayed relatively normal maternal care after the birth of their young after the second and third pregnancies. Such changes in parental behavior, possibly opiate related, thus can and do

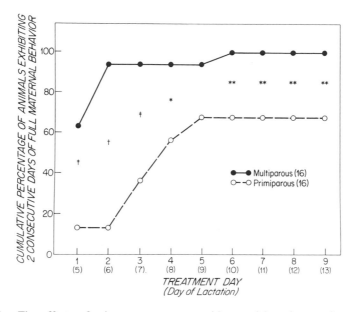

Figure 6.9. The effects of prior pregnancy, parturition, and lactation on the sensitivity of age-matched female rats to the disruptive actions of morphine on maternal behavior. The multiparous, lactating group had previously raised a litter of six young (weaning at 21 days). The primiparous, lactating group was caring for its first litter. All females were injected with morphine sulfate (5 mg/kg) sc 1 hr before the start of behavioral testing. The maternal responsiveness of the multiparous group was significantly less disrupted by morphine from day 5 to 13 of lactation. Full maternal behavior consisted of retrieving all the test young, grouping them in the nest, and crouching over the pups during the 60-min test session. (From Kinsley & Bridges, 1988b)

develop over the course of multiple pregnancies and lactations. Whether similar changes in behavioral and biological states occur in humans has not been examined. Yet such changes, if they do occur, might provide a biological basis that could help explain why a human female might display a large increase in interest in parenting after the birth of a third or even fourth child. The possibility that such changes may occur in humans should be considered in situations of surrogate parenting when a multiparous female serves as the biological mother.

CONCLUDING REMARKS

The present findings demonstrate an important role for hormones, specifically prolactin, in the induction of maternal behavior in female rats. Whereas the induction of maternal behavior at parturition is stimulated by the changes in endocrine activity during gestation in first-time or primiparous mothers, the endocrine system plays a less crucial role in inducing maternal behavior at the birth of a second or third litter or in maintaining ongoing maternal care once it is established. Nevertheless, during lactation, the endocrine system may have more of a modulatory role, adjusting the tone of the female's responses. The induction of maternal behavior, like ongoing maternal care, appears to be under the regulation of endogenous opioids. Opiates inhibit active parenting in female rats. Future studies are needed to define causal links between the opiates and hormones in the induction of maternal care and to establish which factors are key in regulating opiate activity during pregnancy and lactation. It is also important to recognize that a female parenting behavior can be modified throughout development, not just at the time of the birth of her first young. Most recently, changes in neurochemical sensitivity to opiates that may affect the mother's maternal behavior have been found between the first and second lactations. Such findings necessitate that researchers think of the adult mother as a dynamic animal, one that is subject to behavioral change throughout most of her adult life.

Finally, in a broader context, it is important to ask how studying the endocrine regulation of parental behavior in rodents can increase our understanding of the biological regulation of parenting in those mammals, such as primates and humans, that appear to rely less on hormones to stimulate parental responsiveness. The answer to this question is that once those hormones responsible for stimulating maternal behavior are identified, it becomes possible to use them as biochemical probes of the nervous system to identify areas and potential neurochemical systems that mediate these endocrine responses. Whereas not all species may rely equally on hormones to stimulate maternal behavior, those neural regions and neurochemical systems that regulate the expression of parental behavior may indeed be similar in a wide variety of mammals, including humans. By using the information gained from hormonal studies of parental behavior to understand which neural sites and chemical systems are involved in the regulation of parental care, it will be possible to gain greater insights into how the central nervous system functions when parents are not adequately caring for their young—for example, ignoring or abusing them—and it should become possible to develop interventions for those developmental and environmental factors that may adversely affect the parent's capacity or willingness to care for young.

ACKNOWLEDGEMENTS

I would first like to thank all the members of my laboratory who have participated in the studies of parenting and reproductive physiology during the past 10 years. I am also grateful for the funding support provided to me by the NICHD (HD 19789), the NIMH (K02 MH00536), the NIDA (DA 04291) and the March of Dimes Birth Defects Foundation (Grant #1-895), and to the Laboratory of Human Reproduction and Reproductive Biology, Harvard Medical School, by the NIH (Population Center Grant HD 06645). Finally, the support of the National Hormone Pituitary Program, NIADDK, and Dr. Al Parlow for supplying reagents for the prolactin radioimmunoassay and the biologically active hormones used in behavioral studies is deeply appreciated.

REFERENCES

Amenomori, I., Chen, C. L., & Meites, J. (1970). Serum prolactin levels in rats during different reproductive states. *Endocrinology, 86,* 506–510.

Anderson, C. O., Zarrow, M. X., Fuller, G. B., & Denenberg, V. H. (1971). Pituitary involvement in maternal nest building in the rabbit. *Hormones and Behavior, 2,* 183–189.

Baranczuk, R., & Greenwald, G. S. (1974). Plasma levels of oestrogen and progesterone in pregnant and lactating hamsters. *Journal of Endocrinology, 63,* 125–135.

Barbanel, G., Ixar, G., Arancibia, S., & Assenmacher, I. (1986). Probable extrapituitary source of the immunoreactive prolactin measured in the cerebrospinal fluid of un-anesthetized rats by push-pull cannulation of the 3rd ventricle. *Neuroendocrinology, 43,* 476–482.

Baum, M. J. (1978). Failure of pituitary transplants to facilitate the onset of maternal behavior in ovariectomized virgin rats. *Physiology and Behavior, 20,* 87–90.

Beach, F. A., & Wilson, J. R. (1963). Effects of prolactin, progesterone and estrogen on reactions of nonpregnant rats to foster young. *Psychological Reports, 13,* 231–239.

Blake, C. A., Campbell, G. T., Wagoner, J., Rodriguez-Sierra, J. F., Hendricks, S. E., & Elias, K. A. (1983). Dissociation between increased growth hormone and prolactin secretion during the morning hours of early pregnancy in the rat. *Life Sciences, 33,* 1475–1478.

Bridges, R. S. (1975). Long-term effects of pregnancy and parturition upon maternal respon-siveness in the rat. *Physiology and Behavior, 14,* 245–249.

Bridges, R. S. (1977). Parturition: Its role in the long-term retention of maternal behavior in the rat. *Physiology and Behavior, 18,* 487–490.

Bridges, R. S. (1984). A quantitative analysis of the roles of dosage, sequence, and duration of estradiol and progesterone exposure in the regulation of maternal behavior in the rat. *Endocrinology, 114,* 930–940.

Bridges, R. S., DiBiase, R., Loundes, D. D., & Doherty, P. C. (1985). Prolactin stimulation of maternal behavior in female rats, *Science, 227,* 782–784.

Bridges, R. S., & Dunckel, P. T. (1987). Hormonal regulation of maternal behavior in rats: Stimulation following treatment with ectopic pituitary grafts and progesterone. *Biol-ogy of Reproduction, 37,* 518–526.

Bridges, R. S., & Goldman, B. D. (1975). Ovarian control of prolactin secretion during late pregnancy in the rat. *Endocrinology, 97,* 496–498.

Bridges, R. S., & Grimm, C. T. (1982). Reversal of morphine disruption of maternal behavior by concurrent treatment with the opiate antagonist naloxone. *Science, 218,* 166–168.

Bridges, R. S., & Millard, W. J. (1988). Growth hormone is secreted by ectopic pituitary grafts and stimulates maternal behavior in rats. *Hormones and Behavior, 22,* 194–206.

Bridges, R. S., & Ronsheim, P. M. (1986). *Prolactin regulation of maternal behavior in female rats: Bromocriptine treatment delays and prolactin treatment reinstates the rapid onset of behavior.* (Abstract 316.9). Washington, D.C.: Society for Neurosci-ence.

Bridges, R. S., & Ronsheim, P. M. (1987). Immunoreactive beta endorphin concentrations in brain and plasma during pregnancy in rats: Possible modulation by progesterone and estradiol. *Neuroendocrinology, 45,* 381–388.

Bridges, R. S., Rosenblatt, J. S., & Feder, H. H. (1978). Serum progesterone concentrations and maternal behavior in rats after pregnancy termination: Behavioral stimulation following progesterone withdrawal and inhibition by progesterone maintenance. *En-docrinology, 102,* 258–267.

Bridges, R. S., Zarrow, M. X., Goldman, B. D., & Denenberg, V. H. (1974). A develop-mental study of maternal responsiveness in the rat. *Physiology and Behavior, 12,* 149–151.

Brunelli, S. A., Shindledecker, R. D., & Hofer, M. A. (1987). Behavioral responses of juvenile rats *(Rattus norvegicus)* to neonates after infusion of maternal blood plasma. *Journal of Comparative Psychology, 101,* 47–59.

Butcher, R. L., Fugo, N. W., & Collins, W. E. (1972). Semicircadian rhythm in plasma levels of prolactin during early gestation in the rat. *Endocrinology, 90,* 1125–1127.

Challis, J. R. G., Davies, I. J., & Ryan, K. J. (1973). The concentrations of progesterone, estrone and estradiol-17β in the plasma of pregnant rabbits. *Endocrinology, 93,* 971–976.

Clemens, J. A., & Sawyer, B. D. (1974). Identification of prolactin in cerebrospinal fluid. *Experimental Brain Research, 21,* 399–402.

Cosnier, J., & Couturier, C. (1966). Comportement maternal provoqué chez les rattes adultes castrées. *Comptes Rendus des Séances de la Société de Biologie et de Ses Filiales, 160,* 789–791.

Cutting, R., Fitzsimons, N., Gosden, R. G., Humphreys, E. M., Russell, J. A., Scott, S., & Stirland, J. A. (1985). Evidence that morphine interrupts parturition in rats by inhibiting oxytocin secretion. *Journal of Physiology, 371,* 182P.

Diez-Guerra, F. J., Augood, S., Emson, P. C., & Dyer, R. G. (1987). Opioid peptides inhibit the release of noradrenaline from slices of rat medial preoptic area. *Experimental Brain Research, 66,* 378–384.

Dollinger, M. J., Holloway, W. R., & Denenberg, V. H. (1980). Parturition in the rat *(Rattus norvigecus):* Normative aspects and the temporal patterning of behaviours. *Behavioural Processes, 5,* 21–37.

Erskine, M. S., Barfield, R. J., & Goldman, B. D. (1978). Intraspecific fighting during late pregnancy and lactation in rats and effects of litter removal. *Behavioral Biology, 23,* 206–218.

Erskine, M. S., Barfield, R. J., & Goldman, B. D. (1980). Postpartum aggression in rats: I. Effects of hypophysectomy. *Journal of Comparative and Physiological Psychology, 94,* 484–494.

Fahrbach, S. E., & Pfaff, D. W. (1986). Effect of preoptic region implants of dilute estradiol on the maternal behavior of ovariectomized, nulliparous rats. *Hormones and Behavior, 20,* 354–363.

Gintzler, A. R. (1980). Endorphin-mediated increases in pain threshold during pregnancy. *Science, 210,* 193–195.

Grimm, C. T., & Bridges, R. S. (1983). Opiate regulation of maternal behavior in the rat. *Pharmacology, Biochemistry and Behavior, 19,* 609–616.

Gunnet, J. W., & Freeman, M. E. (1983). The mating-induced release of prolactin: A unique neuroendocrine response. *Endocrine Reviews, 4,* 44–61.

Hammer, R. P., Jr., & Bridges, R. S. (1987). Preoptic area opioids and opiate receptors increase during pregnancy and decrease during lactation. *Brain Research, 420,* 48–56.

Kinsley, C. H., & Bridges, R. S. (1986). Opioid mediation of postpartum aggression in rats. *Pharmacology, Biochemistry and Behavior, 25,* 1007–1011.

Kinsley, C. H., & Bridges, R. S. (1988a). *Morphine disruption of maternal behaviors: Possible mediation through alterations in the perception of olfactory stimuli.* Paper presented at the Conference on Reproductive Behavior, Omaha, NE.

Kinsley, C. H., & Bridges, R. S. (1988b). Parity-associated reductions in behavioral sensitivity to morphine. *Biology of Reproduction, 39,* 270–278.

Klindt, J., Robertson, M. C., & Friesen, H. G. (1981). Secretion of placental lactogen, growth hormone and prolactin in late pregnant rats. *Endocrinology, 109,* 1492–1495.

Koller, G. (1956). Hormonale und psychische Steurung bein Nestbau weise Mauser. *Zoologischer Anzeiger*(Suppl.) *19 (Verhandlungen Deutsche Zoologische Gesellschaft,* 1955) 123–132.

Koranyi, L., Lissak, K., Tamasy, V., & Kamaras, L. (1976). Behavioral and electrophysiological attempts to elucidate central nervous system mechanisms responsible for maternal behavior. *Archives of Sexual Behavior, 5,* 503–510.

Leblond, C. P., & Nelson, W. O. (1937). Maternal behavior in hypophysectomized male and female mice. *American Journal of Physiology, 120,* 167–172.

Leon, M., Crosberry, P. G., & Smith, G. K. (1978). Thermal control of mother–young contact in rats. *Physiology and Behavior, 21,* 793–811.

Linkie, D. M., & Niswender, G. D. (1972). Serum levels of prolactin, luteinizing hormone, and follicle-stimulating hormone during pregnancy in the rat. *Endocrinology, 90,* 632–637.

Lott, D. L., & Fuchs, S. S. (1962). Failure to induce retrieving by sensitization or the injection of prolactin. *Journal of Comparative and Physiological Psychology, 65,* 111–113.

Loundes, D. D., & Bridges, R. S. (1986). Length of prolactin priming differentially affects maternal behavior in female rats. *Biology of Reproduction, 34,* 495–501.

Mann, P. E., Kinsley, C. H., & Bridges, R. S. (1988). *Reproductive experience attenuates morphine-induced disruption of maternal behavior.* Paper presented at the Conference on Reproductive Behavior, Omaha, NE.

Martin, G. E., & Morrison, J. E. (1978). Hyperthermia evoked by intracerebral injection of morphine sulfate in the rat. *Brain Research, 145,* 127–140.

Mayer, A. D., & Rosenblatt, J. S. (1979). Ontogeny of maternal behavior in the laboratory rat: Early origins in 18–27 day-old young. *Developmental Psychobiology, 12,* 407–424.

Mayer, A. D., & Rosenblatt, J. S. (1984). Prepartum changes in maternal responsiveness and nest defense in *Rattus norvegicus. Journal of Comparative Psychology, 98,* 177–188.

McCormack, J. T., & Greenwald, G. S. (1974). Progesterone and oestradiol-17β concentrations in the peripheral plasma during pregnancy in the mouse. *Journal of Endocrinology, 62,* 101–107.

Moltz, H., Lubin, M., Leon, M., & Numan, M. (1970). Hormonal induction of maternal behavior in the ovariectomized nulliparous rat. *Physiology and Behavior, 5,* 1373–1377.

Morishige, W. K., Pepe, G. J., & Rothchild, I. (1973). Serum luteinizing hormone, prolactin and progesterone levels during pregnancy in the rat. *Endocrinology, 92,* 1527–1530.

Neill, J. D. (1974). Prolactin: Its secretion and control. In E. Knobil & W. H. Sawyer (Eds.), *Handbook of physiology* (Section 7, Vol. 4. (pp. 469–488). Washington, DC: American Physiological Society.

Nicoll, C. S. (1982). Prolactin and growth hormone: Specialists on one hand and mutual mimics on the other. *Perspectives in Biology and Medicine, 25,* 369–381.

Nicoll, C. S., Mayer, G. L., & Russel, S. M. (1986). Structural features of prolactins and growth hormones that can be related to their biological properties. *Endocrine Reviews, 7,* 169–203.

Numan, M. (1974). Medial preoptic area and maternal behavior in the female rat. *Journal of Comparative and Physiological Psychology, 87,* 746–759.

Numan, M. (1978). Progesterone inhibition of maternal behavior in the rat. *Hormones and Behavior, 11,* 209–231.

Numan, M. (1985). Brain mechanisms and parental behavior. In N. Adler, D. Pfaff, & R. W. Goy (Eds.), *Handbook of behavioral neurobiology: Vol. 7. Reproduction* (pp. 537–605). New York: Plenum Press.

Numan, M., Rosenblatt, J. S., & Komisaruk, B. R. (1977). Medial preoptic area and onset of maternal behavior in the rat. *Journal of Comparative and Physiological Psychology, 91,* 146–164.

Obias, M. D. (1957). Maternal behavior of hypophysectomized gravid albino rats and the development of performance of their progeny. *Journal of Comparative and Physiological Psychology, 50,* 120–124.

Panerai, A. E., Sawynok, J., LaBella, F. S., & Friesen, H. G. (1980). Prolonged hyperprolactinemia influences β-endorphin and met-enkephalin in the brain. *Endocrinology, 106,* 1804–1808.

Pedersen, C. A., & Prange, A. J., Jr. (1979). Induction of maternal behavior in virgin rats after intracerebroventricular administration of oxytocin. *Proceedings of the National Academy of Sciences USA, 76,* 6661–6665.

Perlow, M. J. (1982). Cerebrospinal fluid prolactin: A daily rhythm and response to acute perturbation. *Brain Research, 243,* 382–385.

Petraglia, F., Baraldi, M., Giarre, G., Facchinettii, F., Santi, M., Volpe, A., & Genazzani, A. R. (1985). Opioid peptides of pituitary and hypothalamus: Changes in pregnant and lactating rats. *Journal of Endocrinology, 105,* 239–245.

Riddle, O., Lahr, E. L., & Bates, R. W. (1935). Maternal behavior induced in virgin rats by prolactin. *Proceedings of the Society for Experimental Biology and Medicine, 32,* 730–734.

Robertson, M. C., & Friesen, H. G. (1981). Two forms of placental lactogen revealed by radioimmunoassay. *Endocrinology, 108,* 2388–2390.

Robertson, M. C., Owen, R. E., McCoshen, J. A., & Friesen, H. G. (1984). Ovarian factors inhibit and fetal factors stimulate secretion of rat placental lactogen. *Endocrinology, 114,* 22–30.

Rosenblatt, J. S. (1967). Nonhormonal basis of maternal behavior in the rat. *Science, 156,* 1512–1513.

Rosenblatt, J. S. (1969). The development of maternal responsiveness in the rat. *American Journal of Orthopsychiatry, 39,* 36–56.

Rosenblatt, J. S., & Siegel, H. I. (1975). Hysterectomy-induced maternal behavior during pregnancy in the rat. *Journal of Comparative and Physiological Psychology, 89,* 685–700.

Rubin, B. S., & Bridges, R. S. (1984). Disruption of ongoing maternal responsiveness in rats by central administration of morphine sulfate. *Brain Research, 307,* 91–97.

Rubin, B. S., Ronsheim, P. M., & Bridges, R. S. (1988). *CSF prolactin concentrations in female rats change with endocrine state.* Paper presented at the meeting of the Society for Neuroscience, Toronto.

Ruppenthal, G. C., Arling, G. L., Harlow, H. F., Sackett, G. P., & Suomi, S. J. (1976). A 10-year perspective of motherless-mother behavior. *Journal of Abnormal Psychology, 85,* 341–349.

Sawin, P. B., & Crary, D. D. (1953). Genetic and physiological background of reproduction in the rabbit: I. Some racial differences in the pattern of maternal behavior. *Behaviour, 6,* 128–146.

Selmanoff, M., & Gregerson, K. A. (1986). Suckling-induced prolactin release is suppressed by naloxone and simulated by β-endorphin. *Neuroendocrinology, 45,* 255–259.

Siegel, H. I., & Rosenblatt, J. S. (1975a). Estrogen-induced maternal behavior in hysterectomized-ovariectomized virgin rats. *Physiology and Behavior, 14,* 465–471.

Siegel, H. I., & Rosenblatt, J. S. (1975b). Progesterone inhibition of estrogen-induced maternal behavior in hysterectomized-ovariectomized virgin rats. *Hormones and Behavior, 6,* 223–230.

Siegel, H. I., & Rosenblatt, J. S. (1980). Hormonal and behavioral aspects of maternal care in the hamster: A review. *Neuroscience and Biobehavioral Reviews, 4,* 17–26.

Slotnick, B. M., Carpenter, M. L., & Fusco, R. (1973). Initiation of maternal behavior in pregnant nulliparous rats. *Hormones and Behavior, 4,* 53–59.

Soloff, M. S., Alexandrova, M., & Fernstrom, M. J. (1979). Oxytocin receptors: Triggers for parturition and lactation? *Science,204*, 1313–1314.

Svare, B. D., Mann, M. A., Broida, J., & Michael, S. D. (1982). Maternal aggression exhibited by hypophysectomized parturient mice. *Hormones and Behavior, 16,* 455–461.

Terkel, J., & Rosenblatt, J. S. (1972). Humoral factors underlying maternal behavior at parturition. *Journal of Comparative and Physiological Psychology, 80,* 365–371.

Terry, L. C., Saunders, A., Audet, J., Willoughby, J. O., Brazeau, P., & Martin, J. B. (1977). Physiologic secretion of growth hormone and prolactin in male and female rats. *Clinical Endocrinology, 6,* 19s–28s.

Voci, V. E., & Carlson, N. R. (1973). Enhancement of maternal behavior and nest building following systemic and diencephalic administration of prolactin and progesterone in the mouse. *Journal of Comparative and Physiological Psychology, 88,* 388–393.

Wallis, M. (1978). The chemistry of pituitary growth hormone, prolactin and related hormones, and its relationship to biological activity. In B. Weinstein (Ed.), *Chemistry and biochemistry of amino acids, peptides, and proteins* (Vol. 5, pp. 213–320). New York: Marcel Dekker.

Walsh, R. J., Posner, B. I., Kopriwa, B. M., & Brawer, J. R. (1978). Prolactin binding sites in the rat brain. *Science, 201,* 1041–1043.

Walsh, R. J., Slaby, F. J., & Posner, B. I. (1987). A receptor-mediated mechanism for the transport of prolactin from blood to cerebrospinal fluid. *Endocrinology, 120,* 1846–1850.

Wardlaw, S. L., & Frantz, A. G. (1983). Brain β-endorphin during pregnancy, parturition, and the postpartum period. *Endocrinology, 113,* 1664–1668.

Wardlaw, S. L., Thoron, L., & Frantz, A. G. (1982). Effects of sex steroids on brain β-endorphin. *Brain Research, 245,* 327–331.

Wiest, W. G. (1970). Progesterone and 20 α-hydroxypreg-4-en-3-one in plasma, ovaries and uteri during pregnancy in the rat. *Endocrinology, 87,* 43–48.

Zarrow, M. X., Farooq, A., & Denenberg, V. H. (1962). Maternal behavior in the rabbit: A critical period for nest building following castration during pregnancy. *Proceedings of the Society for Experimental Biology and Medicine, 111,* 537–538.

Zarrow, M. X., Farooq, A., Denenberg, V. H., Sawin, P. B., & Ross, S. (1963). Maternal behaviour in the rabbit: Endocrine control of maternal nest building. *Journal of Reproduction and Fertility, 6,* 375–383.

Zarrow, M. X., Gandelman, R., & Denenberg, V. H. (1971). Prolactin: Is it an essential hormone for maternal behavior in the mammal? *Hormones and Behavior, 2,* 343–354.

7

Maternal Aggression: Hormonal, Genetic, and Developmental Determinants

BRUCE B. SVARE

The study of parental behavior typically involves the analysis of caregiving responses (such as retrieval and nursing behavior) that are directly involved with the nurturing of young. However, other parental responses that are not exhibited toward the young, but are equally important to their survival, tend to be ignored by most biobehavioral scientists in this field of research. The purpose of this chapter is to examine one of the less-well-studied dimensions of parental responding in mammals: the protective behavior called ''maternal aggression,'' which pregnant and lactating females of certain species exhibit toward intruder conspecifics.

The study of maternal aggression in rodents has recently been the subject of experimental attention in several laboratories (for reviews, see Svare, 1977, 1981; Svare & Mann, 1983). The behavior, which has been observed in a variety of mammals in both natural and laboratory settings, function to protect the nest site and young, as well as to disperse adults and juveniles (Rowley & Christian, 1976; Wolff, 1985). It therefore plays an important role in species survival and social organization.

While much has been learned about the physiological determinants of other aspects of parental behavior, far less is known about the biological basis of maternal aggression. This chapter reviews research that examines maternal aggression in the mouse and is divided into five sections: (1) topographic features of maternal aggression; (2) hormonal changes during pregnancy, parturition, and lactation; (3) neuroendocrine determinants of maternal aggression; (4) recent research on individual variation in the behavior; and (5) future tactics in the study of maternal aggression.

TOPOGRAPHY OF MATERNAL AGGRESSION DURING PREGNANCY, PARTURITION, AND LACTATION

Female mice display pronounced changes in their aggressive reactions to male conspecifics as they progress through different reproductive states (Mann, Konen, & Svare, 1984; Mann & Svare, 1982; Svare, Betteridge, Katz, & Samuels, 1981; Svare & Gandelman, 1973) (Figure 7.1). In the Rockland-Swiss (R-S) albino mouse, an outbred stock we have used in all our work, aggression is seldom observed during the virgin state. During pregnancy (18-day duration, with the day of copulatory plug

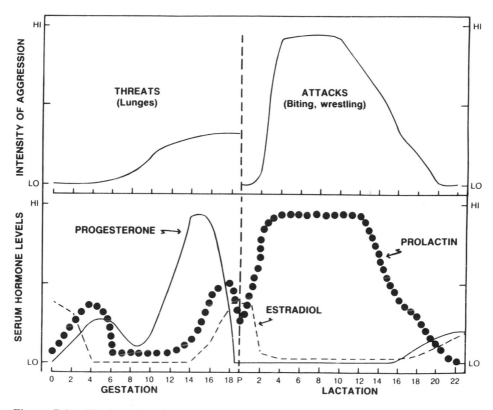

Figure 7.1. The intensity of aggressive behavior (top) and serum hormone levels (bottom) of Rockland-Swiss (R-S) albino mice throughout pregnancy and lactation. The physiological and behavioral data are a composite of information collected in the author's laboratory, as well as other laboratories.

considered gestation day 0), after parturition (evening of the 18th gestation day), and throughout lactation (21 days from the birth to the weaning of young in our colony), dramatic changes in the female's aggressive behavior are exhibited.

Aggression is first observed in the form of threat behavior during mid-pregnancy (gestation day 10), with peak levels of threats seen toward the end of the gestation period (gestation days 14–18) (Mann & Svare, 1982). Aggressive behavior during gestation, referred to as "pregnancy-induced aggression," is much less intense than the agonistic behavior typically observed during lactation (Mann & Svare, 1982). As noted above, late-pregnant females typically exhibit threat behaviors; these responses consist of rapid lunges toward the intruder that fall short of actual physical contact. Attacks (intense biting and wrestling) are also occasionally displayed by late-pregnant females, the primary mode of aggressive responding is threat behaviors.

Immediately following parturition, aggression is absent during the postpartum estrous period when females normally remate with males (Ghiraldi & Svare, unpublished observations; Mann & Svare, 1982). However, after the parturient female has nursed her young for 2 days, intensive aggressive behavior, referred to as "postpartum aggression," emerges and gradually peaks between lactation days 4 and 10 (Svare

& Gandelman, 1973; Svare et al., 1981). This form of aggressive behavior is extremely intense and consists primarily of rapid biting attacks on the intruder. With advancing lactation and the growth of young, lactating females exhibit a dramatic reduction in the intensity of aggressive behavior to a point where there is little postpartum aggression observed around the time when young are weaned (Svare et al., 1981).

HORMONAL CHANGES DURING PREGNANCY, PARTURITION, AND LACTATION

The striking behavioral changes in maternal aggression are associated with changes in hormone profiles. Plasma hormone levels derived from our own work in R-S mice (e.g., Broida, Michael, & Svare, 1981; Mann et al., 1984), as well as from studies in other laboratories using other strains (e.g., Barkley, Michael, Geschwind, & Bradford, 1977; McCormack & Greenwald, 1974; Murr, Stabenfeldt, Bradford, & Geschwind, 1974; Soares & Talamantes, 1982) can be combined to provide a composite picture of maternal hormone secretion during pregnancy, parturition, and lactation (see Figure 7.1).

Progesterone concentrations are low on gestation days 1 and 2, followed by a peak on day 6. Plasma levels of the steroid decline on day 7, but then are dramatically elevated in the second half of pregnancy, beginning around gestation Day 12. Peak serum progesterone is observed between gestation days 14 and 16, with an abrupt decline beginning on gestation day 17, 1 day before parturition. Levels of the steroid reach their lowest point at the time of parturition. In the rat, progesterone apparently remains low during the early postpartum and the lactation periods as a result of the nursing-induced suspension of gonadotropin secretion (Wiest, Kidwell, & Balogh, 1968). A similar situation is thought to be present in the mouse, but we are unaware of peripheral progesterone measurements in this rodent species during the postpartum period.

The level of plasma estradiol is high on the day following the appearance of a copulatory plug, declines to low levels on gestation days 2 and 3, and then remains low until gestation day 14–15. At this time, there is a significant and sustained increase in serum estradiol that continues through the 18th day of pregnancy. At parturition, estradiol levels are reduced and apparently remain low through lactation as a result of suckling-induced suspension of gonadotropin release (Wiest et al., 1968). Once again, however, precise data on estradiol secretion in the mouse during the immediate postpartum period and lactation are not available.

Plasma prolactin shows a brief mating-induced rise, and then decreases and remains low during most of gestation. The hormone begins to increase around gestation day 16 and then declines in the hours just before parturition. Serum prolactin is low on the day of delivery and then dramatically increases, reaching peak levels by the fourth day of lactation. By the 12th postpartum day, plasma levels of prolactin begin to subside, and eventually return to very low concentrations by the time young are weaned.

NEUROENDOCRINE DETERMINANTS OF MATERNAL AGGRESSION

Pregnancy-Induced Aggression

It is well known that steroid hormones play an important role in stimulating some forms of parental behavior in rodents. For example, the rapid onset of pup-directed maternal behavior in rats at the time of parturition is due, in part, to progesterone withdrawal followed by estrogen stimulation (e.g., Bridges, 1984). Our own research, as well as that in other laboratories, suggests that steroid hormones also may be participating in the maternal aggressive behavior exhibited by gravid mice. Specifically, four lines of evidence suggest that elevations in circulating progesterone with advancing gestation may modulate, in part, pregnancy-induced aggression. First, female mice made pseudopregnant by cervical stimulation undergo transient increases in threat behavior and circulating progesterone (Barkley, Geschwind, & Bradford, 1979; Noirot, Goyens & Buhot, 1975). Second, gravid females initiate characteristic threat behaviors when plasma progesterone begins to peak (Mann et al., 1984). Third, although their aggression is not equivalent to that of normal pregnant animals, progesterone-treated virgin female mice exhibit increases in threat behavior when exposed to chronic injections of progesterone or Silastic implants of the steroid (Mann et al., 1984). Fourth, pregnancy termination, a surgical procedure that results in lower serum levels of progesterone, eliminates the aggressive behavior of gravid mice when it is performed on the 15th day of gestation (Svare, Miele, & Kinsley, 1986). Silastic implants of progesterone restore aggression in these mice, although not to the level of fighting normally observed in pregnant animals (Svare et al., 1986).

These results suggest that progesterone plays a modulatory role in stimulating aggression in pregnant mice. Importantly, as noted above, threat behavior is not fully stimulated in virgin or pregnancy-terminated animals by progesterone treatments. The inability to stimulate pregnancy-induced aggressive behavior completely with progesterone treatments suggests that the steroid may not be the only hormone responsible for provoking threat behavior in pregnant animals. Estrogen may be playing a role in the behavior, since progesterone stimulation of aggression in virgin female mice can be achieved only with females that are not ovariectomized (Mann et al., 1984). Interestingly, however, recent work in our laboratory shows that Silastic implants of estradiol inhibit progesterone-stimulation aggression in pregnancy-terminated females (Svare et al., 1986). This preliminary work has to be repeated more systematically, however, since only one dosage of estradiol was used and it may have exceeded the amount that normally circulates in the pregnant mouse. Additional experimentation is clearly needed to explore the extent to which other hormones may synergize with progesterone or even act alone to stimulate aggressive behavior fully.

Another puzzling aspect of these findings concerning progesterone–aggression relationships during gestation is that aggression is still intense late in pregnancy (day 18), when progesterone has reached its lowest concentration (Mann et al., 1984). Perhaps the neural effects of high progesterone concentrations prior to day 18 of gestation persist for 24 hr or more prior to parturition at a time when circulating levels of the steroid have declined. Alternatively, another hormone or combination of hormones may be participating in aggressive behavior, especially late in pregnancy. Testosterone and estrogen are elevated during middle to late pregnancy (e.g., Barkley et al., 1979), but neither hormone has been systematically examined for its role in the development of aggression during gestation. Additionally, it is well known that steroid hormones play a synergistic role in other aspects of parental behavior and other forms of aggression. Testosterone and estrogen are known to stimulate aggression in male mice (Luttge, 1972), and progesterone withdrawal followed by estrogen stimulation can induce pup-directed parental behavior in rats (Bridges, 1984).

The precise brain areas that mediate steroid-hormone effects on pregnancy-induced aggression in the mouse are not known. In rats, progesterone-receptor sites are present in the medial preoptic area (MPOA) of the brain (MacLusky & McEwen, 1980). In mice, progesterone implants in this area are known to facilitate other aspects of maternal behavior, such as nest building (Slotnick, Simoes-Fontes, & Simoes-Fontes, 1977; Voci & Carlson, 1973). In the rat, the MPOA has been implicated in pup-directed maternal behavior (e.g., Numan, 1983), as well as in aggressive behavior (e.g., Potegal & Wagner, 1985). It is interesting to speculate that the MPOA may also mediate aggression in pregnant mice.

Postpartum Docility

Immediately following parturition, aggression is absent during the postpartum estrous period for 24–48 hr, when females normally remate with males (Ghiraldi & Svare, unpublished observations). Research in our laboratory shows that ovarian-hormone stimulation just prior to parturition is probably responsible, in part, for the absence of aggression during the early postpartum period. Ovariectomy just prior to the time of delivery promotes the onset of postpartum aggression earlier (the day of parturition and 24 hr following delivery) than it would normally appear; the administration of estrogen to prepartum, ovariectomized females restores postpartum docility (Ghiraldi & Svare, unpublished observations).

Postpartum Aggression

Intense aggression begins to develop after the postpartum female has received 48 hr of suckling stimulation. It is well documented that the onset of maternal aggression during the postpartum period is contingent on two elements. First, estrogen and progesterone induce growth of the overall size and length of the teats during pregnancy (Svare & Gandelman, 1976a; Svare, Mann, & Samuels, 1980). The growth of the teats, referred to as the "substrate preparation phase" (Svare, 1977), is an important process during pregnancy, since newborns would not be able to attach to and suckle from the mother if it did not occur. Second, the receipt of suckling stimulation from young is essential for the establishment of fighting behavior during the postpartum period (Svare & Gandelman, 1976a, 1976b). Aggressive behavior does not develop in postparturient female mice if either one of these two elements (growth of the nipples or suckling stimulation from young) is missing (Svare & Gandelman, 1976a, 1976b). Thus steroid-hormone exposure during pregnancy stimulates female agonistic behavior in two ways. First, it directly promotes threat behavior during pregnancy;

second, it indirectly stimulates aggression during the postpartum period by preparing the substrate (growth of the nipples) for attachment and suckling by young.

Suckling stimulation activates neuroendocrine relexes involved in hormone release (see chapter 16). Our research has focused on this physiological system in order to probe the factors responsible for the onset of aggression during early lactation. It should be noted that although steroid hormones may play a role in the early inhibition of aggression during postpartum estrus, they do not influence aggression in established lactating mice. Ovariectomy and adrenalectomy do not alter aggression in animals that have been nursing their young for several days (Svare, unpublished observations; Svare & Gandleman, 1976b). Moreover, contrary to our initial speculation, prolactin and other pituitary hormones are not involved in the suckling-induced activation of aggression during the early postpartum period. Aggressive behavior during early lactation is not disrupted by drugs (bromocriptine and ergocornine) that inhibit prolactin release (Mann, Michael, & Svare, 1980), nor is the behavior reduced by hypophysectomy (Svare, Mann, Broida, & Michael, 1982). Instead, the suckling-induced activation of postpartum aggression appears to be due, in part, to elevated serotonergic function. In rats, suckling dramatically increases the hypothalamic turnover rate (5-HT) of serotonin (Kordon, Blake, Terkel, & Sawyer, 1973–1974), and treating postpartum mice with a serotonin depletor ([PCPA]) para-chlorophenylalanine or receptor blocker (methysergide) decreases aggression (Ieni & Thurmond, 1985; Svare, unpublished observations; Svare & Mann, 1983). Finally, Kinsley and Bridges (1986) recently implicated the endogenous-opioid system in rat postpartum aggression. They found that morphine lowered the behavior when administered during lactation, while simultaneous naloxone treatment prevented the effects of morphine. If this also occurs in the mouse, serious consideration must be given to combined 5-HT/opioid system involvement in postpartum aggression.

Although evidence suggests that 5-HT may modulate the onset of postpartum aggression in the mouse, the precise neural areas involved in monoamine regulation of the behavior have not been extensively investigated. Hansen and Ferreira (1986) reported that postpartum aggression in the rat is eliminated by lesions of the peripeduncular region of the lateral midbrain, an area that forms part of the ascending milk-ejection pathway. Also, Flannelly, Kemble, Blanchard, and Blanchard (1986) reported that septal-forebrain lesions abolished postpartum aggression in the rat. In both of these studies, however, maternal behavior was compromised following lesioning, and postpartum females no longer received suckling stimulation from young. Hence, conclusions about septal and peripeduncular-region involvement in postpartum aggression are probably premature.

Once postpartum females have received a sufficient amount of suckling to activate aggression, exteroceptive stimulation from young can maintain the behavior for short periods of time (less than 5 hr). For example, placing pups on the opposite side of a double wire-mesh partition in the dam's home cage maintains the behavior at levels equivalent to those of animals that are left in contact with their young (Svare & Gandelman, 1973). The mechanism responsible for the exteroceptive maintenance of aggression is unknown. However, it is reasonable to suggest that olfactory and/or auditory cues from young may sustain the activity of the physiological systems (serotonergic) ordinarily stimulated by suckling from young.

The precipitous decline in aggression that occurs in lactating animals beginning around postpartum Day 12 has not been extensively researched. The quality and quantity of suckling stimulation and consequent changes in the dam's nursing activity are apparently responsible for changes in aggression. It may be that these behavioral changes, like those occurring early in lactation, are under serotonergic influence in key brain areas that receive stimulation from suckling.

ACCOUNTING FOR INDIVIDUAL VARIATION IN MATERNAL AGGRESSION

The work reviewed thus far has identified a number of biological and behavioral factors that are responsible for the display of aggression during pregnancy and lactation. Progesterone appears to promote aggressive behavior during pregnancy, estrogen suppresses it shortly after parturition, and suckling-induced alterations in serotonergic function facilitate fighting behavior during lactation. To date, however, the findings show that the relationships between these various systems and the display of aggressive behavior are not perfect. For example, individual variation in the display of pregnancy-induced aggressive behavior is not related to individual variation in progesterone titers. Mann et al. (1984) measured circulating progesterone levels in late-pregnant animals that differed widely in the exhibition of aggressive behavior and found no correlation between circulating levels of the hormone and the intensity of the behavior. In fact, some animals that were highly aggressive exhibited plasma progesterone levels that were less than 60 ng/ml, while nonaggressive pregnant mice had serum progesterone titers that exceeded 400 ng/ml.

While similar experimentation is required to elucidate serotonergic function and postpartum aggression, the work to date suggests that other sources of variation should be explored for their role in modulating individual differences in aggressive behavior. Research in our laboratory indicates that individual differences in maternal aggressive behavior can be accounted for, at least in part, by two factors: genotype and prior intrauterine position.

Genotype

For several years, strain differences in maternal aggressive behavior have been studied in C57BL/6J and DBA/2J mice. With respect to maternal aggression, DBA/2J mice exhibit relatively high levels of both pregnancy-induced and postpartum aggression in comparison with C57BL/6J animals (Broida & Svare, 1982a; Svare, 1988) (Figure 7.2). These strain differences cannot be accounted for on the basis of experiential or environmental factors. For example, reciprocally crossed hybrid females exhibit DBA-like patterns of aggressive behavior, and cross-fostered females behave true to their genotype (Broida & Svare, 1982a). As a result, other biologically based factors that could account for strain differences in maternal aggression have been investigated.

As noted earlier, aggression in outbred animals is modulated by increasing progesterone concentrations as pregnancy advances. We examined whether the strain difference in pregnancy-induced aggressive behavior was related to differences in serum progesterone levels; separate groups of pregnant mice from the C57BL/6J and

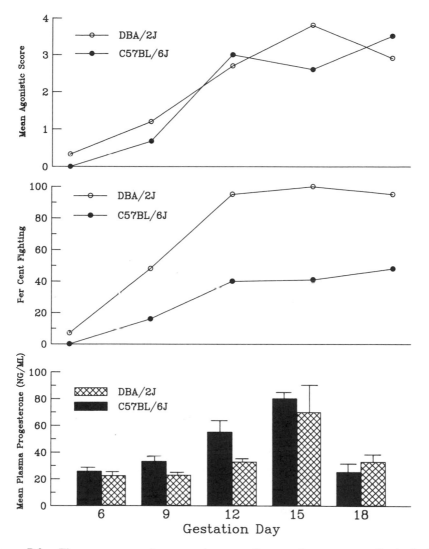

Figure 7.2. The mean composite aggression score (top panel), percentage of animals fighting (middle panel), and serum progesterone (P) levels (bottom panel) during pregnancy in C57BL/6J and DBA/2J mice. A total of 15 animals in each strain were tested for aggressive behavior on gestation days 6, 9, 12, 15, and 18. Separate groups ($N = 10$/group) of animals within each strain were blood sampled on gestation day 6, 9, 12, 15, or 18. Tests for aggression consisted of placing an adult R-S albino male mouse in the home cage of the pregnant female for 3 min. The numbers of attacks and lunges were counted and summed to provide a composite aggression score for each animal. (From Svare, 1988)

DBA/2J strains were tested for aggressive behavior and sampled for blood on gestation day 6, 9, 12, 15, or 18. Although pregnant dams of both strains exhibited the normal rise in serum progesterone with advancing pregnancy and the precipitous drop shortly before parturition, the strains did not differ significantly with respect to serum

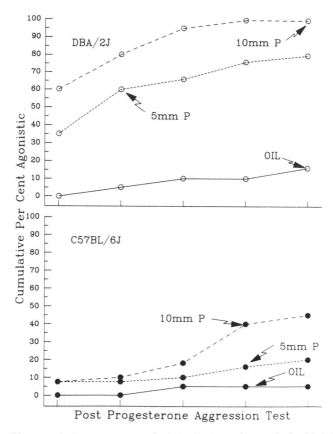

Figure 7.3. The cumulative percentage of adult (60 days of age) virgin CBA/2J and C57BL/6J females that exhibited fighting behavior (attacking or lunging) following exposure to a 5- or 10-mm Silastic implant containing progesterone (P) or oil ($N = 20$/group). Following surgery, the animals were tested for aggression every 3 days for a maximum of five tests (15 days of steroid or oil exposure) or were terminated from testing as soon as fighting (either lunging or attacking) occurred. Adult male R-S Albino mice were used as stimulus opponents for the aggression tests. (From Svare, 1988)

progesterone levels (see Figure 7.2). Thus strain differences in pregnancy-induced aggression cannot be attributed to the amount of this steroid in circulation.

Ongoing work suggests that differential neural sensitivity to progesterone may underlie strain differences in pregnancy-induced aggressive behavior. Virgin C57BL/6J and DBA/2J female mice were subcutaneously implanted with either a 5- or a 10-mm Silastic capsule packed with crystalline progesterone and observed for aggressive behavior toward an intruder male. At both dosage levels, virgin DBA/2J females were more responsive than C57BL/6J females (Figure 7.3). For C57BL/6J females, only animals that were given the 10-mm progesterone implant exhibited a significant elevation in aggression over sesame-oil controls. Also, at the 10-mm dosage, DBA/2J females exhibited a significantly shorter latency (in test days) to begin fighting than did C57BL/6J females.

The mechanism responsible for genetically based differential sensitivity to progesterone is unknown. It is well known that progesterone is concentrated in the MPOA of rodents, where it binds to specific receptors (e.g., MacLusky & McEwen, 1980), and, as stated earlier, progesterone implants in that area are known to increase maternal behaviors in female mice, such as nest building (Voci & Carlson, 1973; Slotnick et al., 1977). Therefore, strain differences in pregnancy-induced aggressive behavior may be related to genotype-dependent variation in progesterone metabolism or uptake, or to progesterone-receptor dynamics in neural areas that are known to concentrate the steroid.

It is interesting to note that other work supports the thesis that C57BL/6J and DBA/2J mice are differentially sensitive to progesterone. Nest-building behavior during gestation in mice is a progesterone-dependent response, and pregnant DBA/2J females exhibit much higher levels of this maternal behavior than do C57BL/6J females (Broida & Svare, 1982b). When administered progesterone in Silastic capsules, virgin DBA/2J females are much more responsive to the hormone and build larger, more fully enclosed maternal nests than do similarly treated C57BL/6J virgins (Broida & Svare, 1983). The utilization of these two inbred strains may be helpful in understanding the etiology of individual variation in progesterone-dependent maternal behaviors.

Finally, DBA/2J mice also exhibit more intense postpartum aggression than do C57BL/6J dams (Broida & Svare, 1982a). If serotonin is found to modulate aggression during this phase of the reproductive cycle in outbred mice, as the present work strongly suggests, then strain differences in the functioning of this monamine ultimately may be found to underlie genotype-based differences in postpartum aggression.

Intrauterine Position

Male and female mice develop in utero adjacent to other fetuses of the same or opposite sex. Such contiguity has dramatic influences on sexually differentiated behavior, physiology, and morphology. Female mice that develop in utero between two males (2M females) have higher amniotic-fluid and fetal-blood titers of testosterone than females that do not develop next to a male fetus (0M females) (vom Saal & Bronson, 1980a). When compared with 0M females with respect to morphology and behavior, 2M females exhibit longer anogenital spacing at birth (an androgen-dependent characteristic) and exhibit higher levels of male-like copulatory behavior (Gandelman, vom Saal, & Reinisch, 1977; vom Saal, 1981; vom Saal & Bronson, 1978). In addition, 2M females are less sexually attractive to males, exhibit longer estrous cycles, achieve puberty later, are less proficient in active avoidance responding, exhibit more locomotor activity, and gain weight more rapidly than 0M females (Hauser & Gandelman, 1983; Kinsley, Miele, Konen, Ghiraldi, & Svare, 1986; Kinsley, Miele, Konen, Ghiraldi, Broida, & Svare, 1986; McDermott, Gandelman, & Reinisch, 1978; vom Saal & Bronson, 1980b). Female mice that develop in utero between one male and one female tend to be intermediate in these reproductive characteristics (e.g., Gandelman et al., 1977).

The extent to which prior intrauterine location influences maternal aggression has been studied in Rockland-Swiss albino mice. Timed-mated female mice were cervically dislocated on the 18th day of gestation. Their fetuses were caesarean delivered, and the respective intrauterine locations were recorded. The fetuses were

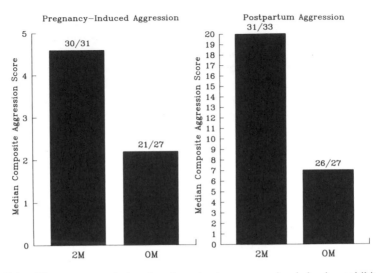

Figure 7.4. The pregnancy-induced and postpartum aggressive behavior exhibited by R-S Albino mice derived from different intrauterine positions. Females located in utero between two males (2M females) were compared with females between no males (0M females) with respect to their aggressive behavior in adulthood on days 15, 16, and 17 of gestation (left panel) and on days 6, 7, and 8 of lactation (right panel). Tests on each day consisted of placing an adult male in the cage of the female for 10 min and counting the number of lunges and attacks. These behaviors were summed over the 3 test days and divided by 3 to produce a composite aggression score for each animal. (The proportions above the histograms represent the proportion of animals fighting on at least one test.) (From Kinsley et al., 1986)

fostered to lactating donor mothers and, at 60 days of age, were mated and examined for maternal aggression. The results indicate that 2M female mice exhibited higher levels of both pregnancy-induced and postpartum aggression than did 0M females (Kinsley, Miele, Ghiraldi, Konen, & Svare, 1986) (Figure 7.4). Although 2M and 0M animals did not differ with respect to the proportion of animals displaying aggression, the former exhibited more intense aggressive behavior than the latter (e.g., a higher composite aggression score).

How intrauterine location modulates maternal aggression is unknown. Exposure of fetuses to testosterone may modify the central neural structures and physiological systems that mediate pregnancy-induced and postpartum aggression (see chapter 5). The sexually differentiated brain areas that receive suckling-induced information and concentrate progesterone (pregnancy-induced aggression) and serotonin (postpartum aggression) are particularly likely candidates for involvement.

CONCLUSIONS AND FUTURE DIRECTIONS

Important progress has been made in understanding the proximate biological causes of maternal aggressive behavior in pregnant and lactating mice. Briefly, progesterone stimulates aggression during pregnancy, estrogen suppresses it in the hours following parturition (postpartum estrus), and suckling-induced changes in serotonergic function modulate the onset of aggressive behavior during the lactation period. The work reviewed here also shows that the dramatic individual variation observed in this aspect of maternal behavior can be traced, at least in part, to genotype and to prior intrauterine position. Genotype may exert its behavioral effects by influencing neural sensitivity to progesterone, while prior intrauterine position may influence aggression by altering testosterone-responsive brain areas during early critical periods of sexual differentiation.

In many ways, the study of maternal aggressive behavior is still in its infancy. More information is needed about the precise neural mechanisms responsible for progesterone, estrogen, and serotonergic modulation of maternal aggression in the mouse. More research is also needed to understand how genotype and prior intrauterine position modulate the physiological systems that control this behavior during pregnancy and lactation. Moreover, recent evidence indicates that other hormones, including prolactin, oxytocin, and the opioids, are implicated in pup-directed maternal care in the rat (e.g., Bridges, DiBiase, Loundes, & Doherty, 1985; Bridges & Grimm, 1982; Pedersen, Ascher, Monroe & Prange, 1982). These hormones need to be explored more systematically to determine their involvement in the maternal aggression displayed by the mouse.

Defense of young from conspecific attacks and predation is as important for species survival as a caregiving maternal behavior that is directly involved in the nurturing of young. Future work should attempt to incorporate findings in both areas in order to provide a cohesive and comprehensive understanding of the proximate and ultimate causes of maternal behavior.

Since an important goal is to extend the reliability and validity of current research on maternal aggression in mice, it is essential that more be learned about it in a much wider variety of species. Currently, there is no systematic information on the extent to which this behavior is represented in mammalian or nonmammalian species. Furthermore, there are no known data on protective aggressive behavior exhibited by male mammals (i.e., paternal aggression). This information is vital for understanding the degree to which the aggression component of maternal behavior has been subject to selection pressures that are crucial to its evolution.

Finally, is there a human analogue to the maternal aggressive behavior that researchers have examined in laboratory rodents? Although several clinical studies have examined the biobehavioral basis of severe mood swings (e.g., postpartum psychosis or postpartum blues) that occur in some peripartum women (Nott, Franklin, Armitage, & Gelder, 1976; Yalom, Lunde, Moos, & Hamburg, 1968), we know of no data that specifically address the issue of maternal protective behavior (or attitude) and its possible biological basis in pregnant or postpartum humans. One can only hope that clinicians will become interested in these questions, since their answers could be critical in understanding the proximate and ultimate causes of this important dimension of maternal behavior.

REFERENCES

Barkley, M. S., Geschwind, I. I., & Bradford, G. E. (1979). The gestational pattern of estradiol, testosterone, and progesterone secretion in selected strains of mice. *Biology of Reproduction, 20*, 733–738.

Barkley, M. S., Michael, S. D., Geschwind, I. I., & Bradford, G. E. (1977). Plasma testosterone during pregnancy in the mouse. *Endocrinology, 100,* 1472–1475.

Bridges, R. S. (1984). A quantitative analysis of the roles of dosage, sequence, and duration of estradiol and progesterone exposure in the regulation of maternal behavior in the rat. *Endocrinology, 114,* 930–940.

Bridges, R. S., DiBiase, R., Loundes, D. D., & Doherty, P. C. (1985). Prolactin stimulation of maternal behavior in female rats. *Science, 227,* 782–784.

Bridges, R. S., & Grimm, C. T. (1982). Reversal of morphine disruption of maternal behavior by concurrent treatment with the opiate antagonist naloxone. *Science, 218,* 166–168.

Broida, J., Michael, S., & Svare, B. (1981). Plasma prolactin is not related to the initiation, maintenance, and decline of maternal aggression in mice. *Behavioral and Neural Biology, 32,* 121–125.

Broida, J., & Svare, B. (1982a). Postpartum aggression in C57BL/6J and DBA/2J mice: Experiential and environmental influences. *Behavioral and Neural Biology, 35,* 76–83.

Broida, J., & Svare, B. (1982b). Strain-typical patterns of pregnancy-induced nestbuilding in mice: Maternal and experiential influences. *Physiology and Behavior, 25,* 153–157.

Broida, J., & Svare, B. (1983). Mice: Progesterone and the regulation of strain differences in pregnancy-induced nestbuilding. *Behavioral Neurosciences, 97,* 994–1004.

Flannelly, K. J., Kemble, E. D., Blanchard, D. C., & Blanchard, R. J. (1986). Effects of septal-forebrain lesions on maternal aggression and maternal care. *Behavioral and Neural Biology, 45,* 17–30.

Gandelman, R., vom Saal, F. S., & Reinisch, J. M. (1977). Contiguity to male foetuses affects morphology and behavior of female mice. *Nature, 266,* 722–724.

Hansen, S., & Ferreira, A. (1986). Food intake, aggression, and fear behavior in the mother rat: Control by neural systems concerned with milk ejection and maternal behavior. *Behavioral Neuroscience, 100,* 64–70.

Hauser, H., & Gandelman, R. (1983). In utero contiguity to males affects level of avoidance responding in adult female mice. *Science, 220,* 437–438.

Ieni, J. R., & Thurmond, J. B. (1985). Maternal aggression in mice: Effects of treatments with PCPA, 5-HTP and 5-HT receptor antagonists. *European Journal of Pharmacology, 111,* 211–220.

Kinsley, C. H., & Bridges, R. S. (1986). Opiate involvement in postpartum aggression in rats. *Pharmacology, Biochemistry, and Behavior, 25,* 1007–1011.

Kinsley, C., Miele, J., Ghiraldi, L., Konen, C., & Svare, B. (1986). Intrauterine position modulates maternal behaviors in female mice. *Physiology and Behavior, 36,* 793–799.

Kinsley, C., Miele, J., Konen, C., Ghiraldi, L., & Svare, B. (1986). Intrauterine contiguity influences regulatory activity in adult female and male mice. *Hormones and Behavior, 20,* 7–12.

Kinsley, C., Miele, J., Konen, C., Ghiraldi, L., Broida, J., & Svare, B. (1986). Prior intrauterine position influences body weight in male and female mice. *Hormones and Behavior, 20,* 201–211.

Kordon, C., Blake, C. A., Terkel, J., & Sawyer, C. H. (1973–1974). Participation of serotonin-containing neurons in the suckling induced rise in plasma prolactin levels in lactating rats. *Neuroendocrinology, 13,* 213–223.

Luttge, W. (1972). Activation and inhibition of isolation induced intermale fighting behavior in castrate male CD-1 mice treated with steroidal hormones. *Hormones and Behavior, 3,* 71–81.

MacLusky, N. J., & McEwen, B. S. (1980). Progestin receptors in rat brain: Distribution and properties of cytoplasmic progestin-binding sites. *Endocrinology, 106,* 192–202.

Mann, M. A., Konen, C., & Svare, B. (1984). The role of progesterone in pregnancy-induced aggression in mice. *Hormones and Behavior, 18,* 140–160.

Mann, M. A., Michael, S., & Svare, B. (1980). Ergot drugs suppress prolactin and lactation but not aggressive behavior in parturient mice. *Hormones and Behavior, 14,* 319–328.

Mann, M. A., & Svare, B. (1982). Factors influencing pregnancy-induced aggression in mice. *Behavioral and Neural Biology, 36,* 242–258.

McCormack, J. T., & Greenwald, G. S. (1974). Progesterone and oestradiol-17β concentrations in the peripheral plasma during pregnancy in the mouse. *Journal of Endocrinology, 62,* 101–107.

McDermott, N. J., Gandelman, R., & Reinisch, J. M. (1978). Contiguity to male fetuses influences anogenital distance and time of vaginal opening in mice. *Physiology and Behavior, 20,* 661–663.

Murr, S. M., Stabenfeldt, G. H., Bradford, G. E., & Geschwind, I. I. (1974). Plasma progesterone during pregnancy in the mouse. *Endocrinology, 94,* 1209–1211.

Noirot, E., Goyens, J., & Buhot, M. C. (1975). Aggressive behavior of pregnant mice towards males. *Hormones and Behavior, 6,* 9–17.

Nott, P., Franklin, M., Armitage, C., & Gelder, M. (1976). Hormonal changes and mood in the puerperium. *British Journal of Psychiatry, 128,* 379–383.

Numan, M. (1983). Brain mechanisms of maternal behavior in the rat. In J. Balthazart, E. Prove, & R. Gilles (Eds.), *Hormones and behavior in higher vertebrates* (pp. 69–85). New York: Springer-Verlag.

Pederson, C. A., Ascher, J. A., Monroe, Y. L., & Prange, A. J. (1982). Oxytocin induces maternal behavior in virgin female rats. *Science, 216,* 648–650.

Potegal, M., & Wagner, H. (1985). Serotonergic control of aggressive "state" by the accumbens/preoptic area (APOA): 3. State-associated changes in 5HT receptors. *Society for Neuroscience Abstracts, 11,* 1176.

Rowley, M. H., & Christian, J. J. (1976). Interspecific aggression between *Peromyscus* and *Microtus* females: A possible factor in competitive exclusion. *Behavioral Biology, 16,* 521–525.

Slotnick, B., Simoes-Fontes, A., & Simoes-Fontes, J. (1977). *Medial preoptic area and stimulation of nestbuilding in mice.* Paper presented at the meeting of the Society for Neuroscience, Anaheim, CA.

Soares, M. M., & Talamantes, F. (1982). Gestational effects on placental and serum androgen, progesterone, and prolactin-like activity in the mouse. *Journal of Endocrinology, 95,* 29–36.

Svare, B. (1977). Maternal aggression in mice: Influence of the young. *Biobehavioral Reviews, 1,* 151–164.

Svare, B. (1981). Maternal aggression in mammals. In D. J. Gubernick & P. Klopfer (Eds.), *Parental care in mammals* (pp. 179–210). New York: Plenum Press.

Svare, B. (1988). Genotype modulates the aggression promoting quality of progesterone in pregnant mice. *Hormones and Behavior, 22,* 90–99.

Svare, B., Betteridge, C., Katz, D., & Samuels, O. (1981). Some situational and experiential determinants of maternal aggression in mice. *Physiology and Behavior, 26,* 253–258.

Svare, B., & Gandelman, R. (1973). Postpartum aggression in mice: Experiential and environmental factors. *Hormones and Behavior, 4,* 323–334.

Svare, B., & Gandelman, R. (1976a). Postpartum aggression in mice: The influence of suckling stimulation. *Hormones and Behavior, 7,* 407–416.

Svare, B., & Gandelman, R. (1976b). Suckling stimulation induces aggression in virgin female mice. *Nature, 260,* 606–608.

Svare, B., & Mann, M. A. (1983). Hormonal influences on maternal aggression in mammals. In B. Svare (Ed.), *Hormones and aggressive behavior* (pp. 91–104). New York: Plenum Press.

Svare, B., Mann, M. A., Broida, J., & Michael, S. (1982). Maternal aggression exhibited by hypophysectomized parturient mice. *Hormones and Behavior, 16,* 455–461.

Svare, B., Mann, M. A., & Samuels, O. (1980). Mice: Suckling but not lactation is important for maternal aggression. *Behavioral and Neural Biology, 29,* 453–462.

Svare, B., Miele, J., & Kinsley, C. (1986). Mice: Progesterone stimulates aggression in pregnancy-terminated females. *Hormones and Behavior, 20,* 194–200.

Voci, V., & Carlson, N. (1973). Enhancement of maternal behavior and nestbuilding following systemic and diencephalic administration of prolactin and progesterone in the mouse. *Journal of Comparative and Physiological Psychology, 83,* 388–393.

vom Saal, F. S. (1981). Variation in phenotype due to random intrauterine positioning of male and female fetuses in rodents. *Journal of Reproduction and Fertility, 62,* 633–650.

vom Saal, F. S. & Bronson, F. H. (1978). In utero proximity of female mouse fetuses to males: Effect on reproductive performance during later life. *Biology of Reproduction, 19,* 842–853.

vom Saal, F. S., & Bronson, F. H. (1980a). Sexual characteristics of adult female mice are correlated with their blood testosterone levels during prenatal development. *Science, 208,* 597–599.

vom Saal, F. S., & Bronson, F. H. (1980b). Variation in length of the estrous cycle in mice due to former intrauterine proximity to male fetuses. *Biology of Reproduction, 22,* 777–780.

Wiest, W. G., Kidwell, W. R., & Balogh, K. (1968). Progesterone catabolism in the rat ovary: A regulatory mechanism for progestational potency during pregnancy. *Endocrinology, 82,* 844–853.

Wolff, J. O. (1985). Maternal aggression as a deterrent to infanticide in *Peromyscus leucopus* and *P. maniculatus. Animal Behaviour, 33,* 117–123.

Yalom, I. D., Lunde, D., Moos, R., & Hamburg, D. (1968). Postpartum ''blues'' syndrome: A description and related variables. *Archives of General Psychiatry, 18,* 16–27.

8
Physiological, Sensory, and Experiential Determinants of Maternal Behavior in Sheep

PASCAL POINDRON AND FRÉDÉRIC LÉVY

By contrast to what is observed in rodents (e.g., the mouse or the maternally experienced female rat), the onset of maternal behavior in sheep is rather strictly dependent on the process of parturition. It is usually only within the last 2 hr preceding birth that pregnant ewes show a characteristic maternal behavior toward lambs (Arnold & Morgan, 1975). This prepartum onset of maternal behavior is accompanied by a strong attraction for amniotic fluids (AF). Females lick the ground where the water bag has ruptured, and this usually determines the site of birth. It is also not uncommon to observe preparturient ewes licking the genital region of other females at the same stage of labor, especially in flocks where reproduction has been highly synchronized, either naturally or artificially. Maternal care for the neonate develops within minutes following expulsion. It consists of avid licking of the neonate associated with attraction for AF, emission of soft or low-pitched bleats, and acceptance of the young at the udder as soon as it is able to stand (usually less than 60 min after birth). The mother can stimulate the activity of the neonate by pawing, and she facilitates access to the udder by proper orientation of the body. In contrast with rodents, mother ewes do not build nests. Whereas in the rat, the maintenance of contact between the dam and the litter is in part ensured through the nest, it is ensured in sheep by the rapid establishment (<2 hr) of an exclusive bond between the mother and her one or two offspring. This is vital, for in many instances the mother moves off the birth site within a few hours after parturition (Alexander, Stevens, Kilgour, et al., 1983). Once this bond is established, mothers suckle only their own lamb(s) and reject, often with aggressive behavior, any alien young that tries to suck. This selective behavior lasts throughout the end of the lactation phase (i.e., 2 to 3 months in most farming conditions). Thus sheep represent an interesting alternative model to that of the rat for the study of both the physiological control of maternal behavior and the formation of an exclusive mother–young bond.

As in many mammals, the onset of maternal behavior in sheep is associated with a sensitive period. Mothers are highly responsive to any neonate for a few hours following birth. During this time, they will spontaneously display maternal behavior as soon as they are in the presence of young. But maternal responsiveness will fade very rapidly (in 50% of females within 4 hr in sheep; Poindron, Martin, & Hooley,

1979) if mothers are totally deprived of stimulation from the neonate during this time. In contrast, deprivation of the lamb at a later time (e.g., 24 hr postpartum) will have no noticeable consequence, provided that mothers have been in contact with their neonate during the sensitive period.

Thus, this time would represent a temporary period of maximum maternal receptivity to some characteristics of the neonate. The very close association between parturition and this temporary emergence of maternal responsiveness led scientists to postulate that the onset of maternal behavior was at least partly controlled by internal factors related to parturition (Hersher, Richmond, & Moore, 1963a). This hypothesis will be examined in the first section of this chapter to underline the primary role played by vaginal stimulation in interaction with estradiol.

Among the characteristics of the neonate that are important to the mother during the sensitive period, olfactory cues rapidly become of major relevance. Not only does exclusive bonding rely on olfaction (Bouissou, 1968), but attraction for the neonate is, at least in part, olfaction dependent through attraction for AF (Lévy, Poindron, & Le Neindre, 1983), and perception of olfactory cues is essential for the maintenance of maternal behavior beyond the sensitive period. These various aspects of the regulation of maternal behavior through olfaction are discussed in the second section of this chapter.

These studies also revealed that close relationships exist between physiological factors that facilitate maternal behavior and the treatment of the olfactory information. For example, attraction to AF depends on genital stimulation and steroid priming, and exclusive bonding can be affected by supplementary vaginal stimulation. These relations are reviewed in the third section of the chapter.

Lastly, experimental studies of the control of maternal behavior have stressed the existence of interactions between previous maternal experience and the factors controlling the immediate development of maternal behavior. Thus the efficacy of steroids in inducing maternal responses in nonpregnant ewes increases with experience (Le Neindre, Poindron, & De Louis, 1979), while the deprivation of sensory cues or internal facilitatory factors is of less consequence in experienced females than in naive ones (Krehbiel, Poindron, Lévy, & Prud'homme, 1987; Lévy & Poindron, 1987). Therefore, any analysis of the determinants of maternal behavior in sheep must take into account this experiential dimension. Evidence on the role of experience is discussed in the fourth section of this chapter.

PHYSIOLOGICAL DETERMINANTS OF THE ONSET OF MATERNAL BEHAVIOR

Parturition in sheep is characterized by various endocrine events that are summarized in Figure 8.1. While circulating progesterone levels fall in the last days of pregnancy, prolactin concentrations rise, as do estrogen levels; estrogens reach a maximum level in the last 24 hr preceding birth. Oxytocin itself is not secreted in large amounts, except in the last stages of labor and at the very time of expulsion. Among the various factors possibly involved, two have been identified so far: estrogens and the mechanical stimulation of the genital tract caused by the expulsion of the fetus.

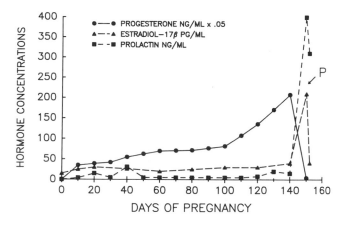

Figure 8.1. Schematic representation of endocrine profiles during pregnancy and parturition (P) in the ewe for progesterone, estradiol 17β, and prolactin. (Progesterone, estradiol, and prolactin data adapted from Stabenfeld, 1974, Terqui, 1974, and Davis et al., 1971, respectively)

Estrogens

Evidence for facilitation of maternal behavior by estrogens have come from three main areas. First, the spontaneous occurrence of maternal behavior in sheep during the reproductive cycle appears limited to times when estrogens are likely to be present in significant quantities in the general circulation (i.e., at estrus and during the last 10 days of pregnancy; Poindron & Le Neindre, 1980).

The second argument for a facilitatory role of estrogens comes from the effects observed on the duration of the sensitive period when lambing is induced by estradiol benzoate (Poindron et al., 1979). Such a treatment, which leads to very high levels of estradiol in maternal blood for at least 24 hr postpartum, also delays significantly the fading of maternal responsiveness. This effect is not observed when lambing is induced with dexamethasone. In this case, there is no enhancement of estradiol concentrations above those normally observed at parturition. Additional experiments have also allowed investigators to rule out a possible indirect action of estradiol via hypersecretion of prolactin (Poindron, Le Neindre, Raksanyi, Trillat, & Orgeur, 1980).

The third and last argument for a facilitation of maternal behavior by estrogen comes from experiments of hormonal induction in nonpregnant ewes. Provided that females have had some experience prior to the experiment, it is possible to induce maternal responses toward neonates in nonpregnant ewes (Le Neindre et al., 1979). This is true regardless of whether the treatment does or does not induce lactation (Poindron & Le Neindre, 1980) and despite the fact that there can be some variability in the proportion of females responding to the treatment (Poindron, Lévy, & Krehbiel, 1988). Indeed, a single injection of 25 mg estradiol-17β can be sufficient to induce maternal responses within 6 to 24 hr (Poindron & Le Neindre, 1980). So far, similar attempts to stimulate maternal behavior with estrogens in goats have failed, according to the literature (Rosenblatt & Siegel, 1981).

Genital Stimulation

Despite the effectiveness of estradiol in inducing maternal behavior in nonpregnant ewes, it would be surprising if an elaborate and biologically important drive such as maternal behavior relied on a single factor. Thus it was logical to investigate the possibility of a multifactorial control of maternal behavior. This hypothesis was further supported by the fact that in many cases an optimal response was not obtained following hormonal treatment alone.

In the first attempts, other behaviors (i.e., male sexual behavior) were also induced by the treatment (Le Neindre et al., 1979). Even when maternal behavior alone was induced, a response was seldom observed in all females, nor was the behavior induced identical to that of parturient ewes. Several causes can be invoked to explain this reaction. It is possible that inappropriate quantities of hormones were used or that the course of injection failed to mimic correctly the hormonal profile necessary to obtain the highest efficacy. This, together with interactions with other hormones (e.g., progesterone), may be important, as it is in the case of female behavior (Fabre-Nys, Lévy, & Poindron, 1986). It is clear that the treatments used so far are very crude and could probably be improved.

Furthermore, in addition to possible interactions between estradiol and progesterone, some other factors could participate in the facilitation of maternal behavior. Indeed, experiments investigating the role of genital stimulation (GS) have led to a new insight in the understanding of the control of maternal behavior (Keverne, Lévy, Poindron, & Lindsay, 1983; Poindron, Lévy, Le Neindre, & Keverne, 1986; Poindron et al., 1988). Experiments have been carried out in nonpregnant and parturient ewes, and in both instances, genital stimulation emerged as a key factor in the rapid onset of maternal behavior (Figure 8.2). In an experiment conducted on nonpregnant Préalpes-du-Sud ewes primed with ovarian steroids, it was found that 80% of the females exhibited immediate (<30 min) maternal behavior, including licking of the neonate, in response to 5 min of genital stimulation. In contrast, only 20% of the females responded maternally in the absence of this stimulation (Keverne et al., 1983) (Figure 8.2A). Similar results were found in aged Merino ewes in an experiment conducted at the University of Western Australia (Poindron et al., 1988). In the second experiment, the response to GS in steroid-primed ewes was compared with that in controls. Whereas GS proved to have little (1/9) effect in unprimed ewes, a significant response, especially concerning licking, was observed in females that had received treatment with progesterone and estradiol (10/26). Also noteworthy was the decrease in aggressive behavior toward the test lamb following genital stimulation, even in ewes that failed to become maternal (Poindron et al., 1988).

There is also increasing evidence that GS facilitates maternal behavior in parturient ewes (Figure 8.2B). The first 5 min of GS 1 hr postpartum greatly enhances licking of an alien neonate (Keverne et al., 1983). Further experiments have been carried out to confirm the importance of the process of fetus expulsion for the onset of maternal behavior. To this end, the effects of peridural anesthesia during labor on the maternal behavior of primiparous and multiparous ewes have been studied. Blocking the sensory input from the genital region led to a highly significant delay in the onset of maternal behavior in most primiparous ewes (7/8) and in a number of multiparous

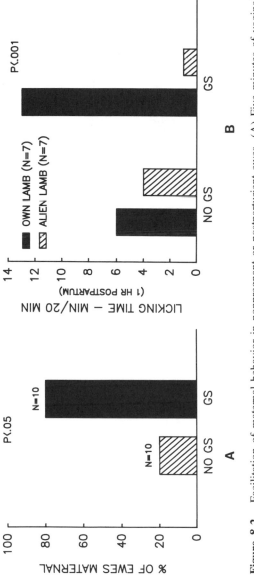

Figure 8.2. Facilitation of maternal behavior in nonpregnant or postparturient ewes. (A) Five minutes of vaginal stimulation in dry ewes primed with steroids. (B) Five minutes of cervicovaginal stimulation 1 hr postpartum. (After Keverne et al., 1983)

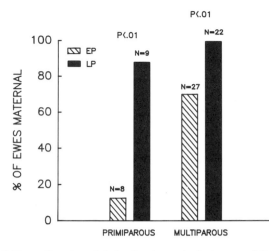

Figure 8.3. Inhibition of maternal behavior in parturient ewes following peridural anesthesia. EP, early peridural (90 min before expulsion—experimental group); LP, later peridural (10 min before expulsion—sham operated group). (After Krehbiel et al., 1987)

sheep (8/27) when peridural anesthesia was administered at the first visible signs of labor (Figure 8.3). Such ewes failed to show any interest in their neonates, in some cases for several hours (6/7 and 7/8 in primiparous and multiparous sheep, respectively; Krehbiel et al., 1987). Thus these ewes not only did not accept their lamb at the udder for a long time, but also failed to lick their lamb or to show interest in AF. This outcome is contrary to what is normally observed in parturient sheep (Collias, 1956; Lévy et al., 1983). Although it could be argued that these effects are mostly due to nonspecific consequences of paralysis associated with peridural anesthesia, this does not seem to be the case. One ewe, although insensitive following peridural anesthesia, was not paralyzed, and yet failed to become maternal. Also, there was no correlation between the time ewes recovered from paralysis and the onset of maternal behavior. Lastly, a sham operated group, which was paralyzed just before expulsion of the fetus, showed no disturbances in maternal behavior.

Thus all available information converges to emphasize the primary role played by GS in the onset of maternal behavior in the ewe. Whether these findings apply to other mammals remains to be established. A role for vaginal distention has been suggested in the goat (Klopfer & Klopfer, 1968; Hemmes, in Rosenblatt & Siegel, 1981). However, no published evidence is available so far. In rats, results obtained following GS (Yeo & Keverne, 1986) suggest that GS may also play some role in this species. If this were confirmed, it would probably lead to an entirely new approach to the study of maternal behavior, as it has in sheep. More particularly, in sheep it has led to entirely new investigations concerning the links between GS and olfaction (see the third section). But before this is discussed, it is necessary to recall the important role played by olfaction in the development of maternal behavior in the parturient ewe.

OLFACTORY REGULATION OF MATERNAL BEHAVIOR IN THE PARTURIENT EWE

During the sensitive period, the neonate provides the mother with cues that allow her to remain maternal past the initial phase of responsiveness controlled by internal factors. Among the sensory cues provided by the lamb, which are important for the normal development of maternal behavior, olfactory ones appear to play a privileged role in several respects.

Maintenance of Maternal Behavior Beyond the Sensitive Period

Two experiments have investigated the cues from the lamb that are necessary for the mother to remain maternal beyond the sensitive period. To this end, lambs were partially separated from their dam at birth, so that only some cues were available. Mothers were tested for retention of maternal responses after 12 (experiment 1) or 8 hr of treatment (exp. 2).

Deprivation of tactile stimulation of suckling by or licking of the neonate had no detrimental effects on maternal behavior. In contrast, placing the lamb 1 m away from its dam—or better, in a smellproof box with a transparent side—led to performances no better than those of mothers totally deprived of their neonate (exp. 1; Poindron, Le Neindre, et al., 1980). These data were confirmed by the results of a second series of experiments. Lambs were placed in smellproof boxes with a ventilation system allowing the smell of the lamb either to return to its dam or to be expelled outside the room. Here again, ewes that could only see and hear their young did not retain maternal behavior more frequently than those totally deprived of a lamb. In contrast, a higher proportion of ewes that had access to the odor of their neonate remained maternal, regardless of whether they could see the young (exp. 2; Poindron et al., 1988).

Thus it is only the deprivation of olfactory cues from the neonate that leads ewes to perform as though they had been totally deprived of their young. This indicates that maternal olfaction plays an important role during the sensitive period for the maintenance of maternal responsiveness beyond this time. However, one must be aware that, at this stage, these data do not allow clarification of whether the maintenance of maternal behavior is due to maintenance of maternal responsiveness or whether actual consolidation (or transition between internal and neurosensory control) has taken place. Also, there is an apparent contradiction between these results and the fact that ewes made anosmic before parturition show proper maternal behavior. An exception exists for exclusive bonding, which is absent in anosmic mothers (Bouissou, 1968; Poindron, 1976). The discrepancies between the effects of olfactory-cue deprivation in intact ewes and those of anosmia suggest that some compensatory process can take place, but only if ewes have lost the sense of smell. These questions bring into focus the mechanism by which the termination of the sensitive period is controlled.

Attraction to AF

Following the results of the experiments on partial separation, the next step was to identify the types of olfactory cues important to the parturient dam. It has long been known that the odor of the lamb is an important parameter in mother–young bonding. But some aspects of the results already obtained suggested that it was not the only important olfactory cue. Other experiments have emphasized the importance of another source of olfactory information: amniotic fluids.

Two sets of experiments have stressed that the newborn lamb has some characteristics that are very attractive to a parturient ewe. These are already absent in lambs as early as 12–24 hr after birth. First, studies of the sensitive period reveal that an alien neonate is more readily accepted by a mother 12 hr postpartum than its own 12-hr-old young (Poindron & Le Neindre, 1980). Second, when swapping or adding lambs at parturition, mothers will accept any alien neonate, whereas they show signs of disturbance with 12- to 24-hr-old alien lambs (Poindron, Le Neindre, et al., 1980; Poindron, Le Neindre, Lévy, & Keverne, 1984). In addition to differences in the behavior of lambs according to their age, the presence of AF on the neonate's coat is a possible factor facilitating its acceptance. Recent studies have largely confirmed the importance of AF in the development of maternal behavior at parturition (Lévy & Poindron, 1984, 1987).

The consumption of amniotic fluids by parturient mothers is the general rule in mammals, including herbivores such as sheep and goats (Collias, 1956; Hersher et al., 1963a). Studies in sheep have emphasized the very close association of AF with parturition. In fact, AF are strongly repulsive to ewes during the whole reproductive cycle, except for a few hours following parturition, during which they become very attractive (Lévy et al., 1983). Such a change in preference does not exist in anosmic females, which are neither repelled by AF outside the peripartum period nor clearly attracted to them during the same period.

Two further experiments provided information on the importance of AF for the development of maternal behavior. In the first case, the acceptance of 24-hr.-old lambs by parturient ewes was studied following application of AF to their coats. In the second experiment, neonates were washed before being presented to their dam. In the first experiment, two lambs were presented to a ewe. In the control group, both were dry, while in the experimental group, one carried AF on its coat. The origin of AF (from the tested ewe or from an alien mother) was also taken into account. Adding AF to the coat of 24-hr-old lambs led to a significant increase in the rate of acceptance of the lambs (Lévy & Poindron, 1984). Interestingly, the facilitation of acceptance was not limited to the lamb carrying AF but also affected the dry lamb of the pair (Figure 8.4). The origin of AF (tested mother or alien mother) had little influence on the outcome of the test: overall, the acceptance rate was similar in both cases. Nevertheless, aggressive behavior was significantly less frequent in tests where the AF were that of the ewe being tested. Thus these results indicate the following:

1. AF facilitate acceptance of lambs by parturient mothers.
2. The facilitation of acceptance by AF is due not only to the fact that the lamb is more attractive, but also to the fact that AF enhance maternal responsiveness.
3. AF stimulate maternal behavior irrespective of their origin (own or alien mother), although they may also contain some individual information. This seems confirmed by studies of partial mother–young separations in which

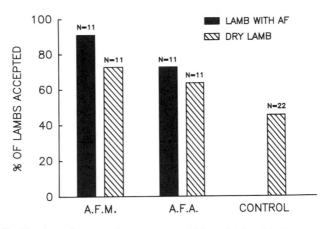

Figure 8.4. Facilitation of maternal acceptance of 12- to 24-hr-old alien lambs by parturient ewes following adjunction of amniotic fluids (AF) on the lambs' coat. AFM, one lamb coated with AF from tested ewe and one lamb dry; AFA, one lamb coated with AF from an alien ewe and one lamb dry; CONTROL: two lambs dry (no adjunction of AF); AFM + AFA vs. CONTROL: $p < .05$. AFM vs. AFA: n.s. (After Lévy & Poindron, 1984)

ewes exposed to lambs with AF tended to be more selective than ewes whose lambs had been cleaned of AF (Alexander et al., 1986).

Depriving parturient ewes of AF by washing the neonate confirmed the first results. In this experiment, the effects of washing the young were studied in primiparous and multiparous ewes. In both groups, washing, with either soap or just water, led to a significant reduction of licking (by 50% in multiparas and by about 90% in primiparas; Lévy & Poindron, 1987). Furthermore, most primiparous ewes whose lambs had been washed failed to accept them. Multiparous mothers, however, were disturbed to a lesser degree, and most established a proper bond with their lamb (Figure 8.5).

Overall, it is clear that olfactory cues provided by AF are important for the normal development of maternal behavior, especially in naive mothers. They orient the mother to its normal target and ensure an immediate onset of care. But AF also have a positive-feedback action on maternal responsiveness as a whole. These results, together with those of Kristal, Thompson, and Abbott (1986) concerning analgesia following AF ingestion in the rat, may lead to a reassessment of the importance of prepartum consumption of AF spilled on the ground. This behavior not only reduces labor pain, but also may help to ensure the full development of maternal responsiveness at the time of birth.

Selective Bonding

Ewes are able to distinguish their young from an alien one very rapidly after parturition. This recognition is associated with an exclusive bond, and as early as 2 hr postpartum, most mothers reject any alien neonate that attempts to suck (Poindron & Le Neindre, 1980). A similar behavior appears to exist in goats, although the dynamics of bonding are not as clearly established in this species (Hersher et al., 1963a;

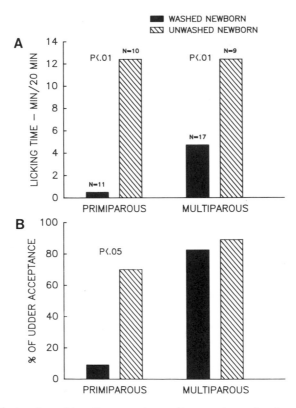

Figure 8.5. Effects of washing the neonate on its acceptance by its parturient mother. (A) Duration of licking of neonates washed with water or unwashed during the first 20 min of contact. (B) Acceptance of these lambs at the udder. (After Lévy & Poindron, 1987)

Klopfer, Adams, & Klopfer, 1964). In both species, olfaction appears to be of primary importance for the early establishment of selective suckling (Bouissou, 1968; Gubernick, 1981a; Klopfer & Gamble, 1966; Morgan, Boundy, Arnold, & Lindsay, 1975). Thus in sheep, anosmia before parturition prevents the formation of selectivity (i.e., exclusive acceptance of own lamb at the udder) (Bouissou, 1968). In contrast, if anosmia is performed when the lambs are about 2 weeks old, the results are more variable. Some mothers fail to remain maternal, others suckle any lamb, and still others show signs of exclusive bonding (Poindron & Le Neindre, 1980). Similar effects have been reported by washing 3-week-old lambs (Alexander, Stevens, & Bradley, 1983a). This suggests the following:

1. The possibility exists at some stage (or in some animals) for a shift to occur between olfactory and other sensory cues in the control of selectivity.
2. Olfaction may participate in both the mediation of selective bonding and the regulation of maternal behavior in a more general sense. This would be more in agreement with the important role that olfaction has been found to play in the development of maternal behavior at parturition (see the preceding paragraphs).

This argument does not exclude the possibility of recognition by other senses, although this probably occurs in a second stage and is not as crucial for the establishment of selective suckling (Alexander & Shillito, 1977; Poindron & Carrick, 1976; Poindron & Le Neindre, 1980; Poindron & Schmidt, 1984).

The nature of the olfactory cues involved in maternal recognition has not been systematically investigated in sheep. It has long been assumed that mothers learn to recognize the individual odor of their young. However, recent results from experiments on goats have led to the statement of an alternative hypothesis in this species. Although it was first hypothesized that does were "imprinted" with the individual smell of their kid (Klopfer & Gamble, 1966), more recent work (see Gubernick, 1981b, for review) suggests that mother goats label their own kids by licking and suckling. There is little evidence that similar labeling occurs in sheep. First, selectivity is established within 2 hr. This duration is too short for labeling to occur. Second, studies of partial mother–young separations have clearly shown that mothers can still learn to discriminate between lambs even if they cannot lick or feed them (case of the double wire-cage group; Poindron & Le Neindre, 1980). Also, ewes do not discriminate between lambs fed with colostrum from an alien mother and those fed from their own following 8 hr of partial separation (Poindron et al., 1988). Lastly, AF facilitated acceptance of alien young, even with 24-hr-old lambs that had been kept with their own dam (and therefore presumably labeled; see the second paragraph of this section). This does not rule out the possibility of labeling in sheep, but it indicates that AF have a facilitatory action on maternal acceptance—a possibility that has been somewhat overlooked in experiments on labeling in goats. Obviously, more experiments are needed, with both sheep and goats, to clarify this issue. Even if labeling proved to occur in only some precise experimental conditions, it would still have important implications for the design of experiments on maternal behavior. It would also illustrate the plasticity of compensatory means used by dams to establish a selective bond.

Another point to be mentioned regarding the intimate mechanisms of recognition of young by the mother is the relationship between the sensitive period, on the one hand, and the establishment of recognition of the neonate and of selective suckling, on the other. In light of the results obtained on imprinting in birds during a critical period, and because of the rapidity of establishment of selective suckling, some confusion has occurred in regard to sheep and goats concerning the sensitive period and selectivity (Hersher, Richmond, & Moore, 1963b; Klopfer et al., 1964). More precisely, it was thought that there was a sensitive period *for* the establishment of a selective bond. Although this is not false, it is clear that the sensitive period during which mothers accept any young and that during which they discriminate between own and alien lamb are separate phenomena (Poindron & Le Neindre, 1980). This is also likely to be the case in goats (Gubernick, 1980, 1981a; Lickliter, 1982).

OLFACTORY AND GENITAL INTERACTIONS IN THE CONTROL OF MATERNAL BEHAVIOR

GS and olfaction are two important factors in the development of maternal behavior at parturition. There are clear indications that these two parameters are closely linked in the control of maternal behavior. Evidence comes from various experiments about the action of GS on the attraction to AF, the establishment of selective suckling, or the induction of maternal behavior in anosmic ewes, as well as from studies on the neurobiological mechanisms mediating GS action.

Genital Stimulation and Attraction to AF

The timing and suddenness of changes in the olfactory reaction to AF rule out a direct involvement of estrogens in the onset of AF attraction. Estradiol rises long before attraction is observed, and the exact timing of this rise probably varies with the individual, while the onset of AF attraction, in contrast, is very stable and precisely related to the last stages of labor. It is therefore reasonable to assume that some relation exists between these two phenomena.

Various studies indicate clearly that genital stimulation facilitates attraction to AF. For example, it has been shown that 2 hr postpartum, 50% of ewes are no longer attracted to AF (Lévy et al., 1983). However, if such ewes undergo 5 min of GS at this time, 70% of them show immediate attraction (Figure 8.6). The situation is similar 4 hr postpartum, but with a lesser degree of success (only 40% of females respond to GS; Lévy, unpublished observations, in Poindron et al., 1986). Eight hours postpartum, GS is without effect (9% of females respond to GS). It may be of some significance that the fading of GS efficacy parallels the decrease in estrogen concentrations in the post-parturient ewe (see Figure 8.1). It may suggest some synergy between estradiol and GS in the control of olfactory processes. This agrees with the fact that such a synergy is known to exist for the induction of maternal behavior (see the second paragraph of the first section), and it is quite reasonable to assume that these two events are closely related. Indeed, induction of attraction to AF has proved very difficult in nonpregnant ewes, and it is only in females receiving both estradiol and GS that licking of a neonate and consumption of AF have been observed (Lévy, unpublished observations). The primary importance of GS is also confirmed by the effects of peridural anesthesia in parturient ewes. In such females, neither licking of the neonate nor interest in AF spilled on the ground appears following expulsion of the fetus (Krehbiel et al., 1987).

Genital Stimulation and Induction of Maternal Behavior in Anosmic Nonpregnant Ewes

Additional evidence for the existence of a relationship between the sense of smell in the ewe and vaginal stimulation comes from experiments in which the facilitation of maternal behavior by GS has been compared in intact and anosmic, nonpregnant ewes primed with steroids. This work, carried out on Merino ewes at the University of Western Australia and reported in detail elsewhere (Poindron et al., 1988), will be briefly summarized here. While anosmia alone had no facilitatory effect on maternal behavior, this was not the case when it was associated with GS. Then the proportion of ewes accepting a lamb was significantly higher in anosmic ewes (45%) than in intact ones (25%; Figure 8.7).

This additional finding suggests an interaction between maternal olfaction and GS in the control of maternal behavior. One interpretation of these results could be

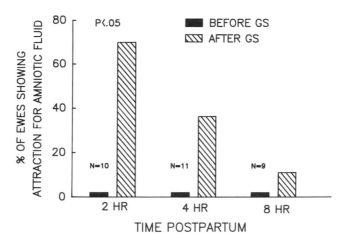

Figure 8.6. Reinduction of attraction for AF by cervicovaginal stimulation in postparturient ewes. Ewes separated from their lamb at birth were selected at various times postpartum for their absence of attraction for AF. They then received 5 min of genital stimulation and were tested again for attraction. (After Lévy, 1985)

that experimental anosmia in nonpregnant ewes helps to mimic some of the olfactory changes normally induced by cervicovaginal stimulation during the process of expulsion. In intact, nonpregnant ewes, artificial GS only partly induces such changes, and anosmia more closely approximates labor.

Thus, as in the rat (Rosenblatt, Siegel, & Mayer, 1979), anosmia in nonpregnant females facilitates the manifestation of maternal behavior, with the restriction that in sheep it does so only when associated with GS. This supports the hypothesis that some physiological anosmia occurs at parturition to facilitate the onset of mater-

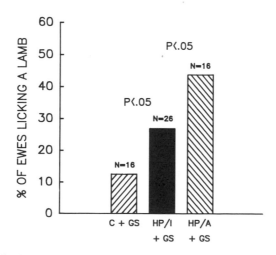

Figure 8.7. Facilitation of maternal behavior by genital stimulation in intact or anosmic, nonpregnant ewes. C + GS: controls + genital stimulation; HP/I + GS: hormonal priming in intact ewes + genital stimulation; HP/A + GS: hormonal priming in anosmic ewes + genital stimulation. (After Poindron et al., 1988)

nal behavior (Keverne, 1988). If this hypothesis is correct, the process would be controlled (in sheep) by the mechanical stimulation caused by the fetal expulsion.

Genital Stimulation and Selective Bonding

Studies on the acceptance of alien neonates in single- and twin-bearing ewes have suggested that expulsion facilitates the acceptance of an alien neonate (Poindron et al., 1988; Poindron, Le Neindre, et al., 1980). This could be explained partly by the facilitating action of GS on licking. However, the facilitation of alien-neonate acceptance is mediated not only by the general action on licking, but also by some more specific influence on the process of selective bonding. Thus 2 hr postpartum, most mothers are selective and reject any alien lamb that is presented to them instead of their own. However, in such selective mothers, 5 min of GS leads to the immediate acceptance of the alien neonate rejected 5 min before (Keverne et al., 1983). This suggests that GS influences the memorization of lamb odor, or at least inhibits rejection of an alien neonate by reducing aggressive behavior, and/or enhances licking of a newborn. Results concerning the action of GS in nonpregnant ewes and the neurobiological mechanisms mediating GS are of crucial importance for elucidating this issue.

Neurobiological Mechanisms Mediating the Control of Maternal Behavior by Genital Stimulation

Recent studies in sheep indicate that GS is undoubtedly an essential factor for the control of maternal behavior in this species. However, it is not clear whether most of this effect is due to the action of GS on the olfactory function or if a more general action on maternal responsiveness is involved. Perhaps internal factors strongly influence olfactory function, and the transition between temporary responsiveness under internal control and a lasting maternal behavior under neurosensory control relies heavily on olfactory cues. This suggests that the olfactory system plays a key role in the control of maternal behavior as a whole.

Studies of the neurobiological mechanisms that mediate the effects of GS will certainly help to clarify this issue. However, current understanding of the neurobiological basis of maternal behavior in sheep is very limited. Two possible mechanisms have been proposed to account for the action of GS, and they are not mutually exclusive. First, it is possible that GS acts on olfaction by specific activation of noradrenergic (NA) fibers coming from the locus coeruleus and projecting to the olfactory bulbs. This appears to be the case for the control of selective behavior by GS. Indeed, lesions of these NA pathways by 6-hydroxydopamine in pregnant ewes prevent the establishment of selectivity in a significant proportion of ewes (Pissonnier, Thiery, Falre-Nys, Poindron, & Keverne, 1985), despite the fact that ewes are not rendered anosmic by the treatment. GS would therefore facilitate the memorization of individual lamb smells by activation of the NA afferents to the olfactory bulbs (OB). But whether the action of GS on NA afferents is direct remains to be established. Also, it is not known if a similar process can account for the interest in AF that is elicited at parturition by GS.

A second hypothesis is that the facilitation of maternal behavior by GS is mediated by the neurohormone oxytocin. Such a case has been demonstrated in rats, at least in a specific context (Pedersen & Prange, 1985; Wamboldt & Insel, 1987). Several facts are in agreement with such a hypothesis in sheep. It has been shown that a rise in the concentration of oxytocin occurs in both the cerebrospinal fluid (CSF) and the OB during parturition (Kendrick, Keverne, Baldwin, & Sharman, 1986; Kendrick, Keverne, Sharman, & Baldwin, 1988a); and following vaginal stimulation (Kendrick et al., 1986; Kendrick, Keverne, Sharman, & Baldwin, 1988b). Moreover, preliminary results indicate that it is possible to induce maternal responses by intracerebroventricular (ICV) injections of oxytocin in nonpregnant ewes. The effects reported so far, though, are weak and transient (Kendrick, Keverne, & Baldwin, 1987).

The level of action and the function of intracerebral oxytocin are still unknown. In contrast to the situation in the rat, the central structures involved in the regulation of sheep maternal behavior remain to be uncovered. Oxytocin could influence the activity of these unknown structures and act to modulate olfactory processing at the level of the OB. This action, in agreement with changes of oxytocin concentration in the OB (Kendrick et al., 1988a, 1988b), could account for attraction to AF and/ or modulation of the memorization of the neonate's odor. Such an endocrine regulation of olfactory processing has been demonstrated for prostaglandin D2, which increases the response of mitral cells to olfactory stimuli in the rabbit (Watanabe, Mori, Imamura, Takagi, & Hayaishi, 1986).

Clearly, further studies, including investigations to localize the distribution of oxytocin, are needed to specify its role in olfactory processing and to clarify the relations between norepinephrine and vaginal stimulation, as well as to reach a better understanding of the neurobiological regulation of maternal behavior in sheep.

INFLUENCES OF EXPERIENCE ON THE PHYSIOLOGICAL CONTROL OF MATERNAL BEHAVIOR

In the preceding sections, most of the results concerned experiments involving multiparous ewes. The action of the various factors identified (hormones, GS, olfactory cues) depends on previous maternal experience (Figure 8.8). Therefore, conclusions drawn from these results apply mainly to experienced mothers. This experimental factor must be taken into account to modulate the description of the physiological control of maternal responsiveness in sheep. Below are reviewed the results already mentioned concerning the role of experience.

Experience and Hormonal Induction of Maternal Behavior

Evidence of an interaction between hormones and experience in the elicitation of maternal behavior in sheep was first obtained by Le Neindre et al. (1979). In 50% of cases or more, multiparous females responded maternally to a hormonal treatment initially designed to induce lactation. In contrast, only 1 nulliparous ewe (out of 13) responded (Figure 8.8A). These results indicate that steroid hormones were ineffective in inducing maternal behavior in naive females. This is confirmed by the fact

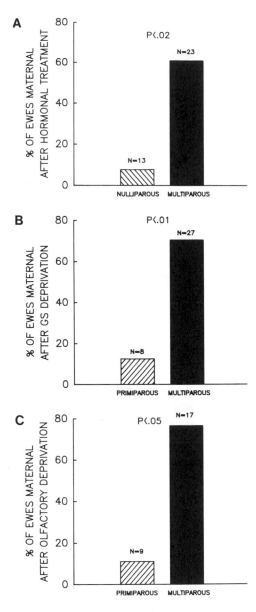

Figure 8.8. Relation between previous maternal experience and the ability of ewes to show maternal behavior in various experimental conditions. (A) Response of nonpregnant ewes to hormonal induction with ovarian steroids (After Le Neindre et al., 1979). (B) Inhibition of maternal behavior in primiparous and multiparous, parturient ewes following deprivation of genital stimulation by peridural anesthesia (After Krehbiel et al., 1987). (C) Perturbation of maternal acceptance of the neonate by its parturient mother in primiparous and multiparous ewes following removal of AF from the lamb's coat by washing. (After Lévy & Poindron, 1987)

that maternal responses to neonates are commonly observed in multiparous sheep 10 to 15 days prepartum (Alexander, 1960; Poindron & Le Neindre, 1980). Responsivity is supposed to be correlated with rising concentrations of estrogen at that time. In contrast, it is rarely seen in primigravid females (Poindron & Le Neindre, 1980). A further experiment tried to clarify whether the facilitatory action of estrogens depended on physiological maturation associated with parturition, or whether the maternal experience of rearing the lamb was necessary (Le Neindre et al., unpublished observations, in Poindron ct al., 1988). Results indicate that a single parental cycle is probably not sufficient to allow later hormonal facilitation of maternal behavior in nonpregnant females. Rather, at least three parental cycles are necessary. In fact, it is in very experienced females (five lactations or more) that hormonal facilitation is best demonstrated (Poindron et al., 1984, 1988). In some instances with very old ewes, maternal responses have even been observed in the absence of any treatment (Poindron & Le Neindre, 1980).

The situation in sheep is not unlike that in the rat. In such animals, experience also appears to determine the facilitation of maternal behavior (Cosnier, 1963). In this species too, there is evidence of an interaction between experience and hormones in the later facilitation of maternal behavior, as best shown in studies by Bridges (1975, 1977). Results demonstrated that neither the physiological experience of parturition alone (mothers deprived of their pups at birth) nor maternal experience alone (virgin females becoming maternal following concaveation) was sufficient to reduce latencies of maternal behavior 25 days later. In contrast, 48 hr of mother–young interactions just after parturition had a lasting effect on later latencies.

These data reveal an interaction between endocrine factors and maternal experience in the later responsiveness of females. However, there appears to be some variation between sheep and rats; ewes must experience several parental cycles for this interaction to give full results, whereas only one parturition is required by rats.

Experience and Facilitation of Maternal Behavior by Genital Stimulation

As in hormonal induction, the action of genital stimulation varies with experience. In sheep, suppression of genital stimulation at parturition by peridural anesthesia in primiparous ewes prevents the normal onset of maternal behavior at the time of fetal expulsion in most females (Krehbiel et al., 1987). By contrast, this deprivation is without effect in more than 50% of experienced mothers (Figure 8.8B). Since it is known that GS also facilitates maternal behavior in multiparous, nonpregnant females (Keverne et al., 1983; Poindron et al., 1988), the deprivation of GS can be compensated for by other factors in experienced mothers. It cannot, however, be as compensated for in naive subjects. The differences between naive and experienced mothers have to be confirmed before this matter is resolved (Krehbiel et al., 1987).

Results obtained in nonpregnant rats by Yeo and Keverne (1986) also indicate some interaction between GS and experience. These authors found that in estrogen-primed subjects, GS significantly improves the proportion of multiparous females showing maternal behavior. Virgins so treated are unaffected. This result provides further evidence that, as in the hormonal induction of maternal behavior in sheep, the ability of GS to induce maternal behavior increases with experience.

Experience and Olfactory Facilitation of Maternal Behavior

Olfaction has been shown to play an important role in the maintenance of maternal behavior beyond the sensitive period and in the normal development of maternal behavior at parturition generally. Among the factors that may be involved, AF have emerged as having primary importance. AF not only orient maternal behavior and facilitate initial contact between the dam and her neonate, but also produce a positive-feedback effect on maternal acceptance (see the second paragraph of the second section). The comparison of results obtained following the removal of AF from the neonate's coat in primiparous and multiparous mothers again indicates the influence of maternal experience. In multiparous mothers, washing the neonate significantly reduces the amount of time the ewe licks her young, but it does not interfere further with acceptance, and mothers readily suckle their lambs (Figure 8.8C). One could conclude, therefore, that AF are of little importance. However, as we have already mentioned, adding AF to the coat of 12-to 24-hr-old lambs facilitates their acceptance by a multiparous, parturient ewe. Moreover, washing the neonate in primiparous ewes not only greatly disturbs licking, but also prevents the acceptance of the young by a majority of ewes (Lévy & Poindron, 1987).

Thus although AF have a positive effect on maternal acceptance in multiparous ewes, such mothers compensate quite easily for their deprivation. In contrast, AF are absolutely necessary for primiparous dams to develop a proper relationship with their young. The situation therefore appears close to that observed for GS. Similar interactions between olfaction and experience have also been reported in rats (Fleming & Rosenblatt, 1974a). The origin of the cues involved remains to be investigated in rats. But results from studies on the behavior of females toward placenta and AF, as well as the influence of these factors on maternal behavior, may suggest that they are involved (Kristal, Whitney, & Peters, 1981).

Neurobiological Implications of the Interactions Between Experience and Internal Factors in the Facilitation of Maternal Behavior

Overall, it is clear that females that have undergone several parental cycles (i.e., pregnancy, parturition, and lactation) develop a greater ability to respond to the presence of the young and to the facilitatory action of internal factors. Results obtained in sheep, as well as in rats (Bridges, 1975, 1977), underline the importance of the temporal association of internal (e.g., hormones) and environmental (e.g., sensory cues from the neonate) factors for the acquisition of what is commonly called "maternal experience."

This brings a new dimension to the study of the neurobiology of memory and learning, for the role of peripheral factors (e.g., E_2 or GS) is usually not taken into account in many studies using conditioning as a model for memory and learning. For example, it would be of interest to know at what level steroids or GS come into play. Is it by an action in associative structures that are classically involved in learning and/or at the level of neural structures such as the hypothalamus, preoptic area, or OB? This possibility should not be overlooked. It is known, for example, that steroids may participate to the neural organization of adult birds (Arnold & Breedlove, 1985). Also, restructuring of hypothalamic synapses occurs in rats following motherhood (Hatton & Ellisman, 1982), and morphological plasticity of the hypothalamo-neurohypophyseal system is influenced by oxytocin (Theodosis, Montagnese, Rodriguez, Vincent, & Poulain, 1986). Lastly, the NA system, which is regarded as playing an important role in neural and behavioral plasticity (Kety, 1970), may act at various levels, including the OB and the hypothalamus (Keverne, 1988; Pissonnier et al., 1985). More studies on the role of experience, related to reproductive behavior, could contribute to an understanding of the neurobiology of memory and learning by taking into account the physiological factors that contribute to these processes.

CONCLUDING REMARKS

Thus three factors appear to be of primary importance in the control of maternal behavior in the parturient ewe: GS, estrogens, and olfactory cues. In addition, their action is closely linked to maternal experience, and this must be as taken into account in the design of experiments. Overall, it appears that as sheep and rats acquire more maternal experience, single factors or determinants are able to stimulate maternal behavior, even though other factors are missing. On the one hand, the ability of any given factor itself to induce maternal behavior increases as the mother experiences more parental cycles. On the other, as experience increases, the mother more easily compensates for the absence of a given factor. In naive females, all the internal and sensory determinants must be present simultaneously for elicitation and development of maternal care. As a consequence, while deprivation of a factor is most likely to provide evidence of its role in naive females, the stimulatory effects of a parameter will be best illustrated in experienced females. Among the three factors identified so far, GS appears to play a special role because it can influence several maternal items (i.e., licking, attraction to AF, exclusive bonding, aggressive behavior). It is clear that further studies at the level of the brain will help to provide us with a better understanding of the relative importance of GS and estrogens in the control of maternal behavior. These studies will have to specify what relations exist between four types of nervous structures: olfactory, estrogen sensitive, oxytocinergic, and noradrenergic. The original and well-established links found between GS, estrogens, and olfaction in sheep render this model very attractive in undertaking such studies.

Comparison of the results obtained in sheep with those found in the rat allows the identification of several similarities between these two somewhat distant species. In both species, there is a sensitive period for the rapid onset of maternal behavior, and in both animals this is under the control of physiological factors. Some of these factors are common to the two species, as has been shown for estrogens. It is only in sheep, though, that the action of cervicovaginal stimulation has been studied extensively. Implications of this work have been demonstrated in the rat as well (Yeo & Keverne, 1986). The action of genital stimulation may also be related to that of intracerebral oxytocin. This hormone also appears to be a common key to maternal-behavior facilitation in rats and sheep. However, the role of this neuropeptide needs to be confirmed in both species. There are few studies in sheep, and the meaning of the interactions existing between anosmia and ICV oxytocin administration is far from being clearly understood in the rat. More studies are needed at both the brain and the behavioral levels to assess the significance of these results. In fact, it is only in sheep, which have been studied extensively that GS is implicated in the control of maternal behavior. Considering the evidence emerging for the control of the olfactory function by internal physiological factors in sheep, it could be of great interest to investigate a possible role of GS on maternal behavior in rats. The important role played by olfaction in rats is a further argument for undertaking such research, especially given the similarity of GS effects in anosmic ewes and of ICV oxytocin in anosmic rats (Poindron et al., 1988; Wamboldt & Insel, 1987).

Prolactin is now also recognized as an important hormone in the facilitation of maternal behavior in the rat (Bridges et al., 1985). No such data exist for sheep. Thus a possible role for this hormone in the ewe cannot be excluded at this stage. Therefore, many points have to be clarified before one can draw conclusions about the possible existence of common control factors for the rat and the sheep. However, while results obtained in other species (e.g., mice, rabbit, hamster) indicate clearly that a physiological facilitation of maternal behavior is a feature common to a wide range of mammalian species (Rosenblatt et al., 1981), they tend to indicate also that the combination of factors involved varies from one species to another.

Another point of comparison between sheep and rats concerns the role of olfaction in these two species. Overall results in sheep indicate a very positive role of olfactory cues in the normal development and control of maternal behavior, especially in the "consolidation" of maternal responsiveness beyond the sensitive period. In contrast, results in the rat identify olfaction as an inhibitory factor, since anosmia facilitates the onset of maternal behavior (Fleming & Rosenblatt 1974a, 1974b). However, this apparently opposite role of olfaction may well be artificial, given the interactions found in sheep between GS and olfaction. Thus while olfactory cues have been found to have an inhibitory action on maternal behavior in nonpregnant rats, this has not been verified in *parturient* females. It may well be that, as in sheep, cues that were found to be aversive in nonpregnant subjects (e.g., odor of pups) were attractive at parturition. This would also agree with the work of Kristal and his colleagues concerning placentophagia and maternal behavior (Kristal et al., 1981).

More generally speaking, olfaction emerges as an important function in the elicitation of maternal behavior in sheep and the rat. It may be of some relevance that these two species are classified as macrosmatic. Whether this is a feature of macrosmatic animals or whether it can also be extended to microsmatic species may be worthy of investigation. Even though olfactory structures are certainly not the most important ones for maternal behavior, they certainly play some role. It would be of interest to learn the extent to which olfaction is a feature common to mammals in general. The fact that AF are attractive at parturition in most mammals (Lehrman, 1961) tends to favor such a hypothesis. The ability of human mothers to establish early olfactory recognition of their infants might reflect some vestigial role of olfaction in mother–infant bonding (Schaal et al., 1980). Variations of olfactory thresholds in women around the time of parturition cannot be excluded (Schaal, personal communication). But so far there is no clear evidence in humans of interactions between internal physiological factors and olfaction at the time of parturition. Here again, results obtained in nonprimate mammals could be used as models to promote new research in primates, including humans.

To conclude, recent findings concerning the roles of GS, intracerebral oxytocin, and amines in the control of maternal behavior may prove to be important for further research and the understanding of the physiological determinants of maternal behavior in mammals. In addition, the constant interaction of these factors with experience raises the question of the neurobiological nature of what is commonly called "maternal experience." Studies of the mechanisms by which the mother improves her ability to respond maternally to internal or sensory cues may prove to be an interesting complement to the classical approach of conditioning in understanding the various mechanisms that account for learning processes.

REFERENCES

Alexander, G. (1960). Maternal behaviour in the Merino ewe. *Proceedings of the Australian Society of Animal Production, 3,* 105–114.

Alexander, G., & Shillito, E. E. (1977). The importance of odour, appearance and voice in maternal recognition of the young in Merino sheep *(Ovis aries). Applied Animal Ethology, 3,* 127–135.

Alexander, G., Stevens, D., & Bradley, L. R., (1983). Washing lambs and confinement as aids to fostering. *Applied Animal Ethology, 10,* 251–261.

Alexander, G., Stevens D., Kilgour, R., de Langen, H., Mottershead, B. E., & Lynch, J. J. (1983). Separation of ewes from twin lambs: Incidence in several sheep breeds. *Applied Animal Ethology, 10,* 301–317.

Alexander, G., Poindron, P., Le Neindre, P., Stevens, D., Lévy, F., & Bradley, L. (1986). Importance of the first hour post-partum for exclusive maternal bonding in sheep. *Applied Animal Behaviour Science, 16,* 295–300.

Arnold, A. P., & Breedlove, S. M. (1985). Organizational and activational effects of sex steroids on brain and behavior: A reanalysis. *Hormones and Behavior, 19,* 469–498.

Arnold, G. W., & Morgan, P. D. (1975). Behaviour of the ewe and lamb at lambing and its relationship to lamb mortality. *Applied Animal Ethology, 2,* 25–46.

Bouissou, M. F. (1968). Effet de l'ablation des bulbes olfactifs sur la reconnaissance du jeune par sa mère chez les Ovins. *Revue de Comportement Animal, 2,* 77–83.

Bridges, R. S. (1975). Long-term effects of pregnancy and parturition upon maternal responsiveness in the rat. *Physiology and Behavior, 14,* 245–249.

Bridges, R. S. (1977). Parturition: Its role in the long-term retention of maternal behavior in the rat. *Physiology and Behavior, 18,* 487–490.

Bridges, R. S., DiBiase, R., Loundes, D. D., & Doherty, P. C. (1985). Prolactin stimulation of maternal behavior in female rats. *Science, 227,* 782–784.

Collias, N. E. (1956). The analysis of socialisation in sheep and goats. *Ecology, 37,* 228–239.

Cosnier, J. (1963). Quelques problèmes posés par le "comportement maternel provoqué" chez la ratte. *Comptes Rendus de la Société de Biologie* (Lyon) *157,* 1611–1613.

Davis, S. L., Reichert, L. E., & Niswender, L. E. (1971). Serum levels of prolactin as measured by radioimmunoassay. *Biology of Reproduction, 4,* 145–147.

Fabre-Nys, Cl., Lévy, F., & Poindron, P. (1986). Analyse du déterminisme du comportement sexuel et maternel de la brebis: Deux études de neuroéthologie. *Comportements, 5,* 111–120.

Fleming, A. S., & Rosenblatt, J. S. (1974a). Olfactory regulation of maternal behavior in rats: I. Effects of olfactory bulb removal in experienced and inexperienced lactating and cycling females. *Journal of Comparative and Physiological Psychology, 86,* 221–232.

Fleming, A. S., & Rosenblatt, J. S. (1974b). Olfactory regulation of maternal behavior in rats: II. Effects of peripherally induced anosmia and lesions of the lateral olfactory tract in pup-induced virgins. *Journal of Comparative and Physiological Psychology, 86,* 233–246.

Gubernick, D. J. (1980). Maternal "imprinting" or maternal "labelling" in goats. *Animal Behaviour, 28,* 124–129.

Gubernick, D. J. (1981a). Mechanisms of maternal "labelling" in goats. *Animal Behaviour, 29,* 305–306.

Gubernick, D. J. (1981b). Parent and infant attachment in mammals. In D. J. Gubernick & P. H. Klopfer (Eds.), *Parental care in mammals* (pp. 243–305). New York: Plenum Press.

Hatton, J. D., & Ellisman, M. H. (1982). A restructuring of hypothalamic synapses is associated with motherhood. *Journal of Neuroscience, 2,* 704–707.

Herscher, L., Richmond, J. B., & Moore, A. U. (1963a). Maternal behavior in sheep and goats. In H. L. Rheingold (Ed.), *Maternal behavior in mammals* (pp. 203–232). New York: Wiley.

Herscher, L., Richmond, J. B., & Moore, A. U. (1963b). Modifiability of the critical period for the development of maternal behavior in sheep and goats. *Behaviour, 20,* 311–319.

Kendrick, K. M., Keverne, E. B., & Baldwin, B. A. (1987). Intracerebroventricular oxytocin stimulates maternal behaviour in the sheep. *Neuroendocrinology, 46,* 56–61.

Kendrick, K. M., Keverne, E. B., Baldwin, B. A., & Sharman, D. F. (1986). Cerebrospinal fluid levels of acetylcholinesterase, monoamines and oxytocin during labour, parturition, vaginocervical stimulation, lamb separation and suckling in sheep. *Neuroendocrinology, 44,* 149–156.

Kendrick, K. M., Keverne, E. B., Sharman, D. F., & Baldwin, B. A. (1988a). Intracranial dialysis measurement of oxytocin, monoamine and uric acid release from the olfactory bulb and substantia nigra of sheep during parturition, suckling, separation from lambs and eating. *Brain Research, 439,* 1–10.

Kendrick, K. M., Keverne, E. B., Sharman, D. F., & Baldwin, B. A. (1988b). Microdialysis measurement oxytocin, aspartate, γ-aminobutyric acid and glutamate release from the olfactory bulb of the sheep during vaginocervical stimulation. *Brain Research, 442,* 171–174.

Kety, S. (1970). The biogenic amines in the central nervous system: Their possible roles in arousal, emotion and learning. In F. O. Schmidt (Ed.), *The neuroscience, second study program* (pp. 324–336). New York: Rockfeller University Press.

Keverne, E. B. (1988). Central mechanisms underlying the neural and neuroendocrine determinants of maternal behaviour. *Psychoneuroendocrinology, 13,* 127–141.

Keverne, E. B., Lévy, F., Poindron, P., & Lindsay, D. R. (1983). Vaginal stimulation: An important determinant of maternal bonding in sheep. *Science, 219,* 81–83.

Klopfer, P. H., Adams, D. K., & Klopfer, M. S. (1964). Maternal imprinting in goats. *Proceedings of the National Academy of Sciences USA, 52,* 911–914.

Klopfer, P. H., & Gamble, J. (1966). Maternal "imprinting" in goats: The role of chemical senses. *Zietschrift für Tierspsychologie, 23,* 588–592.

Klopfer, P. H., & Klopfer, M. S. (1968). Maternal "imprinting" in goats: Fostering of alien young. *Zietschrift für Tierspsychologie, 25,* 862–866.

Krehbiel, D., Poindron, P., Lévy, F., & Prud'homme, M. J. (1987). Effects of peridural anesthesia on maternal behavior in primiparous and multiparous parturient ewes. *Physiology and Behavior, 40,* 463–472.

Kristal, M. B., Thompson, A. C., & Abbott, P. (1986). Ingestion of amniotic fluid enhances opiate analgesia in rats. *Physiology and Behavior, 38,* 809–815.

Kristal, M. B., Whitney, J. F., & Peters, L. C. (1981). Placenta on pup's skin accelerates onset of maternal behaviour in non-pregnant rats. *Animal Behaviour, 29,* 81–85.

Le Neindre, P., Poindron, P., & De Louis, C. (1979). Hormonal induction of maternal behavior in non-pregnant ewes. *Physiology and Behavior, 22,* 731–734.

Lehrman, D. S., (1961). Hormonal regulation of parental behaviour in birds and infrahuman mammals. In W. C. Young (Ed.), *Sex and internal secretions* (pp. 1268–1382). Baltimore: Williams & Wilkins.

Lévy, F. (1985). *Contribution à l'analyse des mécanismes de mise en place du comportement maternel chez la brebis (Ovis aries L.): Étude de la répulsion et de l'attraction vis-à-vis du liquide amniotique, mise en évidence, déterminisme, rôle.* [Analysis of repulsion and attraction towards amniotic fluids in relation with the establishment of maternal behaviour in sheep: evidence, determinants and role]. Unpublished doctoral dissertation, University of Paris VI.

Lévy F., & Poindron, P. (1984). Influence du liquide amniotique sur la manifestation du comportement maternel chez la brebis parturiente. *Biology of Behaviour, 9,* 65–88.

Lévy, F., & Poindron, P. (1987). Importance of amniotic fluids for the establishment of maternal behaviour in relation with maternal experience in sheep. *Animal Behaviour, 35,* 1188–1192.

Lévy, F., Poindron, P., & Le Neindre, P. (1983). Attraction and repulsion by amniotic fluids and their olfactory control in the ewe around parturition. *Physiology and Behavior, 31,* 687–692.

Lickliter, R. E. (1982). Effects of a post-partum separation on maternal responsiveness in primiparous and multiparous domestic goats. *Applied Animal Ethology, 8,* 537–542.

Morgan, P. D., Boundy, C. A. P., Arnold, G. W., & Lindsay, D. R. (1975). The roles played by the senses of the ewe in the location and recognition of lambs. *Applied Animal Ethology, 1,* 139–150.

Pedersen, C. A., & Prange, A. J. (1985). Oxytocin and mothering behavior in the rat. *Pharmacology and Therapeutics, 28,* 287–302.

Pissonnier, D., Thiéry, J. C., Fabre-Nys, C., Poindron, P., & Keverne, E. B. (1985). The importance of olfactory bulb noradrenalin for maternal recognition in sheep. *Physiology and Behavior, 35,* 361–364.

Poindron, P. (1976). Mother–young relationships in intact or anosmic ewes at the time of suckling. *Biology of Behaviour, 2,* 161–177.

Poindron, P., & Carrick, M. J. (1976). Hearing recognition of the lamb by its mother. *Animal Behaviour, 24,* 600–602.

Poindron, P., & Le Neindre, P. (1980). Endocrine and sensory regulation of maternal behavior in the ewe. *Advances in the Study of Behavior, 11,* 75–119.

Poindron, P., Le Neindre, P., Lévy, F., & Keverne, E. B. (1984). Les mécanismes physiologiques de l'acceptation du nouveau-né chez la brebis. *Biology of Behaviour, 9,* 65–88.

Poindron, P., Le Neindre, P., Raksanyi, I., Trillat, G., & Orgeur, P. (1980). Importance of the characteristics of the young in the manifestation and establishment of maternal behaviour in sheep. *Reproduction Nutrition Developpement, 20,* 817–826.

Poindron, P., Lévy, F., & Krehbiel, D. (1988). Genital, olfactory and endocrine interactions in the development of maternal behaviour in the parturient ewe. *Psychoneuroendocrinology, 13,* 99–125.

Poindron, P., Lévy, F., Le Neindre, P., & Keverne, E. B. (1986). The roles of genital stimulation, oestrogens and olfaction in the maternal bonding of sheep and other mammals. In L. Dennerstein & I. Fraser (Eds.), *Hormones and behavior* (pp. 538–548). Amsterdam: Elsevier.

Poindron, P., Martin, G. B., & Hooley, R. D. (1979). Effects of lambing induction on the sensitive period for the establishment of maternal behavior in sheep. *Physiology and Behavior, 23,* 1081–1087.

Poindron, P. Orgeur, P., Le Neindre, P., Kann, G., & Raksanyi, I. (1980). Influence of the blood concentration of prolactin on the length of the sensitive period for establishing maternal behavior in sheep at parturition. *Hormones and Behavior, 14,* 173–177.

Poindron, P., & Schmidt, P. (1984). Distance recognition in ewes and lambs kept permanently indoors or at pasture. *Applied Animal Behaviour Science, 13,* 267–273.

Rosenblatt, J. S., & Siegel, H. I. (1981). Factors governing the onset and maintenance of maternal behavior among nonprimate mammals. In D. J. Gubernick & P. H. Klopfer (Eds.), *Parental care in mammals* (pp. 13–76). New York: Plenum Press.

Rosenblatt, J. S., Siegel, H. I., & Mayer, A. D. (1979). Progress in the study of maternal behavior in the rat: Hormonal, nonhormonal, sensory, and developmental aspects. *Advances in the Study of Behavior, 10,* 225–311.

Schaal, B., Montagner, H., Hertling, E., Bolzoni, D., Moyse, A., & Quinchon, R. (1980). Les stimulations olfactives dans les relations entre l'enfant et la mère. *Reproduction Nutrition Developpement, 20,* 843–858.

Stabenfeldt, G. H. (1974). The role of progesterone in parturition: Premature, normal, prolonged gestation. In M. J. Bosc, P. Palmer, & Cl. Sureau (Eds.), *Avortement et parturition provoqués* (pp. 97–122). Paris: Masson.

Terqui, M. (1974). Les oestrogènes au cours de la gestation et de la parturition chez la truie et la brebis. In M. J. Bosc, R. Palmer, & Cl. Sureau (Eds.), *Avortement et parturition provoqués* (pp. 71–79). Paris: Masson.

Theodosis, D. T., Montagnese, C., Rodriguez, F., Vincent, J. D., & Poulain, D. A. (1986). Oxytocin induces morphological plasticity in the adult hypothalamo–neurohypophyseal system. *Nature, 322,* 738–740.

Wamboldt, M. Z., & Insel, T. R. (1987). The ability of oxytocin to induce short latency maternal behavior is dependent on peripheral anosmia. *Behavioral Neuroscience, 101,* 439–441.

Watanabe, Y., Mori, K., Imamura, K., Takagi, S., & Hayaishi, O. (1986). Modulation by prostaglandin D_2 of mitral cell responses to odor stimulation in rabbit olfactory bulb. *Brain Research, 378,* 216–222.

Yeo, J. A. G., & Keverne, E. B. (1986). The importance of vaginal-cervical stimulation for maternal behaviour in the rat. *Physiology and Behavior, 37,* 23–26.

9

Psychobiology of Maternal Behavior in Nonhuman Primates

CHRISTOPHER L. COE

Anyone who has observed the birth of an animal cannot help but marvel at the seemingly spontaneous and perfect responsiveness of the parturient female to her newborn. Her solicitousness and caregiving appear so ideally suited to the biological and emotional needs of the infant that one usually does not ponder over the complex processes that must underlie her maternal responses. Indeed, the maternal responsiveness of the nonhuman primate female has been so taken for granted that there has been little research on the mechanisms mediating maternal behavior since the pioneering studies by Harry Harlow in the 1960s (Harlow, Harlow, & Hansen, 1963). The paucity of psychobiological research on primates is also due to the strong ethological tradition that has shaped the field of primatology. As a consequence, there have been many descriptive studies of parturient behavior and mother–infant interactions (Shively & Mitchell, 1986a, 1986b), but only a handful of experiments on the neurological and endocrinological processes underlying maternal behavior (reviewed in Capitanio, Weissberg, & Reite, 1985; Higley & Suomi, 1986).

The following review of research on maternal behavior in nonhuman primates reflects the history of this field. The chapter begins with a brief summary of the evolutionary factors that have shaped the expression of maternal behavior in primates. This section is followed by a general description of the behavioral changes observed in the pregnant, parturient, and maternal female. Considered next are the few studies that have attempted to delineate whether maternal responsiveness has a hormonal basis in nonhuman primates. This brief review is followed by a summary of the more extensive literature on the stimulus characteristics of the infant that elicit maternal responses. The final discussion encompasses a somewhat novel approach that can be applied to the study of mother–infant relations. Our laboratory has spent the past 5 years studying aspects of the immunological relationship between the mother and the infant in the squirrel monkey *(Saimiri sciureus)* and the rhesus monkey *(Macaca mulatta)*. We will illustrate the unique perspective that can be brought to bear on the topic of maternal behavior by considering one topic in immunology: the placental transfer of maternal antibody to the fetus.

EVOLUTIONARY CONSIDERATIONS

The Order Primates is composed of a highly successful and diverse group of animals classified into 52 genera and 181 species. Both the tiny 4-ounce, seasonally breeding mouse lemur of Madagascar and the 200-pound, long-lived female gorilla of Africa are primates. It is difficult, therefore, to speak of any one species as being representative of the behavior of all primates. Nevertheless, most of the experimental studies have been conducted on the maternal behavior of only one primate species, the rhesus monkey, with some additional work on the squirrel monkey and the chimpanzee. Fortunately, there has been sufficient research on the evolution of primates, and enough field studies, that one can at least provide an evolutionary perspective for the data on the rhesus monkey.

Certainly, one of the most important changes that affected the evolution of primates was the trend toward a reduction in the number of offspring. Even the prosimians typically give birth to only one infant, or at most two, and this change is reflected in several aspects of the primate female's reproductive physiology. There has been a reduction in the number of mammaries to two, a shift from a bicornate to a unicornate uterus, and, across the Order Primates, the evolution of a hemochorial placenta from the epitheliochorial placenta. The invasive hemochorial placenta permits more direct contact between maternal and fetal blood and symbolizes anatomically the greater investment of the female in her offspring, a process that begins in utero and continues for an extended period after birth. With the reduction in the number of young, it becomes possible for the primate female to carry the infant at birth, and there is a shift away from the nesting behavior observed in the litter-bearing insectivores, such as the tree shrew. In some of the smaller prosimians, like the mouse lemur and the galago, there is still a brief transition period when the neonate is either left in a nest or "parked" on a branch while the mother is foraging, but in most prosimians and all anthropoid primates, the baby is carried continuously from the moment of birth (Klopfer & Boskoff, 1979). Although speculative, the evolution of the opposable thumb probably had as much to do with the infant's need to grasp the mother's fur as it did with locomotion and foraging in adulthood.

The birth of a single infant coincided with a dramatic increase in the life span of the higher primates. All phases of the life span are markedly extended in the monkeys and apes compared with the prosimians. Gestation increases from 3–4 months in the prosimian to 5–6 months in the monkey and 8 months in the great apes. The greater maternal investment made in pregnancy is followed by an equivalent extension of the infancy phase. Maternal care in the prosimians typically lasts for 3–6 months compared with 6–12 months in the monkey and 4–6 years in the great apes. As the number of young declined and the interbirth interval was extended, it became more and more imperative to ensure the survival of each offspring. This constitutes the basis of sociality and the evolution of group living based on female–female bonding (i.e., mother–daughter and sister–sister affiliation). The development of the multi-female group provided greater protection for the infants and, in a number of primate species, led to the evolution of a rearing strategy involving multiple female caregivers (e.g., "aunting" in the squirrel monkey and "baby passing" in the langur and patas monkey) (Quiatt, 1979). In fact, the interest and responsiveness that most monkey females show toward infants, especially distressed or threatened ones, may account for the difficulty in showing a hormonal basis for retrieval behavior in female rhesus monkeys (Holman & Goy, 1980).

There are some exceptions to the statement that infant rearing is exclusively the mother's responsibility in primates, most notably in the monogamous New World monkeys that show paternal behavior (Wamboldt, Gelhart, & Insel, 1988), but certainly the norm for primates is primary caregiving by the mother. Female relatives, usually older daughters, may assist in carrying the infant, but the mother will almost always try to retrieve a distressed infant, and males typically do not interact frequently with their offspring until the juvenile stage (with some exceptions, such as the Gibraltor macaque). This basic sex difference appears to be strongly wired in the primate and is not affected by androgen exposure of female fetuses in utero, a treatment that masculinizes other behaviors (Gibber & Goy, 1985).

The three trends—reduction in infant number, lengthening of the mother–infant phase, and evolution of sociality—served to establish the infancy period as a time of socialization. In monkeys and apes, the socialization process may last for many years, during which time the juvenile primate acquires the necessary social and foraging skills of the species (Lancaster, 1971). Puberty is reached at 3–4 years of age in the monkey and at 8–12 years of age in the great apes. Typically, the developing primate remains associated with its mother until puberty, and adolescents of at least one sex usually remain in the social group after puberty. In monkey species, the daughters usually remain in the natal group, and biological relatives form kinship alliances within the troop (as always, there are some exceptions; in the monogamous species and the leaf-eating monkeys, both sexes may emigrate). In our closest evolutionary relatives, the gorilla and the chimpanzee, there has been an interesting reversal in this developmental pattern. Adolescent males may remain in the natal group, and females emigrate at menarche (Goodall, 1986).

Regardless of the variation in social strategies at the pubertal stage, all monkey and ape females are permitted an extremely long period of time in which to learn the necessary skills for mothering. Juvenile females have many opportunities to observe births and maternal behavior, and in many species, females practice caregiving on their younger siblings (Lancaster, 1971). In some species, such as the squirrel monkey, there may even be the onset of lactation in the aunting adolescent female prior to the birth of her own infant. The importance of this socialization phase may explain why most primate research has emphasized the overriding influence of experience on the expression of maternal behavior. Although one must be careful about drawing definitive conclusions because of the limited research, it is highly probable that experiential and cognitive factors have largely replaced hormone mediation in the induction of maternal behavior in primates. The existing primate studies suggest that hormones do not play an important role in maternal responsiveness, at least in the multiparous female (Gibber, 1986; Holman & Goy, 1980). This question will be addressed in subsequent sections of this chapter.

BEHAVIORAL CHANGES ACROSS PREGNANCY

One other factor that may have influenced researchers not to pursue a detailed study of the induction of maternal behavior in nonhuman primates is that there have been only a few reports of overt changes in maternal responsiveness and behavior across pregnancy (Cross & Harlow, 1963; Gibber, 1986; Rosenblum, 1972). In the absence of brooding and nesting behavior, there is often no clear behavioral sign of the impending delivery date. Typically, there have been reports of increasing lethargy and, in some species, increased intolerance of male proximity in the latter stages of pregnancy (Shively & Mitchell, 1986a, 1986b). This finding of decreased activity was confirmed in an observational study on the behavior of captive female chimpanzees across pregnancy (Coe, Horvat, & Levine, unpublished data). The behavior of six adult females living under semifree ranging conditions (1.5 acres) at the Stanford Outdoor Primate Facility was recorded unobstrusively from an observation tower. All the females had been born in the wild, were socially reared, and were now living in one of two social groups, each consisting of three adult males and three adult females plus infants and juveniles (for further details, see Coe, Connolly, Kraemer, & Levine, 1979). Females were observed in a longitudinal manner for 3 months prior to conception, during the three trimesters of pregnancy (mean length = 241 days), and for 3 months after birth. For two females, there was an opportunity to watch second pregnancies, generating data from a total of eight pregnancies.

As can be seen in Figure 9.1, there was a tendency toward increasing inactivity across the second and third trimesters of pregnancy. The percentage of time spent idle, sleeping, and in self-directed grooming increased noticeably. In six of the pregnancies, lethargy in the final stages was overt, resulting in idle scores 67–80% of the time, although for two females there was only a modest change from baseline levels. During the postpartum period, the females' activity patterns returned toward baseline levels, despite the added burden of nursing and carrying an infant.

Although the percentage of time spent in social behavior did not fluctuate much across pregnancy, typically around 5% of the observation period, there was a dramatic change in social interactions (Figure 9.2). Pregnant females associated more frequently with other females; the percentage of social interactions directed toward females rose from 26% in the baseline period to 45% in the second and third trimesters. Interestingly, there was a concomitant decrease in the percentage of time spent with infants and juveniles, suggesting that there was no general attraction to young chimpanzees across pregnancy. Assessment of the types of social behaviors engaged in by pregnant females revealed that an increased interest in social grooming with adults and a marked decrease in playful behavior accounted for these changes in social affiliation. Association with males continued at a stable level in captivity, whereas they would have decreased in the wild, but sexual interactions ended in the second trimester of pregnancy and did not resume until the females began to cycle 3 years later.

This study indicated that there are several behavioral changes across pregnancy in female chimpanzees and actually underestimates the importance of these changes for chimpanzees under natural conditions. Field research in the Gombe National Park has indicated that female chimpanzees show a major shift in ranging patterns and social affiliation across pregnancy (Goodall, 1986). The pregnant female is less gregarious, is not sexually attractive to males, and appears to be unwilling to keep up with the daily ranging behavior of heterosexual traveling parties, which may average 6–7 miles a day. As a consequence, she withdraws from the heterosexual foraging

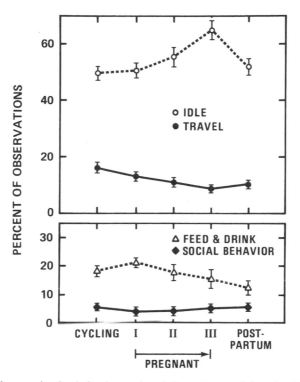

Figure 9.1. Changes in the behavior and activity patterns of female chimpanzees across pregnancy and during the postpartum period.

groups and ultimately travels alone during the days prior to parturition. On the day of parturition, she is usually alone or with only her older offspring.

While most monkey females remain with their troop throughout pregnancy, field researchers have noted that pregnant patas monkeys appear tired by their efforts to forage and keep up with the rest of the group (Chism, personal communication). In the case of the baboon in Kenya, Altmann (1987) reported that the highest incidence of predation occurs in the slower pregnant and lactating females. Further, Marriott (personal communication) found that pregnant rhesus monkeys in Nepal were more reluctant to climb high into trees and, thus, were forced to undergo a major shift in diet. Recently, the results of a longitudinal study of female squirrel monkeys across pregnancy found a similar pattern of behavioral changes. Females living in laboratory groups consisting of one to two males and three to five females were observed across their 160-day pregnancy. During the last 2 months of pregnancy, they became increasingly inactive and withdrew from social interactions. From other studies on pregnant squirrel monkeys, it is known that the latter stage of pregnancy is a time of great nutritional demand when females may consume up to 50% of their body weight in water (Clewe, 1969), and many become anemic without vitamin supplementation (Rosenblum, 1968). Thus even in the absence of overt signs of maternal preparation for the impending birth, there appear to be many significant behavioral changes in

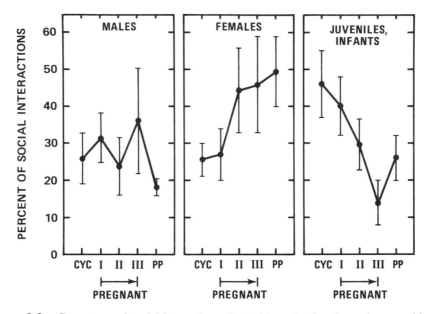

Figure 9.2. Percentage of social interactions directed to male, female, and young chimpanzees across pregnancy and during the postpartum period.

the term-pregnant female. Wasser and Starling (1986) have even offered the provocative suggestion that pregnant female baboons may become extremely aggressive toward other pregnant females in order to cause miscarriages and thereby to decrease food competition subsequently between infants in the following year.

ENDOCRINOLOGY OF PREGNANCY AND PARTURITION

A number of studies have described the hormonal changes associated with pregnancy and parturition in nonhuman primates (Bosu, Johansson, & Gemzell, 1973; Line, Golub, Mahoney, Ford, & Anderson, 1986; Novy & Resko, 1981; Reyes, Winter, Faiman, & Hobson, 1975; Sholl, Robinson, & Wolf, 1979a, 1979b; Weiss, Butler, Hotchkiss, Dierschke, & Knobil, 1976). Basically, the endocrine profile is similar to that of the human female. Depending on the species, there is a moderate to high elevation of the gonadal steroids, estrogen and progesterone, across pregnancy. The predominance of the progesterone influence becomes visually apparent in the third trimester in those primates that have sex skins because the skin becomes flat and loses its swollen, red color. In the chimpanzee, the pregnancy-related changes in the sex skin are associated with the total cessation of sex behavior in the third trimester, which continues throughout the postpartum period (Coe et al., 1979).

The high levels of gonadal hormones appear to be produced primarily by the fetal–placental unit, and endogenous gonadotropin secretion in the mother is low (Chandrashekar et al., 1980; Novy & Resko, 1981; Reyes et al., 1975). This may

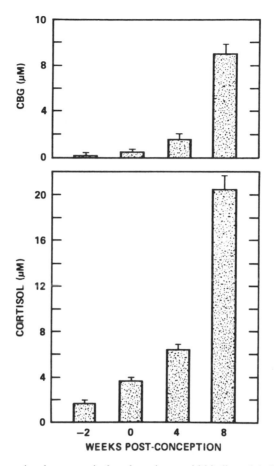

Figure 9.3.   Increases in plasma cortisol and corticosteroid-binding globulin (CBG) soon after conception of the pregnant squirrel monkey. (From Coe et al., 1986)

account for the fact that pregnancy can be maintained in the hypophysectomized female monkey (Smith, 1954). Hypophysectomy also does not appear to disrupt maternal responsiveness after caesarean delivery of the babies, although it was not possible to evaluate maternal behavior in depth in this study because the females could not lactate. The prolactin level in the normal primate female is elevated during the final stages of pregnancy, presumably to prepare the female for lactation (Walsh, Wolf, Meyer, Aubert, & Friesen, 1977). In a study of 97 pigtail macaque females, 47% were lactating prior to delivery (Goodlin & Sackett, 1983).

The pituitary–adrenal axis is also markedly affected by pregnancy. In studies on the squirrel monkey female, cortisol levels began to rise within a few days after conception and continued to increase progressively throughout pregnancy due to estrogenic stimulation of the adrenal (Coe, Murai, Wiener, Levine, & Siiteri, 1986) (Figure 9.3). Initially, the cortisol levels were matched by equivalently high levels of cortisol-binding globulin (CBG) released from the liver, but by the end of pregnancy, cortisol increases exceeded the CBG rise, and there were relatively high levels of free cortisol in circulation. Presumably, these high titers of free cortisol account for some of the immunological changes observed in females during the latter stages of pregnancy (Rocklin, Kitzmiller, & Kaye, 1979). Cortisol probably acts in conjunction with progesterone to suppress the mother's immune system, helping to prevent immunological rejection of the fetus (Siiteri, Febres, & Clemens, 1977). The role of cortisol in parturition is not clear in the primate, although it is not surprising that there is marked activation of the pituitary–adrenal axis, including elevated adrenocorticotropic hormone (ACTH) and endorphins, given the discomfort and sleep disturbance that would be associated with nocturnal labor (Goodlin & Sackett, 1983). Oxytocin is low during gestation and high at parturition (Mitchell, Mountford, Natale, & Robinson, 1980), but its role in the induction of maternal behavior is unclear and it remains an important area for future research (for review, see Capitanio et al., 1985).

The only major deviation from the human endocrine pattern occurs in the New World monkeys, which show a pattern of chorionic gonadotropin release different from that of Old World monkeys, apes, and humans. In contrast to human females, who need human chorionic gonadotropin (HCG) to maintain the corpus luteum during the initial stage of pregnancy, New World primates seem to be able to sustain the ovarian production of progesterone without the release of chorionic gonadotropin from the placenta. Interestingly, the release of chorionic gonadotropin begins to increase in the middle of pregnancy in the New World primate and then subsides at the end of pregnancy.

The similarity between the pregnancy hormone profiles in nonhuman primates and human females would seem to indicate that monkeys are an excellent model for studying the role of hormones in the induction of maternal behavior. However, preliminary studies have suggested that this will not be a fruitful line of research. For example, it has not been possible to show that ovariectomized rhesus monkey females are any less responsive than intact females when presented with young infants (Holman & Goy, 1980). Similarly, menopausal females are extremely responsive to presented infants, and older females frequently carry juveniles in a maternal manner at our laboratory (Deets & Harlow, 1967). Rosenblum's (1972) report of high retrieval behavior in term-pregnant squirrel monkeys provides one of the few suggestions of a hormonal basis for maternal interest in primates, but this work did not generalize to pregnant rhesus monkeys (Gibber, 1986). Based on most studies, it is unlikely that ovarian hormones, at least, play a major role in the induction of maternal responsiveness in primates. Where they may play a greater role is in the establishment of lactation, which becomes important in the behavioral regulation and maintenance of maternal interest by stimulating interactions with the infant. Hormonal changes during parturition may also play a role in the selective focusing of a female's maternal interest on her own infant, but this hypothesis has not been tested directly.

PARTURIENT BEHAVIOR

Detailed reports on the parturient behavior of female primates have existed in the literature for over 50 years (Hartman, 1928; Tinklepaugh & Hartman, 1930, 1932).

For a detailed summary of parturient behavior, the interested reader is referred to several excellent reviews (Brandt & Mitchell, 1971; Goodlin & Sackett, 1983; Lanman, 1977; Rosenblum, 1971; Shively & Mitchell, 1986a, 1986b), and only a brief overview will be presented here.

An increase in spontaneous uterine contractions has been described during the day or two before labor in rhesus monkeys and may account for reports of an increase in behavioral agitation and increased inspection of the vaginal area (Line et al., 1986). Births typically occur at night after a relatively brief labor lasting for 0.5–3 hr. In the pigtailed macaque, 76% of 160 births occurred between 10:00 P.M. and 2:00 A.M. (Goodlin & Sackett, 1983). Although there are exceptions, longer labors, especially during the day, are often an indication of an impending birth complication. In the pigtailed macaque study cited above, longer daytime labors occurred more frequently in high-risk pregnancies. During labor, the female typically sniffs, licks, and touches the vaginal area and bears down in a squatting position. As labor becomes more intense, the female may show signs of discomfort and pain, but often the birth appears to be nontraumatic. In some cases, the female may actually be walking or eating when the baby's head appears. Experienced mothers typically reach down and manually assist the emerging infant onto the ventrum, and begin to inspect and groom it.

The importance of this seemingly innocuous "reach-down" can best be emphasized by describing what can go wrong in an inexperienced, primiparous female. JD, one of the female chimpanzees at the Stanford Outdoor Primate Facility, had spent several years performing in a circus before being reestablished in a chimpanzee social group. Despite having been born in the wild and having witnessed the birth and care of another infant, when JD's turn came, she was inept. As illustrated in Figure 9.4, her labor came during the day instead of at night. As a consequence, she was surrounded by chimpanzees, especially interested juveniles, during the birth. Ultimately, she gave birth in an inappropriate location more than 20 feet above the ground and failed to catch the emerging infant. Although she went to the location where it had fallen, she did not attempt to retrieve it; nor did she prevent the other excited chimpanzees from grabbing it. From this anecdote, it is possible to see that the time and location of the birth process are critical to ensuring the well-being of the neonate. This example also emphasizes the important role of experience for the primiparous female. Despite undergoing the normal hormonal changes associated with pregnancy and parturition, the inexperienced female failed to respond in an appropriate maternal manner. The role of experience is also evident in empirical studies that have evaluated the retrieval patterns of nulliparous rhesus monkey females (Gibber & Goy, 1985; Holman & Goy, 1980). Retrieval of presented infants occurs at a low level in young, inexperienced females compared with experienced adults.

As a postscript, it should be mentioned that the female chimpanzee JD went on to rear infants normally in subsequent pregnancies. This capacity for the normal expression of maternal behavior even after a brief exposure to a previous baby is certainly a topic that warrants further study. Although it is not widely known, even rhesus monkeys that have been socially isolated from birth can eventually learn to become competent mothers after several pregnancies (Ruppenthal, Arling, Harlow, Sackett, & Suomi, 1976). The so-called motherless-mothers improve across successive pregnancies, and 75% are rated as adequate by their fourth pregnancy (Figure

Figure 9.4. Inappropriate parturient behavior shown by an inexperienced female chimpanzee. (A) Poor location and time of birth, and (B) nonresponsive and overly excited reaction to neonate.

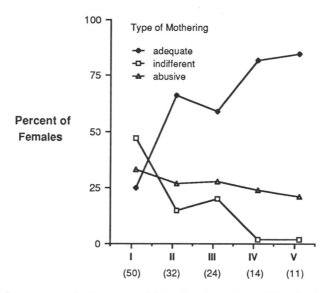

Figure 9.5. Improvement in the maternal behavior shown by socially deprived rhesus monkeys over successive births. Numbers in parentheses are the Ns for each birthing experience. (Modified from Ruppenthal et al., 1976)

9.5). We have noted a similar improvement in squirrel monkey females across first and second pregnancies. In one case, female squirrel monkeys were reared in peer groups in the absence of adult role models after 6 months of age (Coe, Hayashi, & Levine, 1988). When they conceived and gave birth at 3 years of age, they typically failed to care for their first infant. Often the infants were rejected on the day of birth, and so there was little time for exposure to infant stimuli. Nevertheless, 1 year later after a subsequent pregnancy, almost all females responded normally and successfully reared their infants. In the review of motherless-mothers by Ruppenthal et al. (1976), it was noted that only 48-hr exposures to infants were required for an improvement in maternal behavior in subsequent pregnancies. Others have suggested that experiencing parturition is also important in establishing maternal responsiveness in primiparous rhesus monkeys (Gibber, 1986), as suggested in rodent studies (Bridges, 1977; Moltz & Kilpatrick, 1980). In keeping with this view, primiparous rhesus females are much less likely to care for caesarean-delivered babies than are multiparous females (Meier, 1965).

Perhaps the most dramatic aspect of the primate mother's response to her neonate is the almost invariant occurrence of placentophagia (Figure 9.6). While ingestion of the placenta is a fairly routine behavior in mammals, it is a radical departure from the frugivorous and folivorous diet typical of most primates. There have been many descriptive accounts of placentophagia in diverse primate species, and most observers have commented on the ritualized nature of the behavior—seemingly done under compulsion, rather than for its pleasurable rewards (Rosenblum, 1972; Shively & Mitchell 1986a, 1986b). It is often done in association with licking and grooming of the birth fluids from the infant, but the female usually stops her cleaning behavior to devour most or all of the placenta when it is passed within the first minutes or

Figure 9.6. Placentophagia shown by a pigtailed macaque soon after birth of the infant. (Photo by L. Rosenblum, used with permission; from Rosenblum, 1971)

hours after birth (Lindburg & Hzell, 1972). While it may seem intuitively obvious, one must wonder how the female, especially the inexperienced primiparous female, knows where the umbilical cord ends and the neonate begins. The basis of this question becomes more apparent after one has witnessed the birth of a premature or stillborn infant, and the female consumes a part of the fetus's body as well. Indeed, some of the abusive behavior that has been seen in socially deprived rhesus monkeys involves inappropriate biting of the neonate and may reflect misdirected behavioral components of placentophagia.

A number of investigators have proposed plausible but untested explanations for the almost universal performance of placentophagia by primate females. The hypotheses include both proximate and ultimate functions, such as (1) a source of hormones that may stimulate lactation or reduce uterine bleeding, (2) provision of nutrients and fluids to the exhausted female, (3) severing the placenta from the neonate and preventing uncontrolled hemorrhaging from the umbilical cord, and (4) avoidance of predator detection. The last function would be of obvious importance under natural conditions, especially for the smaller primates, and would be in keeping with the birth of infants at night. Although these hypotheses are reasonably deduced, they have not been empirically tested. Placentophagia is certainly an interesting behavioral response that would be amenable to study from the point of view of neurobiological encoding.

STIMULUS CHARACTERISTICS OF THE NEONATE

Most of the research on the induction of maternal behavior in primates has focused on the stimulus qualities of the infant that elicit caregiving behavior. Because of the common belief that visual cues predominate in primates, relatively little attention has been given to the role of olfaction, in contrast to the large literature on olfactory cues in rodents and ungulates (see chapters 8 and 12). One exception is a study by Lundblad and Hodgen (1980) indicating that the presence of vaginal fluids facilitated the acceptance of rhesus infants delivered by caesarean section. A second exception is the research on mother–infant recognition by olfactory cues in the squirrel monkey (Kaplan & Cubiccioti, 1980). The latter study indicated that squirrel monkey females can recognize their own infants by smell within a few days after birth, while tests of visual recognition usually suggest that several weeks are required, at least for visual recognition of the mother by the infant (Rosenblum & Alpert, 1977). Further research on the role of olfaction in nonhuman primates is certainly required, given recent studies on maternal discrimination of human neonates by olfactory cues soon after birth. However, it is clear that the infant presents a multisensory stimulus, and olfaction probably plays a lesser role in the microsmatic higher primates than in other mammals because they have deemphasized olfactory cues in favor of visual input. The evolutionary loss of the vomeronasal system as a source of olfactory information and the absence of scent-marking behavior in the Old World monkeys and apes attest to the relative reduction in the importance of smell in the Order Primates.

Research in primates has focused instead on the visual appearance of the infant at birth and on the neonate's reflexive behavior as stimuli for caregiving. In many primate species, infants can be clearly identified by their distinctive natal coat color. Typically, the color contrast remains until the onset of weaning. In one of the few empirical tests of the stimulus value of infantile color, Higley and Suomi (1986) showed that red dye on the face of juvenile rhesus monkeys facilitated looking behavior in adult females when compared with abnormal green coloration on the face. Other investigations have suggested that infant size is an important feature as well. Sackett and Ruppenthal (1974) found that adult rhesus females show a preference for viewing smaller, young infants compared with larger, older infants in a visual selection paradigm, although the size–age interaction was not specifically controlled. Another widely cited visual characteristic that has not received empirical verification is the foreshortened muzzle of the neonate, the so-called cute response described originally by Lorenz (1971).

While infant size and color are undoubtedly important in eliciting maternal responses, the behavior of the neonate certainly plays a primary role. Primate infants are born with a strong predisposition to cling to the mother's body, and the expression of this reflexive response appears to be essential to the attachment process. As Harlow showed in his seminal research, the object of this clinging becomes imbued with affect, and over a short period of time, either animate or inanimate objects can become the focus of attachment (Harlow & Zimmerman, 1959). The infant's clinging and rooting behavior for the nipple appears to have an important influence on the mother as well, although the reinforcing hormonal and somatosensory aspects of nursing have not been studied in primates.

Female primates show some compensatory behavior for neonates that cannot cling, but weak or stillborn infants are often abandoned by the mother after a few days. This is particularly true in an arboreal species, such as the squirrel monkey, where the neonate is entirely responsible for maintaining itself on the dorsal surface while the mother travels and forages. Equally important is the infant's capacity to elicit caregiving behavior by showing signs of discomfort and distress. Infant primates of virtually all species have a similar type of distress call that elicits maternal responses. In the newborn, calls may be emitted in response to being held upside down, since infants show a strong negative geotropism. The infant may also call if it is held too low on the ventrum or away from ventral contact with the mother. Typically, the mother corrects her carrying behavior and often shifts the rooting infant toward the nipple. In the older infant, the same type of call is used by the distressed monkey to elicit retrieval behavior. Considerable research has been conducted on the stimulus qualities of the "whoo" call in the rhesus monkey and the "isolation peep" of the squirrel monkey (Levine, Wiener, Coe, Bayart, & Hayashi, 1987). Several studies have shown that mothers learn to recognize the distinctive call of their own infant, although all adults usually respond to the distress call of any infant.

This responsiveness may be mediated in part by the fact that distressed infants cause the adults to become disturbed. A number of years ago, a study was done to evaluate whether female squirrel monkeys would be disturbed by the visual presentation of a distressed infant or following observation of the separation of a mother and an infant (Vogt, Coe, Lowe, & Levine, 1980) (Figure 9.7). The levels of adrenal hormones in circulation, as a reflection of arousal, were measured. Significant increases in cortisol levels were found in females, both in response to seeing a separated infant for 30 min and after witnessing a mother–infant separation. It was also of interest that adult males showed signs of behavioral agitation, but not adrenal activation, unless they actually saw the infant held by a human during the test.

During the course of these studies, the authors were struck by an inadvertent observation that the adult squirrel monkeys in adjacent cages seemed to respond only to calls from infants of their own subspecies. In a subsequent experiment, call recognition and responsiveness were shown to be restricted to infants of the monkeys' own genotype (Snowdon, Coe, & Hodun, 1985). Squirrel monkeys of Bolivian and Peruvian origin reacted only to calls recorded from infants of their own genotypes, and Guyanese squirrel monkeys responded only to playback calls from Guyanese infants. These observations concur with the work of Newman (1985) on population variation in infant call structure across closely related species and indicate that the

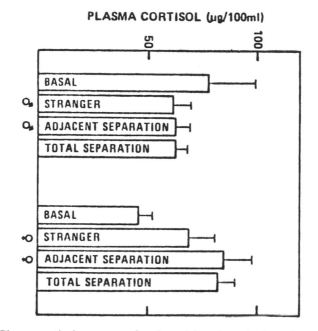

Figure 9.7. Plasma cortisol response of male and female squirrel monkeys to the presence of a disturbed infant or after the removal of an infant from an adjacent group. (From Vogt et al., 1980)

mothers are appropriately predisposed to respond to calls from their own type of infant.

NORMATIVE COURSE OF MOTHER–INFANT INTERACTION

Once the mother–infant bond is established, the normal course of maternal behavior in nonhuman primates can be described as consisting of three stages: (1) a period of support and solicitousness, (2) a period of ambivalence toward infant autonomy, and (3) a period of weaning and active rejection. This transition in maternal attitude can be illustrated by showing the trajectory of mother–infant relations in two species: the squirrel monkey and the chimpanzee.

A number of years ago, the development of eight squirrel monkey infants was evaluated across the first 7 months of life (Coe, Wiener, Rosenberg, & Levine, 1985). As can be seen in Figure 9.8, the infants spent most of the first month in continuous contact with their mothers. By 2 months of age, the infants began to make independent ventures, but were often followed and retrieved by their mothers, especially if there was any sign of disturbance. This maternal concern rapidly diminished by the third month, however, and the mothers began to remove the infants from their backs and actively to encourage independence. After 4–5 months of age, the infants were

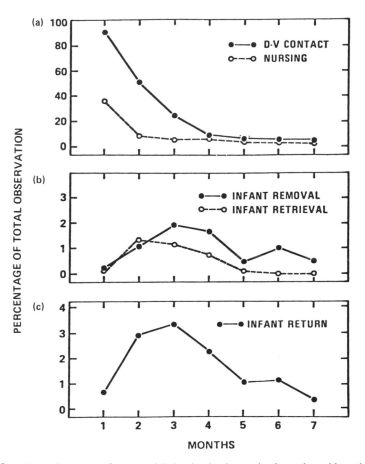

Figure 9.8. Normal pattern of maternal behavior in the squirrel monkey. Note the progressive decline in body contact over the first 3 months (top graph), the onset of infant exploratory behavior and maternal retrieval at 2–3 months (middle graph), and the transition to maternal rejection and decrease in maternal approaches after 3 months of age (bottom graph). (From Coe et al., 1985)

responsible for maintaining contact with their mothers. By 6 months of age, they were weaned and spent most of their waking hours off the mother.

A similar sequence has been described in the Old World monkeys and apes, although the temporal pattern is greatly extended. In the Old World monkeys, the maternal phase usually lasts for 6–12 months, whereas the process typically extends over 3–5 years in the chimpanzee, orangutan, and gorilla (Hansen, 1966; Hinde, 1983). The maternal pattern of the chimpanzee is illustrated in Figure 9.9, based on observations of three mother–infant pairs over a 3-year period (Horvat, Coe, & Levine, 1980). Each pair was observed in a longitudinal manner during 30-min observation sessions, with a total of over 1000 hours per dyad. During the first 6 months, infants were almost continuously in contact with their mothers, and even at 1 year of age, they were spending over 60% of their time in contact with their mothers. Infants began to travel and sit independently in the Near category (<5 m from the mother)

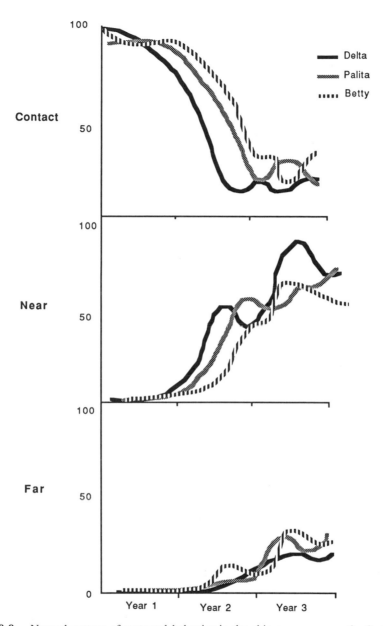

Figure 9.9. Normal pattern of maternal behavior in the chimpanzee across the first 3 years of life. The transition from contact with their mothers to independence for three dyads (near: <5 m; far: >5 m).

more frequently during the second year of life. This transition corresponded to the decline in suckling and the onset of independent eating and play behavior between 12 and 24 months of age. However, it was not until the third year of life that infants were regularly observed in the Far category (>5 m from the mother). Although 3 years of maternal care may seem to be a particularly slow rate of infant maturation,

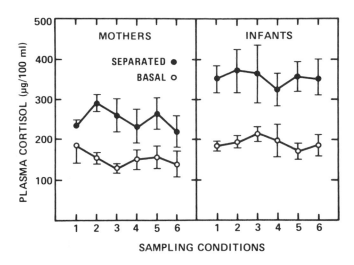

Figure 9.10. Lack of change in the squirrel monkey female's adrenal response to brief separation from her infant, despite its increasing age and the occurrence of weaning. The 30-min separations were conducted at 10-day intervals when the infants were 3–6 months of age. (From Coe et al., 1983)

the developmental pattern in captivity is actually accelerated over that observed in wild chimpanzees at Gombe National Park, Tanzania.

There have been relatively few attempts to analyze the cause of the change in maternal responsiveness during the weaning stage. In our study, active maternal rejection and avoidance were not observed until the third year of life, and for all three infants, the mother's attitude changed with the resumption of the menstrual cycle (for Delta, 37 months; Betty, 32 months; Palita, 28 months of age). These data concur with observations on wild chimpanzees, suggesting that rejections are related more to resumption of the mother's sexual activity than to final weaning efforts. Descriptive studies have suggested that experienced, multiparous females may initiate the process earlier, in part because they have succeeded in rearing offspring that are more independent at an earlier age. Rosenblum and Youngstein (1974) conducted one of the few empirical tests to assess the change in maternal attitude by examining the willingness of macaque mothers to engage in compensatory care. Young and older infants were lightly anesthetized and returned to their mothers. The females readily accepted and carried their immobile infants when they were young but did not attempt to carry their older infants. In the course of extensive breeding of squirrel monkeys, a similar change in maternal responsiveness has been observed when the infants reach 5–6 months of age. With a developmentally retarded or sickly infant, squirrel monkey females may show abnormally high levels of carrying behavior, even past the normal age of weaning. However, once the mother approaches the period when she will resume cycling, she will often actively reject the 5- to 7-month-old infant.

This change in the mother's attitude under nondisturbed conditions does not necessarily mean that she has given up her protective role. Primate mothers do continue to show some evidence of bonding to the older weaned infant, at least in terms

Figure 9.11. Depressive reaction of an older sibling in response to a female chimpanzee's redirection of maternal care to her neonate.

of their reaction to involuntary separation from the infant (Coe, Glass, Wiener, & Levine, 1983). In studies of the adrenal responses of squirrel monkey mothers to 30-min separations from their infants, equivalent responses to 3- and 6-month-old infants have been observed, and no sign of habituation in cortisol responses occurs across repeated separations (Figure 9.10).

Another aspect of the change in maternal attitude toward older infants can be seen in those species where the females continue to suckle or care for the infant throughout the next pregnancy period. As shown in Figure 9.11, there is a rapid redirection of caregiving behavior away from the older sibling and toward the newborn on the day of birth. In these instances, which we have witnessed several times in rhesus monkeys and chimpanzees, the older sibling often shows a transient depression. Quantitative descriptions of this phenomenon have also been made on the vervet monkey (Lee, 1983). Evaluation of this triadic relationship could provide a unique paradigm for assessing the factors underlying selective maternal bonding. The naturalistic features of an extended-family study would also be particularly compelling, especially when one considers that most primate females spend their entire postmenarchal life involved in some aspect of the maternal process.

IMMUNOBIOLOGY OF THE MOTHER–INFANT RELATIONSHIP

Although researchers have brought many evolutionary and physiological perspectives to bear on the study of mother–infant relationships, a different viewpoint that has not received much attention in previous research merits consideration. Over the past several years, our laboratory has begun to examine the immunological relationship between mother and infant, and specifically the role that the mother plays in stimulating the development of infant immunity. The possibilities generated by this approach can be exemplified by considering one topic: the placental transfer of maternal antibody.

This process is part of the considerable prenatal investment that the female makes in her infant. In contrast to the rodent, which can transfer antibody of the IgG class to the immature pup after birth, the primate must transfer this antibody before term because IgG cannot cross the primate infant's gut in substantial amounts. There are several major classes of antibody—IgG, IgM, IgA, and IgE—but IgG is particularly important because it is the predominant antibody class in adults, accounting for about 80% of the antibody in circulation. It is also the predominant memory class, produced during exposure to viruses and bacteria. Thus it makes heuristic sense that IgG should be the only class of antibody to cross the placenta, and it provides an extended period of passive immunity in the infant before it decays between 3 and 6 months of age. This prenatal process is distinct from the more widely known transfer of secretory IgA that occurs in mother's milk. Secretory IgA serves a different role in preventing bacterial infections of the gut.

The transfer of maternal antibody has been examined in detail in both the squirrel monkey and the rhesus monkey (Coe, Cassayre, Rosenberg, & Levine, 1988; Scanlan, Coe, Latts, & Suomi, 1987). Infants in both species are born with relatively high titers of maternal IgG, which decline over the first 3 months of life. Squirrel monkey females are capable of transferring 40–49% of their IgG titers to the infant, while rhesus monkeys can pass up to 100% of their IgG levels to the full-term infant. The difference in the relative levels of maternal antibody reflects an interesting and significant improvement in the placental-transfer process as one looks across the Order Primates. There appears to be a progressive improvement in the transfer process coinciding with the evolution of a hemochorial placenta and longer gestation in the Old World monkey, and eventually the development of an active transport mechanism for IgG in humans (Kohler & Farr, 1966). In humans, the placental-transfer process is so highly developed that infant levels can reach 100–150% of maternal levels (Allansmith, McClellan, Butterworth, & Maloney, 1968), and infants can receive almost normal levels even when the mother's IgG has been pharmacologically suppressed during pregnancy (Cederqvist, Merkatz, & Litwin, 1977).

In contrast, in the monkey infant, a direct correlation was found between the mother's IgG level at term and her infant's titer. Data for 12 mother–infant pairs of squirrel monkeys are shown in Figure 9.12 and demonstrate that the mothers' levels at parturition are highly correlated with their neonates' levels. When infant titers were tracked across the first several months of life, neonatal levels were found to be strongly predictive of the levels for the next 3–4 months. Infants born with low levels, because of either low maternal values or prematurity, showed particularly low levels at 3 months of age. Work is now under way to evaluate whether these infants are at any health risk, as in the human condition of transient hypogammaglobulenemia (Hosking & Roberton, 1983). Three-month-old human infants with markedly low levels of IgG may show an increased incidence of bacterial infection, such as otitis media and bronchitis.

Parallel studies on rhesus monkeys have verified that birth levels of IgG are

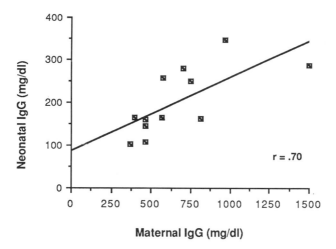

Figure 9.12.   Correlation between maternal and neonatal IgG levels in 12 squirrel monkey dyads.

highly correlated with levels shown over the first 4–6 months of life, prior to the onset of substantial synthesis of IgG by the infant (Scanlan et al., 1987). Further, in collaboration with Joseph Kemnitz, a series of blood samples from fetal rhesus monkeys (samples collected previously for another study) were evaluated to assess the rate of IgG transfer during pregnancy. In keeping with theoretical predictions, there was an accelerating increase in IgG transfer during the last weeks of pregnancy. In fact, the largest increase occurred in the last week of gestation, between days 155 and 162 postconception. Thus any infant born even 1–2 weeks before term would have been significantly deprived of the normal complement of maternal antibody, a deficiency that would have had consequences for the first several months of life. Similar observations have been made after premature births in humans, with significantly lower levels of IgG occurring in infants born before 33 weeks of gestational age or with birth weights below 2000 g (Berg, 1968).

GENERAL CONSIDERATIONS

While these studies on the placental transfer of maternal antibody are not directly related to the induction of maternal behavior, they do indicate the value of bringing different perspectives to bear on the study of mother–infant relationships. In particular, it is necessary to extend evaluations back to the complex relationship between the mother and the fetus during the prenatal period in order to understand the ready predisposition of the parturient female to care for her neonate. Humans clearly have a strong emotional investment in the infant before term, but animals have an equally important physiological investment that establishes the basis for parental care. This investment is particularly important in any species that gives birth to a single infant following a long gestation. Moreover, the placenta provides an important medium of exchange between mother and infant beyond the nutritional role usually ascribed to it. We have touched on only one immunological process, the placental transfer of antibody, but there are many other immunological questions of importance. For example, several hormonal changes the latter part of gestation are directly related to ensuring that the mother will not reject the fetus immunologically. There have been no long-term developmental studies on the possible effect of immunological incompatibility between the mother and her fetus. Further, studies at our laboratory and by others have shown that the nature of postnatal maternal care, the length of nursing, and the occurrence of mother–infant separations may have significant immunological consequences (Coe, Lubach, Ershler, & Klopp, 1988; Coe, Rosenberg, & Levine, 1988; Laudenslager & Reite, 1985).

Immunology is only one of many possible fields that can provide a useful perspective, in addition to the traditional emphasis on endocrinology and the neurosciences. While the study of the hormonal basis of maternal behavior in nonhuman primates has just begun, it is already clear that the simple application of a hormone model will be insufficient to account for maternal responsiveness. The existing studies on infant retrieval patterns of ovariectomized and menopausal rhesus females indicate that ovarian hormones are not a prerequisite for appropriate maternal responding in the primate. Further, hormonal changes associated with pregnancy and parturition are insufficient to override the effects of inexperience or abnormal rearing in the monkey.

It remains to be determined whether hormones serve a priming role in the primiparous female or enhance maternal responses in subsequent pregnancies. It is evident that the process of parturition itself can elicit a dramatic transformation in the responsiveness of a primiparous female monkey. Whether this is due to physiological changes associated with the process of labor or to the stimulus qualities of the neonate is still an open question. Oxytocin is a likely hormonal candidate for further study. It will also be necessary to go back and reevaluate the possible role of olfactory cues, since this factor was largely overlooked in the early primate studies. At a minimum, olfactory cues may facilitate the selective bonding in primates that forms the basis of the attachment relationship.

Another area that cries out for further study is the neurobiology of maternal behavior in nonhuman primates. The existing literature is based entirely on a handful of lesion/ablation studies showing that damage to limbic areas can disrupt normal maternal care (Steklis & Kling, 1985). While there will always be practical and ethical considerations in this type of research with nonhuman primates, modern techniques certainly permit at least correlational studies. Probably one of the greatest hurdles will be the enduring quality of maternal responsiveness once it has been established in the primate. Even in females reared under pathological conditions, there is a resiliency that allows for the improvement of maternal care and, once established, the maintenance of correct behavior over time (Dienske, van Vreeswijk, & Konig, 1979; Meier, 1965; Ruppenthal et al., 1976). This capacity would suggest a strong genetic basis for maternal behavior, with both heritability and penetrance. Under normal circumstances, it is supplemented by experiential factors based on several years of observational learning during the prepubertal stage. Maternal behavior is obviously so important to the survival of a species that it has been "overdetermined"—that is, driven by multiple behavioral and physiological systems. For that reason alone, it would be a mistake to expect any single variable to have exclusive primacy, and the analysis of maternal behavior will probably require the application of transactional models. There may be no overt antecedents prior to emergence of the infant and the expression of maternal behavior. One of the challenges for future researchers will be to develop holistic psychobiological models, rather than attempt to dissect maternal behavior in the traditional manner.

ACKNOWLEDGEMENTS

This research was supported by Grant MH-41659 from the NIMH to C. L. Coe and by Grant HD-02881 from the NICHD to S. Levine. C. L. Coe is partially supported by Grant RR–00167, and S. Levine has a USPHS Research Scientist Award (MH-19936) from the NIMH. This chapter is Publication No. 29-001 of the Wisconsin Regional Primate Center.

REFERENCES

Allansmith, M., McClellan, B. H., Butterworth, M., & Maloney, J. R. (1968). The development of immunoglobulin levels in man. *Journal of Pediatrics, 72*, 276–290.
Altmann, J. (1987). Life span aspects of reproduction and parental care in anthropoid primates. In J. B. Lancaster, J. Altmann, A. S. Rossi, & L. R. Sherrod (Eds.), *Parenting across the life span* (pp. 15–29). New York: Aldine de Gruyter.
Berg, T. (1968) Immunoglobulin levels in infants with low birth weights. *Acta Paediatrica Scandanavica, 57*, 369–376.
Bosu, W. T. K., Johansson, E. D. B., & Gemzell, C. (1973). Patterns of circulating oestrone, oestradiol-17β, and progesterone during pregnancy in the rhesus monkey. *Acta Endocrinologica, 74*, 743–755.
Brandt, E. M., & Mitchell, G. (1971). Parturition in primates: Behavior related to birth. In L. A. Rosenblum (Ed.), *Primate behavior: Developments in field and laboratory research* (Vol. 2, pp. 177–223). New York: Academic Press.
Bridges, R. S. (1977). Parturition: Its role in the long-term retention of maternal behavior in the rat. *Physiology and Behavior, 18*, 487–490.
Capitanio, J. P., Wiessberg, M., & Reite, M. (1985). Biology of maternal behavior: Recent findings and implications. In M. Reite & T. Field (Eds.), *The psychobiology of attachment and separation* (pp. 51–92). Orlando, FL: Academic Press.
Cederqvist, L. L., Merkatz, I. R., & Litwin, S. D. (1977). Fetal immunoglobulin synthesis following maternal immunosuppression. *American Journal of Obstetrics and Gynecology, 129*, 687–690.
Chandrashekar, V., Wolf, R. C., Dierschke, L. J., Sholl, S. A., Bridson, W. E., & Clark, J. R. (1980). Serum progesterone and corpus luteum function in pregnant pigtailed monkeys *(Macaca nemestrina). Steroids, 36*, 483–495.
Clewe, T. H. (1969). Observations on reproduction of squirrel monkeys in captivity. *Journal of Reproduction and Fertility, 6* (Suppl.), 151–156.
Coe, C. L., Cassayre, P., Rosenberg, L. T., & Levine, S. (1988). Effects of age, sex, and psychological disturbance on immunoglobulin levels in the squirrel monkey. *Developmental Psychobiology, 21*, 161–175.
Coe, C. L., Connolly, A. C., Kraemer, H. C., & Levine, S. (1979). Reproductive development and behavior in captive female chimpanzees *(Pan troglodytes). Primates, 20*, 571–582.

Coe, C. L., Glass, J. C., Wiener, S. G., & Levine, S. (1983). Behavioral but not physiological adaptation to repeated separation in mother and infant primates. *Psychoneuroendocrinology, 8,* 401–409.

Coe, C. L., Hayashi, K. T., & Levine, S. (1988). Hormones and behavior at puberty: Activation or concatenation. In M. Gunnar (Ed.), *Development during the transition to adolescence* (pp. 17–41). Hillsdale, NJ: Erlbaum.

Coe, C. L., Lubach, G. R., Ershler, W. B., & Klopp, R. G. (1988). Influence of early rearing on lymphocyte proliferation responses in juvenile rhesus monkeys. *Brain, Behavior, and Immunity 2,* 1–14.

Coe, C. L., Murai, J. T., Wiener, S. G., Levine, S., & Siiteri, P. K. (1986). Rapid cortisol and corticosteroid-binding globulin responses during pregnancy and after estrogen administration in the squirrel monkey. *Endocrinology, 118,* 435–440.

Coe, C. L., Rosenberg, L. T., & Levine, S. (1988). Immunological consequences of psychological disturbance and maternal loss in infancy. In C. Rovee-Collier & L. P. Lipsitt (Eds.), *Advances in infancy research* (pp. 97–134). Norwood, NJ: Ablex.

Coe, C. L., Wiener, S. G., Rosenberg, L. T., & Levine, S. (1985). Endocrine and immune responses to separation and maternal loss in nonhuman primates. In M. Reite & T. Field (Eds.), *The psychobiology of attachment and separation* (pp. 163–199). Orlando, FL: Academic Press.

Cross, H., & Harlow, H. (1963). Observation of infant monkeys by female monkeys. *Perceptual and Motor Skills, 16,* 11–15.

Deets, A. C., & Harlow, H. F. (1967). Adoption of single and multiple infants by rhesus monkey mothers. *Primates, 15,* 193–203.

Dienske, H., van Vreeswijk, W., & Konig, H. (1979). Adequate mothering by partly isolated rhesus monkeys after observation of maternal care. *Annual Report of the Primate Center* TNO, pp. 401–403.

Gibber, J. R. (1986). Infant-directed behavior of rhesus monkeys during their first pregnancy and parturition. *Folia Primatologica, 46,* 118–124.

Gibber, J. R., & Goy, R. W. (1985). Infant-directed behavior in young rhesus monkeys: Sex differences and effects of prenatal androgens. *American Journal of Primatology, 8,* 225–237.

Goodall, J. (1986). *The chimpanzees of Gombe.* Cambridge, MA: Belknap Press.

Goodlin, B. L., & Sackett, G. P. (1983). Parturition in *Macaca nemestrina. American Journal of Primatology, 4,* 283–307.

Hansen, E. W. (1966). The development of maternal and infant behavior in the rhesus monkey. *Behavior, 27,* 107–149.

Harlow, H. F., Harlow, M. K., & Hansen, E. W. (1963). The maternal affectional system of rhesus monkeys. In H. L. Rheingold (Ed.), *Maternal behavior in mammals* (pp. 254–281). New York: Wiley.

Harlow, H. F., & Zimmerman, R. R. (1959). Affectional responses in the infant monkey. *Science, 130,* 421–432.

Hartman, C. G. (1928). The period of gestation in the monkey, *Macacus rhesus,* first description of parturition in monkeys, size and behavior of the young. *Journal of Mammalogy, 9,* 181–194.

Higley, J. D., & Suomi, S. J. (1986). Parental behavior in non-human primates. In W. Sluckin & M. Herbert (Eds.), *Parental behavior* (pp. 152–207). Oxford: Basil Blackwell.

Hinde, R. A. (1983). *Primate social relationships.* Oxford: Blackwell Scientific.

Holman, S. D., & Goy, R. W. (1980). Behavioral and mammary responses of adult female rhesus to strange infants. *Hormones and Behavior, 14,* 348–357.

Horvat, J. R., Coe, C. L., & Levine, S. (1980). Infant development and maternal behavior in captive chimpanzees. In R. W. Bell & W. R. Smotherman (Eds.), *Maternal influences and early behavior* (pp. 285–309), New York: Spectrum.

Hosking, C. S., & Roberton, D. M. (1983). Epidemiology and treatment of hypogammaglob-ulinemia. *Birth Defects, 19,* 223–227.

Kaplan, J. N., & Cubicciotti, D. D. (1980). Early perceptual experience and the development of social preferences in squirrel monkeys. In R. W. Bell & W. P. Smotherman (Eds.), *Maternal influences and early behavior* (pp. 253–270) New York: Spectrum.

Klopfer, P., & Boskoff, K. (1979). Maternal behavior in prosimians. In G. Doyle and R. P. Martin (Eds.), *The study of prosimian behavior* (pp. 123–157). New York: Academic Press.

Kohler, P. F., & Farr, R. S. (1966). Elevation of cord over maternal IgG immunoglobulin: Evidence for an active placental IgG transport. *Nature, 210,* 1070–1071.

Lancaster, J. (1971). Play-mothering: The relations between juvenile females and young infants among free-ranging vervet monkeys *(Cercoptithecus aethiops). Folia Primatologica, 15,* 161–182.

Lanman, J. J. (1977). Parturition in nonhuman primates. *Biology of Reproduction, 16,* 28–38.

Laudenslager, M. L., & Reite, M. L. (1985). Loss and separations: Immunological consequences and health implications. In P. Shaver (Ed.), *Review of personality and social psychology* (pp. 285–311). Beverly Hills, CA: Sage.

Lee, P. C. (1983). Effects of parturition on the mother's relationship with older offspring. In R. A. Hinde (Ed.), *Primate social relationships* (pp. 134–138). Oxford: Blackwell Scientific.

Levine, S., Wiener, S. G., Coe, C. L., Bayart, F. E. S., & Hayashi, K. T. (1987). Primate vocalization: A psychobiological approach. *Child Development, 58,* 1408–1419.

Lindburg, D. G., & Hzell, L. D. (1972). Licking of the neonate and duration of labor in the great apes and man. *American Anthropologist, 74,* 318–325.

Line, S. W., Golub, M. S., Mahoney, C. J., Ford, E. W., & Anderson, J. H. (1986). Parturition: Studies using the conscious rhesus monkey. In P. W. Nathanielsz (Ed.), *Animal models in fetal medicine* (pp. 129–151). New York: Elsevier/North Holland Biomedical Press.

Lorenz, K. (1971). *Studies in animal and human behavior* (Vol. 2). London: Methuen.

Lundblad, E. G., & Hodgen, G. D. (1980). Induction of maternal–infant bonding in rhesus and cynomologous monkeys after caesarean delivery. *Laboratory Animal Science, 30,* 913.

Meier, G. W. (1965). Maternal behavior of feral- and laboratory-reared monkeys following the surgical delivery of their infants. *Nature, 206,* 492–493.

Mitchell, M. D., Mountford, L. A., Natale, R., & Robinson, J. S. (1980). Concentrations of oxytocin in the plasma and amniotic fluid of rhesus monkeys *(Macaca mulatta)* during the latter half of pregnancy. *Journal of Endocrinology, 84,* 473–478.

Moltz, H., & Kilpatrick, S. J. (1980). Pheromonal control of maternal behavior. In R. W. Bell & W. P. Smotherman (Eds.), *Maternal influences and early behavior* (pp. 135–154). New York: Spectrum.

Newman, J. D. (1985). Squirrel monkey communication. In L. A. Rosenblum & C. L. Coe (Eds.), *Handbook of squirrel monkey research* (pp. 99–126). New York: Plenum Press.

Novy, M. J., & Resko, J. A. (1981). *Fetal endocrinology.* New York: Academic Press.

Quiatt, D. (1979). Aunts and mothers: Adaptive implications of allomaternal behavior of nonhuman primates. *American Psychologist, 81,* 310–319.

Reyes, F. I., Winter, J. S. D., Faiman, C., & Hobson, W. C. (1975). Serial serum levels of gonadotrophins, prolactin, and sex steroids in the nonpregnant and pregnant chimpanzee. *Endocrinology, 96,* 1447–1455.

Rocklin, R. E., Kitzmiller, J. L., & Kaye, M. D. (1979). Immunobiology of the maternal–fetal relationship. *Annual Review of Medicine, 30,* 375–404.

Rosenblatt, J. S., & Siegel, H. I. (1980). Maternal behavior in the laboratory rat. In R. W.

Bell & W. P. Smotherman (Eds.), *Maternal influences and early behavior* (pp. 155–199). New York: Spectrum.

Rosenblum, L. A. (1968). Some aspects of female reproductive physiology in the squirrel monkey. In L. A. Rosenblum & R. W. Cooper (Eds.), *The squirrel monkey* (pp. 147–170). New York: Academic Press.

Rosenblum, L. A. (1971). Ontogeny of mother–infant relations in macaques. In H. Moltz (Ed.), *The ontogeny of vertebrate behavior* (pp. 315–368). New York: Academic Press.

Rosenblum, L. A. (1972). Sex and age differences in response to infant squirrel monkeys. *Brain, Behavior, and Evolution, 5,* 30–40.

Rosenblum, L. A.., & Alpert, S. (1977). Response to mother and stranger: A first step in socialization. In S. Chevalier-Skolnikoff & F. E. Poirier (Eds.), *Primate biosocial development* (pp. 463–478). New York: Garland.

Rosenblum, L. A., & Youngstein, K. P. (1974). Developmental changes in compensatory dyadic response in mother and infant monkeys. In M. Lewis & L. A. Rosenblum (Eds.), *The origins of fear: The influence of the infant on the caregiver* (pp. 141–161). New York: Wiley.

Ruppenthal, G. C., Arling, G. L., Harlow, H. F., Sackett, G. P., & Suomi, S. J. (1976). A 10-year perspective of motherless-mother monkey behavior. *Journal of Abnormal Psychology, 85,* 341–349.

Sackett, G. P., & Ruppenthal, G. C. (1974). Some factors influencing the attraction of adult female macaque monkeys to neonates. In M. L. Lewis & L. A. Rosenblum (Eds.), *The effect of the infant on the caregiver* (pp. 163–186). New York: Wiley.

Scanlan, J., Coe, C. L., Latts, A., & Suomi, S. (1987). Effects of age, rearing and separation stress on immunogloblin levels in rhesus monkeys. *American Journal of Primatology, 13,* 11–22.

Shively, C., & Mitchell, G. (1986a). Perinatal behavior of prosimian primates. In G. Mitchell & J. Erwin (Eds.), *Comparative primate biology: Behavior, conservation and ecology* (Vol. 2A, pp. 217–243). New York: Alan R. Liss.

Shively, C., Mitchell, G. (1986b). Perinatal behavior of anthropoid primates. In G. Mitchell & J. Erwin (Eds.), *Comparative primate biology: Behavior, conservation & ecology* (Vol. 2A, pp. 245–294). New York: Alan R. Liss.

Sholl, S. A., Robinson, J. A., & Wolf, R. C. (1979a). Progesterone and 5-pregnane-3,20-dione in serum of peripartum rhesus monkeys. *Endocrinology, 104,* 329–332.

Sholl, S. A., Robinson, J. A., & Wolf, R. C. (1979b). Estrone, 17β-estradiol, and cortisol in serum of peripartum rhesus monkeys. *Endocrinology, 104,* 1274–1278.

Siiteri, P. K., Febres, F., & Clemens, L. E. (1977). Progesterone in the maintenance of pregnancy: Is progesterone nature's immunosuppressant? *Annals of the New York Academy of Science, 286,* 384–397.

Smith, P. E. (1954). Continuation of pregnancy in rhesus monkeys *(Macaca mulatta)* following hypophysectomy. *Endocrinology, 55,* 655–664.

Snowdon, C. T., Coe, C. L., & Hodun, A. (1985). Population recognition of infant isolation peeps in the squirrel monkey. *Animal Behaviour, 33,* 1145–1151.

Steklis, H. D., & Kling, A. (1985). Neurobiology of affiliative behavior in nonhuman primates. In M. Reite & T. Field (Eds.), *The psychobiology of attachment and separation* (pp. 93–134). Orlando, FL: Academic Press.

Tinklepaugh, O. L., & Hartman, K. G. (1930). Behavioral aspects of parturition in the monkey *(Macaca rhesus). Comparative Psychology, 11,* 63–98.

Tinklepaugh, O. L., & Hartman, K. G. (1932). Behavior and maternal care of the newborn monkey *(Macaca mulatta). Journal of Genetic Psychology, 40,* 257–286.

Vogt, J. L., Coe, C. L., Lowe, E. L., & Levine, S. (1980). Behavioral and pituitary-adrenal responses of adult squirrel monkeys to mother–infant separation. *Psychoneuroendocrinology, 5,* 181–190.

Walsh, S. W., Wolf, R. C., Meyer, R. K., Aubert, M. L., & Friesen, H. G. (1977). Chorionic gonadotropin, chorionic somatomammotropin, and prolactin in the uterine vein and peripheral plasma of pregnant rhesus monkeys. *Endocrinology, 100,* 851–855.

Wamboldt, M. Z., Gelhard, R. E., Insel, T. R. (1988). Gender differences in caring for infant *Cebuella pygmaea:* The role of infant age and relatedness. *Developmental Psychobiology, 21,* 187–202.

Wasser, S. K., & Starling, A. K. (1986). Reproductive competition among female yellow baboons. In J. G. Else & P. C. Lee (Eds.), *Ontogeny, cognition and social behavior of primates* (pp. 343–354). Cambridge: Cambridge University Press.

Weiss, G., Butler, W. R., Hotchkiss, J., Dierschke, D. J., & Knobil, E. (1976). Periparturitional serum concentrations of prolactin, the gonadotropins, and the gonadal hormones in the rhesus monkey. *Proceedings of the Society for Experimental Biology Medicine, 151,* 113–116.

10

Hormonal and Experiential Correlates of Maternal Responsiveness in Human Mothers

ALISON S. FLEMING

Considerable research has been done on the psychobiology of maternal behavior in rodents and other mammals. While there is an extensive literature on social and psychological factors mediating human mothering (Belsky, 1984), relatively few studies have focused on the kinds of psychobiological influences known to be important in other species. This chapter delineates some of the hormonal and experiential factors that influence maternal responsiveness in human mothers, highlighting areas of similarity between humans and the animal models. Because of space limitations, relevant animal work will be noted but not described in detail.

While human maternal behavior is considerably more complex in its organization and control than are functionally similar behaviors in other mammals, some important commonalities may also exist. For many mammals, including humans, prior experiences in interacting with young facilitate the subsequent expression of maternal behavior; however, the relevant features of the experience (whether the learning of specific infant cues or of specific movements) and the temporal parameters of that learning (the optimal time, duration, and retention characteristics) may differ widely from species to species. Similarly, although parturitional hormones influence the expression of maternal behavior in a number of species, the hormones involved and their mode of action may differ across species. Thus species-specific behaviors that have similar functions and similar general features or phenotypes may be controlled in quite different ways.

This chapter attempts to evaluate some of the similarities and differences in the factors regulating maternal responsiveness in human and nonhuman mothers. It will be argued that most new mothers are motivated or primed to respond nurturantly to their newborn infants. The priming derives from the mother's physiological state during the periparturitional periods, as well as cultural and background factors that promote a positive attitude toward infants and mothering. If mothers have sufficient interaction with infants during the early postpartum period of heightened receptivity to infants, experiences derived during the interaction increase the likelihood that mothers will continue to respond in a nurturant fashion at a later point when physiological or physical influences associated with the newborn period are no longer present.

Although it is assumed that physiological or physical factors can facilitate the early postpartum expression of maternal responsiveness, these factors are neither necessary nor sufficient for the appearance of these behaviors. Siblings, fathers, adoptive parents, and adults who care for children (nurses, day-care workers, relatives, etc.) can become as attached to infants as biological mothers and, in some cases, as rapidly as or more so. Moreover, in the absence of a positive psychological predisposition to be a mother or respond nuturantly to infants, the hormonal events will not *create* nurturant feelings or behavior in the new mother. Rather, the underlying assumption of this chapter is that the events of pregnancy and the early postpartum period may—given appropriate background and proximal conditions—facilitate the process of attachment to the offspring and make the transition from nonmother to mother somewhat easier.

This chapter first considers the role of pregnancy and early postpartum hormones in the onset of maternal responsiveness. This discussion is followed by an examination of the role of early postpartum experiences with the young in subsequent maternal responsiveness. Finally, the influences of prior maternal experience and parity on the mother's initial and later reactions to infants and infant-related cues are evaluated.

MATERNAL RESPONSIVENESS IN WOMEN: A PROBLEM OF DEFINITION

Perhaps the most fundamental issue concerning human maternal responsiveness is definitional. What measures of maternal responsiveness adequately capture the strong emotional bond or tie that motivates mothers to maintain close proximity to their infants and to nurture and protect them in ways that promote their survival and optimize their healthy development (Ainsworth, 1979; Robson & Moss, 1970)? A cross-cultural survey of mothering practices indicates that although the biological mother is typically the primary caretaker, and in that role usually breast-feeds her infant (Leiderman & Leiderman, 1977), there is enormous variability in how mothers interact with their infants. In some cultures, mothers provide the infants with considerable tactile and kinesthetic stimulation by carrying them on their bodies for most of the day (Brazelton, 1977; Konner, 1977; Leiderman & Leiderman, 1977); in others, infants are left alone in a cradle for much of the time (Caudill & Weinstein, 1969; Moss, 1967); in some colder climates, infants are swaddled and placed on cradle boards, in which case skin-to-skin contact is reduced (Whiting, 1981); in more tropical climates, infants are lightly clothed and free to move their limbs (Whiting, 1981). Although there are some commonalities in how mothers vocalize to their infants (as in the "motherese" pattern described by Fernald, 1984), mothers in some cultures do not address any of their vocalizations directly to the infant (Tronick, 1987), whereas in others, they do not look directly at them or do not look at them "en face" (Brazelton, 1977; Klaus, Kennell, Plumb, & Zuehkle, 1970). Moreover, some of the same types of variability observed across cultures often occur in Western industrialized societies and are likely to be represented in studies of human mothering in our culture.

It appears, then, that there is no single set of behaviors, attitudes, or physiological responses that consistently reflects a high level of maternal motivation, analogous to the species-characteristic retrieval, nursing, and licking responses shown by the

Rattus norvegicus mother (Wiesner & Sheard, 1933). For our studies of human maternal responsiveness, three qualitatively different classes of maternal responses were used: behavioral, self-report, and psychophysiological. Although each has its limitations and they are often uncorrelated, each provides important information about the mother's maternal motivation and represents an index of maternal responsiveness. The behavioral and self-report measures are necessarily culture bound, reflecting modes of responsiveness in women in Western industrialized nations. This chapter focuses primarily on our behavioral and self-report work, although some discussion of the psychophysiological work from other laboratories will be included.

Maternal Behavior

All our studies of maternal behavior involve observations of new mothers interacting with their own infants in a feeding or nursing context. In one study, pregnant mothers were observed while interacting in a nonfeeding situation with an unfamiliar infant. In general, 15-min continuous mother–infant interactions were observed, and the following behaviors were recorded, either continuously with a computer-based event recorder or time-sampled with a 1–0 paper-and-pencil procedure. Affectionate responses include such behaviors as patting, stroking, cuddling, or kissing the baby; vocal responses include talking and cooing to the baby or singing; the en face response consists of visual orientation to the baby. Together, the affectionate and vocal responses constitute approach responses. (The en face trait was excluded from the approach category because there was very little variability in this measure, since most mothers looked at their infants throughout the feeding session.) Mothers' instrumental caretaking responses include burping their infants, changing diapers, adjusting position, and so on. In general, most of the behavioral effects in our studies are evident in relation to approach responses as opposed to caretaking responses, as the former category is felt to be the more sensitive index of underlying maternal responsiveness (e.g., Belsky, Rovine, & Taylor, 1984; Hales, Lozoff, Sosa, & Kennell, 1977).

Maternal Attitudes

In many of our studies, information was obtained on mothers' expressed feelings and attitudes. Mothers filled out extensive questionnaires containing more than 100 attitude items on 7-point Likert-type scales reflecting a range of feelings from "strongly disagree" to "strongly agree." These items have been factor analyzed and separate factors derived reflecting aspects of maternal responsiveness. They include sets of attitude items on the mother's feeling of warmth or closeness to her own fetus or infant (attachment). Another factor is concerned with the mother's feelings about infants in general (children); a third, with her feelings about caretaking activities (caretake); and a final factor, with her feeling of adequacy in the maternal role (maternal adequacy). These attitude clusters tend to be highly intercorrelated in women before the birth of their infants (at 9 months, $rs = .51$ to $.73$) but weakly correlated after some postpartum experience ($rs = .14$ to $.51$). Since these attitudes show a different intercorrelation pattern at the different times, they are assumed to reflect qualitatively different underlying feelings.

Physiological Responses to Infant Cues

The final set of measures are mothers' heart rate and skin conductance responses to a videotape or audiotape of an infant laughing or crying. In some cases, the infant is the woman's own; in others, it is unfamiliar. Although there is some disagreement about the meaning of acceleratory versus deceleratory heart rate responses (Lacey & Lacey, 1970; Obrist, 1976), most of the studies reviewed here have assumed that an acceleratory response to the infant stimulus reflects a heightened state of arousal, emotion, or readiness for action, whereas deceleration reflects more passive attention.

PHYSIOLOGICAL DETERMINANTS OF MATERNAL RESPONSIVENESS

The following two sections describe a series of studies that attempt to identify possible physiological or hormonal correlates of maternal responsiveness. The first approach to this issue was to determine whether nulliparous but pregnant women undergo a growth in mother responsiveness as pregnancy advances and prior to any explicit contact with the young, as has been demonstrated in a number of other species (Lott & Rosenblatt, 1969; Richards, 1966; Siegel, Clark, & Rosenblatt, 1983; Siegel & Rosenblatt, 1975a), including some of the primates (Rosenblum, 1972), although not all (Gibber, 1986). The second approach was to determine whether changes in maternal responsiveness are temporally related to and/or correlated with hormonal changes (Fleming, Krieger, & Wong, in press).

Gestational Changes in Maternal Responsiveness

Self-report data were obtained from a group of 29 Caucasian, middle-class, first-time mothers recruited through prenatal classes or by physician referrals in Ontario, Canada. These women filled out questionnaires four times during pregnancy, immediately after the birth on day 2 or 3 postpartum, and then again at 6 weeks postpartum. At these times, blood samples were also obtained on which radioimmunoassays were done of some of the hormones known to be involved in the periparturitional onset of maternal behavior in other animals (Rosenblatt & Siegel, 1981), including estradiol, progesterone, and testosterone, as well as the binding proteins, sex-hormone-binding globulin, and albumin, which permit an estimate of free versus bound fractions of some of these steroids. Cortisol and some of the protein hormones were also assayed. Unfortunately, plasma oxytocin or the β-endorphins were not assayed, and the prolactin radioimmunoassays are not yet completed.

Using multivariate analysis of variance (ANOVAs), with repeated measures on the time factor, no change was found across pregnancy in women's feelings about other people's children, about caretaking activities, or about maternal adequacy. However, as shown in Figure 10.1, an increase was found from the beginning to the end of pregnancy in women's reported feelings of attachment to their own fetuses/ infants, which occurred primarily between 3 and 5 months; such feelings changed most precipitously, however, after the birth and after mothers had interacted with their infants.

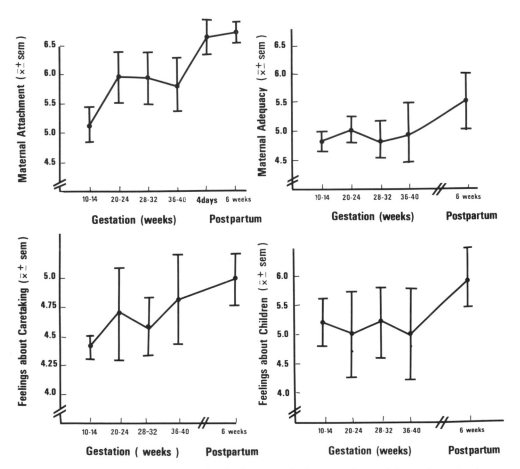

Figure 10.1. Changes in maternal attitude scores during gestation and the early postpartum period in first-time mothers. Ordinates reflect responses on Likert-type scales ranging from 1 to 7. Higher scores reflect more positive attitudes.

These observations are consistent with Leifer's (1977, 1980) observations that most mothers who initially have negative attitudes toward being pregnant come to feel positively at about 5 months, when fetal movements are first detected. These results are also consistent with observations by Bleichfeld and Moely (1984), who found that nulliparous but pregnant women show a more variable heart rate response to the pain cry of an unfamiliar baby than do nonpregnant women, who show either no change or a deceleration; in contrast, under some circumstances, pregnant women show the same pattern of heart rate acceleration as seen in the new mother. Despite these self-report and psychophysiological indications of heightened maternal feelings during pregnancy, there are some behavioral data that suggest that, in comparison with nonpregnant women, pregnant women do not respond more maternally to unfamiliar infants in a waiting-room situation and do not spend more time looking at pictures of infants or other infant-related themes (Feldman & Nash, 1978). What was unclear in this study was the stage of pregnancy at which women were assessed.

Similar assessments very late in the pregnancy may well produce a different outcome from first-trimester assessments.

Hormonal Correlates of Maternal Responsiveness

This section considers the extent to which hormonal changes of pregnancy are associated with the heightened feelings of attachment expressed by pregnant mothers. Most of the hormones that were measured underwent an increase across pregnancy, followed by precipitous declines only at parturition (see Batra & Grundsell, 1978; Fleming & Corter, 1988; Nott, Franklin, Armitage & Gelder, 1976; Willcox, Yovich, McColm, & Schmitt, 1985). In some important respects, these hormonal profiles differ from those reported in rats, which undergo a substantial prepartum decline in progesterone associated with elevations in plasma concentrations of estrogen and prolactin (reviewed by Rosenblatt & Siegel, 1981). Human parturition is also accompanied by large increases in the adrenocorticotropic hormone (ACTH) and adrenal corticoids, as reflected in the large rise in cortisol concentrations, which also remain high for a period during the puerperium (Jolivet, Blanchier, Gautray, & Dhem, 1974).

The decision to concentrate on the association of the pregnancy steroids with maternal responsiveness was based primarily on the extent of the animal literature. Studies on rats, mice, and ewes indicate that the periparturitional rise in maternal behavior is promoted by a rise in estrogen at this time (Moltz, Lubin, Leon, & Numan, 1970; Poindron & Le Neindre, 1980; Siegel & Rosenblatt, 1975a, 1975b). Studies on rats indicate further that a prior progesterone decline (Bridges, Rosenblatt, & Feder, 1978; Siegel & Rosenblatt, 1978) and an estrogen-induced rise in prolactin (Bridges, 1984; see also chapter 6) may also contribute to the onset of maternal behavior. Finally, at least one rat study suggests that removal of the adrenal gland prevents the normal expression of postpartum retrieval behavior (Hennessy, Harney, Smotherman, Coyle, & Levine, 1977).

In order to determine whether individual differences in these hormones are related to differences in maternal attitudes, correlation coefficients between the various maternal constructs and the hormones at each of the five time points were computed. Using the arbitrary criterion of a significant correlation coefficient between the same two variables on at least two time points, no consistent association was found between any hormone or hormone ratio and any of the maternal-attitude constructs during various phases of pregnancy or at 6 weeks postpartum.

These results suggest that there is no relation between hormones and maternal responsiveness in human mothers. When one considers a behavioral measure of responsiveness rather than attitude measures, a somewhat different situation obtains. In a subsequent study (Fleming, Steiner, & Anderson, 1987), 30 first-time mothers filled out the same maternal-attitude questionnaires at the end of pregnancy and again at 3–4 days and 6 weeks postpartum. Mother–infant interactions were observed on days 3 and 4 and at 2 months postpartum while mothers were feeding their infants. Blood samples were obtained on days 3–4 postpartum, immediately before and 1 hr after the feeding session, and were assayed for the same set of hormones described earlier, although in this study prolactin but not the binding globulins were assayed.

Consistent with our earlier observations, no relation was found between puerperal concentrations of any of the hormones and mothers' feelings or attitudes at any

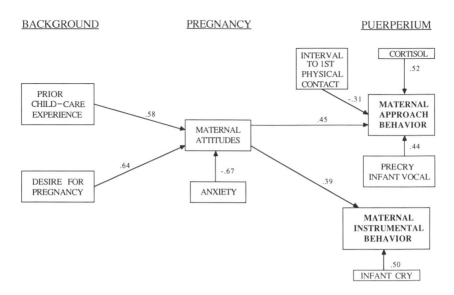

Figure 10.2. Schematic of relations among background factors, pregnancy attitudes, puerperal factors, and maternal behaviors. (Directions of arrows indicate patterns of predictors, established through multiple-regression analyses. Numbers on arrows refer to correlation coefficients between the variables that were significant [rs].) Positive numbers reflect positive relations; negative numbers reflect negative relations.

time point. However, a significant correlation ($r = .53$; $p < .05$) was found between concentrations of plasma cortisol both before and after the nursing bout and mothers' affectionate and vocal interactions with their infants on days 3–4 postpartum. Thus mothers who had higher plasma levels of cortisol spent more time being physically affectionate and talking to their infants than did mothers with lower levels of this hormone.

Of interest, however, were the additional findings that this association was strengthened if mothers also had positive maternal feelings and attitudes during pregnancy. A hierarchical regression analysis in which pregnancy maternal-attitude scores were entered first, followed by cortisol levels, showed that the hormones made a significant contribution to mothers' approach behavior over and above that made by mothers' attitudes. Together, the two variables accounted for 42% of the variance in mothers' approach behaviors. In contrast, neither maternal attitudes nor cortisol concentrations significantly predicted the amount of time mothers spent on caretaking activities during the puerperium. The associations between mothers' pregnancy attitudes, postpartum hormones, and maternal behaviors are schematized in Figure 10.2.

Thus simple correlational procedures do not reveal an association between hormones and any self-report measure of maternal responsiveness. Although an association has been demonstrated between hormones and maternal behavior, without a replication of the study at this stage we do not know whether the correlation reflects a real association or is simply a chance effect. Assuming that the correlation reflects a real relation, we nevertheless have no evidence that cortisol directly influences behavior. It may instead be associated with personality variables or pregnancy or childbirth events that more directly influence behavior. Moreover, the hormonal effect, if it does exist, seems to be limited to the early postpartum period. There were no associations between the concentrations of earlier puerperal hormones and mothers' maternal behavior at 2 months postpartum.

BEHAVIORAL MECHANISMS OF HORMONAL ACTION DURING THE PUERPERIUM

Although the case for a relation between cortisol concentrations and maternal behavior requires substantiation, the following discussion of the variety of ways in which these hormones could influence behavior is included because it illustrates an approach or a way of looking at hormone–behavior relations that could apply more generally. Specifically, the approach assumes that hormones do not activate behavior directly but that they exert multiple and often diffuse effects, which together increase the probability that the behavior will be expressed.

How might cortisol (or, indeed, any hormone) act to augment mothers' maternal responsiveness during the puerperium—that is, the period immediately after birth? In our rat studies, it was suggested that the periparturitional hormones that activate maternal behavior do so not only by acting directly on the maternal behavior substrates, but also by acting indirectly, possibly on other limbic structures (Fleming, Vaccarino, & Luebke, 1980), to reduce the animal's timidity and natural neophobic tendencies that would otherwise be activated by the strangely configured newborn pups (Cohen & Bridges, 1981; Fleming, 1986; Fleming & Luebke, 1981). These hormones also induce an attraction to pup-related odorants (Fleming, 1986) and promote long-term retention of experiences acquired in interaction with pups (Bridges, 1975, 1977; Cohen & Bridges, 1981).

In the same way, it may be argued, cortisol (or possibly other parturitional hormones) released in response to the "stress" of parturition (Ramsay, 1980) does not directly activate positive maternal behaviors but may do so indirectly by increasing the mother's general state of arousal, her responsivity to infant-related cues, and the extent to which she is influenced by interactions with her infant. There are a number of lines of evidence that are consistent with these notions.

It is known, for instance, that at times outside the puerperium, cortisol does have arousal-inducing effects (Mason, 1968; but see Curtis, Buxton, Lippman, Ness, & Wright, 1976), although whether it is experienced as depression or elation depends, in part, on the individual's personality and, presumably, other situational factors (von Zerssen, 1976). Although the underlying hormonal mechanisms have yet to be worked out (Handley, Dunn, Waldron, & Baker, 1980; Nott et al., 1976; Steiner, Fleming, Anderson, Monkhouse, & Boulter, 1986), there is also evidence that new mothers are in a heightened state of arousal and tend to be labile in their moods; some become hypomanic, whereas others experience periods of "blues" or depression (Cutrona, 1982; Hopkins, Marcus, & Campbell, 1984). Although there is no direct evidence to support an arousal interpretation, one could predict that the new mother's heightened arousal state would energize her to show elevated responsiveness to her infant. Such responsivity would be translated into approach in mothers with positive maternal attitudes (which was the case for most of the women in our sample). However, the same heightened arousal could be translated into more negative responses to the infant—as indifference or active rejection for mothers with highly negative attitudes or in at-risk situations. In short, a greater range of maternal attitudes in our sample of women may well have yielded strong interactions between attitudes and hormones on maternal behavior.

There is also indirect support for the contention that cortisol could increase the mother's responsivity to infant-related cues. Exogenous treatment with adrenocorticoids for patients with hypoadrenal conditions increase their ability to recognize and distinguish among different odor and taste qualities (Henken, 1975), and can increase olfactory acuity and recognition performance in patients suffering from olfactory dysfunctions (Goodspeed, Gent, Catalanotto, Cain, & Zagraniski, 1986). Other preliminary evidence indicates, further, that during the first 3 postpartum days (when cortisol levels are still elevated), new mothers also show increased olfactory acuity for neutral odorants (Schaal, 1986), which, one assumes, would also generalize to infant-related odorants. Consistent with this interpretation are observations by Schaal (1986), Schaal et al. (1980), and Porter, Cernock, and McLaughlin (1983) that indicate that new mothers with as little as 2 hr of prior contact with their infants are able to distinguish, at better than chance levels, their own infant's odors from those of other same-age infants. Whether mothers are better than fathers or nonparturient women at this recognition task and whether the effect is specific to infant-related odors is not known. A final possible role for cortisol (or, more likely, ACTH) during the puerperium may be to facilitate mothers' ability to recognize the distinctive odor, vocal, and other characteristics of their young that would promote subsequent interactions with their infants (Dunn, 1984). In this case, mothers would require less exposure to infant-related cues than would fathers or nonparturient women to produce the same level of recognition at a later time. Support of this idea comes from findings that exogenous ACTH 4-10 in animals retards the extinction of both appetitive learning and conditioned avoidance responses (Dunn, 1984). It would be extremely interesting to determine whether new mothers exhibit more rapid learning of infant cues than do fathers or nonparents, whether the learning is easier to infant- than to noninfant-related cues, and whether periparturitional hormones exert an effect on this learning.

POSTPARTUM EXPERIENCE AND THE RETENTION
OF MATERNAL RESPONSIVENESS

Some indirect support exists for the idea that, as in a number of other species (Bridges, 1975, 1977; Orpen & Fleming, 1987; Poindron & Le Neindre, 1980), the early postpartum period is the optimal time for the elicitation of human maternal behavior and that experiences in interacting with young during this time are quickly consolidated by the mother. It is also assumed that it is this experience, rather than hormones, that sustains subsequent responsiveness. Gaulin-Kremer, Shaw, and Thoman (1977) observed maternal and infant behaviors in 28 mother–infant pairs during mothers' first prolonged contact after delivery and found that the closer to delivery such interactions occurred, the more the mother caressed, talked to, and held her infant before actual feeding. Moreover, Schaal (1986), Schaal et al. (1980), Porter et al. (1983), Formby (1967), and others have found that with very little prior experience, mothers are able to identify correctly the odors and vocalizations of their own infants in the context of comparable cues from other newborn infants.

Schaal (1986) and Schaal et al., (1980) tested mothers daily for the first 10 postpartum days on the ability to recognize their own infants body odors (which had accumulated on a newborn T-shirt) from among odorants taken from four other same-age infants. They found that as early as the second postpartum day, many mothers were able to perform at better than chance levels, and they showed marked improvement over days. Porter et al. (1983) found that as little as 2 hr of contact with the infant resulted in successful recognition. Other evidence also indicates that the experience of interacting with infants within the first few hours after birth produces heightened responsiveness, at least for a brief period (Grossman, Thane, & Grossman, 1981; Hales et al., 1977; Hopkins & Vietze, 1977), and may even have longer term effects on such specific behaviors as breast-feeding (Sosa, Klaus, Kennell, & Urrutia, 1976), although not on affectionate behaviors in general (Grossman et al., 1981). In one representative study, Grossman et al. (1981) compared maternal behavior on each of the first 10 postpartum days as a function of the amount of maternal–infant contact during the first few postpartum days. Mothers who had an additional 30 min of contact on the delivery table, either with or without subsequent extended rooming-in, showed heightened affectionate behavior toward their infants during the first 3 days, in comparison with mothers receiving only extended rooming-in starting on the day following birth or with mothers receiving routine hospital contact procedures. By day 10 postpartum, however, after all groups had had additional experience with their infants, early contact advantages were no longer apparent. In many of our own studies, significant associations have been found between the interval to mothers' contact with their infants and the amount of affectionate behavior exhibited during the first few postpartum days (Fleming et al., 1987; Fleming, Ruble, Flett, & Shaul, 1988). Consistent with the Grossman et al. study, however, the correlation does not continue for long and is absent by 6 weeks postpartum, by which time all mothers have had substantial experience.

These studies indicate that very early physical contact with the infant may promote elevated responsiveness during the initial postpartum days, but that the effects are short-lived and can be masked by subsequent interactions with the infant (see Goldberg, 1983; Lamb, 1982). Whether the experience of additional early contact would prevent the decline in responsiveness associated with complete and extended postpartum separation, as has been demonstrated in rats (Bridges, 1975, 1977; Cohen & Bridges, 1981; Jakubowski & Terkel, 1986; Orpen & Fleming, 1987; Stern, 1983), is not known. There is some evidence, however, that maternal experiences acquired during a previous parity can prevent the loss of self-confidence that otherwise occurs during a period of separation from premature infants (Seashore, Leifer, Barnett, & Leiderman, 1973).

Whether women have a postpartum sensitive period analogous to that in sheep (Poindron & Le Neindre, 1980), whether mothers in this period are particularly susceptible to the influence of infant-related cues, and what role hormones play in this process are poorly understood. Part of the limitation in our knowledge arises from the paucity of studies adequately designed to test the sensitive-period hypothesis, including the appropriate controls for time of onset of first contact, duration of first contact, quality of first contact, and time of assessment of effects of first contact (Goldberg, 1983). Other problems plaguing this early contact research, as discussed by Lamb (1982), include the analyses of a large number of outcome measures and results in only a few, the noncomparability of affected measures from study to study, the expectations and biases of the investigators due to the nonblind evaluations, and the nonrandom formation and differential treatment of groups. Despite these constraints, it is striking that no studies have found a reduction in responsiveness with additional contact during and after the postpartum period.

There is also very little information on what constitutes the relevant feature(s) of the early contact. It could be simple exposure to odors, vocalizations, or other exteroceptive cues, as in the ewe (Poindron & Le Neindre, 1980); or it could be that, as in the rat, mothers have to be in actual physical contact with the young and interact with them (Jakubowski & Terkel, 1986; Orpen & Fleming, 1987; Stern, 1983). The weight of current evidence suggests that actual interactive contact is probably important. In most early contact manipulations or situations, the mother either has experienced skin-to-skin contact with her infant (de Chateau & Wiberg, 1977; Grossman et al., 1981; Hales et al., 1977); or has cradled and held her lightly clothed infant (Schaal et al., 1980). In order to give mothers privacy, mother–infant pairs are typically left alone, precluding direct observation of the duration of physical contact or the kinds of interactions that take place (Goldberg, 1983).

DEVELOPMENT OF MATERNAL RESPONSIVENESS
OVER THE POSTPARTUM PERIOD

As mothers gain experience with their infants during the first few postpartum months, most come to feel increasingly attached to them; they express more positive attitudes, become more efficient at tasks such as feeding (Thoman, Barnett, & Leiderman, 1971; Thoman, Turner, Leiderman, & Barnett, 1970), and are more attuned to infant signals (Sagi, 1981). Similar experiential effects on retrieving efficiency (Carlier & Noirot, 1966) and reinduction latencies (Cohen & Bridges, 1981; Fleming & Rosenblatt, 1974) have been reported in mother rats, as well as in sheep (Poindron & Le Neindre, 1980).

As can be seen in Figure 10.1, which is based on our initial pregnancy questionnaire study (Fleming, Krieger, et al., in press), from the end of pregnancy to 6 weeks postpartum, primiparous mothers come to like caretaking activities more, report liking children more, feel more adequate, and feel more attached to their infants. In another study, we have also found that in comparison with their responses during pregnancy, by 3 months postpartum mothers feel themselves to be more knowledgeable, less traditional, and more fun-loving and nurturant as mothers (Fleming et al., 1988). Unfortunately, one problem with these questionnaire measures is that by the 6- to 9-week measurements, women's scores have essentially reached asymptote on a number of the constructs. In a subsequent study with somewhat different self-report measures of maternal responsiveness (i.e., on interviews), Fleming, Ruble, Flett, and

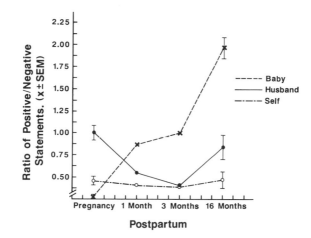

Figure 10.3. Changes in the ratio of positive to negative statements made in reference to the baby, the husband, and self across the first 16 postpartum months among first-time mothers.

Van Wagner (in press) demonstrated that emotional involvement with the infant continues to develop through the first postpartum year and probably longer.

In this study, 68 first-time mothers filled out questionnaires, were observed interacting with their infants, and were interviewed at 9 months of pregnancy, 3 days postpartum, and 1, 3, and 16 months postpartum. The interviews were semistructured, with open-ended questions about the women's feelings about themselves, their infants, and their husbands. They could say as much or as little about each topic as they chose. Using the Gottschalk–Gleser procedures for content analysis (Gottschalk & Gleser, 1969), the interviews were tape-recorded, transcribed verbatim, and coded by a number of independent coders for the frequency of positive and negative phrases made in reference to each of the content categories. All intercoder reliabilities were above 70%. As shown in Figure 10.3, the ratio of positive to negative phrases made in reference to the infant increases substantially across the first postpartum year. Thus by the end of the first year, mothers spend a larger proportion of the interview time talking about their infants, and the underlying affect becomes more and more positive.

Changes in maternal behaviors over this period also occur. In our longitudinal study (Fleming et al., 1988), a subset of mothers was observed during pregnancy in interaction with a 3-month-old infant who was not their own and again during the postpartum period with their own infant (Fleming, Corter, & Ruble, unpublished data). The pregnancy observations took place in a simulated waiting room and were done unobtrusively after the prospective mother agreed to care for a 3-month-old (in nonfeeding interactions) infant for a 5-minute period so that the natural "stooge" mother could briefly leave the room. The prospective mothers were free to leave the infant in the infant seat or to pick it up and hold it. The postpartum observations took place in the mother's home and occurred during a feeding session as well as during a 5-min nonfeeding interaction.

A comparison between pregnancy and 3 months postpartum in the way mothers

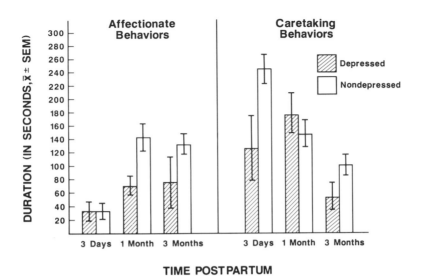

Figure 10.4. Maternal affectionate and caretaking behaviors shown by depressed and non-depressed mothers during the first 3 postpartum months.

interact with 3-month-old infants shows that mothers engage in more caretaking and en face responses with their own infants at 3 months postpartum, but less verbalization compared with their interactions during pregnancy with another 3-month-old. Somewhat surprisingly, however, there are no differences in the amount of affectionate behavior shown in the two situations.

Since the two observation settings were so dissimilar, it is difficult to conclude from these data alone that maternal responsiveness increases from the pregnancy to the postpartum period. However, comparisons of mother–infant interactions under essentially identical conditions at the first three postpartum time points indicate clear changes in behavior (Fleming et al., 1988). As shown in Figure 10.4, from 3 days postpartum to 1 and 3 months postpartum, there are striking increases in the amount of time devoted to affectionate behavior and corresponding reductions in the time for caretaking activities. Thus at 3 days postpartum, mothers are relatively inefficient at nursing and spend most of their time in nursing-related activities. By 3 months postpartum, when they have become more efficient at feeding and are more at ease, they spend more time stroking and interacting affectionately with their infants (Fleming et al., 1988). However, as shown in Figure 10.4 and described below, depression in mothers reduced the affectionate interactions.

In line with changes in maternal attitudes and maternal behavior, more experienced mothers also become more responsive to infant cues. Parents and experienced nonparents (e.g., midwives) are better able than inexperienced women to identify different types of cries (Sagi, 1981; Wasz-Hockert, Partanen, Uorenkowski, Michelsan, & Vallane, 1968); they find infants' cries to be less aversive (Zeskind & Lester, 1978); and they respond to them with more nurturant or caring responses (Boukydis & Burgess, 1982; Sagi, 1981; Zeskind, 1980).

Although these data suggest that in human mothers, as in other animals, the growth of maternal responsiveness over the postpartum period is due to experiential rather than physiological factors, the possible role of hormones during this later post-partum period has not been systematically investigated; however, as indicated earlier, no relation has been found between hormones and maternal attitudes during the early postpartum period (Fleming, Krieger, et al., in press).

While there is no evidence of an association between hormones and behavior at this time, there is some evidence that mothers who are breast-feeding as opposed to bottle-feeding feel and behave differently while feeding their infants and exhibit quite different patterns of autonomic responses to infant stimuli. At both 1 and 3 months postpartum, bottle-feeding mothers show fewer affectionate responses while feeding their infants than do mothers who are breast-feeding, although the former do not feel less close or nurturant (Fleming et al., 1988). Bernal and Richards (1970) and Rich-ards and Bernal (1971) also found that breast-fed babies are left alone less often, are fed for longer periods, and are responded to more contingently during feeding than are bottle-fed babies. These data suggest that in the actual feeding situation, breast-feeders are closer to their infants, but these differences may not be apparent during nonfeeding interactions and may result entirely from physical constraints associated with the two modes of feeding. There is some evidence, however, that differences between feeding modes may have implications beyond the feeding context. Wiesen-feld, Malatesta, Whitman, Granrose, and Uili (1985) report, for instance, that moth-ers who are bottle-feeding, as opposed to breast-feeding, show an augmented and sustained heart rate acceleration arousal response to black-and-white videotapes de-picting their own infant's expressions, as well as overall higher baseline skin con-ductance responses. Although these psychobiological differences may reflect person-ality differences between breast- and bottle-feeding women, these authors argue that the lactational hormones (e.g., prolactin, corticoids, and oxytocin) could modify overall arousal levels. These observations are consistent with evidence from the rodent lit-erature that lactation is associated with reduced responsivity to a variety of environ-mental cues (Stern, Goldman, & Levine, 1973; Stern & Levine, 1974; Thoman, Conner, & Levine, 1970).

According to these data, it may well be that elevated arousal during the puer-perium, which may underlie the heightened maternal responsiveness in motivated mothers, may be disruptive to optimal responsiveness once some maternal experience has been acquired.

EFFECTS OF PRIOR MATERNAL EXPERIENCE ON PRESENT RESPONSIVENESS

The experience of caring for an infant can also influence maternal feelings and atti-tudes at a later time. Primiparous mothers, who have more prior experience in caring for young infants and children prior to pregnancy, express more positive maternal attitudes both during the pregnancy and after the birth. The strength of this relation is underscored by its occurrence in every one of our studies. Similar long-term effects of prior experience during adulthood or during the juvenile and early adolescent pe-riods have been reported in rats (Brunelli, Shindledecker, & Hofer, 1985; Cohen & Bridges, 1981; Fleming & Rosenblatt, 1974; Mayer, Freeman, & Rosenblatt, 1979;

Moretto, Paclick, & Fleming, 1986). Interestingly, in our population of primiparous mothers, no reliable relation has been found between prior maternal experience and mothering behavior; the only consistent relation has been with mothering attitudes.

PARITY EFFECTS AND MATERNAL BEHAVIOR

There is substantial human and nonhuman evidence of strong experiential effects on maternal behavior, as revealed by comparisons of mothers differing in parity (Carlier & Noirot, 1966; Moltz & Wiener, 1966; Poindron & Le Neindre, 1980). In general, the studies of parity effects on maternal responsiveness during the puerperal period indicate that while multiparous mothers are more relaxed and "efficient" at mothering and superior in interpreting infant cues, primiparous mothers tend to be more anxious about and solicitous of their babies and spend more time interacting with them.

Bernal (1972) compared the responses of primiparous and multiparous mothers to their infant's cry during the first 10 postpartum days. She reports that, in comparison with first-time mothers, experienced mothers responded more quickly to crying by picking up and feeding their infants; however, this difference was present only if the infants had not been fed within the past 2–5 hr. Neither group responded rapidly if the infants had just been fed. Bernal interprets these data to mean that multiparous mothers either are more accurate at reading hunger cries or are simply more flexible about feeding their infants on a demand schedule.

Multiparous mothers not only respond more rapidly to the crying infant during the first few postpartum days, but also are more efficient than inexperienced mothers while feeding the infant. Thus in comparison with new primiparous mothers, multiparous mothers spend less time at a feeding session, have fewer feeding episodes within a session, and show less variability in the duration of feeding episodes (Thoman et al., 1970, 1971). In one study (Thoman et al., 1970) but not in the other (Thoman et al, 1971), the second-born infants also took more formula during feedings, despite the reduced time of feeding. Thoman et al. (1970) suggest that experienced mothers are more successful at getting the infant to suck than are less experienced mothers because the former are more inclined to arouse the infant by nipple stimulation to the lip and mouth region, whereas the latter are more inclined to stimulate the infant by talking. Although multiparous mothers are more efficient at feeding their infants, they appear less motivated to spend time with them. Thoman et al. (1970, 1971) report, for instance, that during the first few days, first-time mothers spend more time than experienced mothers even during nonfeeding interactions.

Although many of these parity differences in behavior disappear after additional experience has been acquired (Bernal, 1972; Thoman et al., 1970, 1971), others apparently emerge only at a later point. For instance, in an elegant series of papers on stability and change in parent–infant interactions in a naturalistic home environment, Belsky and his colleagues (Belsky, Gilstrap, & Rovine, 1984; Belsky, Rovine, et al., 1984) assessed primiparous and multiparous parents for their attitudes and parental behaviors at four time points from the end of pregnancy through the first 9 months postpartum. They reported no parity differences in behavior during the first

3 months postpartum, when both groups were extremely attentive to their infants, but substantial differences thereafter, when the more immediate demands were lessened. Compared with the more experienced mothers, the primiparous mothers expressed more positive affect to their infants and showed a higher level of reciprocal interaction with them. This was reflected in increased infant-directed vocalization and general stimulation, as well as more focused attention on the infants. At 10–11 months postpartum, primiparous mothers continued to be more attentive. In a study by Donate-Bartfield and Passman (1985), mothers were led to believe that their infants were alone in a playroom and that the cries heard over an intercom were those of their own infants. Using latency to approach the playroom door as an index of responsiveness, Donate-Bartfield and Passman found that primiparous mothers responded significantly more rapidly than did experienced mothers, in contrast to the longer latencies of response in the earlier postpartum period (Bernal, 1972).

Parity differences during the later postpartum period are also present in mothers' autonomic responses to infant-related cues. For instance, two groups of investigators reported that multiparous mothers exhibit greater heart rate acceleration than inexperienced mothers to both unfamiliar (Bleichfeld & Moely, 1984) and their own (Wiesenfeld & Malatesta, 1982) infants' cries. Using galvanic skin response as a measure, Boukydis and Burgess (1982) compared 2- to 6-month postpartum mothers who varied in parity, as well as nonparents, in their responses to different types of cries recorded from unfamiliar infants. They made the interesting observation that, in this case, primiparous mothers were more likely to show an arousal response to infant cries described as "average," and hence though to reflect hunger, whereas nonparents and multiparous parents showed the highest arousal response to the more "difficult," urgent cries. These data suggest that, in comparison with the other two groups, primiparous mothers are more concerned about hunger-type cries, whereas the other two groups are more aroused by the more stressful stimulus, an interpretation that again is at variance with responses to hunger cries at earlier points (Bernal, 1972). Boukydis and Burgess (1982) argued that the more experienced mothers clearly knew the meaning of hunger cries but had become habituated to them, whereas the nonparents never fully understood the salience or meaning of such cries. Interestingly, when subjects were asked how the cries made them feel, both parent groups indicated an equal desire to nurture, whereas the nonparent group did not.

Although there are some ambiguities in the effects of parity on psychophysiological responses (Boukydis & Burgess, 1982; Wiesenfeld et al., 1985), some of the discrepancies may reflect differences in the particular arousal measure used, in the modality of the eliciting infant stimuli, or in whether women are responding to their own or to an unfamiliar infant.

HORMONE–EXPERIENCE INTERACTIONS IN THE REGULATION OF MATERNAL BEHAVIOR

Under some circumstances, parity effects on maternal responsiveness can be augmented when mothers are in a hormonally (or psychologically?) "primed" state. Poindron and Le Neindre (1980) report, for instance, successful hormonal induction of maternal behavior if ewes have had prior young but not if they are nulliparous. In a recent series of studies with rats, it was found that latencies to the reinduction of maternal behavior 4 weeks after a prior parity were considerably reduced if reinduction tests were done during pregnancy (Fleming, unpublished data). A similar pattern has been reported in women. In their study of women's heart rate responses to the pain cry of unfamiliar infants, Bleichfeld and Moely (1984) tested groups of nonpregnant, pregnant, and newly parturient mothers, half of each group being primiparous and the other half multiparous. They reported a significant interaction between prior maternal experience and parental status on heart rate, such that the elevated heart rate response exhibited by experienced as opposed to inexperienced mothers was accentuated when the mothers in the two parity conditions were tested at the end of pregnancy.

MOOD EFFECTS ON MATERNAL RESPONSIVENESS

How might experience influence mothers' feelings and responses to their infants? Probably in a number of ways, some of which may converge with hypothesized hormonal effects and all of which have been demonstrated in other animals. Simple prolonged exposure to infants could, for instance, increase mothers' attraction to infant cues (Zajonc, 1968); previous practice in caring for infants could influence their actual skill and competence in caring for infants and thereby augment their mothering efficiency. Prior experience with infants could also influence mothers' general affective state and, as a result, affect their feelings and interactions with the infant. There is some direct evidence for this latter point.

In a 1-year longitudinal study, Fleming et al. (1988; also Fleming et al., unpublished manuscript) found that one of the primary predictors of mothers' mood state at all time points was the amount of their prior experience with infants; mood, in turn, was the single most important predictor of maternal attitudes. Strong mood–behavior relations were also found.

As shown in Figure 10.4, mothers who were more contented and less dysphoric showed more affectionate contact interactions with their infants, responded more contingently to them at 1 and 3 months postpartum, and sustained breast-feeding for longer periods.

SUMMARY

Extensive research on the psychobiology of maternal behavior in rodents and other nonhuman mammals has demonstrated that the hormonal changes associated with late pregnancy and the parturitional period prime the primiparous female to show the full complement of nurturant behaviors as soon as the young are born and, in some species, during pregnancy, well before parturition. It also shows that limited experience in interacting with young during the early postpartum period is adequate to sustain maternal responsiveness well beyond the period when hormones are effective.

Finally, it indicates that maternal responsiveness can be induced in the absence of hormones by continuous exposure to pup stimulation and that once responsiveness has been induced, regardless of its mode of induction, it exerts powerful long-term effects, precluding the subsequent need for hormones. In the highly experienced animal, the role of hormones in behavior (as opposed to its obvious need for gestation, parturition, and lactation) may be one of fine-tuning and temporal coordination of early mother–infant interactions.

The purpose of this chapter has been to determine to what extent human mothering shows onset and maintenance mechanisms similar to those described for these other animals. A review of studies concerned with changing maternal responsiveness during pregnancy indicates that for some measures of responsiveness changes do occur, whereas for others they do not. Self-report measures assessing women's feelings of attachment to their fetuses or their acceptance of the pregnancy indicate that many women feel more positively at the end than at the beginning of pregnancy and that the largest increase occurs at the time of quickening. These self-report data are consistent with some psychophysiological data showing heightened autonomic arousal to infant cues during pregnancy. However, these changes in affect were not accompanied by changes in attitudes toward caretaking activities or in mothers' feelings of maternal competence, which may not be surprising, since these primiparous mothers had not yet had the opportunity to practice their mothering skills or to actually care for infants. Prospective mothers are more similar to the monkey model than to the rat model in not showing heightened maternal responsivity during pregnancy to a real infant and to infant pictures or in their expressed feelings about other people's infants. Thus depending on the actual measure of maternal responsiveness used, human mothers either do or do not show a pregnancy-related growth of responsiveness.

The situation after the birth is more clear-cut. Regardless of the measure, new mothers experience a large increase in responsiveness from the end of pregnancy to the postpartum period. They report feeling closer to their infants, and they feel more positively about infants in general, as well as about caretaking and their own competence. They also show a more complete repertoire of maternal behaviors, as well as autonomic responses that indicate heightened arousal and readiness for action. In short, as do most other mammals, human mothers rapidly become attuned to their infant's cues and respond appropriately to them.

The role of hormones in this process is still unclear. Like every other animal that has been studied, human mothers show no direct evidence that the pregnancy-related changes in maternal responsiveness that do occur are associated directly with ongoing hormonal changes. They may, however, come about as a result of cumulative hormonal changes that sensitize (possibly through receptor induction) the relevant underlying substrate; however, such a hormonal effect would clearly not be revealed using correlational techniques.

In contrast, there is some evidence that hormones acting at parturition (e.g., cortisol) may be associated with the early expression of affectionate maternal behaviors; these hormonal effects, if causal, are believed to be nonspecific, their valence dependent on the attitudes and experiences that the mother possesses at the time of hormone stimulation. Although the relevant hormones may be different, the behavioral mechanisms by which hormones "activate" maternal responsiveness in rat and human mothers may show similarities.

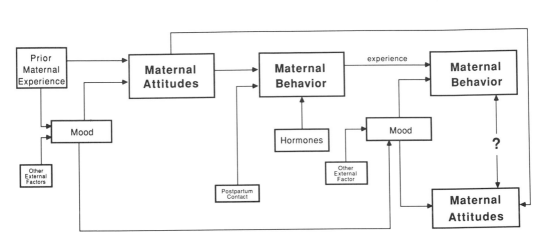

Figure 10.5. Schematic of relations among background, pregnancy, puerperal, and postpartum factors and maternal attitudes and behavior in first-time mothers. (Directions of arrows indicate directions of predictors, established through multiple-regression analyses.)

After birth, human mothers seem to be easily influenced by the types of experiences that they have with their infants. If they are allowed to interact with their infants soon after birth, they feel closer to them and interact more affectionately than if deprived of this opportunity; they also quickly learn the odor and cry characteristics of their infants. Thus human mothers may be similar to rats in that they show elevated responsiveness with early postpartum contact, and they may be similar to ungulates in that they rapidly learn to recognize their own infants. However, they may be different from both of these groups in that the effects of early contact are extremely short-lived and easily masked by subsequent interactive experiences. Also, there is no evidence in humans of a postpartum sensitive period; clearly, one does not have to give birth to become attached to an infant and to experience heightened maternal responsiveness. Adoptive parents as well as biological fathers clearly become attached to young (Silver personal communication; chapter 22). However, whether the events of pregnancy and the early puerperium facilitate this process cannot be determined with the data we now have. It would be extremely interesting—albeit fraught with practical and ethical problems—to study matched groups of adoptive and biological primiparous parents, as well as "surrogate" mothers who do and do not choose to give up their infants for adoption.

In short, maternal responsiveness in humans is multiply determined and may be differentially related to hormones and experience at different time points. As shown schematically in Figure 10.5, during the early postpartum period the response to infants seems to be associated with a coaction of hormones, experiences acquired during early postpartum contact with infants, and the maternal attitudes the mother brings to the situation. Once some experience has been acquired, experiential and mood effects strengthen subsequent responsiveness. Finally, previous parity or prior experiences in caring for young favorably influence the mother's mood state, as well as her reaction to infant cues and her skill in caretaking, thereby promoting easier adjustment to the new baby.

ACKNOWLEDGEMENTS

The research described in this chapter was supported by grants awarded to the author and collaborators from the Laidlaw Foundation, Social Sciences and Humanities Research Council, Health and Welfare Canada, National Sciences Engineering and Research Council.

Many thanks to Brenda Worsnop for her excellent typing; to S. Miltych, G. Owen, N. Pittman, J. Sworn, and V. Vanwagner for their help in carrying out these projects; to all the women who gave so much of their time as participants; and to my collaborators, Professors V. Anderson, C. Corter, G. Flett, H. Krieger, D. Ruble, M. Steiner, and W. Wong.

REFERENCES

Ainsworth, M. (1979). Attachment as related to mother–infant interactions. In J. S. Rosenblatt, R. Hinde, C. Beer, & M. Busnel (Eds.), *Advances in the study of behavior* (Vol. 9, pp. 1–51). New York: Academic Press

Batra, S., & Grundsell, H. (1978). Studies on certain aspects of progesterone and cortisol dynamics before, during and after labour. *Clinical Endocrinology 8*, 403–409.

Belsky, J. (1984). Determinants of parenting: A process model. *Child Development, 55*, 83–96.

Belsky, J., Gilstrap, B., & Rovine, M. (1984). The Pennsylvania Infant and Family Development Project: I. Instability and change in mother–infant interaction in a family setting at 1, 3 and 9 months. *Child Development, 55*, 692–705.

Belsky, J., Rovine, M., & Taylor, D. (1984). The Pennsylvania Infant and Family Development Project: III. The origins of individual differences in infant–mother attachment: Maternal and infant contributions. *Child Development, 55*, 718–728.

Belsky, J., Taylor, D. G., & Rovine, M. (1984). The Pennsylvania Infant and Family Development Project: II. Development of reciprocal interaction in the mother–infant dyad. *Child Development, 55*, 706–717.

Bernal, J. (1972). Crying during the first 10 days of life and maternal responses. *Developmental Medicine and Childhood Neurology, 14*, 362–372.

Bernal, J., & Richards, M. (1970). The effects of bottle and breast feeding on infant development. *Journal of Psychosomatic Research, 14*, 247–252.

Bleichfeld, B., & Moely, B. E. (1984). Psychophysiological responses to an infant cry: Comparisons of groups of women in different phases of the maternal cycle. *Developmental Psychology, 20*, 1082–1091.

Boukydis, C. F., & Burgess, R. L. (1982). Adult physiological response to infant cries: Effects of temperament of infant, parental status and gender. *Child Development, 53*, 1291–1298.

Brazelton, T. B. (1977). Implications of infant development among Mayan Indians of Mexico. In P. H. Leiderman, S. R. Tulkin, & A. Rosenfeld (Eds.), *Culture and infancy variations in human experience* (pp. 151–187). New York: Academic Press.

Bridges, R. S. (1975). Long-term effects of pregnancy and parturition upon maternal responsiveness in the rat. *Physiology and Behavior, 14*, 245–249.

Bridges, R. S. (1977). Parturition: Its role in the long-term retention of maternal behavior in the rat. *Physiology and Behavior, 18*, 487–490.

Bridges, R. S. (1984). A quantitative analysis of the roles of estradiol and progesterone in the regulation of maternal behavior in the rat. *Endocrinology, 114,* 930–940.

Bridges, R. S., Rosenblatt, J. S., & Feder, H. (1978). Serum progesterone concentrations and maternal behavior in rats after pregnancy termination: Behavioral stimulation following progesterone withdrawal and inhibition by progesterone maintenance. *Endocrinology, 102,* 258–267.

Brunelli, S. A., Shindledecker, R. D., & Hofer, M. A. (1985). Development of maternal behaviors in prepubertal rats at three ages: Age-characteristic patterns of responses. *Development Psychobiology, 18,* 309–326.

Carlier, C., & Noirot, E. (1966). Effects of previous experience on maternal retrieving by rats. *Animal Behaviour, 13,* 423–426.

Caudill, W., & Weinstein, H. (1969). Maternal care and infant behavior in Japan and America. *Psychiatry, 32,* 12–43.

Cohen, J., & Bridges, R. S. (1981). Retention of maternal behavior in nulliparous and primiparous rats. Effects of duration of previous maternal experience. *Journal of Comparative and Physiological Psychology, 95,* 450–459.

Curtis, G., Buxton, M., Lippman, D., Nesse, R., & Wright, J. (1976). "Flooding in vivo" during the circadian phase of minimal cortisol secretion: Anxiety and therapeutic success without adrenal cortical activation. *Biological Psychiatry, 11,* 101–107.

Cutrona, C. E. (1982). Nonpsychotic postpartum depression: A review of recent research. *Clinical Psychology Review, 2,* 487–503.

de Chateau, P., & Wiberg, B. (1977). Long-term effect on mother–infant behavior of extra contact during the first hour postpartum: I. First observations at 36 hours. *Acta Paediatrica Scandinavica, 66,* 137–143.

Donate-Bartfield, E., & Passman, H. (1985). Attentiveness of mothers and fathers to their baby's cries. *Infant Behavior and Development, 87,* 385–393.

Dunn, A. J. (1984). Effects of A. C. T. H. β-lipoprotein and related peptides on the central nervous system. In C. B. Nermeroff & A. J. Dunn (Eds.), *Peptides, hormones and behavior* (pp. 273–348). New York: Medical Science Books.

Feldman, S. S., & Nash, S. C. (1978). Interest in babies during young adulthood. *Child Development, 49,* 617–622.

Fernald, A. (1984). Perceptual and affective salience of mothers' speech to infants. In I. Iagans, C. Garvey, & R. Golinkoff (Eds.), *Origins and growth of communication* (pp. 5–29). Norwood, NJ: Ablex.

Fleming, A. S. (1986). Psychobiology of rat maternal behavior. How and where hormones act to promote maternal behavior at parturition. *Annals of the New York Academy of Sciences, 474,* 234–251.

Fleming, A. S., & Corter, C. (1988). Factors influencing maternal responsiveness in humans: Usefulness of an animal model. *Psychoneuroendocrinology, 13,* 189–212.

Fleming, A. S., Krieger, H., & Wong, P. Y. (in press). Affect and nurturance in first-time mothers: Role of psychobiological influences. In B. Lerer & S. Gershon (Eds.), *New directions in affective disorders.*

Fleming, A. S., & Luebke, C. (1981). Timidity prevents the nulliparous female from being a good mother. *Physiology and Behavior, 27,* 863–868.

Fleming, A. S., & Rosenblatt, J. S. (1974). Maternal behavior in the virgin and lactating rat. *Journal of Comparative and Physiological Psychology, 86,* 957–972.

Fleming, A. S., Ruble, D., Flett, G., & Van Wagner, V. (in press). Adjustment in first-time mothers: Changes in mood and mood content during the first postpartum year. *Developmental Psychology.*

Fleming, A. S., Ruble, D., Flett, G., & Shaul, D. (1988). Postpartum adjustment in first-time mothers: Relations between mood, maternal attitudes and mother–infant interactions. *Developmental Psychology, 24,* 71–81.

Fleming, A. S., Steiner, M., & Anderson, V. (1987). Hormonal and attitudinal correlates of maternal behavior during the early postpartum period in first-time mothers. *Journal of Reproductive and Infant Psychology, 5,* 193–205.

Fleming, A. S., Vaccarino, F., & Luebke, C. (1980). Amygdaloid inhibition of maternal behavior in the nulliparous rat. *Physiology and Behavior, 25,* 731–743.

Formby, D. (1967). Maternal recognition of infants' cry. *Developmental Medicine and Child Neurology, 9,* 293–298.

Gaulin-Kremer, E., Shaw, J. L., & Thoman, E. B. (1977, March). *Mother–infant interaction at first prolonged encounter. Effects of variation in delay after delivery.* Paper presented at the biennial meeting of the Society for Research in Child Development, New Orleans.

Gibber, R. (1986). Infant-directed behavior of rhesus monkeys during their first pregnancy and parturition. *Folia Primatologica, 46,* 118–124.

Goldberg, S. (1983). Parent–infant bonding: Another look. *Child Development, 54,* 1355–1382.

Goodspeed, R. B., Gent, J. F., Catalanotto, F. I., Cain, W. S., & Zagraniski, R. T. (1986). Corticosteroids in olfactory dysfunction. In H. L. Mcisclman & R. S. Rivlin (Eds.), *Clinical measurement of taste and smell* (pp. 514–518). New York: Macmillan.

Gottschalk, L., & Gleser, C. (1969). *The measurement of psychological states through the content analysis of verbal behavior.* Los Angeles: University of California Press.

Gray, P. & Chesley, S. (1984). Development of maternal behavior in nulliparous rats *(Rattus norvegicus):* Effects of sex and early maternal experience. *Journal of Comparative Psychology, 98,* 91–99.

Grossman, K., Thane, K., & Grossman, K. E. (1981). Maternal tactual contact of the newborn after various conditions of mother–infant contact. *Developmental Psychology, 17,* 158–169.

Hales, D. J., Lozoff, B., Sosa, R., & Kennell, J. H. (1977). Defining the limits of the maternal sensitive period. *Developmental Medicine and Child Neurology, 19,* 454–461.

Handley, S. L., Dunn, T. L., Waldron, G., & Baker, J. M. (1980). Tryptophan, cortisol and puerperal mood. *British Journal of Psychiatry, 136,* 498–508.

Henken, R. I. (1975). The role of adrenal corticosteroids in sensory processes. In H. Blaschko, G. Savers, & A. D. Smith (Eds.), *The handbook of physiological endocrinology* (Vol. 6, pp. 209–230). Baltimore, Williams & Wilkins.

Hennessy, M. B., Harney, K. S., Smotherman, W. P., Coyle, S., & Levine, S. (1977). Adrenalectomy-induced deficits in maternal retrieval in rat. *Hormones and Behavior, 9,* 222–227.

Hopkins, J. B., Marcus, M., & Campbell, S. B. (1984). Postpartum depression: A critical review. *Psychology Bulletin, 95,* 498–515.

Hopkins, J. B., & Vietze, P. (1977, March). *Postpartum early and extended contact: Quality, quantity or both?* Paper presented at the biennial meeting of the Society of Research in Child Development, New Orleans.

Jakubowski, M., & Terkel, J. (1986). Establishment and maintenance of maternal responsiveness in postpartum wistar rats. *Animal Behaviour, 34,* 256–262.

Jolivet, A., Blanchier, H., Gautray, J. P., & Dhem, N. (1974). Blood cortisol variations during late pregnancy and labor. *American Journal of Obstetrics and Gynecology, 119,* 775–783.

Klaus, M. H., Jerauld, R., Kreger, N., MacAlpine, W., Steffa, M., & Kennell, J. H. (1972).

Maternal attachment: Importance of the first post-partum days. *New England Journal of Medicine, 286,* 460–463.

Klaus, M. H., Kennell, J. H., Plumb, N., & Zuehkle, S. (1970). Human maternal behavior on first contact with her young. *Pediatrics, 46,* 187–192.

Konner, M. (1977). Infancy among the Kalahari Desert San. In P. H. Leiderman, S. R. Tulkin, & A. Rosenfeld (Eds.), *Culture and infancy: Variations in human experience* (pp. 287–328). New York: Academic Press.

Lacey, J. I., & Lacey, B. C. (1970). Some autonomic central nervous system relationships. In P. Black (Ed.), *Physiological correlates of emotion* (pp. 205–227). New York: Academic Press.

Lamb, E. (1982). Early contact and maternal–infant bonding one decade later. *Pediatrics, 70,* 763–777.

Leiderman, P., & Leiderman, M. (1977). Economic change and infant care in an East African agricultural community. In P. H. Leiderman, S. R. Tulkin, & A. Rosenfeld (Eds.), *Culture and infancy: Variations in human experience* (pp. 405–438). New York: Academic Press.

Leifer, M. (1977). Psychological changes accompanying pregnancy and motherhood. *Genetic Psychology Monographs, 95,* 55–96.

Leifer, M. (1980). *Psychological effects of motherhood: A study of first pregnancy.* New York: Praeger.

Lott, D., & Rosenblatt, J. S. (1969). Development of maternal responsiveness during pregnancy in the rat. In B. M. Foss (Ed.), *Determinants of infant behavior,* (Vol. 4, pp. 61–67). London: Metheun.

Mason, J. W. (1968). Review of psychoendocrine research on pituitary-adrenal-cortical system. *Psychosomatic Medicine, 30,* 576–607.

Mayer, A. D., Freeman, N. G., & Rosenblatt, J. S. (1979). Ontogeny of maternal behavior in the laboratory rat: Actors underlying changes in responsiveness from 30 to 90 days. *Developmental Psychobiology, 12,* 425–439.

Moltz, H., Lubin, M., Leon, M., & Numan, M. (1970). Hormonal induction of maternal behavior in the ovariectomized nulliparous rat. *Physiology and Behavior, 5,* 1373–1377.

Moltz, H., & Wiener, E. (1966). Effects of ovariectomy on maternal behavior of primiparous and multiparous rats. *Journal of Comparative and Physiological Psychology, 62,* 382–387.

Moretto, D., Paclik, L., & Fleming, A. S. (1986). The effects of early rearing environments on maternal behavior in adult female rats. *Development Psychobiology, 19,* 581–591.

Moss, H. (1967). Sex, age and state as determinants of mother–infants interaction. *Merrill-Palmer Quarterly, 13,* 19–36.

Nott, P. N., Franklin, M., Armitage, C., & Gelder, M. G. (1976). Hormonal changes and mood in the puerperium. *British Journal of Psychiatry, 128,* 379–383.

Obrist, P. (1976). The cardiovascular behavioral interaction: As it appears today. *Psychophysiology, 13,* 95–107.

Orpen, G., & Fleming, A. S. (1987). Experience with pups sustains maternal responding in postpartum rats. *Physiology and Behavior, 40,* 47–54.

Poindron, P., & Le Neindre, P. (1980). Endocrine and sensory regulation of maternal behavior in the ewe. In J. S. Rosenblatt, R. A. Hinde, C. Beer, & M. C. Bushel (Eds.), *Advances in the study of behavior* (Vol. 2) (pp. 75–119). New York: Academic Press.

Porter, R. H., Cernock, J. M., & McLaughlin, F. (1983). Maternal recognition of neonates through olfactory cues. *Physiology and Behavior, 30,* 151–154.

Ramsay, I. D. (1980). The adrenal gland. In F. Hytten & G. Chamberlain (Eds.), *Clinical physiology in obstetrics* (pp. 411–423). London: Blackwell Scientific.

Richards, M. P. M. (1966). Maternal behavior in the golden hamster: Responsiveness to young in virgin, pregnant and lactating females. *Animal Behaviour, 14,* 310–313.

Richards, M. P. M., & Bernal, J. F. (1971). An observational study of mother–infant interaction. In N. Blurton-Jones (Ed.), *Ethological studies of child behavior* (pp. 175–198). London: Cambridge University Press.

Robson, K. S., & Moss, H. A. (1970). Patterns and determinants of maternal attachment. *Journal of Pediatrics, 77,* 976–985.

Rosenblatt, J. S., & Siegel, H. (1981). Factors governing the onset and maintenance of maternal behavior among nonprimate mammals: The role of hormonal and nonhormonal factors. In D. J. Gubernick & P. H. Klopfer (Eds.), *Parental care in mammals* (pp. 13–76). New York: Plenum Press.

Rosenblum, L. A. (1972). Sex and age differences in response to infant squirrel monkeys. *Brain, Behavior, and Evolution, 5,* 30–40.

Sagi, A. (1981). Mothers' and non-mothers' identification of infant cries. *Infant Behavior and Development, 41,* 37–40.

Schaal, B. (1986). Presumed olfactory exchanges between mother and neonate in humans. In J. LeCamus & R. Campons (Eds.), *Ethologie et psychologie de l'enfant* (pp. 101–110). Toulouse: Privat.

Schaal, B., Montagner, H., Hertling, E., Bolzoni, D., Moyse, A., & Quichon, R. (1980). Les stimulations olfactives dans les relations entre l'enfant et la mère. *Reproduction, Nutrition, Developement, 20,* 843–858.

Seashore, M. J., Leifer, A. D., Barnett, C. R., & Leiderman, P. H. (1973). Effects of denial of early mother–infant interaction on maternal behavior. *Journal of Personality and Social Psychology, 26,* 369–378.

Siegel, H. I., Clark, M. C., & Rosenblatt, J. S. (1983). Maternal responsiveness during pregnancy in the hamster *(Mesocricetus auratus). Animal Behaviour, 31,* 497–502.

Siegel, H. I., & Rosenblatt, J. S. (1975a). Hormonal basis of hysterectomy-induced maternal behavior during pregnancy in the rat. *Hormones and Behavior, 6,* 211–222.

Siegel, H. I., & Rosenblatt, J. S. (1975b). Estrogen induced maternal behavior in virgin rats. *Physiology and Behavior, 14,* 465–471.

Siegel, H. I., & Rosenblatt, J. S. (1978). Duration of estrogen stimulation and progesterone inhibition of maternal behavior in pregnancy-terminated rats. *Hormones and Behavior, 11,* 12–19.

Sosa, R., Klaus, M. H., Kennell, J., & Urrutia, J. J. (1976). The effect of early mother–infant contact on breastfeeding, infection and growth. In *Breastfeeding and the infant,* Ciba Symposium 45. Amsterdam: Elsevier.

Steiner, M., Fleming, A. S., Anderson, V. N., Monkhouse, E., & Boulter, G. E. (1986). Is there a psychoendocrine profile for postpartum blues? In L. Dennerstein & I. Fraser (Eds.), *Hormones and Behavior* (pp. 327–335). Amsterdam: Elsevier.

Stern, J. M. (1983). Maternal behavior priming in virgins and caesarian-delivered Long Evans rats: Effects of brief contact or continuous exteroceptive pup stimulation. *Physiology and Behavior, 31,* 757–763.

Stern, J. M., Goldman, L., & Levine, S. (1973). Pituitary-adrenal responsiveness during lactation in rats. *Neuroendocrinology, 12,* 179–191.

Stern, J. M., & Levine, S. (1974). Psychobiological aspects of lactation in rats. In D. F. Swabb & J. P. Schade (Eds.), *Intergrative hypothalamic activity: Progress in brain research* (Vol. 41, pp. 20–29). Amsterdam: Elsevier.

Thoman, E. B., Barnett, C. R., & Leiderman, P. H. (1971). Feeding behaviors of newborn infants as a function of parity of the mother. *Child Development, 42,* 1471–1483.

Thoman, E. B., Conner, R. L., & Levine, S. (1970). Lactation suppresses adrenal cortico-
    steroid activity and aggressiveness in rats. *Journal of Comparative and Physiological
    Psychology, 70*, 364–369.
Thoman, E. B., Turner, A. M., Leiderman, H., & Barnett, C. R. (1970). Neonate–mother
    interactions: Effects of parity on feeding behavior. *Child Development, 41*, 1103–
    1111.
Tronick, E. Z. (1987, April). *Discussion paper: An interdisciplinary view of human intuitive
    parenting behaviors and their role in interactions with infants.* Paper presented at
    the meetings of the Society for Research in Child Development, Baltimore.
von Zerssen, D. (1976). Mood and behavioral changes under corticosteroid therapy. In T. M.
    Itil, G. Laudahn, & W. M. Herrman (Eds.), *Psychotropic action of hormones* (pp.
    195–222). New York: Spectrum.
Wasz-Hockert, D., Partanen, T., Uorenkowski, V., Michelsson, K., & Vallane, E. (1964).
    Effect of training on ability to identify preverbal vocalizations. *Developmental Med-
    icine and Child Neurology, 6*, 397–402.
Whiting, J. W. (1981). Environmental constraints on infant care practices. In R. H. Munroe,
    R. L. Munroe, & B. N. Whiting (Eds.), *Handbook of cross-cultural human devel-
    opment* (pp. 155–179). New York: Garland.
Wiesenfeld, A. R., & Malatesta, C. Z. (1982). Infant distress: Variables affecting responses
    of caregivers and others. In L. W. Hoffman, R. Gandelman, & H. R. Schiffman
    (Eds.), *Parenting: Its Causes and Consequences* (pp. 123–139). Hillsdale, NJ: Erl-
    baum.
Wiesenfeld, A. R., Malatesta, C. Z., Whitman, P. B., Granrose, C., & Uili, R. (1985).
    Psychophysiological response of breast- and bottle-feeding mothers to their infants'
    signals. *Psychophysiology, 22*, 79–86.
Wiesner, B. P., & Sheard, N. M. (1933). *Maternal behavior in the rat.* Edinburgh: Oliver &
    Boyd.
Willcox, D. L., Yovich, J. L., McColm, S. C., & Schmitt, L. H. (1985). Changes in total
    and free concentrations of steroid hormones in the plasma of women throughout
    pregnancy: Effects of medroxyprogesterone acetate in the first trimester. *Journal of
    Endocrinology, 107*, 293–300.
Zajonc, R. B. (1968). Attitudinal effects of mere exposure. *Journal of Personality and Social
    Psychology, 8*, 1–29.
Zeskind, P. S. (1980). Adult responses to cries of low and high risk infants. *Infant Behavior
    and Development, 3*, 167–177.
Zeskind, P. S., & Lester, B. M. (1978). Acoustic features and auditory perceptions of the
    cries of newborns with prenatal and perinatal complications. *Child Development, 49*,
    580–589.

11

Endocrine Correlates of Human Parenting: A Clinical Perspective

MICHELLE P. WARREN AND BARBARA SHORTLE

The role of hormones in human parenting behavior has received relatively little attention. The so-called maternal, and to some extent the paternal, "instinct" is well characterized in primates and other mammals and is necessary for survival of the young (Velle, 1982). Such instincts undoubtedly exist in the human, yet the link to hormonal influences remains imprecise. Human evolution may have reached a stage at which these biological determinants of behavior play a purely marginal role.

The lack of well-tested and recognized scales of reproducible parenting behavior, and the difficulties of associating these parameters with the very complex hormonal changes observed in pregnancy and lactation, are some of the problems that make investigations in this area difficult.

Moreover, the manifestations of maternal and paternal behavior may be different in humans. Four well-accepted manifestations in rats include nest building, licking, retrieval of the young, and nursing (McCullogh, Quadagno, & Goldman, 1974). Other mammalian species show more stereotyped, species-characteristic behaviors than those exhibited by humans (Zarrow, Denenberg, & Sachs, 1972). Although there are similarities in the parental behavior of humans and other mammals, parental behavior in humans is ill-defined, and, with notable exceptions (see chapter 10), there are few well-standardized measures of parental behavior in use for human research.

Some studies have examined "childhood rehearsal of maternalism" (Ehrhardt & Baker, 1974), and include behaviors such as daydreams and fantasies of pregnancy and motherhood, the importance of marriage as a career, playing with dolls, and interest in infant care. At present, with the exception of Ehrhardt's and Fleming's studies, research examining the hormonal determinants of parenting behavior using reproducible parenting scales and hormonal models has not been performed in humans and, in fact, is sadly lacking in the scientific literature. The postpartum period, an interval in which, according to animal studies, exposure to the young may be a powerful stimulus for both maternal and paternal behavior (Rosenblatt, 1967), is also understudied in humans.

Parenting behavior is strongly influenced by cultural norms, as well as by genetics. Traditionally, the female is more interested in the care of babies than is the male, and this appears to occur whether or not women undergo the birth and nursing process. Thus the female seems genetically predisposed toward the care of young.

This behavior may be a result of hormonal determinants indigenous to the female, or it may be entirely unrelated to hormones. However, it is more likely the result of complex interactions among physiological, cultural, social, and psychological factors.

The purpose of this chapter is to examine the literature in order to ascertain the possible hormonal determinants of parenting behavior in humans. The important cultural and social as well as psychological factors will be addressed elsewhere.

The hormonal changes in pregnancy, parturition, and lactation are extremely complex and interdependent. These changes have been well described in the human, but the complexity of the hormonal changes in pregnancy makes behavioral changes difficult to correlate with hormonal determinants. To separate the cultural, genetic (including biological determinants other than hormonal), and hormonal variables may be possible only in situations where a disease has altered the genetic or hormonal milieu. Since studies on humans are difficult to develop because of ethical problems related to hormonal manipulations, experiments of nature can provide insight into hormonal determinants of parenting behavior, yielding data that are otherwise difficult to generate. The complex hormonal changes seen in pregnancy will be described first to provide a basis for the subsequent discussion of possible hormonal correlates in the parenting behavior of humans.

ENDOCRINOLOGY OF PREGNANCY

Three hormonal units interact in human pregnancy: the mother, the fetus, and the placenta. These provide the setting for the onset and maintenance of pregnancy, fetal growth and development, the initiation of labor, delivery, and the initiation of lactation. The functional hypothalamic–pituitary target organ axis of the fetus becomes an active part of the endocrine system very early in pregnancy and is a unique feature of humans and subhuman primates. This system is best described by its components, with the realization that all three systems cooperate and depend on each other to achieve a successful final outcome (Figure 11.1).

The Placenta

After fertilization and implantation, the trophoblast produces the protein hormone human chorionic gonadotropin (HCG). This is the most important hormone of early pregnancy, since it supports the functioning of the corpus luteum within the mother's ovary beyond its usual life span of 2 weeks. HCG blood levels can be detected as early as 8 days after ovulation and implantation, and they double every 2 days. Concentrations peak at 9–10 weeks, then fall and plateau. After 9–10 weeks, HCG no longer supports the corpus luteum. At this time, HCG's function is unclear, although it is probably important in both placental and ovarian steroid biosynthesis. Interestingly, this placental hormone may stimulate maternal gonadal hormone secretion in early pregnancy, doing the work of the mother's pituitary gland; hypophysectomized patients are in fact able to maintain a pregnancy, indicating that gonadotrophic stimulation is received from a source other than the pituitary (Vande Wiele

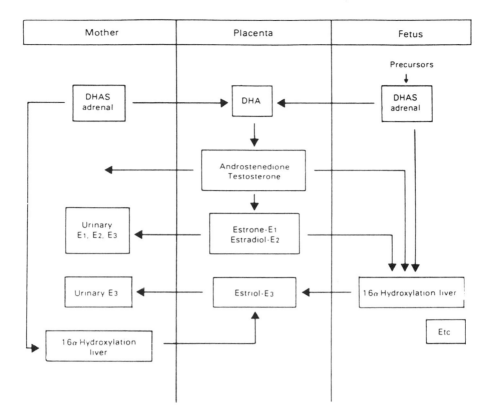

Figure 11.1. Simplified scheme of the biogenesis of placental estrogens in pregnancy in the human female, indicating the predominant fetal contribution in the second trimester and both fetal and maternal contributions in the third trimester. (From Josimovich, 1980)

et al., 1970). In the fetus, HCG stimulates fetal adrenal dehydroepiandrosterone sulfate (DHEAS) synthesis. HCG also stimulates testosterone production in the fetal testes in early pregnancy (Goebelsmann, 1986; Jaffe, 1986a, 1986b). It is noteworthy that in humans, gonadal hormone secretion is not required for pregnancy maintenance after weeks 8–10, being independent of both pituitary and gonadal regulation.

Another placental hormone, human placental lactogen (HPL), also known as "human chorionic somatomammotropin," is synthesized by a part of the placenta known as the "syncytiotrophoblast." HPL is detected at 5–6 weeks and is mostly secreted into the maternal circulation to effect changes that are beneficial to the fetus, such as supplying glucose. In experimental animals, the biological activity of HPL closely resembles that of human growth hormone (HGH) in having both luteotropic and lactogenic properties (Grumbach, Kaplan, Sciarra & Burr, 1968; Jaffe, 1986b). HPL, for example, stimulates milk production in rabbits (Jaffe, 1986b; Josimovich, 1980). In humans, however, HPL is absent during the lactation period, and its role is restricted to pregnancy. Its lactotrophic properties in other species make it an interesting hormone to study with reference to maternal behavior (Jaffe, 1986b).

The placenta secretes progesterone into the maternal and fetal circulations, using maternal low density lipoprotein (LDL)-cholesterol and fetal pregnenolone sulfate as precursors, converting them to pregnenolone and then to progesterone. Some hormones, such as 17-α-hydroxyprogesterone, are manufactured only by using specific ovarian and fetal adrenal enzymes (17-α-hydroxylase and 17,20 desmolase), and are not present in the placenta.

Progesterone (P) is produced during pregnancy, and its function is intriguing. Attention has focused mainly on its effect in maintaining uterine musculature (the myometrium) in a state of relative quiescence during pregnancy (Jaffe, 1986a). It has also been proposed that P produced by the placenta and ovaries is essential in mammalian pregnancy because of its ability to inhibit T-lymphocyte cell-mediated responses involved in tissue rejection (Siiteri et al., 1977). P is called the "hormone of pregnancy" because it appears to be essential for maintaining pregnancy in mammals.

The placenta also synthesizes a number of androgens using precursors from the mother as well as the fetus. DHEAS, for instance, is taken from both to form DHEA. The placenta also has 3-β-ol-dehydrogenase activity and Δ5–Δ4 isomerase activity, and thus it can convert DHEA to androstenedione (A_4) and testosterone (T). Most of these androgens are then converted to estrone (E_1) and estradiol (E_2) rather than released into the circulation. This is an important mechanism because it protects the fetus from excessive androgen exposure.

The important estrogens (E_1 and E_2) are also synthesized from a precursor. However, the fetus and mother must supply the placenta with this precursor. Since the placenta lacks 16-α-hydroxylase activity, it cannot further metabolize E_1 and E_2 to manufacture estriol (E_3).

In human pregnancy, the major urinary estrogen is E_3, which is formed by a unique biosynthetic pathway, demonstrating the interdependence of the fetus, the placenta, and the mother (see below). The maternal and fetal adrenals produce DHEAS, which is then converted to DHEA and eventually to E_2 by the placenta.

In terms of E_1 and E_2 production, the placenta has the ability to convert androstanedione and testosterone to E_1 and E_2. Late in pregnancy, it also produces estrogens from fetal and maternal DHEAS. The net effect is that maternal estrogen levels rise until delivery. The placenta is incapable, however, of converting E_1 or E_2 to E_3, since it lacks 16-α-hydroxylase activity (Jaffe, 1986b).

The Fetus

The fetal glands—adrenal, gonad, thyroid, and pituitary—all provide important hormones for the maintenance and growth of pregnancy. The fetal adrenal is capable of synthesizing steroids de novo from acetate. Importantly, it metabolizes circulating LDL-cholesterol, pregnenolone, and P into cortisol and DHEAs, providing important hormonal precursors as well as the hormones themselves. Fetal DHEA is converted into maternal estrogen, whose levels indicate fetal well-being (Jaffe, 1986c). In early pregnancy, HCG stimulates the fetal adrenal. After 20 weeks, adrenocorticotropic hormone (ACTH) takes over the role of HCG so that the essential production of fetal cortisol and androgens may ensue. The primary job, however, of the fetal adrenal (as discussed above) is to provide DHEAs as a precursor for placental estrogen production (Jaffe, 1986a, 1986b, 1986c).

The fetal gonads, too, contribute hormones that help to coordinate the early stages of pregnancy. In the male fetus, the Leydig cells of the testes appear at 7 weeks and, under the stimulation of HCG, produce T, required for the development of male internal and external genitalia. Fetal T levels and Leydig cell numbers peak at 15–17 weeks and then fall. In the second half of gestation, fetal T production is regulated by fetal gonadotropins, with negative feedback of T on the pituitary and hypothalamus. Fetal follicle-stimulating hormone (FSH) and luteinizing hormone (LH) are detected in fetal pituitaries by 10 weeks of gestation and peak in the second trimester. Their levels are higher in females than in males. In female fetuses, FSH and LH are necessary to facilitate follicle development. Primary follicles develop at 20–25 weeks, with graafian follicles appearing at 28 weeks. The amniotic fluid also reflects the sex of the fetus, with T and androgens (Goebelsmann, 1986; Jaffe, 1986c) being higher in male fetuses and E_2 being higher in female fetuses.

The thyroid gland is present at 9 weeks, and the hypothalamic–pituitary–thyroid axis is fully developed between 30 and 35 weeks. Fetal thyroxin (T_4) rises rapidly between 20 and 30 weeks of gestation, and after 30 weeks, fetal thyroid-stimulating hormone (TSH) falls as free T_4 rises, indicating maturation of feedback. The fetal thyroid is thought to be important for full growth of the brain (Jaffe, 1986c).

During pregnancy, the fetal pituitary begins its production of fetal plasma prolactin (PRL), which starts to increase at the end of the second trimester (Jaffe, 1986a). Fetal PRL levels exceed maternal levels at term but after birth are very low. PRL is also found in amniotic fluid that originates from a third source, decidual placental tissue (Jaffe, 1986c).

The Interdependence of Fetus and Placenta

The interdependence of the fetus and the placenta in the production of steroids is well illustrated by the hormone E_3, which was used in the past as an indication of fetal well-being. The fetal adrenal secretes large quantities of DHEAS, which, in turn, is converted into the compound 16-OH-DHEAS by the fetal liver. This is the principal source of E_3. The placenta, which normally has sulfatase enzyme present, converts this to 16-OH-DHEA, which the placenta next converts to 16-OH-A_4 and 16-OH-T. After aromatization, 16-OH-A_4 becomes 16-OH-E_1 and 16-OH-T becomes 16-OH-E_2, which is E_3. E_3 is then secreted into the maternal and fetal circulations. The fetus sulfuryates it and conjugates it into glucosiduronates. Unconjugated E_3 readily crosses the placenta and is metabolized by the mother into conjugated forms, 90% of which is secreted in the urine (Jaffe, 1986c).

E_3 measurement was once an attractive method of monitoring fetal well-being, since E_3 is almost entirely (90%) of fetal origin and rises until term. The copious amounts of E_3 produced each day raises the question of its function during pregnancy, a question that has caused a great deal of speculation. E_3 is probably active in increasing uteroplacental blood flow and appears to have this estrogenic function only with the placenta; it exerts a relatively weak estrogenic effect on other organ systems (Clark, Paszko, & Peck, 1977; Martucci & Fishman, 1977; Resnik, Killam, Battaglia, Makowski, & Meschia, 1974).

The Mother

The corpus luteum of pregnancy secretes 17α-OH-P and P from the onset of pregnancy. Secretion of these steroids by the corpus luteum in the ovary is required until the seventh week of pregnancy, when the placenta assumes the major role of P synthesis. Ovariectomy prior to the seventh week results in abortion of the fetus. P levels from the corpus luteum decline progressively until the 12th week, at which time P function is taken over by the placenta.

The two- to threefold rise in maternal T in pregnancy is secondary to the increase in sex hormone-binding globulin (SHBG), which is synthesized by the liver. SHBG, in turn, is secondary to the increased E_2 level. Approximately 99% of T is bound to SHBG and presumably is not active.

Cortisol levels rise in pregnancy secondary to the increase in corticoid-binding globulin (CBG), which is stimulated by elevated titers of E_2. The production rate of cortisol is unchanged, and the metabolic clearance rate decreases, while the rate at which cortisol binds to CBG increases. Urinary-free cortisol increases and maternal ACTH stays unchanged. Thus some increase in free active cortisol is present during pregnancy.

Finally, concerning mineralocorticosteroid metabolism, the elevation in E_2 levels cause a rise in angiotensinogen, thereby resulting in a rise in renin and angiotensinogen II, which enhances aldosterone production and excretion in pregnancy.

LABOR AND MATERNAL ENDOCRINOLOGY FOLLOWING BIRTH

Initiation of Labor

To understand the factors that initiate labor, it is important to realize what causes uterine quiescence. P has long been established as a hormone that, in rabbits and rats, causes unresponsiveness of the uterus to oxytocin stimulation. Estrogen (E), on the contrary, increases urine membrane excitability. Other factors that cause uterine relaxation are prostacyclin (PG_{12}), produced by myometrial tissue, and the hormone relaxin, produced by the corpus lectum in pregnancy and probably by the decidua and myometrium. Relaxin stimulates myometrial PG_{12} synthesis. Cortisol, however, has the opposite effect, thereby removing a key factor maintaining uterine quiescence.

What initiates labor in the human is still an enigma. It is known that prostaglandin E_2 and $F_2\alpha$ are important agents causing uterine contractions, but factors leading to their synthesis are still under investigation. Oxytocin plays a significant role in the initiation of labor. One theory proposes that there is a drop in placental P production, along with a rise in E_2 production, just before the onset of labor. A relative decrease in fetal membrane P levels causes the lysosomes to release the enzyme phospholipase A_2, which, in turn, cleaves arachidonic acid from phospholipids and thereby initiates prostaglandin synthesis in membranes and decidua. The synthesis of prostaglandin results in cervical ripening and uterine contractions (Goebelsmann, 1986).

Lactation and the Puerperium

Although many hormones affect the breast, three in particular are important in pregnancy. E_2, P, and PRL; all three are elevated during this period. Both acinar and ductal portions of the breast hypertrophy during pregnancy, and the aveoli become engorged with a liquid substance called "colostrum," a rich source of antibodies for the baby postpartum.

Lactation is initiated by a drop in E_2 levels after delivery. Indeed, E_2 has an inhibitory effect on PRL receptors in alveolar cells. Not until circulating E_2 levels have returned to those found during the follicular phase of the menstrual cycle does lactation begin. This usually occurs the second or third day after delivery. Once the inhibition of E_2 has been removed, lactation depends on the hormones PRL and oxytocin (Mishell & Marrs, 1986).

PRL rises 10- to 20-fold during pregnancy and falls rapidly over a 2-week period postpartum. Circulating PRL levels rise following stimulation of the breast (i.e., suckling), at which point this hormone acts directly on alveolar cells to stimulate lactogenesis. Psychological factors are thought not to be involved in the activation of PRL release during lactation. Instead, suckling or other mechanical stimuli such as the use of a breast pump are the primary stimuli that cause pituitary PRL release. PRL levels are higher in breast-feeding mothers for a 6-week period postpartum, after which all parameters revert to normal (Tyson, Hwang, Guyda, & Friesen, 1972).

Oxytocin is also released as a result of nursing. Other stimuli, such as the sight, sound, or touch of the infant, also elicit oxytocin release. This hormone induces contractions of the myoepithelial cells, which cause "let down" of milk in addition to uterine contractions.

During the puerperium, there is an eventual return to ovarian function. Whether the mother chooses to breast-feed or not determines, to a large extent, the interval of time from delivery to ovulation and menstruation. If the mother does not nurse, the breasts remain engorged for at least 1 week. Ovulation may resume in as little as 4 weeks postpartum, but the average time of ovulation postpartum is 10 weeks.

The resumption of ovulation in the nursing mother is based on the duration and frequency of breast-feeding. For a fully nursing mother, the incidence of ovulation prior to 10 weeks is extremely low. As nursing progresses, by 5 months approximately one-third of women ovulate.

The mechanisms responsible for inhibiting ovulation during lactation are poorly understood. Elevated PRL levels may inhibit ovulation and folliculogenesis, possibly by affecting the secretion of gonadothropin releasing hormone (GnRH), and thus LH and FSH levels (Mishell & Marrs, 1986; Yen, Rebar, & Quesenberry, 1976).

In summary, pregnancy, labor, the puerperium, and lactation are associated with profound and complex endocrine alterations. The factors involved in the regulation of most of these hormonal changes are still under investigation, and their role in affecting maternal behavior in human females has barely been explored.

ENDOCRINOLOGY AND HUMAN PARENTING BEHAVIOR

Whereas the hormonal changes in pregnancy, parturition, and lactation have been well described in humans, the role of hormones in the facilitation of parenting behavior in humans is not yet understood. The research of Fleming, described in chapter 10, addresses some of these questions in a study of 29 white first-time mothers who were studied after birth, on days 2 and 3 postpartum, and again at 6 weeks. The mothers filled out self-report questionnaires four times during pregnancy and underwent venipuncture for assay of hormones known to be involved in the periparturitional onset of maternal behavior in other animals. These included E_2, P, T, and their various ratios, as well as binding proteins, SHBG, and cortisol. Behavioral findings revealed that heightened positive feelings toward pregnancy occurred at about 5 months when fetal movements were first detected. Correlation coefficients were computed between the various maternal constructs and the hormone levels at five points during the testing interval. No consistent associations were found between any hormone or hormone ratio and any of the maternal-attitude constructs during various phases of pregnancy or at 6 weeks postpartum. Fleming concluded that there was no direct relationship between hormone concentrations and maternal responsiveness in human mothers. However, hormone secretion is subject to considerable variation, and one-time sampling may not be sensitive enough to determine these complex cause–effect interactions. Moreover, PRL, a hormone known to facilitate the expression of maternal because in the rat (Bridges, DiBiase, Loundes, & Doherty, 1985), was not included in these computations.

In a subsequent study reported by Fleming and her colleagues (see chapter 10), 30 first-time mothers filled out the same maternal-attitude questionnaire at the end of pregnancy, again at days 3 and 4 postpartum, and then at 6 weeks postpartum. Mother–infant interactions were observed on days 3 and 4 and 2 months postpartum. On days 3 and 4 postpartum, samples were obtained immediately before and 1 hr after the feeding session and were assayed for the same hormones, but this time including PRL. In this case, a highly significant correlation was found between cortisol (both before and after nursing) and mothers' affectionate and vocal interactions with their infants on days 3 and 4 postpartum. These findings were strengthened if the mothers had expressed positive maternal feelings and attitudes during pregnancy. The other hormones measured showed no significant correlation, although the data on PRL measurement are still pending. Thus cortisol levels were correlated with responsiveness rather than attitudes. This finding, then, may be more reflective of inherent personality traits in the mothers, since cortisol is considered a "stress" hormone and may reflect reactivity to environmental situations rather than maternal attitudes. It would be of interest to determine if this correlation occurs in other situations not involving mother–infant interactions. In any case, the correlation between behavioral responsiveness and cortisol is surprising, because corticoids have not been shown to facilitate maternal behavior in other animals. Large increases in ACTH and high cortisol concentrations are, however, found during pregnancy and the puerperium, which in part reflects the large amounts of estrogen secreted in these states, as well as increases in CBG. But free cortisol (the active hormone) also increases in pregnancy, and measurement of the free active form would be of great interest. Conceivably, the cortisol increases observed could be a measure of expressivity that is tied to the maternal response.

Fleming's work is an important study that addresses the questions of hormones and their relevance to maternal behavior; yet, with the exception of cortisol, a hormone not implicated in the regulation of maternal behavior in other species, her findings do not show correlations between the hormones measured and her scales of maternal behavior.

Pregnancy and lactation are times of rapid change when hormonal determinants would be expected to prepare and enhance parenting behavior, and this pattern has been well documented in animals (Bridges, 1984; Moltz, 1974; Rosenblatt & Siegel, 1981). Parturition and milk secretion are synchronized, and yet PRL, the hormone responsible for lactation, has, surprisingly, not been linked to maternal behavior in the human. PRL does rise during pregnancy to approximately 100 times the normal level (Tyson et al., 1972), with the peak level attained in the third trimester. Postpartum levels fall rapidly over a 2-week period (Tyson et al., 1972) and tend to be slightly higher for 6 weeks in breast-feeding mothers; additional spurts of PRL are detected with suckling (Martin, Glass, Chapman, Wilson, & Woods, 1980). After 6 weeks, all parameters return to normal (Tyson et al., 1972, Voogt, 1978) (Figure 11.2).

EXPERIMENTS OF NATURE: CLINICAL ENDOCRINE CONDITIONS

In addition to studying normal pregnancy and lactation in order to understand the relationships between the female's physiology and parenting behavior (see chapter 10), it is possible to examine the behavioral responses associated with a number of aberrant endocrine states. One such state is the syndrome of pseudopregnancy, or pseudocyesis. This syndrome is particularly interesting because it appears to cause altered behavior as well as significant physical changes very similar to those detected in pregnancy. Examining the hormonal parameters of this syndrome may provide insight into the hormonal determinants of pregnancy-related behavior. The syndrome of pseudocyesis, or so-called false pregnancy, is well described in animals, particularly dogs (Horning, 1932). In humans, the condition was described by Hippocrates as early as 300 B.C., and the name is taken from the Greek *cyesis,* meaning "pregnancy." Hippocrates characterized the condition of these women as those "who imagine they are pregnant . . . seeing that their menses are suppressed and their matrices swollen." By the early 1980s, 512 cases of pseudocyesis had been reported in the English literature, with a large number (256) described in the nineteenth century (Bivin & Klinger, 1937; Tulandi, McInness, Mehta, & Tolis, 1982). Interestingly, a review of the recent literature reveals only 17 cases since 1960 (Murray & Abraham, 1978), with 22% being in the menopausal age group. The condition appears to be more common in societies where there is a high premium placed on a women's fertility (Murray & Abraham, 1978) and has been described with more frequency in certain cultures in South Africa where payment to a bride's family depends on the number of children produced (Cohen, 1982).

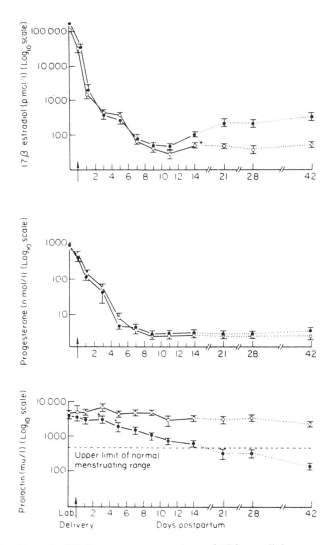

Figure 11.2. Mean (±SEM) serum concentrations of 17β-estradiol, progesterone, and prolactin in lactating and nonlactating subjects throughout the puerperium. Lactating subjects ($N = 10$); nonlactating subjects ($N = 9$). *$p < .01$; +$p < .02$. (From Martin et al., 1980, with permission)

The symptoms of pseudocyesis include amenorrhea, gradual abdominal enlargement, breast changes, and, occasionally, galactorrhea and areolar pigmentation. The patient usually experiences nausea and vomiting, apparent fetal movement, and weight gain. There may be an occasional positive pregnancy test. The distention of the abdomen is very significant and may fool an experienced physician. One unfortunate patient underwent a caesarean section for nonproductive labor. The pregnancy is almost always desired, and the patient becomes hostile and defensive if challenged about her pregnant state (Brown & Barglow, 1971; Murray & Abraham, 1978).

Recent investigations into this syndrome reveal an abnormal hormonal profile that fairly consistently shows elevated LH levels accompanied, in the great majority of cases, by elevated PRL levels. In addition to elevated LH levels, LH pulsatility is increased (Bray, Muneyyirci-Delale, Kofinas, & Reyes, 1987; Devane, Vera, Buhi, & Kalra, 1985; Tulandi, McInness, & Lal, 1983; Starkman, 1984; Tulandi et al., 1982; Yen et al., 1976; Zarate et al., 1974). The pattern of exaggerated pulsations of LH secretion suggests an abnormality of the neuroendocrine system—specifically, the intricate neurotransmitter systems that originate in the arcuate nucleus of the hypothalamus. The arcuate nucleus in the human is thought to act as a pulse generator and determines the pulsation of luteinizing hormone releasing hormone (LHRH). These pulses of LHRH are carried by the pituitary–portal system from the hypothalamus to the anterior pituitary, where they stimulate LH pulsations. Devane and coworkers suggested that endogenous opioid systems that modulate LHRH release may be suppressed in patients with pseudocyesis (Ropert, Quigley, & Yen, 1981). However, dynamic experiments in women using naloxone, an opioid antagonist, had no effect on LH pulsations, suggesting that opioids were not solely involved in producing the pattern of LH found in these women (Devane et al., 1985). The combination of elevated LH and PRL is particularly interesting, since LH is a luteotrophic agent in humans (Vande Wiele et al., 1970), whereas PRL exhibits luteotrophic actions in other animals, including the maintenance of the corpus luteum in the rat (Bishop, Orias, Fawcett, Krulich, & McCann, 1971). PRL is also linked with maternal behavior in these species (Bridges et al., 1985; Zarrow et al., 1972; see also chapter 6). Thus the synergistic effect of LH and PRL elevations may be responsible for the emergence of this syndrome, which involves very dramatic changes in behavior as well as in physiology. Even more dramatic is the reversal of these endocrine patterns with the resolution of these symptoms, suggesting a cause-and-effect relationship between the endocrine and the behavioral symptoms (Ayers & Seiler, 1984; Devane et al., 1985; Tulandi et al., 1983; Yen et al., 1976).

Other authors (Abram, 1969; Ayers & Seiler, 1984; Brown & Barglow, 1971; Cramer, 1971; Devane et al., 1985) have noted a high incidence of depression in patients with pseudocyesis and have suggested that the endocrine alterations associated with depression, which include significant changes in pituitary feedback mechanisms, may in some cases provide the setting for an endocrine dysfunction of the type described and, in turn, cause a neurobehavioral illness (Brown & Barglow, 1971) (Figure 11.3). Moreover, depression is known to be associated with hormonal alterations; these include abnormal neuroendocrine responses to dexamethasone with suppression of cortisol (Carroll et al., 1981), an abnormal response to ACTH (Reus, Joseph, & Dallmann, 1982), and abnormal thyroid releasing factor (TRF) stimulation of TSH (Gold et al., 1979) and GH (Maeda et al., 1975). Hypothalamic dysfunction of hypothalamic pathways in close proximity to those involved in depression is entirely plausible. An exaggerated PRL response to TRH, which has been described in pseudocyesis, is also seen in depression (Maeda et al., 1975; Post et al., 1978). Of particular interest is the report that the drug reserpine, which may induce depression by depleting brain biogenic amines, may also induce pseudocyesis in animals and lactation and amenorrhea in humans (Brown & Barglow, 1971; Cappolla, Leonardi, Lippman, Perrine, & Ringler, 1965). Thus the depletion of important neurotransmitters, such as the biogenic amines in depression, may be a causal event in the development of this syndrome. Other abnormalities suggestive of hypothalamic–pituitary

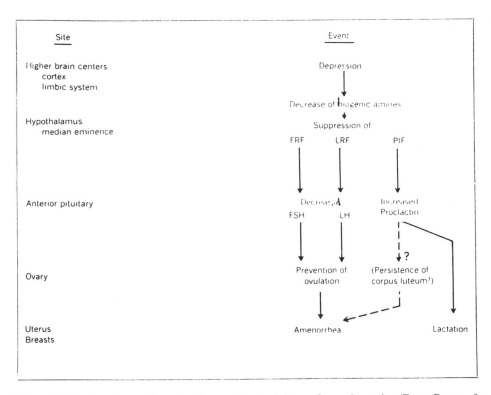

Figure 11.3. A scheme depicting the psychophysiology of pseudocyesis. (From Brown & Barglow, 1971, with permission)

derangement include abnormal LH responses to LHRH, exaggerated PRL responses to TRH, and aberrant LH responses to TRH (Tulandi et al., 1983). Other data have suggested that endogenous depression is associated with a decrease in central nervous system (CNS) catecholamine activity. Specifically, Yen has demonstrated that dopamine inhibits the release of PRL and LH in humans (Leblanc, Lacheliin, Abu-Fadil, & Yen, 1976). Thus reduced dopamine activity could account for the hypersection of PRL and LH in patients with pseudocyesis. A report of a case before and after abdominal deflation suggests that depression in this patient may have induced hormonal changes that led to the pseudopregnant state, although other neuroendocrine mechanisms may have also been involved (Figure 11.4).

In another study, Bray and co-workers (1987) demonstrated an increase in circulating prolactin levels and the amplitude of PRL pulsations in patients with pseudocyesis relative to normal cycling women. In addition, the normal sleep-entrained neuroendocrine rhythm was not preserved in these patients. All these changes suggest an altered neuroendocrine rhythm that probably reflects more than just an increased dopaminergic tone.

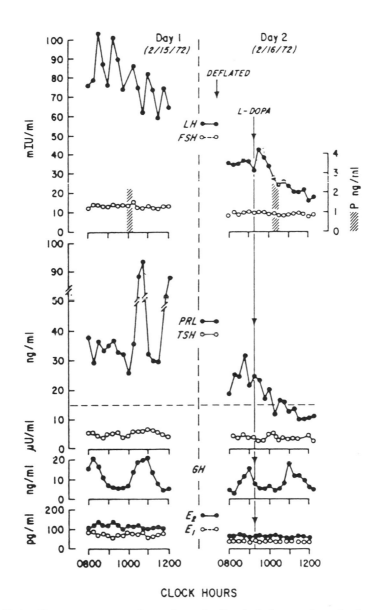

Figure 11.4. Hormone patterns of pseudocyesis. Basal pituitary and ovarian hormone concentrations and their patterns of fluctuation in a patient with pseudocyesis studied before (day 1) and after (day 2) clinical resolution (deflation). L-dopa (500 mg, po) was given on day 2. The hatched bars in the top panels represent progesterone levels. The dotted lines in the second panel denote the mean value of prolactin (PRL) in normal women. (From Yen et al., 1976, with permission)

POTENTIAL MODELS FOR STUDYING ENDOCRINE REGULATION
OF PARENTING IN HUMANS

Another issue of interest when examining parenting behavior is the question of competency in parenting care. Studies in primates have shown that psychosocial development plays a fundamental role in the ability of the mother to care competently for the newborn. With the firstborn, monkey and ape mothers are usually highly incompetent and depend on the close support of female relatives. With each successive birth, there appears to be an increase in maternal competence (Montagu, 1981).

These observations may be of relevance to humans. Statistics have shown that young teenage mothers are in a state of "physiological and psychological unreadiness" (Montagu, 1981) for motherhood. Morbidity, teratology, and mortality rates are significantly higher for both the young teenage mother and the child compared with similar rates in later life. Girls under 15 have only a 50% chance of giving birth to a normal child (Nortman, 1974). The statistics indicate that the adolescent girl not only is unready physiologically for healthy reproduction, but also is a poor mother. The mortality rates of their 1- to 4-year-old infants exceed the average by 4%. Mental retardation and handicap rates also indicate the deficiencies in the children of these young mothers (Nortman, 1974).

Moreover, teenage mothers are thought to have increased medical risks during pregnancy that are tied to their youthful age. The risks are much greater for those who are poor and black, but are observed in all socioeconomic groups (Dryfoos & Lincoln, 1981). Some studies have shown that the problems encountered in teenage pregnancy are as much a result of social-risk factors as of medical risks (Chase, 1973); others have suggested inadequate diet as a persistent element (Weigley, 1975). Still others have reported that physiological and anatomical immaturity (Clark, 1971; Miller & Field, 1985), as well as sociocultural factors, may contribute to the inadequate parenting performance of adolescents. In any case, the outcome of adolescent pregnancy is consistently poor, especially for those under age 14.

The adolescent girl is considered hormonally immature as well. A high incidence of anovulatory cycles occur in women under age 19 (Collett, Wertenberger, & Riske, 1954; Apter, Viinikka, & Wihko, 1978; Venturoli et al., 1986), especially in the first 2 years after menarche. If hormonal factors affect maternal competence, one might predict that hormones would be less effective in the adolescent. One hormone that might affect the caretaking behavior of adolescents is PRL. PRL levels seem to rise with each successive birth, especially when an infant is breast-fed, and may account for the improved maternal competence of mothers who have already had a child. A rise in the concentration of PRL has also been observed in adolescents during puberty, and the presence of this hormone may be necessary for maternal competence. The role of PRL is complex, but in view of its importance as a luteotropic and nurturing factor in animals, its role in preparing women for caretaking behavior deserves investigation. Unfortunately, the relationship between maternal caretaking behavior and the hormonal changes that take place during pregnancy, parturition, and lactation has not been extensively studied in humans. On a statistical level, however, maternal interest and breast-feeding have been correlated in the majority of studies (Newton, 1973). The hormonal basis for triggering of caretaking behavior, including maternal aggression (Svare, Mann, Broida, & Michael, 1982), in humans in the postpartum period is still unchartered territory. Breast-feeding has in fact powerful physiological effects, including causing uterine contractions due to a release of oxytocin from the posterior pituitary. In animals, experiments suggest that maternal care appears heavily dependent on hormonal conditions normally present around the time of parturition (Zarrow et al., 1972).

CONCLUSION

Although there is little direct evidence at present for the existence of hormonal determinants of parental behavior in humans, the degree of conformity observed in the natural situation with respect to maternal and paternal behavior, as well as the behavioral changes associated with the endocrine states of pseudocyesis and the state of teenage pregnancy, suggests that hormones may play some role in affecting parental behavior in the human female. The clinical condition of pseudopregnancy presents researchers with a model for examining the role of endocrine factors in regulating maternal behavior. Together with studies of pregnant and postparturient women and possibly the examination of the relationships between the endocrine system and behavior in teenage mothers, it will be possible to gain a better understanding of the relationships between endocrine functions associated with normal pregnancy, parturition, and lactation and parenting behaviors in humans. Whether and how hormonal determinants may be involved in appropriate caretaking behavior and survival of the young are important issues that can be addressed by examining clinical endocrine conditions as well as normative physiological and behavioral states.

REFERENCES

Abram, H. S. (1969). Pseudocyesis followed by true pregnancy in the termination phase of an analysis. *British Journal of Medical Psychology, 42,* 255–262.

Apter, D., Viinikka, L., & Wihko, R. (1978). Hormonal pattern of adolescent menstrual cycles. *Journal of Clinical Endocrinology and Metabolism, 47,* 944–954.

Ayers, J. W. T., & Seiler, J. C. (1984). Neuroendocrine indices of depression in pseudocyesis: A case report. *Journal of Reproductive Medicine, 29,* 67–70.

Bishop, W., Orias, R., Fawcett, C. P., Krulich, L., & McCann, S. M. (1971). Plasma gonadotropins and prolactin in pseudopregnancy in the rat. *Proceedings of the Society for Experimental Biology and Medicine, 137,* 1411–1414.

Bivin, G. D., & Klinger, M. P. (1937). *Pseudocyesis.* Bloomington, IN: Principia.

Bray, M. A., Muneyyirci-Delale, O., Kofinas, G. D., & Reyes, F. I. (1987). Prolactin pulsatility is pseudocyesis. *Endocrinology* (Suppl.) *120,* 405.

Bridges, R. S. (1984). A quantitative analysis of the roles of dosage, sequence, and duration of estradiol and progesterone exposure in the regulation of maternal behavior in the rat. *Endocrinology, 114,* 930–940.

Bridges, R. S., DiBiase, R., Loundes, D. D., & Doherty, P. C. (1985). Prolactin stimulation of maternal behavior in female rats. *Science, 227,* 782–784.

Brown, E., & Barglow, P. (1971). Pseudocyesis: A paradigm for psychophysiological interactions. *Archives of General Psychiatry, 24,* 221–229.

Cappolla, J. A., Leonardi, R. G., Lippman, W., Perrine, J. W., & Ringler, I. (1965). Induction of pseudopregnancy in rats by depletors of endogenous catecholamines. *Endocrinology, 77,* 485–490.

Carroll, B. J., Feinberg, M., Gredan, J. F., Tarika, J., Albala, A. A., Haskett, R. F., James, N. M., Kronfol, Z., Lohr, N., Steiner, M., de Vigne, J. P., & Young, E. (1981). A specific laboratory test for the diagnosis of melancholia. *Archives of General Psychiatry, 38,* 15–22.

Chase, H. C. (1973). A study of risks, medical care and infants mortality: 1. Selected substantive results. *American Journal of Public Health, 63* (Suppl.), 3–16.

Clark, J. F. J. (1971). Adolescent obstetrics: Obstetric and sociologic implications. *Clinical Obstetrics Gynecology, 14,* 1026–1036.

Clark, J. H., Paszko, Z., & Peck, E. J., Jr. (1977). Nuclear binding and retention of the receptor–estrogen complex: Relation to the agonistic and antagonistic properties of estriol. *Endocrinology, 100,* 91–96.

Cohen, L. (1982). A current perspective of pseudocyesis. *American Journal of Psychiatry, 139,* 1140–1144.

Collett, M. E., Wertenberger, G. E., & Riske, V. M. (1954). The effect of age upon the pattern of the menstrual cycle. *Fertility and Sterility, 5,* 437–438.

Cramer, B. (1971). Delusions of pregnancy in a girl with drug-induced lactation. *American Journal of Psychiatry, 127,* 136–139.

Devane, G. W., Vera, M. I., Buhi, W. C., & Kalra, P. S. (1985). Opioid peptides in pseudocyesis. *Obstetrics and Gynecology, 65,* 183–187.

Dryfoos, J. G., & Lincoln, R. (1981). *Teenage pregnancy: The problem that hasn't gone away.* New York: Alan Guttmacher Institute.

Ehrhardt, A. A., & Baker, S. W. (1974). Fetal androgens, human central nervous system differentiation and behavior sex differences. In R. C. Friedman, R. M. Richard, R. L. Vande Wiele (Eds.), *Sex differences in behavior* (pp. 33–51). New York: Wiley.

Goebelsmann, U. (1986). Endocrinology of pregnancy. In D. R. Mishell & V. Davajan (Eds.), *Infertility, contraception and reproductive endocrinology* (pp. 113–141). Oradell, NJ: Medical Economics.

Gold, M. S., Pottash, A. L. G., Davies, R. K., Ryan, N., Sweeney, D. R., & Martin, D. M. (1979). Distinguishing unipolar and bipolar depression by thyrotropin release test. *Lancet, 2,* 411–412.

Horning, J. C. (1932). Nervous pregnancy in the dog. *Veterinarian Medicine* (Praha), *27,* 24–31.

Jaffe, R. B. (1986a). Endocrine physiology of the fetus and feto-placental unit. In S. S. C. Yen & R. B. Jaffe (Eds.), *Reproductive endocrinology* (pp. 737–757). Philadelphia: Saunders.

Jaffe, R. B. (1986b). Protein hormones of the placenta, decidua and fetal membranes. In S. S. C. Yen & R. B. Jaffe (Eds.), *Reproductive endocrinology* (pp. 758–769). Philadelphia: Saunders.

Jaffe, R. B. (1986c). Integrative maternal–fetal endocrine control systems. In S. S. C. Yen & R. B. Jaffe (Eds.), *Reproductive endocrinology* (pp. 770–778). Philadelphia: Saunders.

Josimovich, T. B. (1980). Hormonal physiology of pregnancy: Steroid hormones of the placenta, and polypeptide hormones of the placenta and pituitary. In J. J. Gold & T. B. Josimovich (Eds.), *Gynecologic endocrinology.* Hagerstown, MD: Harper & Row.

Leblanc, H., Lacheliin, G. C., Abu-Fadil, S., & Yen, S. S. C. (1976). Effects of dopamine infusion on pituitary hormone secretion in humans. *Journal of Clinical Endocrinology and Metabolism, 43,* 668–674.

Maeda, K., Kato, Y., Ohgo, S., Chihara, K., Yoshimoto, Y., Yamaguchi, N., Kuromaru,

S., & Imura, H. (1975). Growth hormone and prolactin release after injection of thyrotropin-releasing hormone in patients with depression. *Journal of Clinical Endocrinology and Metabolism, 40,* 501–505.

Martin, R. H., Glass, M. R., Chapman, C., Wilson, G. D., & Woods, K. L. (1980). Human α-lactalbumin and hormonal factors in pregnancy and lactation. *Clinical Endocrinology, 13,* 223–230.

Martucci, C., & Fishman J. (1977). Direction of estradiol metabolism as a control of its hormonal action-uterotrophic activity of estradiol metabolites. *Endocrinology, 101,* 1709–1715.

McCullogh, J., Quadagno, D. M., & Goldman, B. D. (1974). Neonatal gonadal hormones: Effect on maternal and sexual behavior in the male rat. *Physiology and Behavior, 12,* 183–184.

Miller, K. A., & Field, C. S. (1985). Adolescent pregnancy: Critical review for the clinician. *Seminars in Adolescent Medicine, 1,* 195–212.

Mishell, D. R., & Marrs, R. P. (1986). Endocrinology of lactation and the puerperium. In D. R. Mishell & V. Davajan (Eds.), *Infertility, contraception and reproductive endocrinology* (pp. 143–162). Oradell, NJ: Medical Economics.

Moltz, H. (1974). Some mechanisms governing the induction, maintenance and synchrony of maternal behavior in the laboratory rat. In W. Montagna & W. A. Sadler (Eds.), *Reproductive behavior* (pp. 77–96). New York: Plenum Press.

Montagu, A. (1981). The adolescent's unreadiness for pregnancy and motherhood. *Pediatric Annals, 10,* 597–511.

Murray, J., & Abraham, G. (1978). After office hours: Pseudocyesis, a review. *Obstetrics and Gynecology, 51,* 627–631.

Newton, N. (1973). Interrelationships between sexual responsiveness, birth, and breast feeding. In J. Zubin & J. Money (Eds.), *Contemporary sexual behavior: Critical issues in the 1970's* (pp. 77–98). Baltimore: Johns Hopkins University Press.

Nortman, D. (1974, August). Parental age as a factor in pregnancy outcome and child development. *Reports on Population Planning* (No. 16). New York: Population Council.

Post, R. M., Gerner, R. H., Carman, J. S., Gillin, J. C., Jimerson, D. C., Goodwin, F. K., & Bunney, W. E., Jr. (1978). Effects of a dopamine agonist piribedil in depressed patients: Relationship of pretreatment homovanillic acid to antidepressant response. *Archives of General Psychiatry, 35,* 609–615.

Resnik, R., Killam, A. P., Battaglia, F. C., Makowski, E. L., & Meschia, G. (1974). The stimulation of uterine blood flow by various estrogens. *Endocrinology, 94,* 1192–1196.

Reus, V. I., Joseph, M. S., & Dallman, M. F. (1982). ACTH levels after the dexamethasone suppression test in depression. *New England Journal of Medicine, 306,* 238–239.

Ropert, J. F., Quigley, M. E., & Yen, S. S. C. (1981). Endogenous opiates modulate pulsatile luteinizing hormone release in humans. *Journal of Clinical Endocrinology and Metabolism, 52,* 583–585.

Rosenblatt, J. S. (1967). Nonhormonal basis of maternal behavior in the rat. *Science, 156,* 1512–1514.

Rosenblatt, J. S., & Siegel, H. I. (1981). Factors governing the onset and maintenance of maternal behavior among nonprimate mammals: The role of hormonal and nonhormonal factors. In D. J. Gubernick & P. H. Lopfer (Eds.), *Parental care in mammals* (pp. 13–76). New York: Plenum Press.

Siiteri, P. K., Febres, F., Clemens, L. E., Chang, R. J., Gondos, B., & Stites, D. (1977). Progesterone and maintenance of pregnancy: Is progesterone nature's immunosuppressant? *Annals of the New York Academy of Sciences, 286,* 384–397.

Starkman, M. N. (1984). Impact of psychodynamic factors on the course and management of patients with pseudocyesis. *Obstetrics and Gynecology, 64,* 142–145.

Svare, B., Mann, M. A., Broida, J., & Michael, S. D. (1982). Maternal aggression exhibited by hypophysectomized parturient mice. *Hormones and Behavior, 16,* 455–461.

Tulandi, T., McInnes, R. A., & Lal, S. (1983). Altered pituitary hormone secretion in patients with pseudocyesis. *Fertility and Sterility, 40,* 637–641.

Tulandi, T., McInnes, R. A., Mehta, A., & Tolis, G. (1982). Pseudocyesis: Pituitary function before and after resolution of symptoms. *Obstetrics and Gynecology, 59,* 119–121.

Tyson, J. E., Hwang, P., Guyda, H., & Friesen, H. G. (1972). Studies of prolactin secretion in human pregnancy. *American Journal of Obstetrics and Gynecology, 113,* 14–20.

Vande Wiele, R. L., Bogumil, J., Dyrenfurth, J., Ferin, M., Jewelewicz, R., Warren, M., Rizkallah, T., & Mikhail, G. (1970). Mechanisms regulating the menstrual cycle in women. In G. Pincus (Ed.), *Recent progress in hormone research* (Vol. 26, pp. 63–103). New York: Academic Press.

Velle, W. (1982). Sex hormones and behavior in animals and man. *Perspectives in Biology and Medicine, 25,* 295–315.

Venturoli, S., Porcu, E., Fabbri, R., Paradisi, R., Ruggeri, S., Bolelli, G., Orsini, L. F., Gabbi, D., & Flamigni, C. (1986). Menstrual irregularities in adolescents: Hormonal pattern and ovarian morphology. *Hormone Research, 24,* 269–279.

Voogt, J. L. (1978). Control of hormone release during lactation. *Clinics in Obstetrics and Gynaecology, 5,* 435–455.

Weigley, E. S. (1975). The pregnant adolescent: A review of instructional research and programs. *Journal of the American Dietetic Association, 66,* 588–592.

Yen, S. S. C., Rebar, R. W., & Quesenberry, W. (1976). Pituitary function in pseudocyesis. *Journal of Clinical Endocrinology and Metabolism, 43,* 132–136.

Zarate, A., Canales, E. S., Soria, J., Jacobs, L. S., Daughaday, W. H., Kastin, A. J., & Schally, A. V. (1974). Gonadotropin and prolactin secretion in human pseudocyesis: Effect of synthetic luteinizing hormone-releasing hormone (LH-RH) and thyrotropin releasing hormone (TRH). *Annales d'endocrinologie, 35,* 445–450.

Zarrow, M. X., Denenberg, V. H., & Sachs, B. D. (1972). Hormones and maternal behavior in mammals. In S. Levine (Ed.), *Hormones and behavior* (pp. 105–134). New York: Academic Press.

III
NEUROBIOLOGICAL CORRELATES OF PARENTING

The work of this part focuses on the central nervous system and its role in parenting. The chapters review neuroanatomical, neurochemical, neurohumoral, and neurophysiological factors involved in the regulation of maternal behavior in female rodents and sheep.

The first chapter by Michael Numan, presents a review of the literature on the neuroanatomical bases for maternal behaviors (e.g., nest building, retrieval of pups, anogenital licking) observed in female rats. In addition to analyzing the contributions made by sensory factors (tactile sensitivity and olfaction) that have been shown to be correlated with maternal behavior, Numan provides an extensive overview of the literature on the neural structures that have been shown by electrolytic, chemical, and knife-cut lesions to subsume the behavioral repertoire associated with parenting in laboratory studies. He describes in detail the known efferent and afferent connections of the medial preoptic area (MPOA) to other brain regions and structures thought to be necessary for maternal behavior. The evidence for involvement of the septal area, amygdala, hypothalamus, and ventral tegmentum in circuits that impinge on the MPOA is also reviewed. Finally, Numan discusses the neural sites at which hormones act to regulate aspects of maternal behavior.

This chapter is followed by Thomas Insel's, which elucidates the role of the pituitary hormone oxytocin in the initiation of maternal behavior in the rat. Insel discusses experimental evidence that demonstrates how high concentrations of oxytocin administered centrally will, under appropriate environmental circumstances, elicit full maternal behavior in estrogen-primed female rats. Next, he summarizes the putative mechanisms of action for the effects of oxytocin and provides evidence that brain oxytocin receptors are increased in female rats that have given birth. Finally, Insel relates the action of oxytocin to that of endogenous opiates and prolactin, two other factors directly implicated in the onset of maternal behavior.

The next chapter, by E. B. Keverne and K. M. Kendrick, reviews evidence on the neurochemical changes that occur in the cerebrospinal fluid (CSF) during labor, vaginal-cervical stimulation, lamb separation, and suckling in the female sheep. Measurements made in CSF oxytocin concentrations and in the levels of monoamines and their metabolites are described. The authors then summarize studies on microdialysis measures of neurochemical release in specific brain loci during parturition. The results described focus on dopamine, homovanillic acid, MHPG, and oxytocin. Also covered are experiments that measure transmitter turnover in CSF. Such turnover is considered an important index of the degree of neurochemical activity within the nervous system, since there is thought to be a widespread distribution throughout the brain of terminals that arise from relatively few neurons. Like Insel, Keverne and Kendrick discuss the functional significance of central oxytocin release and parturition. They also summarize the evidence on what they believe is the functional significance of noradrenergic activation that occurs at parturition. Similar to the other authors in this volume, they recognize the role of sensation (particularly olfaction) in the initiation of parenting behaviors. However, Keverne and Kendrick link sensory stimuli specifically with subsystems of the brain's noradrenergic projections.

In the following chapter, Barbara Modney and Glenn Hatton describe new experimental evidence that indicates that the brains of animals that exhibit maternal behavior are significantly different from those of animals that do not express such behavior. The observations leading to the conclusion are based on studies of the magnocellular hypothalamo-neurohypophyseal system, which makes and secretes the neuropeptides oxytocin and vasopressin. The data support the conclusion that cell-to-cell interactions are characterized by the formation of new specialized synapses on the dendrites of magnocellular neurons during pregnancy and just after birth. Further, neuronal cell bodies receive new synaptic input, but unlike the dendrites, these cells become synaptically connected during lactation. Modney and Hatton suggest that electrical coupling among the neurons of interest may be facilitated by olfactory and/or vomeronasal stimulation. Also of great interest is the authors' report that their data support the conclusion that the changes observed return to their prepregnancy status once the females' pups are weaned.

The last chapter of this part, by Clark Grosvenor, Girish Shah, and Bill Crowley, focuses on the tactile and sensory stimuli emitted by the pup that influence the release of prolactin in the lactating female rat. The authors examine the hypothalamic mechanisms that transform neural stimuli into secretory changes in the pituitary cells that secrete prolactin. Next, they review the evidence concerning afferent control of prolactin secretion during the female's lactation phase. The authors provide an overview of the data on how the litter, acting as an exteroceptive stimulus, influences the secretion of prolactin. They also discuss the specific studies that demonstrate induction of prolactin release by pups when they suckle the dam. In addition, they review studies on the effects of deficiencies in milk-derived prolactin upon the subsequent development of neuroendocrine function. Their results suggest that decreases in the availability of milk-derived prolactin during a brief postnatal period result in reduced dopaminergic inhibition of prolactin secretion, which can continue into maturity. The authors conclude by describing additional studies on the effects of early prolactin deficiency on the regulation of prolactin secretion.

12

Neural Control of Maternal Behavior

MICHAEL NUMAN

The purpose of this chapter is to summarize what is known about the neural basis of maternal behavior. Since most of the work in this area has been done on the rat, the review is restricted to that species. This approach allows for an integrative presentation.

A complex of behavioral changes is associated with the maternal condition, and these changes distinguish the postpartum female from her estrous cycling, nonmaternal counterpart. Maternal behavior is defined as those behaviors shown by the postpartum female that increase the likelihood that her offspring will survive. Such behaviors can be classified into those that are infant directed (retrieving, nursing, anogenital licking) and those that are not (nest building, maternal aggression, lactational hyperphagia). Rats give birth to young that are helpless, essentially immobile, incapable of temperature regulation, and completely dependent on maternal care for their growth and survival. As a result of such care, the young develop and are weaned between 3 and 4 weeks of age. The mother rat builds an elaborate nest around her young from available materials, such as paper strips. These materials are carried to the nest site in the female's mouth; she also uses her mouth (and forepaws) to mold the nest material into a compact structure. The nest serves to insulate the young and therefore to keep them warm in the mother's absence. A maternal female can also be observed to retrieve or transport her young by carrying them in her mouth. Such behavior occurs when a pup becomes displaced from the nest or if the mother changes her nesting area. In the nest, the mother can be seen to lick the anogenital region of the young and to nurse the young. Anogenital licking stimulates urination and defecation; interestingly, the consumption of urine by the mother returns to her a significant proportion of the water that she loses as a result of lactation (Gubernick & Alberts, 1983). During nursing, the mother crouches over the young in order to expose her mammary region. Such crouching behavior not only provides milk, but also serves to warm the young. Rodent mothers also show a high level of aggression toward intruders at the nest site (Ostermeyer, 1983; Svare, 1981). This maternal aggression presumably protects the young from possible injury. Finally, lactating females are hyperphagic and show an increased appetite for salt (Denton & Nelson, 1980; Fleming, 1976). The increased food intake appears to be important for meeting the nutritional demands associated with lactation.

Most studies on the neural basis of maternal behavior have examined the neural control of retrieving, nest-building, and nursing behavior; therefore, this chapter is, for the most part, restricted to those behaviors. However, the above description of maternal behavior stresses its complexity and raises the question of whether a unitary neural system underlies the entire complex. Some of the work reviewed below is relevant to this issue.

The approach taken in this chapter can be termed a "functional neuroanatomical" one. An attempt is made to present an analysis of the neural circuits underlying maternal behavior and the interactions among neural structures underlying this response. Given that something is known about the functions served by the elements of the circuits, hypotheses can be developed about the requisite sensory, motor, or other changes that occur in the maternal organism.

To conclude this introduction, the role of hormones in the control of maternal behavior in the rat should be briefly mentioned. A large body of evidence indicates that the endocrine changes associated with the termination of pregnancy activate pup-directed maternal behaviors in rats (see chapter 6). Interestingly, although the initiation of maternal behavior at parturition is dependent on hormonal stimulation, its continuance during the postpartum period appears to be free from hormonal control (Numan, Leon, & Moltz, 1972; Rosenblatt, Siegel, & Mayer, 1979). The current view is that the continuance of maternal behavior during the postpartum period is solely dependent on the stimulative properties of the young acting on the relevant neural substrate.

SENSORY BASIS OF MATERNAL BEHAVIOR

In 1956, Beach and Jaynes invoked the concept that the retrieving response in the rat is under multisensory control. By this, they meant that no one sensory modality is essential for the performance of retrieving by postpartum rats. They found that the surgical elimination of either vision, olfaction, or tactile sensitivity of the snout and perioral region leaves retrieving intact. Subsequent work has confirmed that anosmia or blindness does not eliminate the maternal behavior of postpartum rats (Benuck & Rowe, 1975; Herrenkohl & Rosenberg, 1972) and has added the finding that the surgical removal of auditory input also leaves pup-directed maternal responses intact (Herrenkohl & Rosenberg, 1972).

Recent work by Kenyon, Cronin, and Keeble (1981, 1983) has shown that tactile input from the perioral region may be more important for retrieving behavior in the rat than the work of Beach and Jaynes (1956) originally suggested. It should be noted that Beach and Jaynes severed the sensory branches of the trigeminal nerve in order to eliminate tactile input from the perioral region. However, they did not report the time interval between transection and retrieval tests, nor did they verify that the deafferentation procedure resulted in an absence of perioral tactile sensitivity. Therefore, peripheral regeneration of trigeminal input to perioral regions may have been responsible for the lack of behavioral effect they observed. Kenyon et al. (1981, 1983), in contrast, found that injection of a local anesthetic into the mystacial pads (which desensitizes the snout region) or a section of the infraorbital branch of the trigeminal nerve (which blocks tactile input from the upper lip and snout) interferes with the retrieval response of postpartum rats. The females approach displaced pups and nose them but do not retrieve them. Importantly, however, retrieving behavior recovers in spite of continued perioral desensitization of the upper lip and snout. The mechanism mediating this recovery has not been elucidated. Perhaps a more complete desensitization of the perioral region, including the mandibular region (lower lip and jaw), would result in a more long-lasting deficit. The findings of Kenyon et al. have been confirmed by Stern and Kolunie (1987). In addition, these investigators (Kolunie & Stern, 1987) report that desensitization of the snout region eliminates maternal aggression toward an intruder male rat.

These results on the role of tactile input from the perioral region in the control of retrieving and aggression are important because they suggest that central neural mechanisms might operate to influence maternal responsiveness by affecting trigeminal sensorimotor mechanisms (see MacDonnell & Flynn, 1966; Numan, 1987). That is, tactile input from the perioral region may be ineffective in triggering certain maternal responses in a nonmaternal organism, but, as a result of the activation of the appropriate central mechanisms, may be quite effective in the postpartum female.

Although olfaction is not essential for pup-directed maternal responses in the rat, a recent line of research indicates that olfaction plays an *inhibitory* role in the control of maternal behavior. Naive virgin female rats, unlike parturient females, are not immediately responsive to young (Rosenblatt, 1967). This is what should be expected if the endocrine events associated with the termination of pregnancy promote maternal behavior. However, the experimental production of anosmia in virgin females promotes maternal responsiveness in the *absence* of hormonal stimulation (Fleming & Rosenblatt, 1974; Mayer & Rosenblatt, 1977). This suggests that pup odors may inhibit maternal behavior in virgins and that one of the ways hormonal stimulation may facilitate maternal responsiveness is by modifying the parturient female's reaction to pup odors. Two major chemoreceptive structures exist within the nasal cavity: the primary olfactory epithelium and the vomeronasal organ (Raisman, 1972). Fleming, Vaccarino, Tambosso, and Chee (1979) have shown that both of these sources of olfactory input inhibit maternal behavior in virgin female rats, and that the surgical removal of input from both the vomeronasal system and the primary system facilitates maternal behavior in virgins to a greater degree than elimination of either system alone.

One sensory modality that has received very little attention with respect to its role in maternal behavior control is taste. Since maternal females lick their young at parturition and during the postpartum period, gustatory input may provide important information to central mechanisms underlying maternal behavior.

FOREBRAIN STRUCTURES AND MATERNAL BEHAVIOR

Medial Preoptic Area

Research from several laboratories has indicated that the medial preoptic area (MPOA) of the basal forebrain is one of the most important structures underlying the neural control of maternal behavior. The MPOA lies just rostral to the anterior hypothalamus and just caudal to the diagonal band–septal area. Lateral to the MPOA lies the lateral preoptic area (LPOA), and dorsal to the MPOA lies the bed nucleus of the stria terminalis (BNST) and the anterior commissure (refer to Figure 12.1 for the

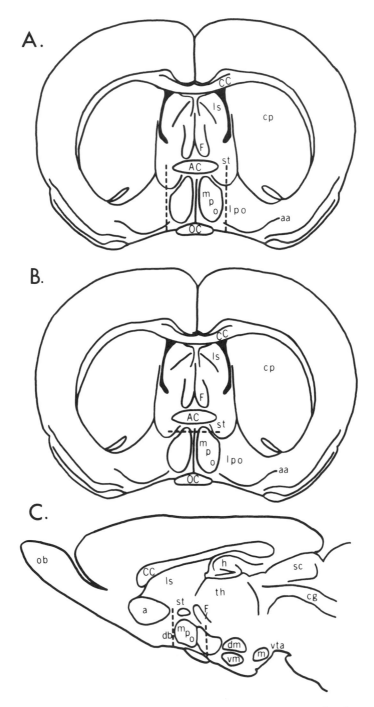

Figure 12.1. (A) A frontal section through the medial preoptic area showing a knife cut (dashed lines) severing the lateral connections of the medial preoptic area. (B) A frontal section through the medial preoptic area showing a knife cut (dashed line) severing the dorsal connections of the medial preoptic area. (C) A sagittal section through the level of the medial

Table 12.1 Some important abbreviations

BNST	Bed nucleus of the stria terminalis
HRP	Horseradish peroxidase
LH	Lateral hypothalamus
LMT	Lateral midbrain tegmentum
LPOA	Lateral preoptic area
MBH	Medial basal hypothalamus
MLR	Mesencephalic locomotor region
MPOA	Medial preoptic area
NE	Norepinephrine
NMA	N-methyl-aspartic acid
NST	Nucleus of the solitary tract
PCPA	Parachlorophenylalanine
PVN	Paraventricular hypothalamic nucleus
VTA	Ventral tegmental area

general location of the MPOA). (See Table 12.1 for a list of important abbreviations used in this chapter.)

Electrolytic or radiofrequency lesioning of the MPOA, performed on postpartum lactating rats, has been found to disrupt retrieving, nest-building, and nursing behavior (Jacobson, Terkel, Gorski, & Sawyer, 1980; Numan, 1974; Numan, Corodimas, Numan, Factor, & Piers, 1988). Most importantly, the selective destruction of MPOA cell bodies with N-methyl-DL-aspartic acid (NMA), a neurotoxic amino acid that spares fibers of passage, has also been found to disrupt nest-building, retrieving, and nursing behavior (Numan et al., 1988). Some of these results are shown in Table 12.2. It is clear, therefore, that MPOA neurons are essential for the display of maternal behavior.

Evidence concerning the MPOA efferents that may be important for maternal behavior has been provided by studies employing the knife-cut technique (Franz, Leo, Steuer, & Kristal, 1986; Miceli, Fleming, & Malsbury, 1983; Numan, 1974; Numan & Callahan, 1980; Numan & Corodimas, 1985; Terkel, Bridges, & Sawyer, 1979). These studies suggest that it is the lateral efferent projections from the MPOA that are most important for maternal behavior. The Numan and Callahan study provides the clearest results. Female rats received knife cuts that severed either the lateral, dorsal, anterior, or posterior connections of the MPOA. The various cuts are shown in Figure 12.1. The basic finding was that only females that received knife

preoptic area showing the placement of a knife cut severing the anterior connections of the medial preoptic area (dashed line between db and mpo) and a knife cut severing the posterior connections of the medial preoptic area (dashed line below F). Abbreviations: a, nucleus accumbens; aa, anterior amygdaloid area; AC, anterior commissure; ah, anterior hypothalamic area; CC, corpus callosum; cg, central gray; cp, caudate-putamen; db, nucleus of the diagonal band; dm, dorsomedial nucleus of the hypothalamus; F, fornix; h, hippocampus; lpo, lateral preoptic area; ls, lateral septal nucleus; m, mammillary bodies; mpo, medial preoptic area; ob, olfactory bulb; OC, optic chiasm; sc, superior colliculus; st, bed nucleus of the stria terminalis; th, thalamus; vm, ventromedial hypothalamic nucleus; vta, ventral tegmental area. (From Numan, 1988)

Table 12.2 Preoperative and postoperative retrieval, nest-building, and nursing scores in postpartum rats[a]

Group	N	Percent retrieving		Median nest-building score[b]		Mean ($\pm$ SE) nursing time (min)[c]	
		Preop.	Postop.	Preop.	Postop.	Preop.	Postop.
NMA-MPOA	9	100	0*	3.3	0.2*	15.8 ± 1.5	3.3 ± 1.0*
				(2.3–3.7)	(0–1.1)		
PB-MPOA	10	100	100	3.1	2.5	15.8 ± 0.9	12.9 ± 0.9
				(2.5–3.7)	(1.0–3.0)		

Note: NMA = *N*-methyl-DL-aspartic acid lesion; PB = phosphate buffer vehicle solution; MPOA = medial preoptic area. Preoperative tests were conducted on days 1 to 3 postpartum. NMA injections or control injections of PB into the MPOA were performed on day 4 postpartum. Postoperative tests were conducted on days 5 to 12 postpartum.
[a] Modified from Numan et al. (1988).
[b] Nests were rated on a 4-point scale, with 0 = no nest and 4 = excellent nest. Numbers in parentheses indicate the range.
[c] Nursing observations consisted of daily 20-min observations.
* Significantly different from the PB-MP control group; $p < .01$.

cuts severing the lateral connections of the MPOA showed a selective disruption of maternal behavior.

In all the studies cited above, the *acute* effects of preoptic damage to maternal responsiveness were examined. That is, maternal behavior tests began within 24 hr of the production of brain damage and continued for 1 or 2 weeks. When appropriate measures were taken, these studies indicated that retrieving, nest-building, and nursing behavior were depressed and that the young did not gain weight normally. Importantly, however, many of these studies found that the *oral* components of maternal behavior (retrieving and nest building) were depressed more than crouching behavior. In a study that examined the *long-term* effects of preoptic damage on the various components of maternal behavior, this differential effect was even more extreme (Numan, unpublished findings; see also Jakubowski & Terkel, 1986). Female rats received knife cuts that severed the lateral connections of the MPOA or sham knife cuts. After a 2-week postoperative recovery period, these rats were mated. Following a 22-day pregnancy, the rats gave birth normally, and their maternal behavior was studied for 2 weeks. All females cleaned the placentas and membranes off their young. Retrieving was completely eliminated in all the females with lateral MPOA cuts, and although such females built nests, these nests were inferior to those built by sham females. In contrast to retrieving and nest building, the time spent nursing the young did not differ between the two groups. The lateral-cut females were also capable of lactating, and their pups gained weight daily, showing increases in body weight of about 10% per day. This pup weight increase, however, was significantly inferior to that shown by the pups of the sham females (16% per day). These results indicate that the lateral connections of the MPOA are most important for the display of retrieval behavior and that the disruption of retrieval behavior as a result of severing these connections is relatively permanent.

An important issue that arises at this point is whether the retrieval deficit represents a general oral motor deficit rather than a deficit that is specific to maternal behavior. That the retrieval deficit is not a general oral motor deficit has been shown by Numan and Corodimas (1985). They found that nonretrieving females with lateral MPOA cuts are capable of hoarding candy that approximates the size and weight of rat pups.

The effects of preoptic damage on aspects of maternal behavior other than retrieving, nest-building, and crouching behavior have not been examined. In this regard, it would be interesting to determine the level of maternal aggression in postpartum females with preoptic damage that are nursing but not retrieving their young. Recall that both retrieving and maternal aggression are dependent on perioral tactile sensory input. Perhaps the lateral connections of the MPOA are involved in facilitating the transmission of trigeminal sensory input in response to maternally relevant stimuli (pups, intruder at the nest site).

Septal Area

Lesions of the septal area (which lies rostral and dorsal to the preoptic area) severely disrupt nursing, nest building, and retrieving behavior in postpartum rats (Fleischer & Slotnick 1978). These deficits, however, are distinct from those caused by preoptic damage. Destruction of the septal region does not lead to a disruption in the willingness of the lesioned females to care for their young. Instead, the disruption involves the organization of individual maternal responses over time and space. Septal mothers persistently carry their pups around the cage and drop them at random locations. The retrieving is so disorganized, and yet so persistent, that the females rarely initiate nest-building and nursing behavior.

Amygdala

The amygdala has been shown to exert inhibitory control over maternal behavior in the rat. Fleming, Vaccarino, and Luebke (1980) found that lesions of the medial amygdala facilitated the response of virgin female rats to test pups. Importantly, the medial amygdala receives input from the vomeronasal organ, and the facilitation of maternal behavior in virgins with amygdala lesions is equivalent to that seen in virgin females with selective damage to the vomeronasal nerves (Fleming et al., 1979). The relevant anatomical relationships between the vomeronasal organ, amygdala, and MPOA are shown in Figure 12.2 (De Olmos & Ingram, 1972; Krettek & Price, 1978; Scalia & Winans, 1975; Simerly & Swanson, 1986). The vomeronasal organ projects to the accessory olfactory bulb, which, in turn, projects to the medial amygdala. The medial amygdala projects, via the stria terminalis, to both the BNST and the MPOA. The BNST also has strong projections to the MPOA. The hypothesis that emerges is that vomeronasal input to the BNST/MPOA region inhibits maternal behavior, and when the MPOA is released from this inhibition, maternal behavior is facilitated. This hypothesis is supported by the finding that stria terminalis lesions facilitate maternal behavior in virgins to the same extent as vomeronasal or medial amygdala lesions (Fleming et al., 1980). Also consistent is the finding that MPOA lesions block the amygdala lesion–induced facilitation of maternal behavior in virgins (Fleming, Miceli, & Moretto, 1983).

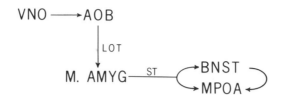

Figure 12.2. Diagram of pathways through which vomeronasal input reaches the preoptic region. Abbreviations: VNO, vomeronasal organ; AOB, accessory olfactory bulb; LOT, lateral olfactory tract; M. AMYG, medial amygdala; ST, stria terminalis; BNST, bed nucleus of the stria terminalis; MPOA, medial preoptic area.

NEURAL CIRCUITRY ANALYSIS

Output Circuits

Given that evidence points to the involvement of laterally projecting MPOA neurons in the control of maternal behavior, the question that arises is the site of termination of these neurons in their influence over maternal behavior. One of the principles that has guided the search for this termination site (or sites) was the relation of anatomy to function. That is, does the MPOA project to sites that are involved in functions that might be related to maternal behavior?

Paraventricular Nucleus of the Hypothalamus

Anatomical evidence indicates that the MPOA projects to the paraventricular nucleus of the hypothalamus (PVN) and that some of this input to the PVN is carried by the lateral efferents of the MPOA (Conrad & Pfaff, 1976; Sawchencko & Swanson, 1983). The PVN gains importance for maternal behavior because it is the main source of oxytocinergic neural pathways within the brain (Sofroniew, 1985) and because several lines of research have indicated that central oxytocinergic neural pathways play a positive role in promoting maternal responsiveness (Pedersen & Prange, 1987; see also chapter 13). Therefore, perhaps MPOA projections to the PVN activate oxytocinergic neural pathways in order to promote maternal behavior.

Numan and Corodimas (1985) presented evidence that preoptic projections to the PVN are *not* involved in the control of maternal behavior in postpartum rats. Females received lesions during the postpartum period after maternal behavior was clearly established. Although knife cuts severing the lateral connections of the MPOA severely disrupted maternal behavior, lesions of the entire PVN left maternal behavior intact. Therefore, the disruption of maternal behavior in postpartum rats resulting from preoptic damage is not mediated by PVN.

Although oxytocinergic neural pathways arising from the PVN do not appear essential for the continuance of maternal behavior that is already established in postpartum rats, the findings of Numan and Corodimas (1985) do not rule out the possibility that such pathways are important for the onset or initial occurrence of maternal behavior at parturition. Indeed, research has recently indicated that oxytocin's role in maternal behavior pertains to the onset, but not the maintenance, of maternal behavior (Fahrbach, Morrell, & Pfaff, 1985; Van Leengoed, Kerker, & Swanson,

1987). It should be noted that neurons that bind oxytocin are found in brain regions that receive primary olfactory and/or vomeronasal input. These regions include the BNST, the anterior olfactory nucleus, and the olfactory tubercle (De Kloet, Voohuis, Boschma, & Elands, 1986; Freund-Mercier et al., 1987; Insel, 1986; Ravid, Swaab, Van der Wounde, & Boer, 1986; Van Leeuwen, Herrikhuize, Van Der Meulen, & Wolters, 1985). Therefore, it is entirely possible that oxytocinergic systems, perhaps arising from the PVN, modify olfactory/vomeronasal function at parturition, in this way facilitating maternal behavior.

Ventral Tegmental Area of the Midbrain

Research from my laboratory has explored the possibility that MPOA influences on the ventral tegmental area (VTA) of the ventromedial midbrain are important for maternal behavior. Figure 12.3 shows the two ways that the lateral efferents of the MPOA can reach the VTA (Chiba & Murata, 1985; Conrad & Pfaff, 1976; Swanson, 1976). First, there is a direct MPOA-to-VTA projection. Second, the MPOA projects to the LPOA, which then projects to the VTA. Both MPOA and LPOA efferents travel through the lateral hypothalamus (LH) on their descent to the VTA. Severing the lateral connection of the MPOA would interfere with both the MPOA-to-VTA and the MPOA-to-LPOA-to-VTA pathways. Damage to the LH would also interfere with both circuits. Avar and Monos (1969) have found that LH lesions disrupt maternal behavior in rats in a manner similar to that observed after preoptic area damage.

The VTA has diverse ascending and descending projections (Beckstead, Domesick, & Nauta, 1979; Simon, Le Moal, & Calas, 1979; Swanson, 1982). Most importantly, some of the ascending projections of the VTA reach the striatum, which includes the nucleus accumbens, medial caudate-putamen, and olfactory tubercle. Since the olfactory tubercle receives input from the olfactory bulb (Scott, McBride, & Schneider, 1980), and since VTA projections to the striatum have been shown to be involved in the control of oral motor responses (Jones & Mogenson, 1979; Kelley & Stinus, 1985), it is possible that preoptic projections to the VTA might influence maternal behavior by altering olfactory processes and/or oral motor processes.

Evidence supporting the idea that preoptic projections to the VTA are critical for maternal behavior has been provided by Numan and Smith (1984). First, they found that bilateral electrolytic lesions of the VTA severely disrupted maternal behavior in postpartum rats. Then, using an asymmetrical lesion design, they found that bilateral damage to the preoptic-to-VTA projection at a different point in the circuit on each side of the brain severely depressed retrieving and nest-building behavior. Specifically, they found that the maternal behavior of postpartum rats that received a unilateral knife cut severing the lateral connections of the MPOA paired with a contralateral electrolytic lesion of the VTA was disrupted in a manner similar to that observed in females with either bilateral MPOA lateral cuts or bilateral VTA lesions, and that this disruption was much more severe than that observed in females that had only unilateral damage to the preoptic-to-VTA pathway.

This evidence supports the view that preoptic projections to the VTA are important for maternal behavior, but it provides no information with respect to whether an MPOA-to-VTA or an MPOA-to-LPOA-to-VTA circuit may be involved. This is be-

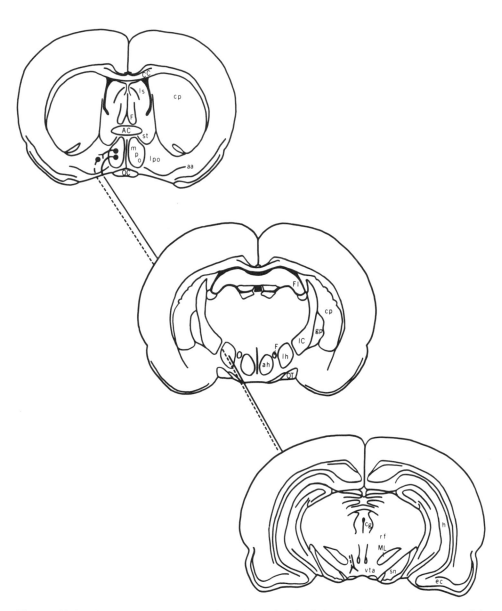

Figure 12.3. Frontal plane sections through the level of the medial preoptic area, caudal anterior hypothalamic area, and ventral tegmental area showing the two major routes through which the lateral efferents of the medial preoptic area can reach the ventral tegmental area. For clarity, the pathways are shown on only one side of the brain. Medial preoptic axons are shown as a solid line. Lateral preoptic axons are shown as a dashed line. Abbreviations: ec, entorhinal cortex; FI, fimbria; gp, globus pallidus; IC, internal capsule; lh, lateral hypothalamus; ML, medial lemniscus; OT, optic tract; rf, reticular formation; sn, substantia nigra. For additional abbreviations see Figure 12.1. (From Numan, 1988)

cause the preoptic-VTA lesions would destroy both circuits. Numan et al. (1988) examined this issue by determining whether LPOA neurons are important for maternal behavior. Fully maternal lactating rats received one of the following treatments. (1) bilateral injections of NMA into the LPOA, which would selectively destroy LPOA neurons while sparing fibers of passage; (2) bilateral injections of phosphate buffer into the LPOA; (3) bilateral radiofrequency lesions of the LPOA; (4) bilateral injections of NMA into the LH. The results indicated that LPOA neurons are important for maternal behavior in that both NMA and radiofrequency lesions of the LPOA severely disrupted retrieving, nest-building and nursing behavior. Control injections of phosphate buffer into the LPOA or of NMA into the LH were without effect on maternal behavior.

Another approach to examining the preoptic efferents descending to the brain stem that may be important for maternal behavior was taken by Numan, Morrell, and Pfaff (1985). Coronal knife cuts were made through the dorsal lateral hypothalamus at a point caudal to the preoptic area in fully maternal lactating rats. Supporting previous work (Avar & Monos, 1969), the cuts disrupted all aspects of maternal behavior. One of the novel aspects of this work was that the wire knife used to make the cuts was coated with horseradish peroxidase (HRP), which would be taken up by the cut axons and retrogradely transported to their cell bodies. HRP-labeled cell bodies, therefore, would serve as a marker indicating the location of neurons damaged by the knife cuts. A diagram indicating the LH knife cuts and the resultant labeling of cell bodies in the preoptic region with HRP is shown in Figure 12.4. LH knife cuts that disrupt maternal behavior label few cells in the MPOA with HRP, but label many cells in the LPOA and BNST with HRP. This correlational evidence supports the view that LPOA axons descending to the brain stem via the LH are important for maternal behavior. The results also point to the possible importance of descending BNST efferents, and this needs further investigation. Recall that the BNST receives vomeronasal input. Also worth mentioning is that the BNST is a recipient of gustatory input from the pontine taste area (Norgren, 1976). Another interesting aspect of the anatomical evidence shown in Figure 12.4 is that LH knife cuts that disrupt maternal behavior damage many neurons at the junction point between the MPOA, the LPOA, and the BNST.

To summarize, the evidence reviewed so far supports the view that an MPOA-to-LPOA-to-VTA circuit underlies maternal behavior. The evidence is also consistent with the possibility that a select group of neurons at the LPOA–MPOA–BNST border descend to or through the VTA to influence maternal behavior. The overall evidence can be summarized as follows:

1. Destruction of MPOA or LPOA neurons with NMA disrupts maternal behavior.
2. Knife cuts severing the lateral connections of the MPOA disrupt maternal behavior.
3. Electrical lesions of the VTA disrupt maternal behavior.
4. A unilateral knife cut of the lateral MPOA connections paired with a contralateral VTA lesion disrupts maternal behavior.
5. Coronal knife cuts through the LH disrupt maternal behavior and sever the axons of many descending LPOA neurons.

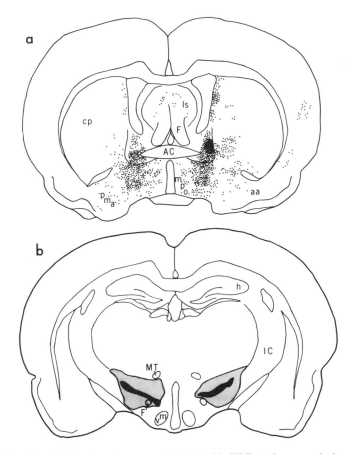

Figure 12.4. The distribution of neurons labeled with HRP at the preoptic level (a) follow-ing knife cuts of the lateral hypothalamus with an HRP-coated knife. Each dot represents a labeled cell. (b) The knife cuts through the dorsal lateral hypothalamus (black) and the HRP deposit site (stipple). Abbreviations: MT, mammillothalamic tract; pma, magnocellular preop-tic nucleus. For additional abbreviations, see Figures 12.1 and 12.3. (Data from Numan et al., 1985. Copyright © 1985 by Alan R. Liss, Inc.)

Although the evidence in favor of an MPOA-to-LPOA-to-VTA circuit in mater-nal behavior control is strong, it is not conclusive (see Franz et al., 1986; Miceli et al., 1983). One of the most important issues that remains is whether VTA neurons are important for maternal behavior. The work showing that VTA lesions disrupt maternal behavior (Gaffori & Le Moal, 1979; Numan & Smith, 1984) produced the lesions with electric current; therefore, the lesions damaged both VTA neurons and fibers of passage. Importantly, preoptic neurons not only terminate in the VTA, but also pass through the VTA to reach other brain stem regions (Conrad & Pfaff, 1976; Swanson, 1976; Swanson, Mogenson, Gerfen, & Robinson, 1984; Swanson, Mogen-son, Simerly, & Wu, 1987).

In an attempt to examine the role of VTA neurons in maternal behavior control, NMA was injected into the VTA of postpartum rats (Numan, unpublished findings).

NMA injections into the VTA, at twice the dosage level found to disrupt maternal behavior when this neurotoxin was injected into the preoptic region, did not disrupt maternal behavior. The fact that NMA injections into the VTA are ineffective, while electrical lesions of the VTA are effective in disrupting maternal behavior, suggests the involvement of fibers of passage. However, negative results with NMA are difficult to interpret. NMA is believed to act on specific receptor sites on neuronal cell bodies, and there is regional variation within the brain in the density of NMA receptors, with the VTA showing a small number of such receptors (Monaghan & Cotman, 1985). Since the lesions of the VTA that were produced with NMA were very small, it is possible that NMA did not destroy a significant number of VTA neurons important for maternal behavior. Obviously, more research is needed on this issue. The fact that many VTA neurons with projections to the striatum are dopaminergic (Swanson, 1982) suggests another research strategy. The effects on maternal behavior of injection of the catecholaminergic neurotoxin 6-hydroxydopamine into the VTA would allow us to determine the role of VTA dopamine neurons in maternal behavior control. This work remains to be done.

Alternatives to the VTA

Since preoptic neurons descending to the brain stem via the LH not only terminate in the VTA, but also pass through it to terminate in other brain stem sites, it may be that brain stem regions other than or in addition to the VTA are critical for maternal behavior. At present, other than the VTA, at least two regions possess anatomical and functional characteristics suggesting that they may be possible terminal brain stem sites for preoptic influences on maternal behavior.

The MPOA, LPOA, and BNST send projections through the LH and the VTA to terminate in a region of the brain stem caudal to the VTA, lateral to the central gray, at the midbrain–pontine border, that is called the "pedunculo-pontine" nucleus or "mesencephalic locomotor region" (MLR) (Swanson et al., 1984, 1987). The MLR, in turn, has ascending and descending projections to various components of the motor system (Garcia-Rill, Skinner, Conrad, Mosley, & Campbell, 1986). It has been suggested that the MLR plays an important role in limbic–motor integration (Mogenson, 1987), and therefore, preoptic projections to the MLR may play a role in activating certain aspects of maternal behavior. At present, no evidence is available on this point, but the possibility is certainly worth investigating (see Steele, Rowland, & Moltz, 1979).

Another brain stem region that may be influenced by preoptic efferents relevant to maternal behavior is the lateral midbrain tegmentum (LMT). This region, which lies dorsal to the lateral aspect of the substantia nigra and medial to the ventral aspect of the medial geniculate nucleus, includes the peripeduncular nucleus. Importantly, the LMT receives input from descending preoptic efferents (Arnault & Roger, 1987; Chiba & Murata, 1985). Equally important, ascending trigeminal sensory pathways, carrying tactile input from the perioral region, pass through the VTA region before terminating in the LMT area (Dubois-Dauphin, Armstrong, Tribollet, & Dreifuss, 1985; Erzurumlu & Killackey, 1980; Peschanski, 1984; Smith, 1973). Therefore, it is very likely that the preoptic-VTA lesions of Numan and Smith (1984) interfered with aspects of these pathways bilaterally. Hansen and his colleagues (Hansen &

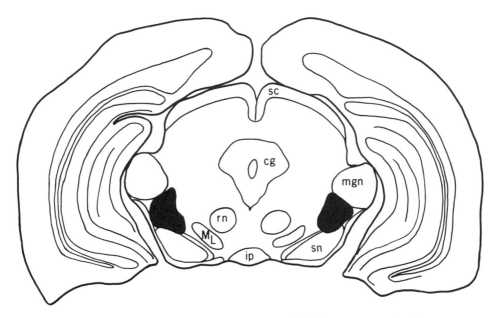

Figure 12.5. An example of a lesion of the lateral midbrain tegmentum that disrupts maternal aggression and milk ejection in postpartum rats. The lesion is shown in solid black. Abbreviations: cg, central gray; ip, interpeduncular nucleus; mgn, medial geniculate nucleus; ML, medial lemniscus; rn, red nucleus; sc, superior colliculus. (Modified from Hansen & Ferreira, 1986)

Ferreira, 1986; Hansen & Gummesson, 1982; Hansen & Kohler, 1984) have explored the effects of LMT lesions on aspects of maternal behavior in postpartum lactating rats. Their basic findings are that LMT lesions abolish maternal aggression toward a male rat intruder and that LMT lesions do not interfere with nursing behavior, but do block the milk-ejection reflex. Figure 12.5 shows a diagram of an effective LMT lesion. Hansen and his co-workers suggest that LMT lesions do not interfere with pup-directed maternal activities but do block maternal aggression. These investigators, however, took very superficial observations of nursing behavior, and they did not test for retrieval behavior. Since both retrieval behavior and maternal aggression are influenced by trigeminal input, it is possible that LMT lesions would disrupt retrieving and maternal aggression while sparing nursing behavior. Since the LMT projects to brain regions involved in motor control (globus pallidus, zona incerta; Arnault & Roger, 1987; Veazey & Severin, 1980), it is possible that the LMT passes on perioral tactile input to the motor system, in this way influencing the occurrence of oral motor responses. Perhaps preoptic efferents to the LMT are involved in gating this sensorimotor system.

Afferent Input to the MPOA that May Be Important for Maternal Behavior

An important research goal, in addition to uncovering sites of termination of preoptic efferents involved in maternal behavior, is to uncover sources of afferent input to the

MPOA necessary for maternal behavior. Recently, a research program (Numan, McSparren, & Numan, in preparation) was initiated to get some information on this issue. This research is based on the findings of Numan and Callahan (1980) that knife cuts that sever the lateral connections of the MPOA, but not those that sever the anterior, posterior, or dorsal connections of the MPOA, selectively disrupt maternal behavior. These findings suggest that knife cuts lateral to the MPOA should interfere not only with efferents of the MPOA that are important for maternal behavior, but also with any afferent input to the MPOA that is essential for the occurrence of maternal behavior.

Several studies have examined the afferents to the MPOA (Berk & Finkelstein, 1981; Chiba & Murata, 1986; Kita & Oomura, 1982; Simerly & Swanson, 1986), but none of these reports has related such inputs to maternal behavior mechanisms. To this end, knife cuts were performed that severed the lateral connections of the MPOA with an HRP-coated knife. After the maternal behavior of the lesioned females was studied, the brains were processed for the localization of HRP-filled cell bodies. Although neurons in several brain regions were found to be labeled with HRP following lateral cuts that disrupted maternal behavior, indicating that such neurons were damaged by the cuts, the work presented here concentrates on two brain stem regions that have been reliably labeled in all the nonmaternal lesioned females and that other research has implicated in maternal behavior control.

The knife cuts that disrupted maternal behavior labeled neurons in the raphe nuclei, particularly the median raphe nucleus, and in the posteromedial nucleus of the solitary tract (NST). In addition to this work, ample evidence exists that these regions project to the MPOA (Azmitia & Segal, 1978; Berk & Finkelstein, 1981; Conrad, Leonard, & Pfaff, 1974; Day, Blessing, & Willoughby, 1980; Ricardo & Koh, 1978; Simerly & Swanson, 1986; Simerly, Swanson, & Gorski, 1984). The importance of this behavioral-neuroanatomical work is that such input is related, on the one hand, to severing of the lateral connections of the MPOA, and, on the other, to a disruption of maternal behavior. Figure 12.6 shows the lateral MPOA cuts and associated HRP-filled cell bodies in the raphe nuclei and in the NST. The issue is whether this relationship between damage to raphe and NST input to the MPOA and maternal behavior deficit is a cause–effect relationship.

Concerning the raphe nuclei, there is evidence that both serotonergic and nonserotonergic neurons from the raphe nuclei project to the MPOA (Simerly et al., 1984). Importantly, there is also evidence that serotonin neurons in the median raphe play some role in regulating maternal behavior. First, systemic injections of parachlorophenylalanine (PCPA), an inhibitor of serotonin synthesis, result in large decreases in whole brain levels of serotonin (85–95% decrease) and cause a large increase in the incidence of infanticide in postpartum rats (Copenhaver, Schalock, & Carver, 1978; Moore & Hampton, 1974). Second, Barofsky, Taylor, Tizabi, Kumar, and Jones-Quartey (1983) have reported that the injection of 5,7-dihydroxytryptamine, a serotonergic neurotoxin, into the median raphe nucleus disrupted the maternal behavior of lactating rats and caused a partial depletion (55% decrease) of hypothalamic serotonin levels.

Concerning the nature of the deficits observed after damage to median raphe serotonin neurons, there was a relatively permanent but partial disruption of crouching behavior, a temporary disruption of retrieving behavior, and a temporary increase

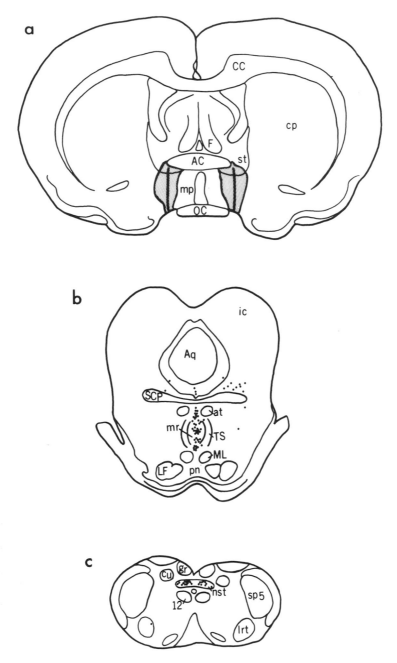

Figure 12.6. Neurons in the brain stem labeled with HRP following knife cuts lateral to the medial preoptic area. The wire knife used to make the cuts was coated with HRP. (a) The preoptic knife cuts in solid black and the HRP deposit site in stipple. (b) Labeled cells at the level of the raphe nuclei. (c) Labeled cells at the level of the nucleus of the solitary tract. Each dot represents a labeled cell. Abbreviations: Aq, aqueduct; at, anterior tegmental nucleus; cu, cuneate nucleus; gr, gracile nucleus; ic, inferior colliculus; LF, longitudinal fasciculus of the pons; lrt, lateral reticular nucleus; ML, medial lemniscus; mr, median raphe nucleus; nst, nucleus of the solitary tract; pn, pontine nuclei; SCP, superior cerebellar peduncle; sp5, spinal trigeminal nucleus; TS, tectospinal tract; l2, hypoglossal nucleus. For additional abbreviations see Figure 12.1.

in the incidence of infanticide (Barofsky et al., 1983). These effects are clearly different from the more severe deficits observed after lateral MPOA cuts, where complete and permanent elimination of retrieval behavior occurs but where infanticide does not occur. The following should be noted, however: the destruction of median raphe serotonin neurons depletes serotonin from widespread brain regions, not just from the MPOA; the depletion of brain serotonin after median raphe lesions is partial. Therefore, future research should explore the specific role of serotonergic input to the MPOA in maternal behavior control. The role of nonserotonergic median raphe neurons should also be explored. The best conclusion to reach at this time is that there is suggestive evidence that median raphe input to the MPOA modulates certain aspects of maternal behavior.

Concerning the posteromedial NST, this area contains norepinephrine (NE) and non-NE neurons that project to the preoptic region (Day et al., 1980; Pickel, Joh, Chan, & Beaudet, 1984; Ricardo & Koh, 1978; Van der Kooy, Koda, McGinty, Gerfen, & Bloom, 1984). With respect to our findings suggesting a correlation between damage to NST input to the MPOA and maternal behavior deficits, we should note that there is conflicting evidence with respect to the involvement of NE neural systems in maternal behavior. Rosenberg, Halaris, and Moltz (1977) reported that injections of 6-hydroxydopamine into the lateral ventricle of pregnant rats disrupted subsequent postpartum maternal behavior and partially depleted the hypothalamus of NE (60% decrease). Such females showed inferior nursing and nest building compared with controls, and their pups did not gain weight normally, suffering a high mortality rate. In contrast to these findings, Bridges, Clifton, and Sawyer (1982) reported that knife cuts in the brain stem of pregnant rats, which disrupted ascending NE systems, partially depleted the hypothalamus of NE (70% depletion) but did not interfere with nursing behavior in postpartum rats. The primary deficit they observed was lactational. Clearly, more work has to be done to explain these differences, particularly in light of our findings. It would be interesting to observe the effects of direct NE antagonism within the preoptic region on maternal behavior in rats.

The posteromedial NST receives afferent visceral input primarily from the vagus nerve (Hamilton & Norgren, 1984). Therefore, its main function appears to be to relay autonomic input to other parts of the brain. A good possibility is that NE input to the preoptic region from the NST is involved in temperature-regulatory mechanisms (Day, Willoughby, & Geffen, 1979; Millan, Millan, & Herz, 1983). Interestingly, there are important relationships between body-temperature regulation and maternal behavior, particularly with respect to nursing behavior (Leon, Croskerry, & Smith, 1978). Therefore, it is possible that temperature-relevant input to the MPOA modulates the output of MPOA neurons concerned with nursing behavior.

OPIATES AND MATERNAL BEHAVIOR

Research has suggested that endogenous opiates might act to depress maternal responsiveness in rats. Systemic treatment of rats with morphine has been found to inhibit maternal behavior, and, importantly, central application of morphine to the MPOA duplicates this inhibitory effect (Bridges & Grimm, 1982; Grimm & Bridges, 1983; Rubin & Bridges, 1984).

The question that concerns us here is whether morphine's inhibition of maternal behavior is indicative of endogenous central neural opiate modulation of maternal behavior. There are three major central neural opioid systems—the β-endorphin system, the enkephalin system, and the neoendorphin system—and axons from each of these are found in the preoptic region (Finley, Lindstrom, & Petrusz, 1981; Khachaturian, Lewis, Schafer, & Watson, 1985; Khachaturian, Lewis, Tsou, & Watson, 1985; Petrusz, Merchenthaler, & Maderdrut, 1985). In addition, opiate receptor sites are located in the MPOA (Hammer, 1984). Future studies, by examining the effects on maternal behavior of direct application of enkephalins, endorphins, and neoendorphins to the MPOA, will provide information on whether endogenous opiates inhibit maternal behavior and, if so, which system is involved.

Although clearly speculative, some ideas about opiate neural pathways reaching the preoptic region that might inhibit maternal behavior will be presented. First, an enkephalinergic pathway travels from the amygdala to the BNST via the stria terminalis (Rao, Yamano, Shiosaka, Shinohara, & Tohyama, 1987; Uhl, Kuhar, & Snyder, 1978). Perhaps this pathway underlies amygdaloid (and ultimately vomeronasal) inhibition of maternal behavior. The problem with this view, however, is that the behavioral evidence indicates that the medial amygdala is inhibitory for maternal behavior (Fleming et al., 1980), while the anatomical evidence indicates that the central nucleus of the amygdala provides enkephalin input to the BNST. Interestingly, although the central nucleus does not receive vomeronasal input (Scalia & Winans, 1975), it is a site of termination of taste pathways that originate in the pons (Fulwiler & Saper, 1984; Norgren, 1976).

Concerning endorphins, there are two groupings of β-endorphin cell bodies in the brain. One lies in the medial basal hypothalamus (MBH), and the other is located in the posteromedial NST (Finley et al., 1981; Khachaturian et al., 1985). There is direct evidence that the MBH system projects to the MPOA. Although there is no direct evidence that β-endorphin neurons in the NST project to the MPOA, the fact that posteromedial NST neurons project to the MPOA (reviewed in the previous section) suggests that some of these neurons may contain β-endorphin. Interestingly, a recent study has found that β-endorphin is capable of inhibiting the release of NE from axon terminals within the MPOA (Diez-Guerra, Augood, Emson, & Dyer, 1987). Given that we have reviewed the suggestive evidence that NE input from the NST to the MPOA may be important for maternal behavior, an idea worth pursuing is that endorphin action at the level of the MPOA is capable of modulating NE's effect on maternal behavior systems.

NEURAL SITE OF HORMONE ACTION

The activation of maternal responsiveness at parturition in the rat is triggered by hormones. Given that we now know something about the neural basis of maternal behavior, we can ask whether hormones act on the brain to stimulate maternal behavior, and if so, what sites are involved. On this issue, there is evidence only with respect to estradiol. The findings show that given the appropriate hormonal background, implants of small amounts of estradiol into the MPOA stimulate maternal

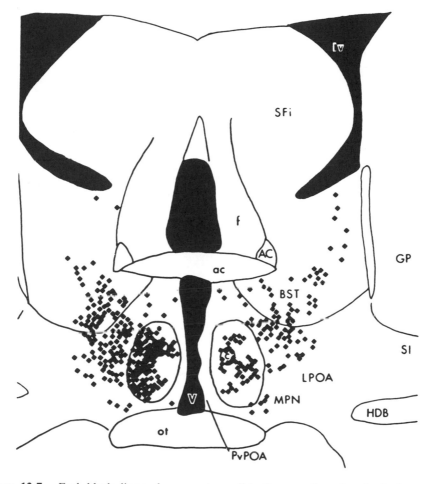

Figure 12.7. Each black diamond represents a cell in the preoptic region that both concentrates estradiol and projects to or through the ventromedial midbrain. Abbreviations: SF, septofimbrial nucleus; f, fornix; ac, anterior commissure; AC, nucleus of the anterior commissure; BST, bed nucleus of the stria terminalis; GP, globus pallidus; MPN, medial preoptic nucleus; LPOA, lateral preoptic area; SI, substantia innominata; PvPOA, periventricular preoptic area; HDB, horizontal division of the nucleus of the diagonal band; ot, optic tract. (From Fahrbach et al., 1986. Copyright © Alan R. Liss, Inc. Reprinted by permission)

behavior, while similar implants into other neural regions are ineffective (Fahrbach & Pfaff, 1986; Numan, Rosenblatt, & Komisaruk, 1977).

A recent finding related to the neural circuits affected by estradiol action on the preoptic region has been presented by Fahrbach, Morrell, and Pfaff (1986). This study combined a fluorescent dye retrograde tracing with estradiol autoradiography to determine whether estrogen-concentrating preoptic neurons project to the ventromedial midbrain (an area that includes the VTA). Figure 12.7 shows the distribution of neurons in the preoptic region that both concentrate estradiol and project to the ventromedial midbrain. Each black diamond represents a double-labeled cell. Note that such cells are concentrated in the medial LPOA, lateral MPOA, and ventral BNST. Since the fluorescent dyes that were injected into the ventromedial midbrain can be taken up by fibers of passage as well as by axon terminals (Sawchenko & Swanson, 1981), the findings of Fahrbach et al. (1986) indicate that estradiol-concentrating preoptic neurons project to or through the ventromedial midbrain. The fact that estrogen acts on the preoptic region to stimulate maternal behavior, and that a neural pathway that extends between the preoptic region and the VTA is important for maternal behavior, suggests that estrogen might stimulate maternal behavior by activating this pathway. Finally, it is worthwhile to compare Figure 12.7 with Figure 12.4. Note that there is an overlap between cells that are damaged by LH knife cuts that disrupt maternal behavior and cells that concentrate estradiol and project to the midbrain. Perhaps LH cuts are effective in disrupting maternal behavior because they damage these estradiol-concentrating neurons.

SUMMARY

Our knowledge of the neural basis of maternal behavior in the rat can be summarized as follows:

1. Neurons in both the MPOA and the LPOA are important for retrieving, nest-building, and nursing behavior.
2. A neural pathway that travels between the preoptic region and the ventro-medial midbrain is important for maternal behavior.
3. It has been suggested that preoptic projections to the VTA are important for maternal behavior, although at present there is no firm evidence that VTA neurons are involved in maternal behavior control. Other brain stem sites that may receive preoptic efferents relevant to maternal behavior are the MLR and the LMT.
4. Estrogen action on the preoptic area stimulates maternal behavior. Evidence was reviewed that suggests that estrogen may produce this effect by activating neurons that descend to or through the ventromedial midbrain via the LH.
5. Concerning afferent input to the preoptic region relevant to maternal behavior, evidence suggests that inhibitory olfactory input reaches the preoptic region by way of projections from the amygdala. Evidence was also presented that emphasized the possible facilitatory effects of afferent input to the MPOA from the median raphe nucleus and from the NST.
6. One of the functions of preoptic output with respect to maternal behavior might be to facilitate transmission across trigeminal sensorimotor pathways. This view is related to the facts that tactile input from the perioral region is important for retrieving and that preoptic damage disrupts retrieving and nest building to a greater extent than nursing behavior.

Figure 12.8 is a summary diagram that incorporates some of these conclusions.

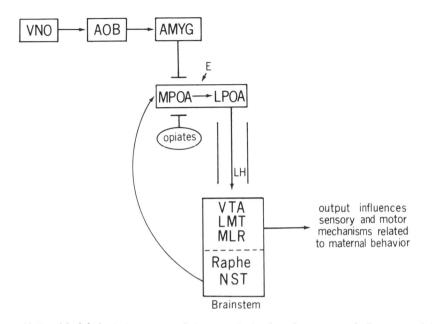

Figure 12.8. Model depicting some of the neural circuitry that may underlie maternal behavior in the rat. ⊣ signifies inhibitory relationships. The areas in boxes represent brain nuclei.
Opiates are shown within ovals to signify that neural pathways containing these compounds
act on the indicated region, although the source of such input is not known. Abbreviations:
AMYG, amygdala; AOB, accessory olfactory bulb; E, estrogen; LH, lateral hypothalamus;
LMT, lateral midbrain tegmentum; LPOA, lateral preoptic area; MLR, mesencephalic locomotor region; MPOA, medial preoptic area; NST, nucleus of the solitary tract; VNO, vomeronasal organ; VTA, ventral tegmental area.

GENERAL DISCUSSION

Several important issues are worth discussing. First, most of the work on the neural
basis of maternal behavior has been done on the rat. This raises the question of
whether the results can be generalized to other species, including primates. In this
regard, the following should be noted:

1. The neural circuitry mapped out for the rat involves the phylogenetically
   older parts of the brain.
2. There is evidence that the preoptic area is involved in the control of parental
   responses not only in rats, but also in hamsters (Miceli & Malsbury, 1982),
   ring doves (Komisaruk, 1967), and bluegill sunfishes (Demski & Knigge,
   1971).
3. A large body of evidence demonstrates that the preoptic region is involved
   in nonmaternal aspects of the reproductive function in a variety of vertebrates, including primates (Crews & Silver, 1985; Morrell & Pfaff, 1978).
   Therefore, throughout the course of vertebrate evolutionary history, cells in
   the preoptic region have been involved in the control of reproduction. It is

therefore reasonable to suggest that this region may influence parental motivation in humans.

Related to the issue of the generality of neural mechanisms underlying maternal behavior, we should note that olfactory input appears to play an important role in regulating aspects of maternal behavior in a variety of mammalian species (Numan, 1988; see chapter 8, this volume). Since, as indicated in this chapter, the preoptic area is a recipient of olfactory input, it is certainly possible that olfactory–preoptic–brain stem circuits form the foundation of the neural circuitry of maternal behavior in a variety of mammalian species (for a hypothetical neural model, see Numan, 1988).

In all mammalian species, the primary parental figure is the mother, although paternal behavior sometimes occurs (Kleiman & Malcolm, 1981). Work on the neural basis of paternal behavior is quite limited. (see Fleischer et al., 1981; see also chapter 20). It would certainly be interesting to explore the involvement of the preoptic area in naturally occurring paternal behavior.

An issue related to paternal-behavior mechanisms is the evidence that suggests that perinatal testosterone secretion depresses parental responsiveness in males (see chapter 17). The fact that anatomical differences exist in the bed nucleus of the stria terminalis–preoptic area organization of male and female rodents and primates, and that these differences can be influenced by perinatal testosterone (see chapter 5), suggests that differences in preoptic area neural circuitry may contribute to differences in the parental inclinations of males and females. It would be interesting to compare the degree of anatomical sex differences in the preoptic area of species that show paternal behavior with those that do not.

A final issue is whether various aspects of maternal behavior (i.e., retrieving, nursing, maternal aggression) are regulated by a common neural mechanism. At present, this does not appear to be the case, particularly when maternal aggression is compared with pup-directed maternal activities. First, retrieving, nest-building, and nursing (crouching) behavior can be observed in maternally behaving virgin female rats in the absence of maternal aggression (for a review, see Numan, 1988). Second, as reviewed in this chapter, brain stem lesions that disrupt maternal aggression do not disrupt nursing behavior. Although the reader is referred to the main body of this chapter for cautionary notes and additional analyses, the available evidence suggest that pup-directed maternal activities (particularly nursing behavior) and maternal aggression have at least partially independent neural regulatory mechanisms. How can this be reconciled with the data showing that the postpartum female shows high levels of both pup-directed behaviors and maternal aggression? The most appropriate answer is that maternal behavior is "turned on" by a complex group of external and internal stimuli, which may include several hormones (see chapter 6). These stimuli may active different neural circuits at the same time, creating the impression of a unitary neural system.

In conclusion, this chapter has outlined what is known about the neural bases of maternal behavior. It has been argued that the findings should have wide generality. Such research could ultimately provide insights into the neuroanatomical substrate for abnormal behaviors associated with the maternal condition in humans, such as postpartum depression and child abuse.

REFERENCES

Arnault, P., & Roger, M. (1987). The connections of the peripeduncular area studied by retrograde and anterograde transport in the rat. *Journal of Comparative Neurology, 258*, 463–476.

Avar, Z., & Monos, E. (1969). Biological role of lateral hypothalamic structures participating in the control of maternal behavior in the rat. *Acta Physiologica Academiae Scientiarum Hungaricae, 35*, 285–294.

Azmitia, E. C., & Segal, M. (1978). An autoradiographic analysis of the differential ascending projections of the dorsal and median raphe nuclei in the rat. *Journal of Comparative Neurology, 179*, 641–668.

Barofsky, A., Taylor, J., Tizabi, Y., Kumar, R., & Jones-Quartey, K. (1983). Specific neurotoxic lesions of the median raphe serotonergic neurons disrupt maternal behavior in the lactating rat. *Endocrinology, 113*, 1884–1893.

Beach, F. A., & Jayens, J. (1956). Studies of maternal retrieving in rats: III. Sensory cues involved in the lactating female's response to her young. *Behaviour, 10*, 104–125.

Beckstead, R. M., Domesick, V. B., & Nauta, W. J. H. (1979). Efferent connections of the substantia nigra and ventral tegmental area in the rat. *Brain Research, 175*, 191–217.

Benuck, I., & Rowe, F. A. (1975). Centrally and peripherally induced anosmia. Influences on maternal behavior in lactating female rats. *Physiology and Behavior, 14*, 439–447.

Berk, M. L., & Finkelstein, J. A. (1981). Afferent projections to the preoptic area and hypothalamic regions in the rat brain. *Neuroscience, 6*, 1601–1624.

Bridges, R. S., Clifton, D. K., & Sawyer, C. H. (1982). Postpartum luteinizing hormone release and maternal behavior in the rat after late-gestational depletion of hypothalamic norepinephrine. *Neuroendocrinology, 34*, 286–291.

Bridges, R. S., & Grimm, C. T. (1982). Reversal of morphine disruption of maternal behavior by concurrent treatment with the opiate antagonist naloxone. *Science, 218*, 166–168.

Chiba, T., & Murata, Y. (1985). Afferent and efferent connections of the medial preoptic area in the rat: A WGA-HRP study. *Brain Research Bulletin, 14*, 261–272.

Conrad, L. C. A., Leonard, C. M., & Pfaff, D. W. (1974). Connections of the median and dorsal raphe nuclei in the rat. An autoradiographic and degeneration study. *Journal of Comparative Neurology, 156*, 179–206.

Conrad, L. C. A., & Pfaff, D. W. (1976). Efferents from the medial basal forebrain and hypothalamus in the rat: I. Medial preoptic area. *Journal of Comparative Neurology, 169*, 185–220.

Copenhaver, J. H., Schalock, R. L., & Carver, M. J. (1978). Para-chloro-D, L-phenylalanine induced filicidal behavior in the female rat. *Pharmacology, Biochemistry and Behavior, 8*, 263–270.

Crews, D., & Silver, R. (1985). Reproductive physiology and behavior interactions in non-mammalian vertebrates. In N. Adler, D. Pfaff, & R. W. Goy (Eds.), *Handbook of behavioral neurobiology: Vol. 7. Reproduction* (pp. 101–182). New York: Plenum Press.

Day, T. A., Blessing, W., & Willoughby, J. O. (1980). Noradrenergic and dopaminergic projections to the medial preoptic area of the rat. A combined horseradish peroxidase/catecholamine fluorescence study. *Brain Research, 193*, 543–548.

Day, T. A., Willoughby, J. O., & Geffen, L. B. (1979). Thermoregulatory effects of preoptic injections of noradrenaline in restrained and unrestrained rats. *Brain Research, 174*, 175–179.

De Kloet, E. R., Voorhuis, D. A. M., Boschma, Y., & Elands, J. (1986). Estradiol modulates density of putative oxytocin receptors in discrete rat brain regions. *Neuroendocrinology, 44*, 415–421.

Demski, L. S., & Knigge, K. M. (1971). The telencephalon and hypothalamus of the bluegill *(Lepomis macrochirus):* Evoked feeding, aggressive and reproductive behavior with representative frontal sections. *Journal of Comparative Neurology, 143*, 1–16.

Denton, D. A., & Nelson, J. F. (1980). The influence of reproductive processes on salt appetite. In M. R. Kare (Ed.), *Biological and behavioral aspects of salt intake* (pp. 229–246). New York: Academic Press.

De Olmos, J. S., & Ingram, W. R. (1972). The projection field of the stria terminalis in the rat brain: An experimental study. *Journal of Comparative Neurology, 146*, 303–334.

Diez-Guerra, F. J., Augood, S., Emson, P. C., & Dyer, R. G. (1987). Opioid peptides inhibit the release of noradrenaline from slices of rat medial preoptic area. *Experimental Brain Research, 66*, 378–384.

Dubois-Dauphin, M., Armstrong, W. E., Tribollet, E., & Dreifuss, J. M. (1985). Somatosensory systems and the milk-ejection reflex in the rat: I. Lesions of the mesencephalic lateral tegmentum disrupt the reflex and damage mesencephalic somatosensory connections. *Neuroscience, 15*, 1111–1129.

Erzurumlu, R. S., & Killackey, H. P. (1980). Diencephalic projections of the subnucleus interpolaris of the brainstem trigeminal complex in the rat. *Neuroscience, 5*, 1891–1901.

Fahrbach, S. E., Morrell, J. I., & Pfaff, D. W. (1985). Possible role for endogenous oxytocin in estrogen-facilitated maternal behavior in rats. *Neuroendocrinology, 40*, 526–532.

Fahrbach, S. E., Morrell, J. I., & Pfaff, D. W. (1986). Identification of medial preoptic neurons that concentrate estradiol and project to the midbrain in the rat. *Journal of Comparative Neurology, 247*, 364–382.

Fahrbach, S. E., & Pfaff, D. W. (1986). Effects of preoptic region implants of dilute estradiol on the maternal behavior of ovariectomized, nulliparous rats. *Hormones and Behavior, 20*, 354–363.

Finley, J. C. W., Lindstrom, P., & Petrusz, P. (1981). Immunocytochemical localization of β-endorphin-containing neurons in the rat brain. *Neuroendocrinology, 33*, 28–42.

Fleischer, S., Kordower, J. H., Kaplan, B., Dicker, R., Smerling, R., & Ilgner, J. (1981). Olfactory bulbectomy and gender differences in maternal behaviors of rats. *Physiology and Behavior, 26*, 957–959.

Fleischer, S., & Slotnick, B. M. (1978). Disruption of maternal behavior in rats with lesions of the septal area. *Physiology and Behavior, 21*, 189–200.

Fleming, A. S. (1976). Control of food intake in the lactating rat: Role of suckling and hormones. *Physiology and Behavior, 17*, 841–848.

Fleming, A. S., Miceli, M., & Morretto, D. (1983). Lesions of the medial preoptic area prevent the facilitation of maternal behavior produced by amygdaloid lesions. *Physiology and Behavior, 31*, 502–510.

Fleming, A. S., & Rosenblatt, J. S. R. (1974). Olfactory regulation of maternal behavior in rats: II. Effects of peripherally induced anosmia and lesions of the lateral olfactory tract in pup-induced virgins. *Journal of Comparative and Physiological Psychology, 86*, 233–246.

Fleming, A. S., Vaccarino, F., & Luebke, C. (1980). Amygdaloid inhibition of maternal behavior in the nulliparous female rat. *Physiology and Behavior, 25*, 731–743.

Fleming, A. S., Vaccarino, F., Tambosso, L., & Chee, P. (1979). Vomeronasal and olfactory system modulation of maternal behavior in the rat. *Science, 203*, 372–374.

Franz, J. R., Leo, R. J., Steuer, M. A., & Kristal, M. B. (1986). Effects of hypothalamic knife cuts and experience on maternal behavior in the rat. *Physiology and Behavior, 38*, 629–640.

Freund-Mercier, M. J., Stoeckel, M. E., Palacios, J. M., Pazos, A., Reichhart, J. M., Porte, A., & Pichard, Ph. (1987). Pharmacological characteristics and anatomical distribution of [³H]oxytocin-binding sites in the Wistar rat brain studied by autoradiography. *Neuroscience, 20,* 599–614.

Fulwiler, C. E., & Saper, C. B. (1984). Subnuclear organization of the efferent connections of the parabrachial nucleus in the rat. *Brain Research Reviews, 7,* 229–259.

Gaffori, O., & Le Moal, M. (1979). Disruption of maternal behavior and appearance of cannibalism after ventral mesencephalic tegmentum lesions. *Physiology and Behavior, 23,* 317–323.

Garcia-Rill, E., Skinner, R. D., Conrad, C., Mosley, D., & Campbell, C. (1986). Projections of the mesencephalic locomotor region in the rat. *Brain Research Bulletin, 17,* 33–40.

Grimm, C. T., & Bridges, R. S. (1983). Opiate regulation of maternal behavior in the rat. *Pharmacology, Biochemistry, and Behavior, 19,* 609–616.

Gubernick, D. J., & Alberts, J. R. (1983). Maternal licking of young: Resource exchange and proximate controls. *Physiology and Behavior, 31,* 593–601.

Hamilton, R. B., & Norgren, R. (1984). Central projections of gustatory nerves in the rat. *Journal of Comparative Neurology, 222,* 560–577.

Hammer, R. P. (1984). The sexually dimorphic region of the preoptic area in rats contains denser opiate receptor binding sites in females. *Brain Research, 308,* 172–176.

Hansen, S., & Ferreira, A. (1986). Food intake, aggression, and fear behavior in the mother rat: Control by neural systems concerned with milk ejection and maternal behavior. *Behavior Neuroscience, 100,* 64–70.

Hansen, S., & Gummesson, B. M. (1982). Participation of the lateral midbrain tegmentum in the neuroendocrine control of sexual behavior and lactation in the rat. *Brain Research, 251,* 319–325.

Hansen, S., & Kohler, C. (1984). The importance of the peripeduncular nucleus in the neuroendocrine control of sexual behavior and milk ejection in the rat. *Neuroendocrinology, 39,* 563–572.

Herrenkohl, L. R., & Rosenberg, P. A. (1972). Exteroceptive stimulation of maternal behavior in the naive rat. *Physiology and Behavior, 8,* 595–598.

Insel, T. R. (1986). Postpartum increases in brain oxytocin binding. *Neuroendocrinology, 44,* 515–518.

Jacobson, C. D., Terkel, J., Gorski, R. A., & Sawyer, C. H. (1980). Effects of medial preoptic area lesions on maternal behavior. *Brain Research, 194,* 471–478.

Jakubowski, M., & Terkel, J. (1986). Female reproductive function and sexually dimorphic prolactin secretion in rats with lesions in the medial preoptic–anterior hypothalamic continuum. *Neuroendocrinology, 43,* 696–705.

Jones, D. L., & Mogenson, G. J. (1979). Oral motor performance following central dopamine receptor blockade. *European Journal of Pharmacology, 59,* 11–21.

Kelley, A. E., & Stinus, L. (1985). Disappearance of hoarding behavior after 6-hydroxydopamine lesions of the mesolimbic dopamine neurons and its reinstatement with L-dopa. *Behavioral Neuroscience, 99,* 531–545.

Kenyon, P., Cronin, P., & Keeble, S. (1981). Disruption of maternal behavior by perioral anesthesia. *Physiology and Behavior, 27,* 313–321.

Kenyon, P., Cronin, P., & Keeble, S. (1983). Role of the infraorbital nerve in retrieving behavior in lactating rats. *Behavioral Neuroscience, 97,* 255–269.

Khachaturian, H., Lewis, M. E., Schafer, M. K. H., & Watson, S. J. (1985). Anatomy of CNS opioid systems. *Trends in Neuroscience, 8,* 111–119.

Khachaturian, H., Lewis, M. E., Tsou, K., & Watson, S. J. (1985). β-Endorphin, α-MSH, ACTH, and related peptides. In A. Bjorklund & T. Hokfelt (Eds.), *Handbook of chemical neuroanatomy: Vol. 4, Part I. GABA and neuropeptides in the CNS* (pp. 216–272). Amsterdam: Elsevier.

Kita, H., & Oomura, Y. (1982). An HRP study of the afferent connections of the rat medial hypothalamic region. *Brain Research Bulletin, 8,* 53–62.

Kleiman, D. G., & Malcolm, J. R. (1981). The evolution of male parental investment in mammals. In D. J. Gubernick & P. H. Klopfer (Eds.), *Parental care in mammals* (pp. 347–387). New York: Plenum Press.

Kolunie, J. M., & Stern, J. M. (1987). *Perioral desensitization reduces maternal aggression in postpartum Norway rats independent of prior experience.* Paper presented at the Conference on Reproductive Behavior, Tlaxcala, Mexico.

Komisaruk, B. R. (1967). Effects of local brain implants of progesterone on reproductive behavior in ring doves. *Journal of Comparative and Physiological Psychology 64,* 219–224.

Konig, J. F. R., & Klippel, R. A. (1963). *The rat brain.* Baltimore: Williams & Wilkins.

Krettek, J. E., & Price, J. L. (1978). Amygdaloid projections to subcortical structures within the basal forebrain and brainstem in the rat and cat. *Journal of Comparative Neurology, 178,* 225–254.

Leon, M., Croskerry, P. G., & Smith, G. K. (1978). Thermal control of mother–young contact in rats. *Physiology and Behavior, 21,* 793–811.

MacDonnell, M., & Flynn, J. P. (1966). Control of the sensory fields by stimulation of the hypothalamus. *Science, 152,* 1406–1408.

Mayer, A. D., & Rosenblatt, J. S. (1977). Effects of intranasal zinc sulfate on open field and maternal behavior in female rats. *Physiology and Behavior, 18,* 101–109.

Miceli, M. O., Fleming, A. S., & Malsbury, C. W. (1983). Disruption of maternal behavior in virgin and postparturient rats following sagittal plane knife cuts in the preoptic area–hypothalamus. *Behavioral Brain Research, 9,* 337–360.

Miceli, M. O., & Malsbury, C. W. (1982). Sagittal knife cuts in the near and far lateral preoptic area–hypothalamus disrupt maternal behavior in female hamsters. *Physiology and Behavior, 28,* 857–867.

Millan, M. J., Millan, M. H., & Hertz, A. (1983). The role of the ventral noradrenergic bundle in relation to endorphins in the control of core temperature, open-field and ingestive behavior in the rat. *Brain Research, 263,* 283–294.

Mogenson, G. J. (1987). Limbic–motor integration. In A. N. Epstein & A. R. Morrison (Eds.), *Progress in psychobiology and physiological psychology* (Vol. 12, pp. 117–170). Orlando, FL: Academic Press.

Monaghan, D. J., & Cotman, C. W. (1985). Distribution of N-methyl-D-aspartate-sensitive-L-[^{3}H] glutamate binding sites in rat brain. *Journal of Neuroscience, 5,* 2909–2919.

Moore, W. J., & Hampton, J. K. (1974). Effects of parachlorophenylalanine on pregnancy in the rat. *Biology of Reproduction, 11,* 280–287.

Morrell, J. I., & Pfaff, D. W. (1978). A neuroendocrine approach to brain function: Localization of sex steroid concentrating cells in vertebrate brains. *American Zoologist, 18,* 447–460.

Norgren, R. (1976). Taste pathways to hypothalamus and amygdala. *Journal of Comparative Neurology, 166,* 17–30.

Numan, M. (1974). Medial preoptic area and maternal behavior in the female rat. *Journal of Comparative and Physiological Psychology, 87,* 746–759.

Numan, M. (1987). Preoptic area neural circuitry relevant to maternal behavior in the rat. In N. A. Krasnegor, E. M. Blass, M. A. Hofer, & W. P. Smotherman (Eds.), *Perinatal development: A psychobiological perspective* (pp. 275–298). Orlando, FL: Academic Press.

Numan, M. (1988). Maternal behavior. In E. Knobil & J. D. Neill (Eds.), *The physiology of reproduction* (pp. 1569–1645). New York: Raven Press.

Numan, M., & Callahan, E. C. (1980). The connections of the medial preoptic region and maternal behavior in the rat. *Physiology and Behavior, 25*, 653–665.

Numan, M., & Corodimas, K. P. (1985). The effects of paraventricular hypothalamic lesions on maternal behavior in rats. *Physiology and Behavior, 35*, 417–425.

Numan, M., Corodimas, K. P., Numan, M. J., Factor, E. M., & Piers, W. D. (1988). Axon-sparing lesions of the preoptic region and substantia innominata disrupt maternal behavior in rats. *Behavioral Neuroscience, 102*, 381–396.

Numan, M., Leon, M., & Moltz, H. (1972). Interference with prolactin release and the maternal behavior of female rats. *Hormones and Behavior, 3*, 29–38.

Numan, M., Morrell, J. I., & Pfaff, D. W. (1985). Anatomical identification of neurons in selected brain regions associated with maternal behavior deficits inducted by knife cuts of the lateral hypothalamus in rats. *Journal of Comparative Neurology, 237*, 552–564.

Numan, M., Rosenblatt, J. S., & Komisaruk, B. R. (1977). Medial preoptic area and onset of maternal behavior in the rat. *Journal of Comparative and Physiological Psychology, 91*, 146–164.

Numan, M., & Smith, H. G. (1984). Maternal behavior in rats: Evidence for the involvement of preoptic projections to the ventral tegmental area. *Behavioral Neuroscience, 98*, 712–727.

Ostermeyer, M. C. (1983). Maternal aggression. In R. W. Elwood (Ed.), *Parental behavior of rodents* (pp. 151–179). Chichester: Wiley.

Pedersen, C. A., & Prange, A. J. (1987). Evidence that central oxytocin plays a role in the activation of maternal behavior. In N. A. Krasnegor, E. M. Blass, M. A. Hofer, & W. P. Smotherman (Eds.), *Perinatal development: A psychobiological perspective* (pp. 299–320). Orlando, FL: Academic Press.

Peschanski, M. (1984). Trigeminal afferents to the diencephalon in the rat. *Neuroscience, 12*, 465–487.

Petrusz, P., Merchenthaler, I., & Maderdrut, J. L. (1985). Distribution of enkephalin-containing neurons in the central nervous system. In A. Bjorklund & T. Hokfelt (Eds.), *Handbook of chemical neuroanatomy: Vol. 4, Part I. GABA and neuropeptides in the CNS* (pp. 273–334). Amsterdam: Elsevier.

Pickel, V. M., Joh, T. H., Chan, J., & Beaudet, A. (1984). Serotoninergic terminals: Ultrastructure and synaptic interaction with catecholamine-containing neurons in the medial nuclei of the solitary tracts. *Journal of Comparative Neurology, 225*, 291–301.

Raisman, G. (1972). An experimental study of the projection of the amygdala to the accessory olfactory bulbs and its relationship to the concept of a dual olfactory system. *Experimental Brain Research, 14*, 395–408.

Rao, Z. R., Yamano, M., Shiosaka, S., Shinohara, A., & Tohyama, M. (1987). Origin of leucine-enkephalin fibers and their two main afferent pathways in the bed nucleus of the stria terminalis in the rat. *Experimental Brain Research, 65*, 411–420.

Ravid, R., Swaab, D. F., Woude, V., & Boer, G. J. (1986). Immunocytochemically stained binding sites for oxytocin and melanocyte-stimulating hormone in rat brain following ventricular administration. *Brain Research, 379*, 404–408.

Ricardo, J. A., & Koh, E. J. (1978). Anatomical evidence of direct projections from the nucleus of the solitary tract to the hypothalamus, amygdala, and other forebrain structures in the rat. *Brain Research, 153*, 1–26.

Rosenberg, P., Halaris, A., & Moltz, H. (1977). Effects of central norepinephrine depletion on the initiation and maintenance of maternal behavior in the rat. *Pharmacology, Biochemistry and Behavior, 6*, 21–24.

Rosenblatt, J. S. (1967). Nonhormonal basis of maternal behavior in the rat. *Science, 156,* 1512–1514.

Rosenblatt, J. S., Siegel, H. I., & Mayer, A. D. (1979). Progress in the study of maternal behavior in the rat: Hormonal, nonhormonal, sensory and developmental aspects. In J. S. Rosenblatt, R. A. Hinde, C. G. Beer, & M. C. Bunsel (Eds.), *Advances in the study of behavior* (Vol. 10, pp. 225–311). New York: Academic Press.

Rubin, B. S., & Bridges, R. S. (1984). Disruption of ongoing maternal responsiveness by central administration of morphine sulfate. *Brain Research, 307,* 91–97.

Sawchenko, P. E., & Swanson, L. W. (1981). A method for tracing biochemically defined pathways in the central nervous system using combined fluorescence retrograde transport and immunohistochemical techniques. *Brain Research, 210,* 31–51.

Sawchenko, P. E., & Swanson, L. W. (1983). The organization of forebrain afferents to the paraventricular and supraoptic nuclei of the rat. *Journal of Comparative Neurology, 218,* 121–144.

Scalia, F., & Winans, S. S. (1975). The differential projections of the olfactory bulb and accessory bulb in mammals. *Journal of Comparative Neurology, 161,* 31–55.

Scott, J. W., McBride, R. L., & Schneider, S. P. (1980). The organization of projections from the olfactory bulb to the piriform cortex and olfactory tubercle in the rat. *Journal of Comparative Neurology, 194,* 519–534.

Simerly, R. B., & Swanson, L. W. (1986). The organization of neural inputs to the medial preoptic nucleus of the rat. *Journal of Comparative Neurology, 246,* 312–342.

Simerly, R. B., Swanson, L. W., & Gorski, R. A. (1984). The cells of origin of a sexually dimorphic serotonergic input to the medial preoptic nucleus of the rat. *Brain Research, 324,* 185–189.

Simon, H., Le Moal, M., & Calas, A. (1979). Efferents and afferents of the ventral tegmental area–A10 region studied after local injection of [^{3}H] leucine and horseradish peroxidase. *Brain Research, 178,* 17–40.

Smith, R. L. (1973). The ascending fiber projections from the principal sensory trigeminal nucleus in the rat. *Journal of Comparative Neurology, 148,* 423–446.

Sofroniew, M. V. (1985). Vasopressin, oxytocin, and their related neurophysins. In A. Bjorklund & T. Hokfelt (Eds.), *Handbook of chemical neuroanatomy: Vol. 4, Part I. GABA and neuropeptides in the CNS* (pp. 93–165). Amsterdam: Elsevier.

Steele, M., Rowland, D., & Moltz, H. (1979). Initiation of maternal behavior in the rat: Possible involvement of limbic nonrepinephrine. *Pharmacology, Biochemistry and Behavior, 11,* 123–130.

Stern, J. M., & Kolunie, J. (1987). *Perioral stimulation from pups: Role in Norway rat maternal retrieval, licking, crouching and aggression.* Paper presented at the Conference on Reproductive Behavior, Tlaxcala, Mexico.

Svare, B. B. (1981). Maternal aggression in mammals. In D. J. Gubernick & P. H. Klopfer (Eds.), *Parental care in mammals* (pp. 179–210). New York: Plenum Press.

Swanson, L. W. (1976). An autoradiographic study of the efferent connections of the preoptic region in the rat. *Journal of Comparative Neurology, 167,* 227–256.

Swanson, L. W. (1982). The projections of the ventral tegmental area and adjacent regions: A combined retrograde fluorescent tracer and immunofluorescence study in the rat. *Brain Research Bulletin, 9,* 321–353.

Swanson, L. W., Mogenson, G. J., Gerfen, C. R., & Robinson, P. (1984). Evidence for a projection from the lateral preoptic area and substantia innominata to the mesencephalic locomotor region in the rat. *Brain Research, 295,* 161–178.

Swanson, L. W., Mogenson, G. J., Simerly, R. B., & Wu, M. (1987). Anatomical and electrophysiological evidence for a projection from the medial preoptic area to the mesencephalic and subthalamic locomotor regions in the rat. *Brain Research, 405,* 108–122.

Terkel, J., Bridges, R. S., & Sawyer, C. H. (1979). Effects of transecting lateral neural connections of the medial preoptic area on maternal behavior in the rat: Nest building, pup retrieval and prolactin secretion. *Brain Research, 169,* 369–380.

Uhl, G. R., Kuhar, M. J., & Snyder, S. H. (1978). Enkephalin-containing pathway: Amygdaloid efferents in the stria terminalis. *Brain Research, 149,* 223–228.

Van der Kooy, D., Koda, L. Y., McGinty, J. F., Gerfen, C. R., & Bloom, F. E. (1984). The organization of projections from the cortex, amygdala, and hypothalamus to the nucleus of the solitary tract in the rat. *Journal of Comparative Neurology, 224,* 1–24.

Van Leengoed, E., Kerker, E., & Swanson, H. H. (1987). Inhibition of post-partum maternal behavior in the rat by injecting an oxytocin antagonist into the cerebral ventricles. *Journal of Endocrinology, 112,* 275–282.

Van Leeuwen, F. W., Van Heerikhuize, J., Van Der Meulen, B., & Wolters, P. (1985). Light microscopic autoradiographic localization of [3H]oxytocin binding sites in the rat brain, pituitary and mammary gland. *Brain Research, 359,* 320–325.

Veazey, R. B., & Severin, C. M. (1980). Efferent projections of the deep mesencephalic nucleus (pars lateralis) in the rat. *Journal of Comparative Neurology, 190,* 231–244.

13

Oxytocin and Maternal Behavior

THOMAS R. INSEL

It was known early in this century that a posterior pituitary hormone was critical for stimulating uterine contractions during labor and milk letdown during nursing (Dale, 1909; Ott & Scott, 1910). In 1953, Du Vigneaud and his colleagues (Du Vigneaud et al., 1953) isolated this hormone, oxytocin, and determined its nine amino acid sequence. In addition to its well-known effects on uterine and mammary tissues, oxytocin may also play a role within the central nervous system in initiating maternal behavior. Several other neuropeptides, such as corticotropin-releasing factor (Koob & Bloom, 1985), angiotensin (Simpson, Epstein, & Camando, 1978), and cholecystokinin (Dockray, 1983), appear to function within the brain to integrate behaviors consistent with their peripheral effects. Whether this consistency of central and peripheral effects also characterizes oxytocin remains unclear. This chapter reviews the evidence implicating oxytocin in the neural integration of maternal behavior and then describes some recent studies of changes in extrahypothalamic oxytocin pathways at the time of parturition. But first, a brief description of oxytocin is in order.

OXYTOCIN

Oxytocin is one of a family of structurally related peptides including arginine vasopressin (AVP) and arginine vasotocin (AVT), which are often called "neurohypophyseal peptides" because they are stored in the neurohypophysis or posterior pituitary (Figure 13.1). Each of these peptides contains nine amino acids, including two cysteine residues that are linked together to form a six amino-acid ring (for oxytocin, this ring is called "tocinoic acid"), leaving a three amino acid tail. Although these peptides are stored in the pituitary, they are synthesized primarily in three regions of the hypothalamus: the paraventricular nucleus (PVN), the supraoptic nucleus (SON), and the anterior commissural nucleus (ACN).

Since the time of the classic studies of the Sharrers (1954), it has been clear that oxytocin is transported via neurosecretory axons to the posterior pituitary, from which it is released during parturition and nursing. Recent immunohistochemical studies have defined additional nonpituitary, extrahypothalamic projections for oxytocin in the rat brain (Buijs, 1978; Sawchenko & Swanson, 1982; Sofroniew & Weindl, 1981). Oxytocin cells within the PVN and ACN (and, to a lesser extent, the SON) project to the olfactory nuclei, tenia tectis, ventral septum, and central nucleus of the amyg-

```
              1   2   3   4   5   6   7   8   9
```

Arginine Vasotocin	(AVT)	Cys-Tyr-Ile-Gln-Asn-Cys-Pro-Arg-Gly-NH2
		$\qquad\qquad\;\star$
Arginine Vasopressin	(AVP)	Cys-Tyr-Phe-Gln-Asn-Cys-Pro-Arg-Gly-NH2
		$\qquad\qquad\qquad\qquad\qquad\qquad\star$
Oxytocin	(OXY)	Cys-Tyr-Ile-Gln-Asn-Cys-Pro-Leu-Gly-NH2

Figure 13.1. Nine amino acid sequence of related peptides: oxytocin, arginine-vasopressin, and arginine-vasotocin. Note that peptides differ only at positions 3 and 8. Arginine-vasotocin is the evolutionary precursor; oxytocin and arginine-vasopressin are found almost exclusively in mammals.

dala, and to several sites within the medulla and spinal cord. One should note that these projections appear to represent both collaterals of cells secreting oxytocin to the posterior pituitary and a different population of neurons that may not project to the posterior pituitary, so that the central and peripheral release of oxytocin may be uncoupled. Indeed, under various experimental conditions, oxytocin content in cerebrospinal fluid (CSF) does not correlate with plasma levels of the peptide (Jones, Robinson & Harris, 1983; Kendrick, Keverne, Baldwin, & Sharman, 1986).

Although oxytocin is found only in mammals, AVT and related peptides can be traced extensively in phylogeny (Acher, 1974). AVT, believed to be the ancestral peptide for both oxytocin and AVP, has been associated with reproductive or nesting behavior in the rough-skinned newt (Moore, 1987; Moore & Miller, 1983), the leopard frog (Diakow & Raimondi, 1981), and a variety of geckoes (Guillette & Jones, 1982; Jones & Guillette, 1982). There is also some evidence linking AVT to imprinting behavior in Pekin ducks (Martin & van Wimersma Greidamus, 1979). While it is tempting to view these behavioral effects of AVT as precursors to parental or sexual behaviors in mammals, a comparative analysis is still premature.

OXYTOCIN INDUCTION OF SHORT-LATENCY MATERNAL BEHAVIOR

Like most other rodents, nulliparous rats are typically not maternal, at least not unless they are exposed to pups for 5–7 days (Rosenblatt, 1967). If the female is ovariectomized and injected with estrogen, this latency period for the onset of maternal behavior can be reduced to 1–3 days (Siegel & Rosenblatt, 1975). A generation of research has confirmed that the decrease in progesterone associated with ovariectomy and the increase in estrogen produced by exogenous administration simulate the physiological changes of parturition and thus may be necessary (if not sufficient) for maternal behavior.

Pedersen and Prange (1979) were the first to report that intracerebroventricular (ICV) administration of oxytocin (400 ng) to an ovariectomized, estrogen-primed virgin female resulted in full maternal behavior within 1 hr. Full maternal behavior, including nest building, grouping, licking, carrying, and crouching over pups, was observed in 11/13 rats following oxytocin administration, but in only 2/11 estrogen-primed females given ICV saline and in no animals given ICV oxytocin without estrogen priming. In a subsequent study, Pedersen, Ascher, Monroe, and Prange (1982) demonstrated that this effect on maternal behavior could be elicited at lower doses (100 ng) and was specific for oxytocin or its ring structure, tocinoic acid. AVP administration also increased maternal behavior, although with a longer latency (2 hr). Eighteen other peptides were not effective. Although these results with oxytocin were largely corroborated by Fahrbach and co-workers (1984), Rubin and associates (1983) and Bolwerk and Swanson (1984) were unable to find any effect of oxytocin on maternal behavior in estrogen-primed rats.

Why these differing results among studies? In part, these differences reflect varying methodologies. Fahrbach, Morrell, and Pfaff (1986) have shown that habituation to the test apparatus decreases the oxytocin effect on maternal behavior. A novel environment (or stress) might be expected to increase endogenous oxytocin (Lang et al., 1983; although see also Gibbs, 1984) and possibly for this reason is associated with a higher frequency of maternal behavior. Rubin et al. (1983) tested females in their home cage, thus reducing the chance of finding an oxytocin effect. Bolwerk and Swanson (1984) used rats of the Wistar strain, which may require a different treatment schedule for estrogen priming. A further difference between the studies mentioned above was that those that found a positive effect for oxytocin used Sprague-Dawley females from the same supplier. This colony was infected with pulmonary pathogens throughout the period of these experiments (Pedersen, personal communication, 1985). Wamboldt and Insel (1987) found that oxytocin interacts with anosmia (produced by zinc sulfate lesions of the olfactory epithelium) to induce maternal behavior in ovariectomized, estrogen-primed rats. One explanation for the positive findings in the Pedersen et al. and Fahrbach et al. experiments is that these rats were anosmic, stressed, or both in sufficient amounts for an oxytocin effect to emerge.

Whatever the ultimate explanation for the different results among studies, it appears that high (i.e., supraphysiological) concentrations of the peptide administered centrally to an estrogen-primed female under the right set of environmental conditions may be associated with the onset of full maternal behavior within a matter of minutes. This effect is dose dependent and specific to oxytocin (Pedersen et al., 1982) but appears fickle, probably because of interactions with a number of environmental stimuli. Furthermore, such a pharmacological effect may have little relevance to the normal mechanisms underlying maternal behavior.

More direct evidence that oxytocin is involved in the physiological mediation of maternal behavior comes from studies with a selective oxytocin antagonist, d (CH$_2$)$_5$-8-ornithine vasotocin, and with oxytocin antisera (Fahrbach, Morrell, & Pfaff, 1985a; Pedersen, Caldwell, Fort, & Prange, 1985; van Leengoed, Kerker, & Swanson, 1987). Fahrbach and co-workers used a pregnancy termination model of maternal behavior: rats on day 16 of pregnancy were hysterectomized, ovariectomized, and injected with 100 μg/kg estrogen. Following this procedure, at least 80% of the rats appeared maternal within the first day of pup contact. However, after ICV administration of either the oxytocin antagonist or the oxytocin antisera, less than 40% appeared maternal on the first day of testing. Administration of the antagonist to lactating females had little effect, suggesting that oxytocin may be particularly important to the initiation, but not the maintenance, of maternal behavior (Fahrbach, Morrell, & Pfaff, 1985b). Concurrently, Pedersen and co-workers (1985) reported that anti-oxytocin antiserum decreased the incidence of steroid-induced maternal behavior. More recently, Van Leengoed and co-workers (1987) demonstrated a delay in the onset of maternal behavior following administration of an oxytocin antagonist to Wistar rats undergoing natural parturition. In this study, females injected with the oxytocin antagonist following the birth of the first pup showed no change in the course of parturition but had a 10-fold longer latency for retrieval 1 hr following delivery of the last pup. The antagonist did not alter the females' activity level. Maternal behavior was noted to be normal 24 hr later with continuous pup exposure.

Finally, lesioning the PVN, which removes the majority of oxytocin neurons, appears to confer a profound deficit on maternal behavior. Initial results with primigravid Sprague-Dawley females following bilateral electrolytic stimulation (12 mAmp for 20 sec) on day 15 of pregnancy reveal slower retrieval and increased cannibalization postpartum (Figure 13.2). These lesion studies defy a single interpretation, as the PVN contains AVP, corticotropin-releasing factor, norepinephrine, and several other neurotransmitters or neuromodulators in addition to oxytocin (Swanson & Sawchenko, 1980). Which of these constituents, if any, is responsible for the observed deficit in maternal behavior following ablation of the PVN remains to be studied. Of note, however, no deficit in maternal behavior appears following PVN lesions during lactation (Numan & Corodimas, 1985), so that the effect observed following lesioning during pregnancy may be conferred specifically on the initiation, and not the maintenance, of maternal behavior.

Does oxytocin have a similar role in other species? Studies in nonhuman primate species are extremely important but unfortunately will be very difficult to interpret because of the high rate of short-latency maternal behavior evident in the nontreated females of most primates. This is not a problem with most rodent species other than inbred strains of mice. Studies in wild mice have shown that high doses of oxytocin administered peripherally can decrease spontaneous pup killing and induce maternal behavior (McCarthy, Bare, & vom Saal, 1986).

Aside from rodent studies, the most significant evidence implicating oxytocin in maternal behavior comes from studies in sheep. Sheep normally show little response to newborns unless they are exposed within 2 hr of parturition (see chapter 8). Keverne and his co-workers (1982) have demonstrated that maternal bonding of the ewe to the newborn lamb can be facilitated by vaginocervical stimulation. Vaginocervical stimulation of the ewe is an extremely potent releaser of oxytocin into ventricular CSF as well as blood (Kendrick et al., 1986). In addition, ICV administration of oxytocin (5, 10, or 20 μg) stimulates maternal behavior in estrogen-primed, ovariectomized, multiparous ewes tested either in their home pens or in a semi-naturalistic outdoor arena (Kendrick, Keverne, & Baldwin, 1987). Thus far, oxytocin has not been effective in nulliparous or primiparous ewes, suggesting that experience may be more important in sheep than in rodents. Studies with oxytocin antagonists or antisera remain to be done in sheep.

MECHANISMS FOR OXYTOCIN'S EFFECTS

How might oxytocin induce maternal behavior? Two observations should guide us in answering this question. First, the behavioral effects of oxytocin relevant to maternal behavior are entirely sex steroid dependent. Without ovariectomy and estrogen priming, oxytocin administration is not associated with maternal behavior. The second

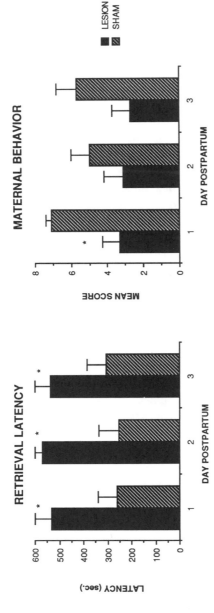

Figure 13.2. Postpartum maternal behavior in primigravid females following either electrolytic lesions of the PVN or sham surgery on day 15 of pregnancy. Means from seven animals in each group show significant ($p < .05$) group differences by repeated measures ANOVA in retrieval latency and maternal behavior. Retrieval latency was tested twice daily by placing three pups in corners of the female's home cage and timing the interval for grouping into a single clump (maximum = 600 sec). Maternal behavior reflects a composite of scores for nest building (1–4), grouping (0–2), licking (0–1), and crouching (0–2) during three daily 5-min observations in the home cage. Higher scores reflect more maternal behavior. Animals with histologically incomplete lesions were deleted from data analysis.

observation, based on an examination of peripheral target tissues for this peptide, is that oxytocin's effects are mediated by membrane-bound receptors. In uterine and mammary tissue, these receptors appear to increase markedly at parturition, suggesting that, at least in the periphery, oxytocin may use increased receptor synthesis to amplify its signal (Soloff, Alexandrova, & Fernstrom, 1979). This increase, like the behavioral effects, appears to be sex steroid dependent. Increased estrogen and/or increased estrogen receptors are associated with the induction of oxytocin receptors. Increased progesterone, on the contrary, blocks the induction of both estrogen and oxytocin receptors in the uterine myometrium (Fuchs, Periyasamy, Alexandrova, & Soloff, 1983). Using this change in peripheral tissue as a guide, we hypothesized that increases in oxytocin receptors in brain might accompany parturition.

BRAIN OXYTOCIN RECEPTORS ARE INCREASED POSTPARTUM

That specific receptors for oxytocin should be found in brain is, by itself, remarkable, for this is suggestive evidence that the brain is a target organ for this peptide. Unfortunately, oxytocin receptors are too sparse in brain to be characterized using conventional homogenate-binding techniques. With slide-mounted frozen sections, specific binding of ^{3}H-Tyr-oxytocin can be demonstrated with in vitro receptor autoradiography (Brinton, Wamsley, Gee, Wan, & Yamamura, 1984; De Kloet, Rotteveel, Vorhuis, & Terlou, 1985; Insel, 1986; Van Leeuwen, Heerikhuize, Vander Meulen, & Wolters, 1985). Briefly, fresh-frozen brains are cut in a cryostat, 12- to 20-micron (μ) thick sections are thaw mounted on slides, and then these slide-mounted sections are prewashed to remove endogenous oxytocin. Sections are then incubated in 0.1 M Tris HCl containing divalent cations, protease inhibitors, and 5 nM ^{3}H-Tyr-oxytocin for 90 min at room temperature. Adjacent sections are incubated under identical conditions with an excess of unlabeled oxytocin to define nonspecific binding. All sections are then washed, dried under a stream of cool air, and apposed to ^{3}H-sensitive film for at least 28 days to generate the autoradiogram.

Using this technique, oxytocin receptors were evident in the following areas: anterior olfactory nucleus (AOP), tenia tectis (TT), bed nucleus of the stria terminalis (BNST), central nucleus of the amygdala (AmC), ventromedial nucleus of the hypothalamus (VMN), and ventral subiculum (VS) (Figure 13.3).

To determine if brain oxytocin receptors increase with parturition, brains from 15-day pregnant (primigravid) rats were compared with brains from females between 1 and 6 days postpartum (for a full description of the methods used, see Insel, 1986). Postpartum rats showed a selective 43% increase in oxytocin binding in the BNST, without evidence of a change in any other receptor region (Figure 13.4). This increase in BNST binding appeared to be due to an increase in receptor number, not a change in affinity, as increasing the concentration of ^{3}H-Tyr-oxytocin to 50 nM, above the level of receptor saturation, did not reduce the difference between the groups. Nor was this increased binding an artifact due to decreased quenching of tritium in the postpartum period, as transmission of tritium through 4-μ-thick brain sections from both groups was equal.

What might cause this focal increase in oxytocin receptors? The profound in-

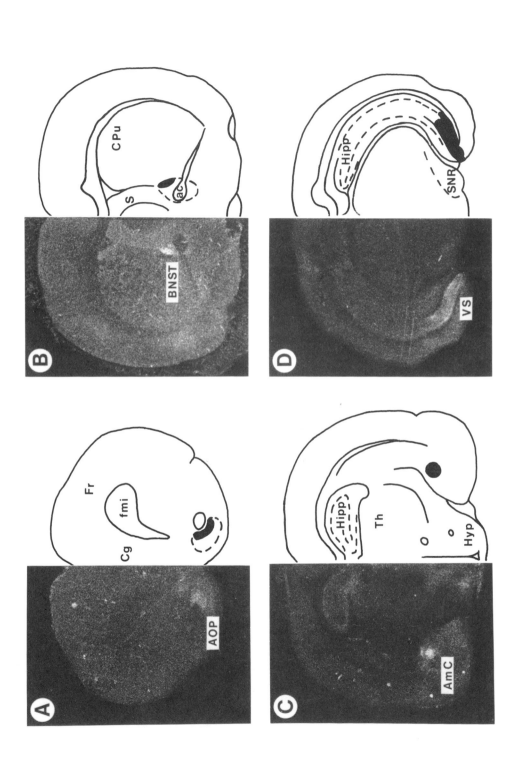

crease in estrogen and the decrease in progesterone that precede parturition (and experimentally decrease the latency to maternal behavior) seemed a likely possibility. Ovariectomized virgin females were injected with estradiol benzoate (100 μg/kg) or oil vehicle, and sacrificed 48 hr later for receptor analysis. Once again, oxytocin receptors in the BNST increased in the estrogen-primed group (Figure 13.5), as seen with the postpartum females in the previous experiment. In addition, oxytocin receptors in the VMN were greater in the estrogen-primed group (as previously described by De Kloet, Voorhuis, & Elands, 1986). These VMN changes were not evident in the postpartum females. Comparing the two studies (Figures 13.4 and 13.5), it appears that the ovariectomized/vehicle females had very few receptors in the VMN, in contrast to the three other groups, which were roughly equivalent. One interpretation of these results in the VMN is that this group of oxytocin receptors is estrogen dependent. Presumably, in the pregnant, postpartum, or estrogen-replaced conditions, there is enough estrogen available to maintain these receptors in the VMN. In contrast, the BNST appears more finely tuned to the endocrine changes in the postpartum condition: decreased progesterone *and* increased estrogen. It is not yet known if progesterone will block the induction of oxytocin receptors in either site.

Why aren't these sex steroid effects conferred on other populations of oxytocin receptors? Estrogen binds to cytosolic receptors that are distributed unevenly within brain (Pfaff & Keiner, 1973; Stumpf, Sar, & Keefer, 1975). Of those brain regions expressing oxytocin receptors (AOP, BNST, AmC, VMN, VS), it is only the BNST and VMN that contain estrogen receptors. Furthermore, in these two regions, oxytocin receptors and estrogen receptors appear in the same subnuclei, although double-labeling studies are still needed to define whether these receptors are indeed within the same cells.

So far, then, it is apparent that oxytocin receptors increase selectively in the BNST sometime after day 15 of pregnancy, probably concurrent with the physiological changes in sex steroids accompanying parturition. What does this receptor change mean? An increase in receptor number generally reflects either an "up-regulation" to compensate for decreased neurotransmitter release or an independent increased synthesis of receptors to amplify the effects of neurotransmitter release. The latter condition obtains in mammary and uterine tissue at parturition: increased surges of oxytocin release are accompanied by increased numbers of oxytocin receptors in these target tissues. The question that remains is, how does brain oxytocin change with parturition?

Figure 13.3. Distribution of oxytocin receptors in female rat brain. Binding, assessed by in vitro receptor autoradiography, is most apparent in the posterior aspects of the anterior olfactory nucleus (AOP), lateral band of the bed nucleus of the stria terminalis (BNST), central nucleus of the amygdala (AmC), and ventral subiculum of the hippocampus (VS). Other abbreviations: fmi, fornix major; Cg, cingulate cortex; Fr, frontal cortex; CPu, caudate-putamen; ac, anterior commissure; S, septum; Hyp, hypothalamus; Th, thalamus; Hipp, hippocampus; SNR, substantia nigra.

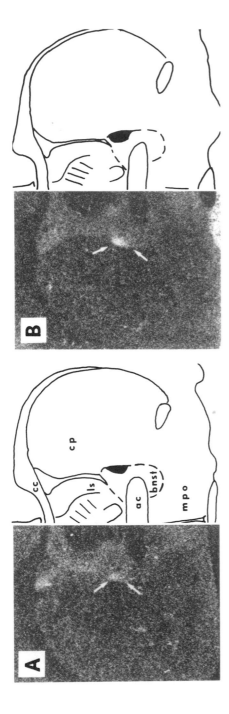

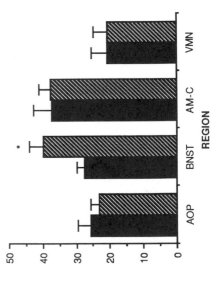

Figure 13.4. ^{3}H-Tyr-oxytocin binding to 16-μ brain sections from (A) 15-day pregnant and (B) 1- to 6-day postpartum rats. Quantitative analysis of autoradiograms using computerized densitometry with ^{3}H-labeled standards reveals a significant ($p < .05$) increase in binding to BNST of postpartum females, as shown in (B), but no difference in the other regions studied (C). Analysis was based on at least four readings per region per animal, with seven animals in each group.

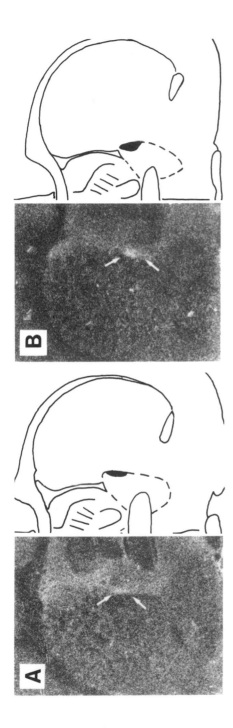

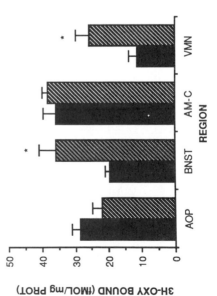

Figure 13.5. ³H-Tyr-oxytocin binding as in Figure 13.4, this time comparing females 48 hr postovariectomy with either (A) oil or (B) estradiol benzoate (100 μg/kg) treatment. Again, computerized densitometry reveals a significant (p < .05) increase in binding to BNST (C) of the same magnitude as seen in Figure 13.4. In addition, binding in the VMN for the ovariectomized-oil group is significantly less than in either the estrogen-treated group or either group studied in Figure 13.4C.

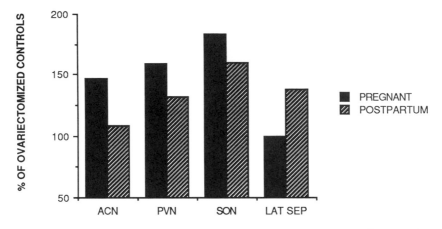

Figure 13.6. Oxytocin measured by radioimmunoassay of brain punches from ovariectomized controls, 16-day pregnant, or 1-day postpartum rats. Oxytocin appears to increase during pregnancy and decrease postpartum in the anterior commissural nucleus (ACN). Concurrently, oxytocin content increases in the ventral septal region (including the BNST), which receives projections from the ACN. (Data adapted from Caldwell et al., 1987)

OXYTOCIN CONTENT IN POSTPARTUM BRAIN

Levels of oxytocin in plasma (Chard, Boyd, Forsling, McNeilly, & Landon, 1970; Glatz et al., 1981; Otsuki, Tanizawa, Yamaji, Fujita, & Kurachi, 1983) and CSF (Kendrick et al., 1986) increase during labor in a variety of species. Caldwell and co-workers (1987) have recently shown, using radioimmunoassay of brain punches, that oxytocin increases in the PVN (159%) and SON (183%) in pregnant rats compared with ovariectomized controls. Oxytocin levels in these nuclei decrease slightly around the time of parturition, probably reflecting decreased storage and increased release. Elevated levels of oxytocin are also evident in the ACN (147%) on day 16 of pregnancy, but they decrease to ovariectomized control levels on day 1 postpartum. Concurrent with this postpartum decrease in the ACN, oxytocin content increases in the "ventral septal region" receiving the ACN projection (Figure 13.6). The ventral septal region in this study includes not only the ventral septum, but the BNST and dorsal preoptic area as well. These data are roughly consistent with an earlier immunohistochemical study suggesting decreases in oxytocin staining in the ACN following estrogen treatment (Rhodes, Morrell, & Pfaff, 1981).

Techniques such as immunohistochemistry or radioimmunoassay, especially when combined, are powerful tools for describing quantitative changes in the regional distribution of a peptide. However, they cannot define whether a decrease in oxytocin content in the ACN reflects a decrease in synthesis or an increase in release.

To investigate this question directly, the amount of messenger RNA (mRNA) for oxytocin was examined using a ^{35}S-labeled synthetic oligonucleotide probe. This probe, with complementary nucleotide pairs to oxytocin mRNA, binds selectively to those cells synthesizing oxytocin. Using in situ hybridization, 12-μ slide-mounted sections are exposed to the oxytocin probe for 16 hr, then coated with emulsion, and counterstained. Differences in message level can be assessed by counting the grains over labeled cells in each nucleus synthesizing oxytocin. Using this method, Lightman and Young (1987) have shown a fivefold increase in oxytocin message level in the PVN during lactation.

A preliminary comparison of sections through the ACN in 15-day pregnant and 1-day postpartum females reveals no difference in the number of cells expressing the oxytocin message. A further analysis to determine the number of copies per cell will be necessary to state definitively that there is no change in the synthesis of oxytocin with parturition. But from these preliminary results, the decreased oxytocin content reported by Caldwell and co-workers would most likely reflect increased release of oxytocin, not decreased synthesis. Of interest is that in both groups of animals the distribution of labeled cells extends anterior to the ACN to include cells along either side of the third ventricle and laterally through the medial preoptic area (Figure 13.7). Although these cells have not been previously visualized with oxytocin immunohistochemistry, very recent work with females in the periparturitional period has noted increased immunostaining of perikarya in these regions (Jirikowski, Caldwell, Pedersen, & Pilgrim, in press). Identification of oxytocin-producing neurons in the medial preoptic area is of particular interest because lesions in this area have been shown to disrupt maternal behavior (Numan, 1974).

OXYTOCIN AND MATERNAL BEHAVIOR: HOW SPECIFIC?

Thus far, we have summarized the evidence that oxytocin induces short-latency maternal behavior in estrogen-primed females; that oxytocin receptors increase in the BNST around parturition, possibly due to the concurrent changes in sex steroids; and, finally, that oxytocin appears to be released in increased amounts to both pituitary and extrahypothalamic sites at about the time of parturition. While the data are currently inconclusive, they are at least consistent in describing a role for oxytocin in the estrogen-primed brain to initiate maternal behavior.

Figure 13.8 summarizes a hypothetical model of oxytocin's effects on maternal behavior. Note that oxytocin release can be increased by numerous inputs, including suckling, labor, vaginocervical stimulation, and stress. In the presence of increased estrogen and decreased progesterone, receptors in the BNST, mammary myoepithelium, and uterine myometrium increase. A surge of oxytocin in the presence of increased receptors in these target fields leads to a coordinated maternal response. This model may have some heuristic value, but it is too simplistic.

In particular, the specific role of the BNST in the initiation of maternal behavior has to be more carefully investigated. Earlier it was noted that anosmia, produced by zinc sulfate lesions of the olfactory epithelium, interacts with oxytocin to induce maternal behavior (Wamboldt & Insel, 1987). This lesion affects not only the olfactory tract, but also the vomeronasal system, which projects via the accessory olfactory system to the BNST (Scalia & Winans, 1975). This may be a critical input, as lesions of the vomeronasal system are associated with maternal behavior even in nonoxytocin-treated females (Fleming & Rosenblatt, 1974). If decreasing this input to the BNST increases maternal behavior, what about lesions of the BNST? The BNST is a vital crossroad for messages from the amygdala, hypothalamus, brain

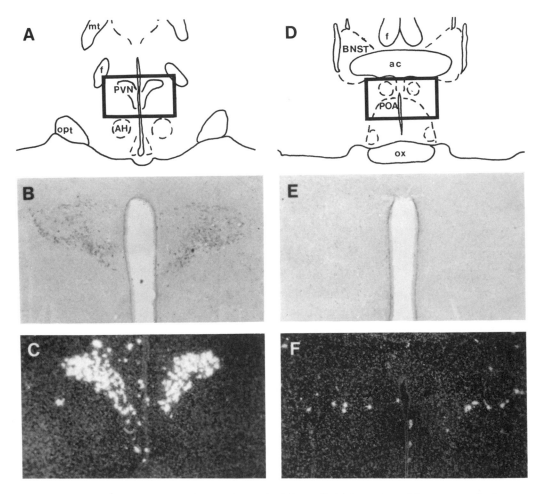

Figure 13.7. Visualization of oxytocin-synthesizing cells in the region of the paraventricular nucleus (A–C) and the preoptic area (D–F). Immunohistochemistry, which uses an antibody to oxytocin in fixed tissue, is effective for demonstrating oxytocin-containing cells in the PVN (B), but in this case fails to reveal cells in the preoptic region (E). In situ hybridization with a ^{35}S-labeled oligonucleotide probe for oxytocin mRNA reveals multiple cells with oxytocin message in both regions (C and F). Abbreviations: mt, mammillothalamic tract; f, fornix; AH, anterior hypothalamus; opt, optic tract; ac, anterior commissure; POA, preoptic area; ox, optic chiasm.

stem, and medial basal forebrain (Figure 13.9). In our own studies using electrolytic lesions of the lateral BNST in 15-day pregnant females, mean retrieval latency during the first day postpartum was actually 27% more rapid in lesioned primigravid rats compared with sham operated controls. How could a lesion in such a vital area, an area with increased oxytocin binding at parturition, improve rather than disrupt maternal behavior? One explanation may be that oxytocin has an inhibitory effect on the normal functioning of this area. The BNST, when uninhibited, might lead to

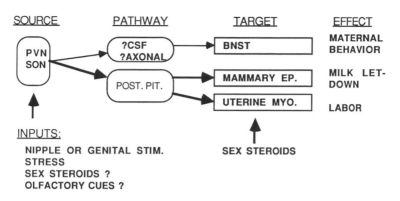

Figure 13.8. Integrative model of oxytocin effects on maternal behavior in the rat. Note how coordinated behavioral and physiological effects could result from increased oxytocin receptors in both central and peripheral target tissues. In this model, estrogen increases gene expression of oxytocin receptors. Oxytocin itself may be increased by suckling, vaginocervical stimulation, and stress, as well as by olfactory cues and changes in sex steroids.

cannibalization, aggression, or avoidance of the pup. Oxytocin might function to turn off these responses, thus "disinhibiting" maternal behavior.

There is another, more general way in which the model suggested by Figure 13.8 is too simplistic. Oxytocin has a number of other behavioral effects, some of which can be demonstrated at doses below those required for inducing maternal behavior. For instance, oxytocin has been implicated in grooming (Caldwell, Hruby, Hill, Prange, & Pedersen, 1986) and reproductive behavior (Caldwell, Prange, & Pedersen, 1986; Hughes, Everitt, Lightman, & Todd, 1987).

It may be useful to consider that grooming and reproductive behavior, along with maternal behavior, mediate social bonds. Some years ago, Newton (1978) suggested a similar connection in a paper entitled "The Role of the Oxytocin Reflexes in Three Interpersonal Reproductive Acts: Coitus, Birth, and Breastfeeding." Perhaps oxytocin in mammals (and AVT in reptiles and birds) functions within the brain to initiate affiliative bonds, either between a parent and offspring or between two sexually mature adults.

Two relevant pieces of clinical evidence are worth noting. First, high plasma levels of oxytocin have been measured during orgasm (Carmichael et al., 1987; Murphy, Seckl, Burton, Checkley, & Lightman, 1987). These levels, comparable with the plasma levels in nursing females, have been demonstrated in males and, in addition to any affiliative function, may be relevant to oxytocin's effects on sperm transport. The second point of note is that in nursing women, oxytocin increases in response to the cry of the infant several minutes prior to suckling (McNeilly, Robinson, Houston, & Howie, 1983). This response, independent of suckling, suggests the importance of affective as well as physiological stimuli in oxytocin release.

Finally, one nonclinical study provides evidence for a curious role of oxytocin in social bonding. Rat pups will not attach to the nipple of a nursing mother following a thorough washing of her ventrum. If the mother is injected with oxytocin immediately after washing, pups continue to attach (Singh & Hofer, 1978). The mech-

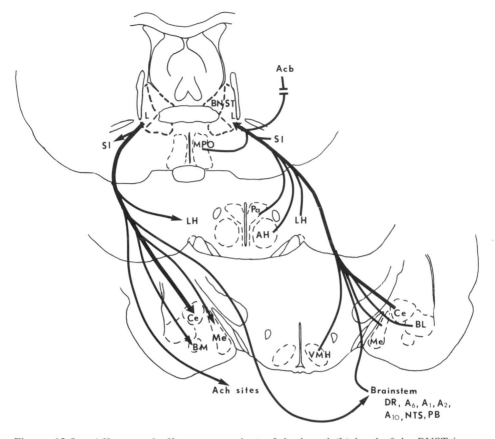

Figure 13.9.   Afferent and efferent connections of the lateral (L) band of the BNST in rat
brain. Abbreviations for afferents: Acb, nucleus accumbens; SI, substantia innominata; MPO,
medial preoptic area; Pa, paraventricular; AH, anterior, LH, lateral, VMH, ventromedial nuclei of the hypothalamus; Ce, central, BL, basolateral, Me, medial nuclei of the amygdala;
DR, dorsal raphe; NTS, nucleus tractus solitarius; PB, parabrachial nuclei in the brain stem.
A_6 (locus coeruleus), A_1, A_2, and A_{10} represent catecholamine cell groups in the brain stem.
Efferent connections, in addition to reciprocal innervation to SI, LH, and Ce, include Me
(medial) and BM (basomedial) nuclei of the amygdala and cholinergic sites in the brain stem.

anism for this effect is not related to lactation, as attachment occurs even to ligated
nipples. Presumably oxytocin increases the secretion of some chemical attractant from
the areolar tissue, which literally induces attachment.

   There are other reported behavioral effects of oxytocin that are less clearly related to a ''hormone for social bonding.'' For instance, extensive evidence suggests
that oxytocin has amnestic effects (reviewed by Kovacs & Telegdy, 1985). It is
counterintuitive that a peptide important for bonding would block memory consolidation, and yet the cumulative evidence from pharmacological studies, injection of
antisera, and local injections of peptide reveal that oxytocin, in contrast to AVP,
facilitates the extinction of an active avoidance response. Closer analysis of these
studies, however, suggests that the specific memory tests used—extinction of active

and passive avoidance—may not be salient to the stimuli for maternal behavior. One might even argue that for rodents and sheep, the onset of maternal behavior requires "forgetting" the species-typical avoidance of newborns. Amnesia, in this sense, might be better characterized as disinhibition, as mentioned above. In addition to oxytocin's effects on cognition, increasing evidence implicates this peptide in the stress response (Bruhn, Sutton, Plotsky, & Vale, 1986; King, Brown, & Kusnekov, 1985; Lang et al., 1983) and in salt and water balance (Forsling & Brimble, 1985). The relationship of these effects to maternal behavior is not clear.

OXYTOCIN AND THE NEUROENDOCRINE CONTEXT

It would certainly be an oversimplification to assume that one peptide mediates maternal behavior. Several monoamines and other neuropeptides have also been implicated in maternal care, and undoubtedly many more remain to be discovered. Currently, two neuropeptides besides oxytocin have received special attention: prolactin and the endogenous opiates. Data on each of these are reviewed elsewhere in this volume (see chapter 6), but here one should note that oxytocin interacts with both systems. Oxytocin increases prolactin secretion both in vitro (Lumpkin, Samson, & McCann, 1983) and in vivo (Samson, Lumpkin, & McCann, 1986), possibly via a direct action either in the external layer of the median eminence or in the anterior pituitary itself. And oxytocin secretion is affected by administration of endogenous opiates (Bicknell & Leng, 1982; Hartman, Rosella-Dampman, Emmert, & Summy-Long, 1986). Enkephalin, β-endorphin, and δ-endorphin (5 μg ICV in each case) decrease oxytocin release into plasma (van Wimersma Greidanus & ten Haaf, 1985). These effects are not naloxone reversible, suggesting that they may not be mediated entirely via opiate receptors. As magnocellular neurons of the hypothalamus contain opioid peptides, this opiate–oxytocin interaction may occur at the level of peptide synthesis, although no direct studies of this interaction have been published. In addition, the oxytocin effects of central opiate release have not yet been reported.

CONCLUSIONS

In summary, data from rat and sheep studies suggest that exogenous oxytocin given under specific environmental conditions can induce short-latency maternal behavior within minutes. These effects are entirely dependent on ovariectomy and estrogen priming. Blockade of endogenous oxytocin inhibits experimentally induced maternal behavior. Oxytocin release and oxytocin receptors are increased at parturition in the forebrain as well as in the periphery. In addition to inducing maternal behavior, exogenous administration of oxytocin has been associated with grooming and reproductive behavior, raising the possibility that this peptide's apparent effects on parental care are part of a larger class of effects on affiliative behavior.

Several issues remain to be explored. Is oxytocin involved in the initiation of parental care in humans and other primates? The relationship of oxytocin to male parental care is of particular interest. Several nonhuman primate species in which males provide much of the parental care will be a valuable resource for exploring this relationship. On a more anatomical level, the fate of increased oxytocin receptors in the BNST should be examined. Do these receptors remain increased following lactation, do they decrease with age, and will they respond to ovariectomy? And on a more molecular level, we have to investigate the second messenger system for transducing oxytocin binding into an intracellular effect. Identification of this second messenger will complete only the first step in the cascade of mechanisms by which oxytocin functions in the brain.

REFERENCES

Acher, R. (1974). Chemistry of the neurohypophyseal hormones: An example of molecular evolution. In E. Knobil & W. Sawyer (Eds.), *Handbook of physiology* (Vol. 4, pp. 119–130). Washington, DC: American Physiological Society.

Bicknell, R. J., & Leng, G. (1982). Endogenous opiates regulate oxytocin but not vasopressin secretion from the neurohypophysis. *Nature, 298,* 161–162.

Bolwerk, E. L. M., & Swanson, H. H. (1984). Does oxytocin play a role in the onset of maternal behaviour in the rat? *Journal of Endocrinology, 101,* 353–357.

Brinton, R. E., Wamsley, J. K., Gee, K. W., Wan, Y. P., & Yamamura, H. I. (1984). ^{3}H-oxytocin binding sites demonstrated in the rat brain by quantitative light microscopic autoradiography. *European Journal of Pharmacology, 102,* 365–367.

Bruhn, T. O., Sutton, S. W., Plotsky, P. M., & Vale, W. W. (1986). Central administration of corticotropin-releasing factor modulates oxytocin secretion in the rat. *Endocrinology, 119,* 1558–1563.

Buijs, R. (1978). Intra- and extrahypothalamic vasopressin and oxytocin pathways in the rat: Pathways to the limbic system, medulla oblongata and spinal cord. *Cell Tissue Research, 252,* 355–365.

Caldwell, J. D., Greer, E. R., Johnson, M. F., Prange, A. J., Jr., & Pedersen, C. A. (1987). Oxytocin and vasopressin immunoreactivity in hypothalamic and extrahypothalamic sites in late pregnant and post-partum rats. *Neuroendocrinology, 46,* 39–47.

Caldwell, J. D., Hruby, V. J., Hill, P., Prange, A. J., Jr., & Pedersen, C. A. (1986). Is oxytocin-induced grooming mediated by uterine-like receptors? *Neuropeptides, 8,* 77–86.

Caldwell, J. D., Prange, A. J., Jr., & Pedersen, C. A. (1986). Oxytocin facilitates the sexual receptivity of estrogen-treated female rats. *Neuropeptides, 7,* 175–189.

Carmichael, M. S., Humbert, R., Dixen, J., Palmisano, G., Greenleaf, W., & Davidson, J. (1987). Plasma oxytocin increases in the human sexual response. *Journal of Clinical Endocrinology and Metabolism, 64,* 27–31.

Chard, T., Boyd, N. R. H., Forsling, M. L., McNeilly, A. S., & Landon, J. (1970). The development of a radioimmunoassay for oxytocin: The extraction of oxytocin from plasma and its measurement during parturition in human and goat blood. *Journal of Endocrinology, 48,* 223–234.

Dale, H. H. (1909). The action of extracts of the pituitary body. *Biochemical Journal, 4,* 427–447.

De Kloet, E. R., Rotteveel, E., Voorhuis, T. A. M., & Terlou, M. (1985). Topography of binding sites for neurohypophyseal hormones in rat brain. *European Journal of Pharmacology, 110,* 113–119.

De Kloet, E. R., Voorhuis, T. A. M., & Elands, J. (1986). Estradiol induces oxytocin binding sites in rat hypothalamic ventromedial nucleus. *European Journal of Pharmacology,* *118,* 185–186.

Diakow, C., & Raimondi, D. (1981). Physiology of *Rana pipiens* reproductive behavior: A proposed mechanism for inhibition of the release call. *American Zoology, 21,* 296–304.

Dockray, G. J. (1983). Cholescystokinin and behavior. In D. T. Krieger, M. J. Brownstein, & J. B. Martin (Eds.), *Brain peptides* (pp. 851–869). New York: Wiley.

Drago, F., Pedersen, C. A., Caldwell, J. D., & Prange, A. J. Jr., (1986). Oxytocin potently enhances novelty-induced grooming behavior in the rat. *Brain Research, 368,* 287–295.

Du Vigneaud, V., Ressler, C., Swann, J. M., Roberts, C. W., Katsoyannis, P. C., & Gordon, S. (1953). The sequence of amino acids in oxytocin. *Journal of the American Chemistry Society, 75,* 4879–4880.

Fahrbach, S. E., Morrell, J. J., & Pfaff, D. W. (1984). Oxytocin induction of short-latency maternal behavior in nulliparous, estrogen-primed female rats. *Hormones and Behavior, 18,* 267–286.

Fahrbach, S. E., Morrell, J. I., & Pfaff, D. W. (1985a). Possible role for endogenous oxytocin in estrogen-facilitated maternal behavior in rats. *Neuroendocrinology, 40,* 526–532.

Fahrbach, S. E., Morrell, J. I., & Pfaff, D. W. (1985b). Role of oxytocin in the onset of estrogen-facilitated maternal behavior. In J. A. Amico & A. G. Robinson (Eds.), *Oxytocin: Clinical and laboratory studies* (pp. 372–388). New York: Elsevier.

Fahrbach, S. E., Morrell, J. I., & Pfaff, D. W. (1986). Effect of varying the duration of pre-test cage habituation on oxytocin induction of short-latency maternal behavior. *Physiology and Behavior, 37,* 135–139.

Fleming, A. S., & Rosenblatt, J. S. (1974). Olfactory regulation of maternal behavior in rats: II. Effects of peripherally induced anosmia and lesions of the lateral olfactory tract in pup-induced virgins. *Journal of Comparative and Physiological Psychology, 86,* 233–246.

Forsling, M. L., & Brimble, M. J. (1985). The role of oxytocin in salt and water balance. In J. Amico & A. G. Robinson (Eds.), *Oxytocin: Clinical and laboratory studies* (pp. 167–178). New York: Elsevier.

Fuchs, A.-R., Periyasamy, S., Alexandrova, M., & Soloff, M. S. (1983). Correlation between oxytocin receptor concentration and responsiveness to oxytocin in pregnant rat myometrium: Effects of ovarian steroids. *Endocrinology, 113,* 742–749.

Gibbs, D. M. (1984). Dissociation of oxytocin, vasopressin, and corticotropin secretion during different types of stress. *Life Sciences, 35,* 487–491.

Glatz, T. H., Weitzman, R. E., Eliot, R. J., Klein, A. H., Nathaneilsz, P. W., & Fisher, D. A. (1981). Ovine maternal and fetal plasma oxytocin concentrations before and during parturition. *Endocrinology, 108,* 1328–1332.

Guillette, L. J., & Jones, R. E. (1982). Further observations on arginine vasotocin-induced oviposition and parturition in lizards. *Journal of Herpetology, 16,* 140–144.

Hartman, R. D., Rosella-Dampman, L. M., Emmert, S. E., & Summy-Long, J. Y. (1986). Inhibition of release of neurohypophyseal hormones by endogenous opioid peptides in pregnant and parturient rats. *Brain Research, 382,* 352–358.

Hughes, A. M., Everitt, B. J., Lightman, S. L., & Todd, K. (1987). Oxytocin in the central nervous system and sexual behaviour in male rats. *Brain Research, 414,* 133–137.

Insel, T. R. (1986). Postpartum increases in brain oxytocin binding. *Neuroendocrinology, 44,* 515–518.

Jirikowski, G. F., Caldwell, J. D., Pedersen, C. A., & Pilgrim, C. (in press). Changes in oxytocin immunostaining in the mouse forebrain across late pregnancy, parturition and early lactation. *Cell and Tissue Research.*

Jones, P. M., Robinson, I. C. A. F., & Harris, M. C. (1983). Release of oxytocin into blood and cerebrospinal fluid by electrical stimulation of the hypothalamus or neural lobe in the rat. *Neuroendocrinology, 37,* 454–458.

Jones, R. E., & Guillette, L. J. (1982). Hormonal control of oviposition and parturition in lizards. *Herpetologica, 38,* 80–93.

Kendrick, K. M., Keverne, E. B., & Baldwin, B. A. (1987). Intracerebroventicular oxytocin stimulates maternal behaviour in the sheep. *Neuroendocrinology, 46,* 56–61.

Kendrick, K. M., Keverne, E. B., Baldwin, B. A., & Sharman, D. F. (1986). Cerebrospinal fluid levels of acetylcholinesterase, monoamines and oxytocin during labour, parturition, vaginocervical stimulation, lamb separation and suckling in sheep. *Neuroendocrinology, 44,* 149–156.

Keverne, E. B., Levy, F., & Lindsay, D. R. (1982). Vaginal stimulation: An important determinant of maternal bonding in sheep. *Science, 219,* 81–83.

King, M. G., Brown, R., & Kusnekov, A. (1985). An increase in startle response in rats administered oxytocin. *Peptides, 6,* 567–568.

Koob, G. F., & Bloom F. O. (1985). Corticotropin-releasing factor and behavior. *Federation Proceedings, 44,* 259–263.

Kovacs, G. L., & Telegdy, G. (1985). Role of oxytocin in memory, amnesia, and reinforcement. In J. Amico & A. G. Robinson (Eds.), *Oxytocin: Clinical and laboratory studies* (pp. 359–371). New York: Elsevier.

Lang, R. E., Heil, J. W. E., Ganten, D., Hermann, K., Unger, T., & Raschen, W. (1983). Oxytocin, unlike vasopressin, is a stress hormone in the rat. *Neuroendocrinology, 37,* 314–316.

Lightman, S. L., & Young, W. S. III (1987). Vasopressin, oxytocin, enkephalin, dynorphin, corticotrophin releasing factor mRNA stimulation in the rat. *Journal of Physiology, 394,* 23–39.

Lumpkin, M. D., Samson, W. K., & McCann, S. M. (1983). Hypothalamic and pituitary sites of action of oxytocin to alter prolactin secretion in the rat *Endocrinology, 112,* 1711–1714.

Martin, J. T., & van Wimersma Greidanus, T. B. (1979). Imprinting behavior—influence of vasopressin and ACTH analogues. *Psychoneuroendocrinology, 3,* 261–270.

McCarthy, M. M., Bare, J. E., & vom Saal, F. S. (1986). Infanticide and parental behavior in wild female house mice: Effects of ovariectomy, adrenalectomy, and administration of oxytocin and prostaglandin F_2 alpha. *Physiology and Behavior, 36,* 17–23.

McNeilly, A. S., Robinson, I. C. A. F., Houston, M. J., & Howie, P. W. (1983). Release of oxytocin and prolactin in response to suckling. *British Medical Journal, 286,* 207–209.

Moore, F. L. (1987). Behavioral actions of neurohypophyseal peptides. In D. Crews (Ed.), *Psychobiology of reproductive behavior* (pp. 61–87). Englewood Cliffs, NJ: Prentice-Hall.

Moore, F. L., & Miller, L. J. (1983). Arginine vasotocin induces sexual behavior of newts by acting on cells in the brain. *Peptides, 4,* 97–102.

Murphy, M. R., Seckl, J. R., Burton, S., Checkley, S. A., & Lightman, S. L. (1987). Changes in oxytocin and vasopressin secretion during sexual activity in men. *Journal of Clinical Endocrinology and Metabolism, 65,* 738–741.

Newton, N. (1978). The role of the oxytocin reflexes in three interpersonal reproductive acts: Coitus, birth, and breastfeeding. In L. Carenza, P. Pacheri, & L. Vichella (Eds.), *Clinical psychoneuroendocrinology in reproduction* (pp. 411–418). New York: Academic Press.

Numan, M. (1974). Medial preoptic area and maternal behavior in the female rat. *Journal of Comparative and Physiological Psychology, 87,* 746–759.

Numan, M., & Corodimas, K. P. (1985). The effects of paraventricular hypothalamic lesions on maternal behavior in rats. *Physiology and Behavior, 35,* 417–425.

Otsuki, Y., Tanizawa, O., Yamaji, K., Fujita, M., & Kurachi, K. (1983). Serial plasma oxytocin levels during pregnancy and labor. *Acta Obstetricia et Gynecologica Scandinavica, 62,* 235–237.

Ott, I., & Scott, J. C. (1910). *Proceedings of the Society for Experimental Biology and Medicine, 8,* 48–49.

Pedersen, C. A., Ascher, J. A., Monroe, Y. L., & Prange, A. J., Jr. (1982). Oxytocin induces maternal behavior in virgin female rats. *Science, 216,* 648–649.

Pedersen, C. A., Caldwell, J. D., Fort, S. A., & Prange, A. J., Jr. (1985). Oxytocin antiserum delays onset of ovarian steroid-induced maternal behavior. *Neuropeptides, 6,* 175–182.

Pedersen, C. A., & Prange, A. J., Jr. (1979). Induction of maternal behavior in virgin rats after intracerebroventricular administration of oxytocin. *Proceedings of the National Academy of Sciences USA, 76,* 6661–6665.

Pfaff, D. W., & Keiner, M. (1973). Atlas of estradiol-concentrating cells in the central nervous system of the female rat. *Journal of Comparative Neurology, 151,* 121–158.

Rhodes, C. H., Morrell, J. I., & Pfaff, D. W. (1981). Changes in oxytocin content in the magnocellular neurons of the rat hypothalamus following water deprivation and estrogen treatment. *Cell Tissue Research, 216,* 47–55.

Rosenblatt, J. S. (1967). Nonhormonal basis of maternal behavior in the rat. *Science, 156,* 1512–1514.

Rosenblatt, J. S., & Siegel, H. I. (1981). Factors governing the onset and maintenance of maternal behavior among nonprimate mammals. In D. J. Gubernick & P. H. Klopfer (Eds.), *Parental care in mammals* (pp. 1–76). New York: Plenum Press.

Rubin, B. S., Menniti, F. S., & Bridges, R. S. (1983). Intracerebral administration of oxytocin and maternal behavior in rats after prolonged and acute steroid pretreatment. *Hormones and Behavior, 17,* 45–53.

Samson, W. K., Lumpkin, M. D., & McCann, S. M. (1986). Evidence for a physiological role for oxytocin in the control of prolactin secretion. *Endocrinology, 119,* 554–560.

Sawchencko, P. E., & Swanson, L. W. (1982). Immunohistochemical identification of neurons in the paraventricular nucleus of the hypothalamus that project to the medulla or to the spinal cord in the rat. *Journal of Comparative Neurology, 205,* 260–272.

Scalia, F., & Winans, S. S. (1975). The differential projections of the olfactory bulb and accessory olfactory bulb in mammals. *Journal of Comparative Neurology, 161,* 31–56.

Sharrer, E., & Sharrer, B. (1954). Hormones produced by neurosecretory cells. *Recent Progress in Hormone Research, 10,* 183–240.

Siegel, H. I., & Rosenblatt, J. S. (1975). Estrogen induced maternal behavior in hysterectomized-ovariectomized virgin rats. *Physiology and Behavior, 14,* 465–471.

Simpson, J., Epstein, J. N., & Camando, J. S., Jr. (1978). The localization of receptors for the dipsogenic action of angiotensin II in the subfornical organ of rat. *Journal of Comparative and Physiological Psychology, 92,* 581–608.

Singh, P. J., & Hofer, M. A. (1978). Oxytocin reinstates maternal olfactory cues for nipple orientation and attachment in rat pups. *Physiology and Behavior, 20,* 385–389.

Slotnick, B. M., & Nigrosh, B. J. (1975). Maternal behavior of mice with cingulate cortical, amygdala, or septal lesions. *Journal of Comparative and Physiological Psychology, 88,* 118–127.

Sofroniew, M. V., & Weindl, A. (1981). Central nervous system distribution of vasopressin, oxytocin and neurophysin. In J. L. Martinez, R. A. Jensen, R. B. Mesing, H. Rigter, & J. L. McGaugh (Eds.), *Endogenous peptides and learning and memory processes* (pp. 327–369). New York: Academic Press.

Soloff, M. S., Alexandrova, M., & Fernstrom, M. J. (1979). Oxytocin receptors: Triggers for parturition and lactation? *Science, 204,* 1313–1314.

Stumpf, W. E., Sar, M., & Keefer, D. A. (1975). Atlas of estrogen target cells in rat brain. In W. E. Stumpf & L. D. Grant (Eds.), *Anatomical neuroendocrinology* (pp. 104–119). Basel: Karger.

Swanson, L. W., & Sawchencko, P. E. (1980). Paraventricular nucleus: A site for the integration of neuroendocrine and autonomic mechanisms. *Neuroendocrinology, 31,* 410–417.

Terkel, J., Bridges, R. S., & Sawyer, C. H. (1979). Effects of transecting lateral neural connections of the medial preoptic area on maternal behavior in the rat: Nest building, pup retrieval and prolactin secretion. *Brain Research, 169,* 369–380.

van Leengoed, E., Kerker, E., & Swanson, H. H. (1987). Inhibition of postpartum maternal behaviour in the rat by injecting an oxytocin antagonist into the cerebral ventricles. *Journal of Endocrinology, 112,* 275–282.

Van Leeuwen, F. W., Heerikhuize, J. V., Van der Meulen, G., & Wolters, P. (1985). Light microscopic autoradiographic localization of ^{3}H-oxytocin binding sites in the rat brain, pituitary, and mammary gland. *Brain Research, 359,* 320–325.

van Wimersma Greidanus, T. B., & ten Haaf, J. A. (1985). The effects of opiates and opioid peptides on oxytocin release. In J. Amico & A. G. Robinson (Eds.), *Oxytocin: Clinical and laboratory studies* (pp. 145–154). New York: Elsevier.

Wamboldt, M. Z., & Insel, T. R. (1987). The ability of oxytocin to induce short latency maternal behavior is dependent on peripheral anosmia. *Behavioral Neuroscience, 101,* 439–441.

Yeo, J. A. G., & Keverne, E. B. (1986). The importance of vaginal-cervical stimulation for maternal behaviour in the rat. *Physiology and Behavior, 37,* 23–26.

14

Neurochemical Changes Accompanying Parturition and Their Significance for Maternal Behavior

E. B. KEVERNE AND K. M. KENDRICK

Very few mammals spontaneously exhibit maternal responses when exposed to neonates, although steroid priming of the kind that occurs during pregnancy greatly facilitates such behavior. In rats, maternal behavior can be elicited by the presence of pups, but this may take several days, especially in nulliparous females (Rosenblatt & Siegel, 1981). In sheep, ewes are responsive to lambs only around the time of parturition, and even with hormone priming, nulliparous ewes are unlikely to foster lambs. The difficulty with inducing maternal behavior in sheep may be related to the fact that parturient ewes rapidly establish an exclusive bond with their offspring, and, although very maternal, violently reject any suckling attempt from other neonates within 2 hr of parturition (Poindron & Le Neindre, 1980). Such rejection of alien lambs may be reversed in the early postpartum period by 5 min of vaginal-cervical stimulation, a procedure that also induces the rapid onset of maternal behavior in both estrogen-primed rats (Yeo & Keverne, 1986) and sheep (Keverne, Lévy, Poindron, & Lindsay, 1983). Clearly, neural mechanisms other than those called on by steroid hormones and the sensory cues emitted by neonates are essential for the immediate onset of maternal behavior, and hence offspring survival. The purpose of this chapter is to emphasize the importance of parturition itself as the physiological event for synchronizing a multiplicity of neural mechanisms that coordinate rapid and successful maternal care. The neurochemical changes that occur at parturition or during vaginal-cervical stimulation will be considered and attributed significance according to three contexts; behavior, neuroendocrine responsiveness, and sensory recognition. These contexts are not, of course, mutually exclusive.

Sheep have a relatively large brain, which greatly facilitates neurochemical analysis, since cerebrospinal fluid may be sampled at regular intervals, and dialysis probes can be monitored at a number of different sites simultaneously in the same animal. Since the neurochemical systems of interest to us are either brain stem or diencephalic in origin, there is no reason to assume that the changes measured are functionally very different across the mammalian phyla. However, those parts of the brain that have access to and address these systems are certainly very different, especially in human and nonhuman primates. Nevertheless, a knowledge of the changing neurochemical activity in the limbic brain of parturient sheep may provide some

fundamental understanding of those mechanisms subserving maternal care at a subcortical level.

NEUROCHEMICAL CHANGES DURING PARTURITION
OR VAGINAL-CERVICAL STIMULATION

Cerebrospinal Fluid

Although it is well established, in *Homo sapiens* and a number of mammalian species, that peripheral levels of oxytocin (OXY) are raised during parturition (Chard,
Boyd, Forsling, McNeilly, & Landon, 1970), vaginal-cervical stimulation (Flint,
Forsling, Mitchell, & Turnbull, 1975), and suckling (Robinson & Jones, 1982), little
is known about the central effects of these events on the cerebrospinal fluid (CSF)
levels of this peptide. The source of OXY in the CSF would appear to be from the
intracerebral terminals, since this peptide does not pass the blood–brain barrier in
appreciable quantities (Jones, Robinson, & Harris, 1983), and it is therefore unlikely
that feedback from the peripheral circulation is responsible for elevations in CSF
OXY. There are a number of OXY-containing neurons in the hypothalamus and
terminals in the limbic system, septum, midbrain, brain stem, and spinal cord (Sofroniew 1980, 1985). Although the precise origin of the OXY cell bodies for these
widely distributed terminals is known in only a few cases, it is clear that many of
these neuronal systems are anatomically distinct from those implicated in the release
of OXY from the posterior pituitary.

Changes in CSF concentrations of OXY were recently measured during labor,
parturition, vaginal-cervical stimulation, lamb separation, and suckling in the sheep
(Kendrick, Keverne, Baldwin, & Sharman, 1986). In addition, measurements were
made of CSF levels of monamines and their metabolites, since dopamine (Numan,
1983), noradrenaline (Gaffori & le Moal, 1979; Steele, Rowland & Moltz, 1979),
and serotonin (Barofsky, Taylor, Tizabi, & Kumar, & Jones-Quartey, 1983) have
been implicated in the control of maternal behavior. Concentrations of the dopamine
(DA) metabolites (DOPAC) and homovanillic acid (HVA), the noradrenaline metabolite 3-methoxy-4-hydroxyphenylglycol (MHPG), and OXY in the CSF varied significantly among samples taken in late pregnancy, labor, parturition, and 24 hr postpartum. Labor and parturition significantly increased mean concentrations of OXY,
whereas concentrations of DOPAC and HVA were significantly lower during pregnancy than at parturition, 24 hr postpartum, and during the remaining postpartum
period. Vaginal-cervical stimulation given to estrogen-treated, nonparturient animals
also induced significant increases in CSF OXY, which declined immediately after
stimulation was stopped. The levels of OXY during parturition were usually higher
than those induced by vaginal-cervical stimulation and in some individuals were as
high as or higher than the plasma concentrations. Concentrations of OXY in the CSF
were, however, unlikely to be influenced by peripheral levels, since it was found that
0.01% of the peptide given intravenously crossed the blood–brain barrier.

The above results provided evidence that significantly raised levels of OXY are
present in the CSF of sheep during parturition and after vaginal-cervical stimulation
when increased concentrations of this peptide are also found in blood. CSF levels of

OXY were not normally detectable (< 20 pg/ml) at other times by the radioimmunoassay used. Significant increases in the CSF levels of the noradrenaline metabolite MHPG, but not DA or serotonin, were also found during parturition.

Microdialysis Measurement of Neurochemical Release in Discrete Regions of the Brain During Parturition

Although OXY has been shown to stimulate maternal behavior in rats (Pedersen & Prange, 1979), little is known about the precise brain regions or pathways where OXY might act to stimulate maternal behavior. Lesion and knife-cut studies in rodents have shown that sites in the amygdala (Fleming, Vaccarino, & Luebke, 1980), midbrain (Numan & Nagle, 1983), preoptic area (Numan, Rosenblatt, & Komisaruk, 1977), and olfactory bulb (Fleming & Rosenblatt, 1974) are important for the normal expression of maternal behavior, and some of these areas overlap with the known distribution of OXY-containing fibers. The olfactory bulbs are also thought to play a role in maternal recognition in the sheep (Baldwin & Shillito, 1974; Poindron, 1976). A study of sheep therefore aimed, using intracranial dialysis, to measure the release of OXY from two brain regions implicated in the control of maternal behavior—the olfactory bulb (OB) and the substantia nigra, (SN)—during parturition, suckling, and separation from lambs.

Figure 14.1 shows the mean concentrations of OXY, DA, and MHPG found in 50-min microdialysis samples taken from the SN, taking into account in vitro recoveries of the individual probes. Significant differences in the concentrations of OXY, DA, and MHPG, but not the serotonin metabolite 5-hydroxyindole acetic acid (5-HIAA) were found in the SN across treatments. Mean OXY concentrations rose significantly during parturition and suckling. DA concentrations were significantly raised during suckling. MHPG was significantly increased during parturition and separation of the lambs from the ewes.

Figure 14.2 also shows the mean concentrations of OXY, DA, and MHPG in dialysis samples from OB, taking into account individual probe recoveries. Analysis of variance revealed significant differences for OXY and DA, these concentrations being significantly raised during parturition and suckling (Figure 14.2).

These results provide the first demonstration that OXY is released in the OB and SN of freely mobile sheep during parturition and suckling (Figure 14.3). OXY was not detectable in dialysis samples taken from the cortex, and a 20-μg intravenous OXY infusion had no effect on the concentration of this peptide measured in dialysis samples taken from the SN of two sheep.

These findings reveal that although there are similarities in the release of OXY, DA, and MHPG in the OB and SN at parturition, there are also some differences, particularly in OXY. This OXY release, although increasing during parturition in both the OB and the SN, increases more in the former during suckling and more in the latter during parturition. Lamb separation increased MHPG levels in the SN but not in the OB. These differences for the same transmitter system, probably originating in the same population of neurons, would suggest that differential control over release is exerted at terminal areas.

The OB have been implicated in the control of maternal behavior in rodents (Fleming & Rosenblatt, 1974) and in maternal recognition in sheep (Poindron, Kev-

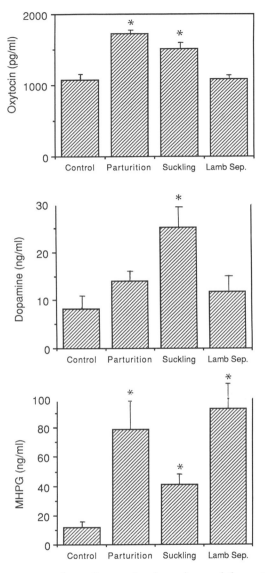

Figure 14.1. Mean concentrations of oxytocin, dopamine, and the metabolite of noradrenaline (MHPG) in the substantia nigra, as revealed by microdialysis during the prepartum period, parturition, suckling, and lamb separation (significantly different from controls).

erne, Chapman, & Baldwin, 1976). A further study (Kendrick, 1988b) employed the technique of microdialysis to investigate whether OXY released in the OB of sheep during parturition is due to the vaginal-cervical stimulation that occurs at this time. Additionally, the release of amino acids was measured, since aspartate and glutamate are thought to function as excitatory transmitters in the efferents of the OB to the piriform cortex (Godfrey, Ross, Carter, Loury, & Matschinsky, 1980; Macrides & Davis, 1985) and γ-aminobutyric acid (GABA) as an inhibitory transmitter at the

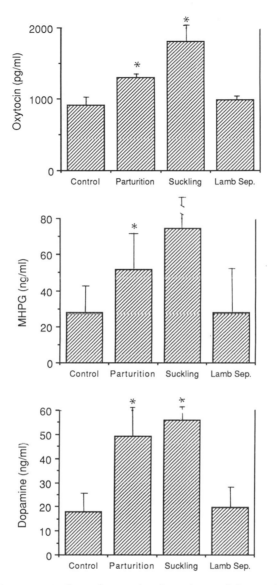

Figure 14.2. Mean concentrations of oxytocin, dopamine, and the metabolite of noradrenaline (MHPG) in the olfactory bulb, as revealed by microdialysis during the prepartum period, parturition, suckling, and lamb separation (significantly different from controls).

dendrodendritic synapses of the granule cells onto mitral and tufted cells (Quinn & Kagan, 1981).

Figure 14.4 shows that vaginal-cervical stimulation given to estrogen-treated, nonparturient ewes significantly increased mean concentrations of OXY, aspartate, GABA, and glutamate compared with prestimulation levels. The figure also shows that the mean concentrations of these substances decreased significantly poststimulation. Concentrations of other amino acids did not change significantly following vaginal-

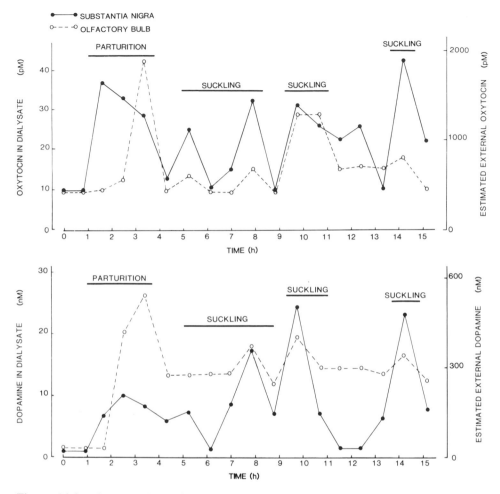

Figure 14.3. Concentrations of oxytocin and dopamine, as measured in microdialysis samples taken simultaneously from the olfactory bulb and substantia nigra of sheep over a 15-hr period, including parturition. Periods of parturition and suckling are indicated by solid lines above the graphs (for suckling, the lines do not indicate that this behavior was continuous but that it was seen for at least 10 min during the sampling period of 50 min). The scale on the left shows the actual concentrations of oxytocin and dopamine measured in the dialysis samples and that on the right, the oxytocin and dopamine concentrations estimated in the brain tissue outside the probes (based on in vitro recoveries).

cervical stimulation. This vaginal-cervical stimulation is probably the stimulus that causes OXY release in the OB during parturition, and produces profound neurochemical changes and amino acids transmitters as well.

Turnover

In addition to measuring changes in a transmitter or its metabolite in CSF, the measurement of enhanced synthesis or turnover of transmitter provides direct evidence

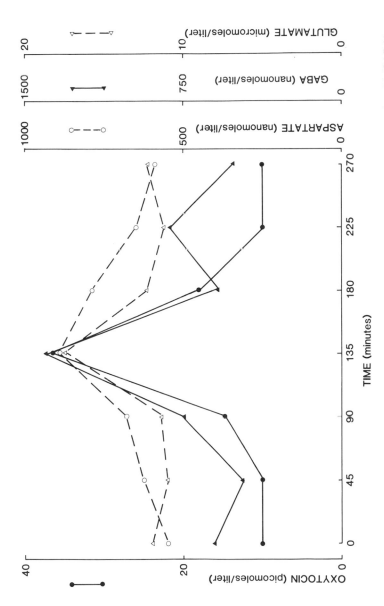

Figure 14.4. Data showing the mean ±SEM concentrations of oxytocin, aspartate, γ-aminobutyric acid (GABA), and glutamate in microdialysis samples taken from the olfactory bulbs of estrogen-treated sheep before, during, and after a 10-min period of vaginocervical stimulation (80–90 minutes). **$p < .02$ (oxytocin, GABA) and *$p < .05$ (aspartate, glutamate) (paired t-test) compared with either pre- or poststimulation samples.

on which brain regions are active. This is a particularly important approach for neural systems that have a widespread distribution of terminals throughout the brain arising from a relatively sparse collection of neurons. Such is the case for the noradrenergic system, which has relatively few cell bodies in the locus coeruleus (A6) and ventral tegmental region (A1 and A2) that are distributed to the whole of the forebrain and limbic brain, respectively (Hokfelt, Johansson, & Goldstein, 1984). The question arises as to whether these small populations of noradrenergic neurons are simultaneously active during parturition, or whether the increased levels of noradrenaline metabolites at this time reflect regulatory controls over release in restricted terminal areas. This can be assessed by microdissection of brain regions following administration of α-methylparatyrosine (α-MPT), an analogue of tyrosine hydroxylase, which prevents the conversion of tyrosine to DOPA and hence blocks the synthesis of noradrenaline. If parturition or vaginal-cervical stimulation enhances noradrenergic activity in a given region of the brain, levels of this transmitter would be expected to decrease as a result of increased turnover of transmitter following synthesis blockade. Such an approach requires a comparison of vaginal-cervically stimulated animals with a group of nonstimulated controls and is clearly not suitable for studying in sheep. For this reason, these studies have been conducted in rodents, from which the brain can be removed rapidly for punch microdissection and transmitter analysis.

Female mice given vaginal-cervical stimulation contingent upon α-MPT administration showed significantly greater noradrenaline depletion of the OB as measured by electrochemical detection after high-pressure liquid chromatography (HPLC) separation. Noradrenaline depletion was found to be greater at 1, 2, and 3 hr after stimulation compared with that in mice given α-MPT alone (Figure 14.5), but was no different at 6, 12 and 24 hr after cervical stimulation. Tissue removed from the cortex did not show enhanced depletion of noradrenaline when compared with that of controls. It appears, therefore, that vaginal stimulation enhances locus coeruleus activity, presumably as a result of local regulation in the terminals in the OB, but not the cortex. Turnover of transmitter is active for some hours as a consequence of olfactory input coincident with somatosensory stimulation (Rosser & Keverne, 1985).

Using a different dosage of α-MPT (100 mg/kg), similar studies on the rat have shown that vaginal-cervical stimulation enhances noradrenaline turnover in the hypothalamus (Figure 14.6). This increased turnover is found only at the 2-hr point after drug treatment and vaginal-cervical stimulation. Taken together, these results suggest that prolonged vaginal-cervical stimulation of the kind that occurs during parturition increases noradrenergic activity synchronously in the A6 and A1, A2 cell groups. However, turnover is high in the OB but not in other telencephalic cortical areas, and outlasts that occurring in the hypothalamus, which suggests that local regulation of release, possibly by presynaptic events (Gervais 1987), plays an important part in enhancing OB noradrenergic activity.

WHAT SIGNIFICANCE MAY BE ATTACHED
TO THESE NEUROCHEMICAL CHANGES?

Although a number of neurochemical changes have been measured in the brain at parturition, one can be sure that there are many more that remain to be measured.

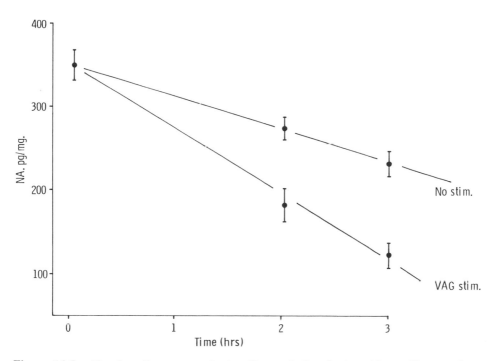

Figure 14.5.   Noradrenaline turnover in the olfactory bulbs of mice with or without vagino-cervical stimulation and α-methylparatyrosine (α-MPT, 50 mg/kg) injections.

Nevertheless, making functional sense of the small number of changes that have already been found is sufficiently perplexing. Even within the relatively short time frame of 2 hr around the period of parturition, when these neurotransmitter events were monitored, there are numerous changes, not only in behavior, but also in neuroendocrine function. Sheep, which are normally gregarious, isolate themselves: their irritable circling and pawing of the ground revert to complete immobility, and even the most timid animals become oblivious to the approach of the experimenter, permitting blood and CSF sampling without restraint. A loss of appetite in the immediate prepartum period is followed by a voracious appetite in the postpartum period. Maternal behavior is immediate at parturition, and high-pitched bleats revert to low-pitched bleats; the odors of amniotic fluids that impregnate the lamb's coat, which are normally repellent to ewes, suddenly become attractive (Lévy & Poindron, 1987). Neuroendocrine changes result in milk letdown, and sexual receptivity and ovulation give rise to postpartum estrus. Hence, there are a multitude of behavioral and neuroendocrine changes at parturition, all of which characterize maternal behavior, but few of which are necessarily specific to maternal behavior. For this reason, it is convenient to distinguish the neurochemical events that occur in the context of maternal behavior as falling into two main categories: (1) the "general synchronizers" and (2) the "specific coders."

   Those systems that simultaneously address wide areas of the brain and lack any specific "coding" for maternal behavior, but are nevertheless essential for it to occur, may be considered general synchronizers. The steroid hormones and rostrally

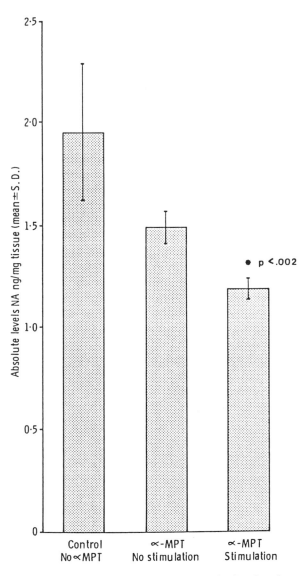

Figure 14.6. Rats given α-MPT (100 mg/kg) and either 5 min of vaginocervical stimulation or 5 min of handling. Noradrenaline levels in the hypothalamus were significantly lower in the stimulated group, suggesting an increase in turnover of noradrenaline in the hypothalamus.

projecting catecholamine systems fall into this category. The steroid hormones may be viewed as primers, but not just for maternal behavior; the amines, specifically noradrenaline, act to synchronize a variety of neural systems associated with maternal behavior. These include the activation of neuroendocrine mechanisms important for the peripheral changes associated with maternal behavior; an interaction with the OXY peptidergic system, which is a more specific addressing system; and the enhancement of sensory signals and learning contingent on parturition. Indeed, parturition is the physiological event that activates the noradrenergic system in the context of maternal behavior.

Those systems that address restricted areas of the brain and can be viewed as specific to maternal behavior may be considered specific coders. Their action is dependent on estrogen priming and the noradrenergic synchronization of other neural events in order for complete maternal behavior to ensue. The oxytocinergic peptidergic system falls into this category, promoting maternal behavior. To ensure specificity, other peptidergic systems (e.g., β-endorphin) may be inhibitory to maternal behavior (see chapter 6), to potentially competing behaviors such as sexual behavior, and to the reproductive neuroendocrine response associated with sexual behavior.

FUNCTIONAL SIGNIFICANCE OF CENTRAL OXY RELEASE AT PARTURITION

OXY exists in two major cell groups within the brain: the so-called magnocellular neurosecretory neurons, which project from the paraventricular and supraoptic nuclei to the posterior pituitary; and the parvocellular neurons of the paraventricular nucleus (PVN), which project to the median eminence and have widespread intracerebral projections. The terminals of the magnocellular neurons are closely associated with the capillary plexus of the neurohypophysis, and the release of their OXY has an important role in the control of uterine contractions at parturition and in milk ejection in response to suckling. The centrally projecting oxytocinergic neurons of the parvocellular PVN are located principally in the lateral parvocellular area, with ascending fibers to the septum and the medial preoptic area and descending projections to the brain stem and spinal cord (Swanson & Cowan, 1979; Swanson & Kuypers, 1980). The descending terminal areas include autonomic centers (tractus solitarius and vagal nucleus), SN, locus coeruleus, raphe magnus, and ventral tegmental area, as well as the intermediolateral cell column of the spinal cord (Swanson & Sawchenko, 1980). The lateral parvocellular neurons of the PVN take up estradiol; this steroid is also known to increase OXY receptors in terminal areas (de Kloet, Voorhuis, & Elands, 1986; Insel, 1986).

In 1979, Pedersen and Prange turned their attention to the possible involvement of OXY in maternal behavior. It is an attractive idea that both the central and the peripheral components of a neuroendocrine system may be activated simultaneously as part of a coordinated neuroendocrine response. However, OXY does not cross the blood–brain barrier in appreciable quantities in rats (Jones & Robinson, 1982) or sheep (Kendrick et al., 1986), and there is no correlation of CSF levels with those in blood following peripheral infusions. It seems clear, therefore, that even if the magnocellular and parvocellular systems releasing OXY work in concert, they do so only to the extent that both may be regulated by the same central mechanisms.

Pedersen and Prange (1979) were the first to test the idea that central OXY has a role in maternal behavior. Their experiments involved infusion of this and other peptides via indwelling cannulas into the ventricular system of virgin females that had been ovariectomized and hormonally primed. These nulliparous females were then presented with 1- to 2-day-old pups and observed for maternal behavior. The

animals became fully maternal within 2 hr following injections of OXY and its analogue, tocinoic acid. These findings were followed up by studies that revealed a dose-
dependent effect of OXY given intraventricularly on promoting maternal behavior
(Pedersen, Asher, Monroe, & Prange, 1982). Some authors were unable to repeat
these findings (Bolwerk & Swanson, 1984; Rubin, Menniti & Bridges, 1983) while
others have confirmed them (Fahrbach, Morrell, & Pfaff, 1984). Indeed, the experimental paradigm used may be an important determinant of the positive results obtained (Fahrbach, Morrell, & Pfaff, 1986; Wamboldt & Insel, 1987; see chapter 13).

Clearly, the strain of rats in which positive results were obtained was prone to
display spontaneous maternal behavior (Wamboldt & Insel, 1987). Other findings
that show that prepartum changes in maternal responsiveness are associated with
uterine contractions (Mayer & Rosenblatt, 1984) and that cervical stimulation induces
rapid maternal responsiveness in estrogen-primed, multiparous females (Yeo & Keverne, 1986) support the proposition that OXY is implicated in the induction of maternal behavior. Also, the onset of maternal behavior in estrogen-treated, pregnancy-
terminated (Fahrbach et al., 1985) and normal postpartum rats (van Leengoed,
Kerher, & Swanson, 1987) can be delayed by intracerebroventricular (ICV) treatment
with OXY antisera or a synthetic OXY antagonist.

In the sheep, vaginal-cervical stimulation of multiparous, estrogen-primed ewes
induces a rapid onset of maternal behavior (Keverne et al., 1983). Moreover, CSF
concentrations of OXY are significantly increased following both vaginal-cervical
stimulation and parturition (Kendrick et al., 1986), even though the blood–brain
barrier of sheep is relatively impermeable to OXY. In sheep, ICV administration of
OXY stimulates a rapid (by 1 min postinfusion) onset of maternal behavior in the
animal's home pen (Figure 14.7) and induces the ewe to maintain proximity to the
lambs in a novel open field (Figure 14.8) even when another attractive stimulus,
food, is present (Kendrick, Keverne, & Baldwin, 1987). The duration of this behavioral effect is approximately 1 hr, and it is noteworthy that selective maternal behavior is established within 1 hr in normal parturient ewes (Poindron & Le Neindre,
1980). The 30-min half-life of OXY in CSF (Jones & Robinson, 1982) indicates that
any behavioral actions of OXY may not outlast the presence of its high concentration
in CSF found at parturition. Moreover, the comparatively short duration of OXY
stimulation of maternal behavior in sheep suggests that its central release may be
important only for the immediate induction of maternal behavior, unless additional
mechanisms are available for maintaining central OXY release. In this context, it is
interesting to note that in sheep, not only parturition but also suckling increases CSF
OXY (Kendrick et al., 1986), as well as OXY release in the OB and SN (see Figure
14.3).

A persuasive story is thus emerging to link OXY with maternal behavior, and it
would appear that parturition simultaneously evokes both central and peripheral components of the paraventricular oxytocinergic system as part of a coordinated neuroendocrine response that is as vital to maternal behavior as to parturition itself. Lesion
and knife-cut studies in the rat have shown sites in the hypothalamus (Jacobson,
Terkel, Gorski, & Sawyer, 1980), septum (Cruz & Beyer, 1972), midbrain (Gaffori
& le Moal, 1979), and OB (Fleming & Rosenblatt, 1974) to be important in maternal
behavior, and it is surely no coincidence that these areas heavily overlap with the
distribution of oxytocinergic terminals (Sofroniew, 1983). Moreover, the magnocel-

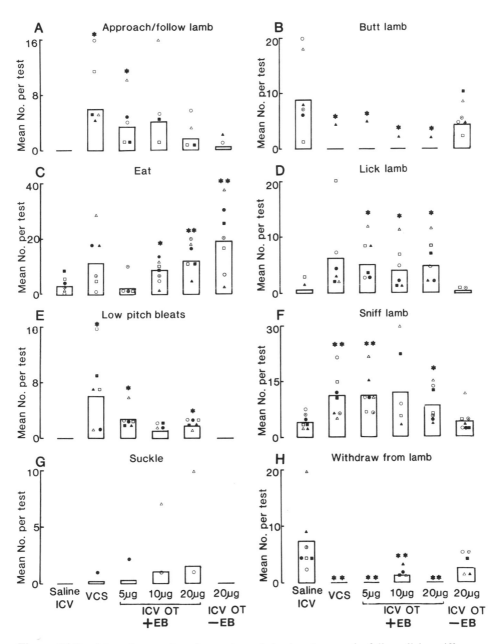

Figure 14.7. Mean frequencies of acceptance behaviors (approach, follow, lick, sniff, suckle, and low-pitched bleats) and rejection behaviors (butt lamb, withdraw from lamb) following intracerebroventricular (ICV) oxytocin or saline infusions, with ($+$) or without ($-$) estradiol benzoate (EB) treatment of ewes, and compared with vaginocervical stimulation (VCS). $*p < .05$, $**p < .01$.

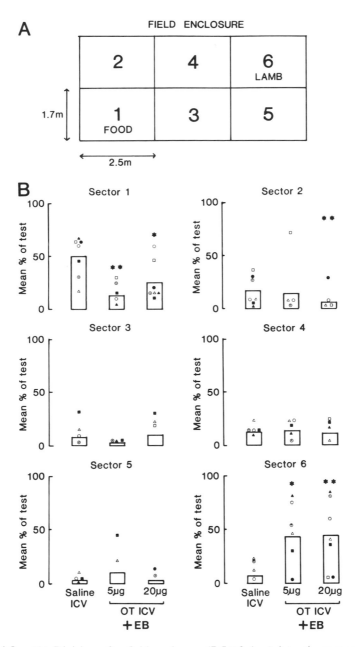

Figure 14.8. (A) Division of a field enclosure (7.5 × 3.4 m) into six sectors, with caged lambs and food in opposite sectors. (B) The relative amounts of time spent by ewes (individual values represented by small symbols) in each sector of the field enclosure following intracerebral or saline infusions. *$p < .05$, **$p < .01$.

lular oxytocinergic neurons have, in common with dorsolateral parvocellular oxytocinergic neurons, inputs from the medial preoptic area, a brain region of considerable
importance to maternal behavior. While there are obvious pitfalls in drawing conclusions from neuroanatomical findings alone, taken together with the functional findings, considerable evidence is accumulating to support an important role for centrally
released OXY in maternal behavior.

FUNCTIONAL SIGNIFICANCE OF NORADRENERGIC ACTIVATION AT PARTURITION

Most behavioral (sexual, maternal, feeding, exploratory) and neuroendocrine (ovulation, lactation, pseudopregnancy, and parturition) functions have at some stage implicated the involvement of the brain's noradrenergic projections. And yet, in neural
terms, the noradrenergic system is poorly defined, with relatively few noradrenergic
cell bodies—so few that they can be counted (about 1500 in the rat's locus coeruleus). These cell bodies are located in the medullary and pontine reticular formation,
and project rostrally via the ventral and dorsal noradrenergic bundles supplying most
limbic and forebrain areas, respectively (Hokfelt et al., 1984). Clearly, so few neurons with such a widespread terminal distribution have insufficient carrying capacity
to code for any complex behavior, let alone all the vital functions mentioned above.
Such a diffuse neural system does, however, lend itself to integration, and any subset
of specific functions does require coordination. This would imply that activation of
the ascending noradrenergic system is significant only in those terminal areas that are
themselves active; that is, while it is capable of regulating "attention" to a wide
variety of cues in the environment, selectivity is achieved by the context—more
specifically, by those cues that are being processed at the time (Mason & Iverson,
1979).

If we examine maternal behavior, what evidence do we have that noradrenergic
mechanisms are implicated, and what might their function be in this particular context? One plausible expectation would be that the brainstem's noradrenergic perikarya
receive a spinal input that is activated by parturition or vaginal-cervical stimulation.
In this context, it is interesting to note that a large proportion of the ascending fibers
in the anterolateral columns of the spinal cord, through which somatosensory stimuli
from the perineal area reach the brain, have collaterals that terminate in the reticular
formation in the vicinity of the adrenergic and noradrenergic cell groups (Zelman,
Leonard, Kow, & Pfaff, 1978). Moreover, a proportion of the noradrenergic cell
bodies in those cell groups concentrate and retain estradiol (Heritage, Grant, & Stumpf,
1977), providing a potential site of action for the hormone implicated in maternal
behavior to amplify or facilitate information transfer at times when maternal responsiveness is most needed.

A number of studies have shown changes in the metabolism of hypothalamic
noradrenaline associated with the onset of maternal behavior in the nulliparous rat
(Rosenberg Leidahl, Halaris, & Moltz, 1976) and shortly after parturition (Moltz,
Rowland, Steele, & Halaris, 1975) or following vaginal-cervical stimulation, a pro-

cedure that facilitates maternal behavior in the rat (Yeo & Keverne, 1986) and the sheep (Keverne et al., 1983). Moreover, CSF noradrenergic metabolites (MHPG) increase significantly at parturition in the ewe (see the previous section). Other studies in the mouse have demonstrated that vaginal-cervical stimulation enhances noradrenaline turnover in the OB (Keverne & de la Riva, 1982), but not in other cortical areas that receive locus coeruleus noradrenergic projections (Rosser & Keverne, 1985). Taken together, these studies provide clear evidence that the brain's noradrenergic projections are indeed activated by somatosensory stimuli associated with parturition.

The question now arises as to what this activation of noradrenergic systems may be accomplishing in the context of maternal behavior and, more pertinently, how this is achieved. One possibility is that the rostral projections of the noradrenergic system are activated at parturition and serve as a signal for coordinating the cascade of events associated with maternity. These include hypothalamic mechanisms of importance for the later stages of parturition, maternal behavior and milk letdown, which would be signaled via the ventral noradrenergic bundle, while dorsal bundle projections are associated with the regulation of attention to cues in the environment by way of hippocampal and olfactory terminations.

NEUROENDOCRINE FUNCTION

It has long been established that noradrenaline has an important role in neuroendocrine function, including prolactin release from the anterior pituitary (Hohn & Wuttke, 1978) and lactation (Clarke, Lincoln, & Merrick, 1979). More recently, noradrenaline has been implicated in the release of peripheral OXY (Bicknell & Leng, 1983). In sheep and goats, it has been known for many years that vaginal or cervical distention releases OXY from the pituitary gland (Ferguson reflex). This facilitates expulsion of the fetus from the birth canal by stimulating and coordinating uterine contractions. This would explain why OXY levels are generally low at the onset of labor, but rise with the passage of the fetus through the birth canal. There is also considerable evidence that noradrenergic neurons are important in relaying visceral afferent information to the PVN in the hypothalamus (Sawchenko & Swanson, 1981). Most of these noradrenergic terminals give rise to a rich innervation of the anterior and medial parvocellular divisions of the PVN. These terminals may provide the signal for the intracerebral OXY release that is important in the context of maternally motivated behavior. An action of noradrenaline on the magnocellular oxytocinergic neurons is probably indirect, since no noradrenergic synapses are made directly onto these neurons.

SENSORY RECOGNITION

The brain's noradrenergic projections have also been described as a neural subsystem that identifies situations of great survival value and hence instructs the storage of relevant incoming sensory information (Kety, 1970). A functional role for such a subsystem is now recognized with respect to the visual, olfactory, and auditory systems (Keverne, 1983). In the case of maternal behavior, a mother's recognition of her offspring as opposed to something alien or an object for eating, or, more selectively, a mother's recognition of her own offspring as opposed to those of strangers, depends heavily on olfaction.

In virgin rats, pup odors are aversive. The nulliparous female may cannibalize pups, but more commonly she shows a pattern of rapidly alternating withdrawal and approach responses. She approaches and intensively sniffs the pups and then withdraws, and may approach again more cautiously, but ultimately she settles at a distance from the pups, actively avoiding them (Fleming & Luebke, 1981). Lesions to the olfactory receptors or their primary projections in the OB (Fleming & Rosenblatt, 1974) or to their subsequent projections to the amygdala (Fleming, Vaccarino, & Luebke, 1980) predictably have the effect of facilitating maternal behavior in nonparturient rodents. Rat mothers, of course, uniformly commence licking their pups as they emerge from the birth canal. Why, it might be questioned, do they likewise not find pup odors aversive? The answer is that they do if selective lesions are made to the ascending noradrenergic terminals in the OB. Inexperienced females given infusions of the neurotoxin 6-hydroxydopamine (6-OHDA) into the medial olfactory stria prior to their first parturition fail to respond to pups normally and cannibalize them or desert them. However, given subsequent exposures to many pups, they eventually show perfectly normal maternal behavior (Yeo & Keverne, personal observations).

Group-living mothers, particularly those that undergo synchronized seasonal breeding, require some means of recognition to enable them to restrict their maternal care to their offspring. Sheep, goats, and pigs all show selective behavior toward their own offspring, and this recognition is achieved by olfactory cues. This selectivity in maternal responsiveness is achieved during a restricted period immediately after birth but is prevented by bulbectomy (Baldwin & Shillito, 1974), after which ewes will accept any lamb. In sheep, therefore, maternal behavior proceeds perfectly normally in the absence of olfactory information, but the ability to distinguish the offspring of other ewes is diminished or lost altogether, and with it the ewe's selective behavior. The consequence of depleting the OB of noradrenaline in sheep, therefore, is very different from the result in rats. Instead of being impaired, maternal behavior proceeds normally, but the ability to be selectively maternal is reduced (Pissonier, Thierry, Fabre-Nys, Poindron, & Keverne, 1985).

This situation is dependent on the neural circuitry of the OB, since the noradrenergic supply terminates on the intrinsic neurons, particularly granule cells and, to a lesser extent, the periglomerular cells (Shipley, Halloram, & de la Torre, 1985). The granule cells inhibit the mitral cell relay neuron by releasing GABA (Halasz & Shepherd, 1983). The mitral cells release the neurotransmitter N-acetyl aspartylglutamate (Blakely, Ory-Lavollee, Grzanna, Koller, & Coyle, 1987), which is excitatory to the granule cells via reciprocal dendrodendritic synapses (Shepherd, 1982). Intracellular recordings have revealed that noradrenaline reduces the inhibition exerted by GABAergic granule cells on mitral cells (Nicoll & Jahr, 1982), and in vitro studies have demonstrated that GABA significantly enhances the evoked release of newly accumulated, tritiated noradrenaline without affecting spontaneous release (Gervais, 1987). This implies that the presence of presynaptic GABA receptors on noradrenaline terminals could establish a functional link between mitral cell activity and local noradrenaline release throughout the population of bulbar noradrenaline terminals. These combined findings have been interpreted as showing that release of noradrenaline in the OB results in the amplification of the mitral cell response to odors, thereby enhancing the signal-to-noise ratio (Jahr & Nicoll, 1982). In order to interpret these findings in the context of what is known about olfactory and maternal behaviors, it is necessary to propose that activation of noradrenaline release has both short- and long-term effects. The immediate effects are to enhance the signal that enables the mother rat to distinguish the odor of pups form that of the placenta, which is eaten. Hence, anosmic, maternally inexperienced rodents or those with a bulbar noradrenaline lesion show cannibalism. The long-term effect of this noradrenaline activation is the establishment of neural circuitry for subsequent amplification of this pup signal, an event that may require activation of second messengers as a consequence of long-term potentiation (LTP) similar to that described for the hippocampus (Lynch & Baudry, 1984). Once this is achieved, the simultaneous noradrenergic activation is not required, since pup olfactory signals themselves act on the now established neural circuitry that effects their signal. It is possible that at some point in the circuitry this centrally transmitted signal is recognized and gated, thus providing a plausible explanation for the finding that anosmia enhances the maternal behavior of experienced, nonparturient rodents (Fleming & Rosenblatt, 1974).

CONCLUDING REMARKS

This chapter has focused on the neurochemical changes that occur during the few hours around parturition. Parturition is an event of considerable significance for the rapid onset of maternal behavior in sheep and, interestingly, has much in common neurochemically with the changes that also occur during suckling. It would therefore come as no surprise to find that the neural pathways from the brain stem rostrally that are implicated in suckling have much in common with the pathways carrying vaginal-cervical somatosensory information. It might also explain why a failure to suckle results in a rapid waning of maternal behavior in sheep.

The findings reported here on the neurochemical changes in amines, peptides, and excitatory amino acids have established both similarities and differences in the release of these compounds in at least two distinct regions of the brain: the SN and the OB. These findings, together with the behavioral studies also reported here on selective neurotoxic lesions to the noradrenergic system and the intracerebral infusions of the peptide OXY, allow us to draw some conclusions about what these systems may be doing in the context of maternal behavior.

Consider first the noradrenergic system. This system shows enhanced activity at the time of parturition (MHPG in CSF) synchronously in widely separated neural regions (OB, hypothalamus, SN). Other arousing stimuli, such as lamb separation, enhance activity in the SN, but in the absence of sensory odor stimuli from the lamb, there is no increased release in the OB. Likewise, turnover studies reveal enhanced activity in the OB, but not in other telencephalic areas that receive an innervation from the same population of noradrenergic neurons. This would suggest that any specificity in this system is determined in the terminal regions, and when the noradrenergic neurons are excited, they serve to enhance neural activity synchronously in these terminal regions. Electrophysiological studies in many regions of the brain have revealed that this enhanced neural activity is significant in facilitating the signal and reducing the noise normally associated with it. An immediate consequence of this is a synchronous enhancement of all the neural subsystems that are active in the context of maternal behavior. Hence, a potentiation of maternal behavior occurs. A long-term consequence may be the facilitation of activity among the neural networks established in those neural subsystems that are active in the context of maternal behavior. This might account for the differences found in maternal behavior between nulliparous and multiparous animals, and helps to explain why multiparous animals were used in our infusion studies.

If OXY is representative of a peptidergic system activated during parturition (and there are others, including corticotropin releasing factor [CRF], vasopressin, and β-endorphin), then it has a relatively discrete distribution—mainly to regions of the brain that have been implicated in maternal behavior. OXY is released intracerebrally at times when maternal behavior is called on (parturition, suckling); when infused intracerebrally, it stimulates maternal behavior, and antagonists or antibodies to OXY block maternal behavior. It also has an important role in producing milk letdown during suckling. It is tempting to say that OXY is chemically coded for maternal behavior, but in doing so, it is important to remember that such coding depends on the priming of the brain by steroids and the synchronous activity in many other transmitter systems.

It is tempting to conclude that studies on the physiological factors underlying the stimulation for maternal behavior in animals around the time of parturition have little relevance to our understanding of human maternal behavior. After all, women who adopt children have no apparent difficulty in exhibiting maternal behavior, even though they have not actually been pregnant and given birth. Equally, some women may function imperfectly as mothers and reject their children even after a normal pregnancy and birth. However, it should also be stressed that many animals can be induced to show maternal behavior without undergoing pregnancy and birth, simply by constant exposure to young. Thus physiological events occurring at parturition may stimulate or enhance the display of maternal behavior, but they are not the exclusive medium through which this behavior can be induced.

So, why are humans apparently so different from other animals in their ability to respond readily in a parental manner? Undoubtedly, growth of the neocortex and emancipation of behaviors from endocrine determinants have played a large part in the evolution of human behavior. The development of a complex social structure first seen in many nonhuman primates has further resulted in shared roles. The family unit has become extended and integrated into the social group, with a contribution to social care from other group members. Socially living primates have extensive opportunities for interacting with young, and it is not unusual to see a nonpregnant monkey pick up another females' infant and nurse it. Through constant social exposure to young, both human and nonhuman primates are potentially maternal throughout their adult lives. Of course, pregnancy and parturition are not without significance. Not only does pregnancy serve as a social and even status signal to the rest of the group, but the endocrine changes of pregnancy and parturition provide the mother with milk and a direct means of feeding her offspring. Moreover, it is extremely likely that parturition in women may bring about an increase in noradrenergic activation. Women rapidly learn to recognize the cry and smell of their own babies, a possible consequence of noradrenergic enhancement of the signal-to-noise ratio in these sensory systems, together with its hippocampal effects on improved learning. Hence, parturient women are probably better equipped to learn the critical sensory signals associated with their own babies.

Finally, the establishment of the neural and neurochemical mechanisms of importance to maternal behavior in animals may be of fundamental significance in understanding the basis of human maternal behavior. Many of the neurochemical changes examined at parturition occur in the limbic, not the neocortical brain, which is remarkably similar in both primate and nonprimate mammals. Hence, the neural basis for maternally motivated behaviors may be very similar across mammalian species, but the mechanisms available for addressing them are clearly very different.

REFERENCES

Baldwin, R. A., & Shillito, E. (1974). The effects of ablation of the olfactory bulbs on parturition and maternal behaviour in Soay sheep. *Animal Behaviour 22*, 220–223.

Barofsky, A. L., Taylor, J., Tizabi, Y., Kumar, R., & Jones-Quartey, K. (1983). Specific neurotoxin lesions of the median raphe serotonergic neurons disrupt maternal behaviour in the lactating rat. *Endocrinology 113*, 1884–1893.

Bicknell, R. J., & Leng, G. (1983). Differential regulation of oxytocin- and vasopressin-secreting nerve terminals. *Progress in Brain Research, 60*, 331–341.

Blakely, R. D., Ory-Lavollee, R., Grzanna, R., Koller, K. J., & Coyle, D. T. (1987). Selective immunocytochemical staining of mitral cells in rat olfactory bulb with affinity purified antibodies against N-acetyl-aspartate-glutamate. *Brain Research, 402*, 373–378.

Bolwerk, E. L. M., & Swanson, H. H. (1984). Does oxytocin play a role in the onset of maternal behaviour? *Journal of Endocrinology, 101*, 353–357.

Chard, T., Boyd, N. R. H., Forsling, M. L., McNeilly, A. S., & Landon, J. (1970). The development of a radioimmunoassay for oxytocin: The extraction of oxytocin from plasma and its measurement during parturition in human and goat blood. *Journal of Endocrinology, 48*, 223–234.

Clarke, G., Lincoln, D. W., & Merrick, L. P. (1979). Dopaminergic control of oxytocin release in lactating rats. *Journal of Endocrinology, 83*, 409–420.

Cruz, M. L., & Beyer, C. (1972). Effects of septal lesions on maternal behavior and lactation in the rabbit. *Physiology and Behavior, 9*, 361–365.

DeKloet, E. R., Voorhuis, T. A. M., & Elands, J. (1986). Estradiol induces oxytocin binding sites in rat hypothalamic ventromedial nucleus. *European Journal of Pharmacology, 118*, 185–186.

Fahrbach, S. E., Morrell, J. I., & Pfaff, D. W. (1984). Oxytocin induction of short latency maternal behavior in nulliparous estrogen primed female rats. *Hormones and Behavior, 18*, 267–286.

Fahrbach, S. E., Morrell, J. I., & Pfaff, D. W. (1985). Possible role for endogenous oxytocin in estrogen facilitated maternal behavior in rats. *Neuroendocrinology, 40*, 526–532.

Fahrbach, S. E., Morrell, J. I., & Pfaff, D. W. (1986). Effect of varying duration of pre-test cage habituation on oxytocin induction of short latency maternal behavior. *Physiology and Behavior, 37,* 135–139.

Fleming, A. S., & Luebke, C. (1981). Timidity prevents the virgin female rat from being a good mother: Emotionality differences between nulliparous and parturient females. *Physiology and Behavior, 27,* 863–868.

Fleming, A. S., & Rosenblatt, J. S. (1974). Olfactory regulation of maternal behavior in rats: I. Effects of olfactory bulb removal in experienced and inexperienced lactating and cycling females. *Journal of Comparative and Physiological Psychology, 86,* 221–232.

Fleming, A. S., Vaccarino, F., & Luebke, C. (1980). Amygdaloid inhibition of maternal behavior in the nulliparous female rat. *Physiology and Behavior, 25,* 731–743.

Flint, A. P. F., Forsling, M. L., Mitchell, M. D., & Turnbull, A. C. (1975). Temporal relationship between changes in oxytocin and prostaglandin F levels in response to vaginal distention in the pregnant and puerperal ewe. *Journal of Reproduction and Fertility, 43,* 551–554.

Gaffori, O., & le Moal, M. (1979). Disruption of maternal behavior after ventral mesencephalic tegmental lesions. *Physiology and Behavior, 23,* 317–323.

Gervais, R. (1987). Local GABAergic modulation of noradrenaline release in the rat olfactory bulb measured on superfused slices. *Brain Research, 400,* 151–154.

Godfrey, D. A., Ross, C. D., Carter, J. A., Lowry, O. H., & Matschinsky, F. M. (1980). Effect of intervening lesions on amino acid distributions in rat olfactory cortex and olfactory bulb. *Journal of Histochemistry and Cytochemistry, 28,* 1157–1169.

Grossman, S. P. (1979). The biology of motivation. *Annual Review of Psychology, 30,* 209–242.

Halasz, N., & Shepherd, G. M. (1983). Neurochemistry of the vertebrate olfactory bulb. *Neuroscience, 10,* 570–619.

Heritage, A. S., Grant, L. D., & Stumpf, W. E. (1977). H[3] estradiol in catecholamine neurons of the rat brain stem. *Journal of Comparative Neurology, 176,* 607–630.

Hohn, K. G., & Wuttke, W. D. (1978). Changes in catecholamine turnover in the mediobasal hypothalamus and the medial pre-optic area in response to hyperprolactinaemia in ovariectomized rats. *Brain Research, 156,* 241–252.

Hokfelt, T., Johansson, O., & Goldstein, M. (1984). Central catecholamine neurons as revealed by immunocytochemistry with special reference to adrenaline neurons. In A. Bjorklund & T. Hokfelt (Eds.), *Handbook of chemical neuroanatomy* (Vol 2, pp. 157–276). Amsterdam: Elsevier.

Insel, T. R. (1986). Postpartum increases in brain oxytocin binding. *Neuroendocrinology, 44,* 515–518.

Jacobson, C. D., Terkel, J., Gorski, R. A., & Sawyer, (1980). Effects of small preoptic area lesions on maternal behavior: Retrieving and nest building in the rat. *Brain Research, 194,* 471–478.

Jahr, C. E., & Nicoll, R. A. (1982). Noradrenergic modulation of dendrodendritic inhibition of the olfactory bulb. *Nature, 297,* 227–228.

Jones, P. M., & Robinson, I. C. A. F. (1982). Differential clearance of neurophysin and neurohypophyseal peptides from the cerebrospinal fluid in conscious guinea pigs. *Neuroendocrinology 34,* 297–302.

Jones, P. M., Robinson, I. C. A. F., & Harris, M. C. (1983). Release of oxytocin into blood and cerebrospinal fluid by electrical stimulation of the hypothalamus or neural lobe in the rat. *Neuroendocrinology, 37,* 454–458.

Kendrick, K. M., Keverne, E. B., & Baldwin, B. A. (1987). Intracerebroventricular oxytocin stimulates maternal behaviour in the sheep. *Neuroendocrinology, 46,* 56–61.

Kendrick, K. M., Keverne, E. B., Baldwin, B. A., & Sharman, D. F. (1986). Cerebrospinal fluid levels of acetylcholinecterase, monoamines and oxytocin during labour, parturition, vagino-cervical stimulation, lamb separation and suckling in sheep. *Neuroendocrinology, 44*, 148–156.

Kendrick, K. M., Keverne, E. B., Chapman, C., & Baldwin, B. A. (1988a). Intracranial dialysis measurement of oxytocin, monoamines and uric acid release from the olfactory bulb and substantia nigra of sheep during parturition, suckling, separation from lambs and eating. *Brain Research, 439*, 1–10.

Kendrick, K. M., Keverne, E. B., Chapman, C., & Baldwin, B. A. (1988b). Microdialysis measurement of oxytocin, aspartate, γ-aminobutyric acid and glutamate release from the olfactory bulb of the sheep during vaginocerival stimulation. *Brain Research, 442*, 171–177.

Kety, S. S. (1970). In F. O. Schmidt (Ed.), *The neurosciences, second study program* (pp. 324–336). New York: Rockefeller University Press.

Keverne, E. B. (1983). Pheromonal influences on the endocrine regulation of reproduction. *Trends in Neuroscience, 6*, 381–384.

Keverne, E. B., & de la Riva, C. (1982). Pheromones in mice: Reciprocal interaction between the nose and brain. *Nature, 296*, 148–150.

Keverne, E. B., Lévy, F., Poindron, P., & Lindsay, D. (1983). Vaginal stimulation: An important determinant of maternal bonding in sheep. *Science, 219*, 81–83.

Lévy, F., & Poindron, P. (1987). The importance of amniotic fluids for the establishment of maternal behaviour in experienced and non-experienced ewes. *Animal Behaviour, 35*, 1188–1192.

Lynch, G., & Baudry, M. (1984). The biochemical basis for learning and memory: A new and specific hypothesis. *Science, 224*, 1057–1062.

Macrides, F., & Davis, B. J. (1985). The olfactory bulb. In P. C. Emson (Ed.), *Chemical neuroanatomy* (pp. 391–426). New York: Raven Press.

Mason, S. T., & Iversen, S. D. (1979). Theories of the dorsal bundle extinction effect. *Brain Research Review, 1*, 107–137.

Mayer, A. D., & Rosenblatt, J. S. (1984). Preparative changes in maternal responsiveness and nest defense in *Rattus norvegicus*. *Journal of Comparative Psychology, 98*, 177–188.

McCaleb, M. I., & Myers, R. D. (1980). Cholecystokinin acts on the hypothalamic "noradrenergic" system involved in feeding. *Peptides, 1*, 47–49.

Moltz, H., Rowland, D., Steele, M., & Halaris, A. (1975). Hypothalamic norepinephrine: Concentration and metabolism during pregnancy and lactation in the rat. *Neuroendocrinology, 19*, 252–258.

Nicoll, R. A., & Jahr, C. E. (1982). Self-excitation of the olfactory bulb neurons. *Nature, 296*, 441–443.

Numan, M. (1983). Brain mechanisms of maternal behavior in the rat. In J. Balthazart, E. Prove, & R. Gilles (Eds.), *Hormones and behavior in higher vertebrates* (pp. 69–85). Berlin: Springer-Verlag.

Numan, M., & Nagle, D. S. (1983). Preoptic area and substantia nigra interact in the control of maternal behavior in the rat. *Behavioral Neuroscience, 97*, 120–139.

Numan, M., Rosenblatt, J. S., & Komisaruk, B. R. (1977). Medial preoptic area and onset of maternal behavior in the rat. *Journal of Comparative and Physiological Psychology, 91*, 146–164.

Pedersen, C. A., Asher, J. A., Monroe, Y. L, & Prange, A. J. (1982). Oxytocin induces maternal behavior in virgin female rats. *Science, 216*, 648–649.

Pedersen, C. A., & Prange, A. J. (1979). Induction of maternal behavior in virgin rats after intracerebroventricular administration of oxytocin. *Proceedings of the National Academy of Sciences USA, 76*, 6661–6665.

Pissonier, D., Thiery, J. C., Fabre-Nys, C., Poindron, P., & Keverne, E. B. (1985). The importance of olfactory bulbs and noradrenaline for maternal recognition in sheep. *Physiology and Behavior 35,* 361–363.

Poindron, P. (1976). Effets de la suppression de l'odorat, sans lésion des bulbes olfactifs, sur la sélectivité du comportement maternel de la Brebis. *Comptes Rendus Hebdomadaires des Séances de l'Académie des Sciences, Series D, 282,* 489–491.

Poindron, P., & Le Neindre, P. (1980). Endocrine and sensory regulation of maternal behavior in the ewe. *Advances in the Study of Behavior, 11,* 75–119.

Quinn, M. R., & Cagan, R. H. (1981). Neurochemical studies of the γ-aminobutyric acid system in the olfactory bulb. In R. H. Cagan & M. R. Kare (Eds.), *Biochemistry of taste and olfaction,* (pp. 395–415). New York: Academic Press.

Robinson, I. C. A. F., & Jones, P. M. (1982). Oxytocin and neurophysin in plasma and CSF during suckling in the guinea pig. *Neuroendocrinology, 34,* 59–63.

Rosenberg, P., Leidahl, L., Halaris, A., & Moltz, H. (1976). Changes in the metabolism of hypothalamic norepinephrine associated with the onset of maternal behavior in the nulliparous rat. *Pharmacology, Biochemistry and Behavior, 4,* 647–649.

Rosenblatt, J. S., & Siegel, H. I. (1981). Factors governing the onset and maintenance of maternal behavior among nonprimate mammals: The role of hormonal and nonhormonal factors. In D. J. Gubernick & P. H. Klopfer (Eds.), *Parental care in mammals* (pp. 2–76). New York: Plenum Press.

Rosser, A. E., & Keverne, E. B. (1985). The importance of central noradrenergic neurons in the formation of an olfactory bulb memory in the prevention of pregnancy block. *Neuroscience, 15,* 1141–1147.

Rubin, B. S., Menniti, F. J., & Bridges, R. S. (1983). Intracerebroventricular administration of oxytocin and maternal behavior in rats after prolonged and acute steroid pretreatment. *Hormones and Behavior, 17,* 45–53.

Sawchenko, P. E., & Swanson, L. W. (1981). Central noradrenergic pathways for the integration of hypothalamic neuroendocrine and endocrine responses. *Science, 214,* 685–687.

Shepherd, G. M. (1982). Synaptic organization of mammalian olfactory bulb. *Physiological Reviews, 52,* 864–917.

Shipley, M. T., Halloram, F. J., & de la Torre, J. (1985). Surprisingly rich projection from locus coeruleus to the olfactory bulb in the rat. *Brain Research, 329,* 294–299.

Sofroniew, M. V. (1980). Projections from vasopressin, oxytocin, and neurophysin neurons to neural targets in the rat and human. *Journal of Histochemistry and Cytochemistry, 28,* 475–478.

Sofroniew, M. V. (1983). Morphology of vasopressin and oxytocin neurons and their central and vascular projections. *Progress in Brain Research, 60,* 121–134.

Sofroniew, M. V. (1985). Vasopressin and oxytocin in the mammalian brain and spinal cord. In D. Bousfield (Ed.), *Neurotransmitters in action* (pp. 329–337). Amsterdam: Elsevier.

Steele, M. K. Rowland, D., & Moltz, H. (1979). Initiation of maternal behavior in the rat: Possible involvement of limbic norepinephrine, *Pharmacology, Biochemistry and Behavior, 11,* 123–130.

Swanson, L. W., & Cowan, W. M. (1979). The connections of the septal region in the rat. *Journal for Comparative Neurology, 186,* 621–656.

Swanson, L. W., & Kuypers, H. G. J. M. (1980). The paraventricular nucleus of the hypothalamus: Cytoarchitectonic subdivisions and organization of projections to the pituitary, dorsal vagal complex and spinal cord as demonstrated by retrograde fluorescence double labeling methods. *Journal of Comparative Neurology, 194,* 555–570.

Swanson, L. W. & Sawchenko, P. E. (1980). The paraventricular nucleus: A site for the integration of neuroendocrine and autonomic mechanisms. *Neuroendocrinology, 31,* 410–417.

Van Leengoed, E., Kerker, K., & Swanson, H. H. (1987). Inhibition of post-partum maternal behaviour in the rat by injecting an oxytocin antagonist into the cerebral ventricles. *Journal of Endocrinology, 12,* 275–282.

Wamboldt, M. Z., & Insel, T. R. (1987). The ability of oxytocin to induce short latency maternal behavior is dependent on peripheral anosmia. *Behavioral Neuroscience, 101,* 439–441.

Yeo, J. A. G., & Keverne, E. B. (1986). The importance of vaginal-cervical stimulation for maternal behavior in the rat. *Physiology and Behavior, 37,* 23–26.

Zemlan, F. P., Leonard, C. M., Kow, L. M., & Pfaff, D. W. (1978). Ascending tracts of the lateral columns of the rat spinal cord: A study using the silver impregnation and horseradish peroxidase techniques. *Experimental Neurology, 62,* 298–334.

15

Motherhood Modifies Magnocellular Neuronal Interrelationships in Functionally Meaningful Ways

BARBARA K. MODNEY AND GLENN I. HATTON

This chapter is concerned with rather remarkable recent discoveries demonstrating that the brains of maternally behaving rats are structurally different from those of other animals. The differences discussed here have been observed in the magnocellular hypothalamo-neurohypophysial system, but similar brain alterations may occur in other areas as well. Neurons in this system manufacture and release the peptide hormones oxytocin and vasopressin. Oxytocin is of particular interest because it is involved in producing the uterine contractions preceding and during parturition, as well as in producing the milk-ejection reflex in response to the suckling stimuli of the young. A new conceptualization of brain reorganization is supported by the findings reviewed here. The brains of maternally behaving animals can no longer be thought of as similar to those of naive virgin animals that simply have some pathways differentially activated by the circumstances of motherhood. Rather, the data support the idea of extensively reorganized cell–cell interactions. For example, new specialized synapses form on the dendrites of magnocellular neurons during or immediately after parturition. The neuronal cell bodies similarly receive new synaptic inputs, but in contrast to the dendrites, the cell bodies receive these novel synapses during lactation. Electrical coupling among these neurons appears to be inducible by olfactory and/or vomeronasal organ stimulation that directly activates oxytocinergic neurons. It is hypothesized that the anogenital licking of the pups in the period immediately prior to suckling is a powerful and crucially important stimulus to these two olfactory systems. It is felt that the consequent electrical coupling "fine-tunes" the neuronal interactions, maximizing the efficiency of the milk-ejection reflex, which occurs in response to suckling stimuli. Among the more intriguing recent findings in this system is the fact that virgin rats that have been induced by the continuous presence of young rat pups to show maternal behavior also display some of the same brain modifications found in the lactating, real mothers. Finally, nearly all the changes that have so far been found to occur in the brains of maternal rats disappear about a month after the pups are weaned.

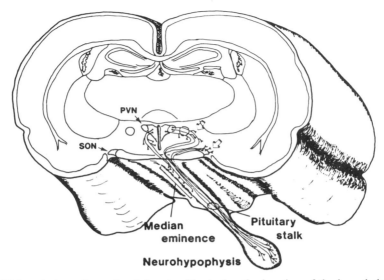

Figure 15.1. A three-dimensional drawing illustrating the location of the hypothalamo-neurohypophysial system within the brain. Many cells in the supraoptic nucleus (SON) and paraventricular nucleus (PVN) have axons that give off local collaterals in areas adjacent to the nuclei (open arrows). Parent axons continue and course through the median eminence and pituitary stalk to terminate in the neurohypophysis. Some axons from portions of the PVN also project to the median eminence.

ANATOMY OF THE HYPOTHALAMIC MAGNOCELLULAR SYSTEM

Among the many physiological events that accompany pregnancy, birth, and subsequent lactation in mammals is a significant increase in the synthesis and release of the peptide oxytocin. Along with vasopressin, oxytocin is synthesized in magnocellular neuroendocrine cells (MNCs) located in the hypothalamus, the two hormones being manufactured by separate cell populations. Although several accessory nuclei that manufacture these peptides have been identified (Peterson, 1966), the predominant nuclei of this system are the supraoptic nuclei (SON) and the magnocellular division of the paraventricular nucleus. These nuclei, their relative locations in the brain, and some of their axonal projections are diagrammed in Figure 15.1. Several anatomical features distinguish MNCs from surrounding areas of the hypothalamus. As shown at the light microscopic level in Figure 15.2, the nuclei are conspicuous because of their large, densely packed, densely staining somata (15–30 μm in diameter) and their locations lateral to the third ventricle and at the ventral surface of the brain just lateral to the optic chiasm and optic tracts. Simply branching dendrites of paraventricular MNCs project medially toward the third ventricle (Armstrong, Warach, Hatton, & McNeill, 1980; van den Pol, 1982), while those of the SON initially project ventrally and turn to course in a parallel fashion rostrocaudally along the ventral surface of the brain (Armstrong, Schöler, & McNeill, 1982). In this chapter, the discussion will be confined largely to the SON (Figure 15.3).

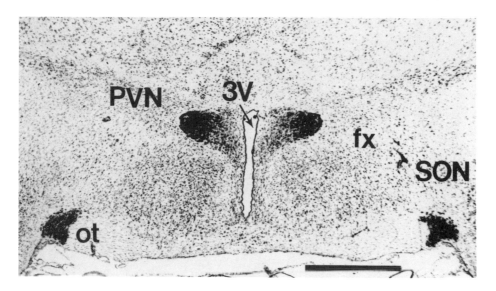

Figure 15.2. Thionin-stained coronal section of the hypothalamus. Magnocellular nuclei are easily identified by their densely stained and packed cell bodies. Abbreviations: PVN, paraventricular nucleus; SON, supraoptic nucleus; ot, optic tract; fx, fornix; 3V, third ventricle. Bar = 1 mm.

Morphological Characteristics of Magnocellular Neurons

Electron microscopic observations have demonstrated that even though somata are densely packed, under basal conditions neighboring cells are isolated from one another by thin astrocytic processes (Hatton & Tweedle, 1982; Tweedle & Hatton, 1976, 1977) (Figure 15.4A). Studies of the ventral dendritic zone of the SON have shown that thin astrocytic processes separate dendrites in a manner similar to that seen in the somatic region (Perlmutter, Tweedle, & Hatton, 1984, 1985) (Figure 15.4B). These dendrites, again under basal conditions, also possess an ultrastructural

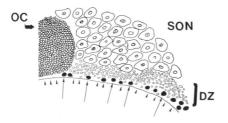

Figure 15.3. Schematic diagram of the supraoptic nucleus (SON), which is located immediately lateral to the optic chiasm (OC) and dorsal to the pial surface (marked by arrowheads). Dendrites from these cells form a distinct dendritic zone (DZ) ventral to the cell bodies. Astrocytic cell bodies (some of which are indicated by arrows) line the most ventral portion of this area.

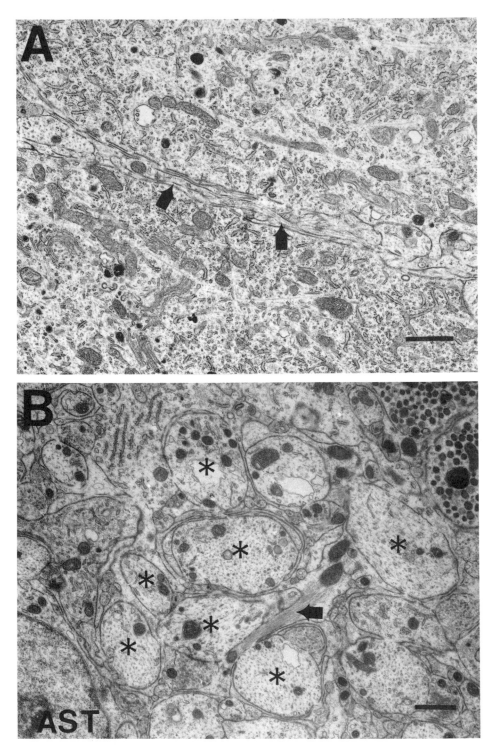

Figure 15.4. Ultrastructural appearance of adjacent SON magnocellular neurons and dendrites in normal animals. (A) Tightly packed magnocellular cell bodies are isolated from one another by astrocytic processes (arrows). Bar = 1 μm. (B) Individual dendrites (asterisks) in the dendritic zone are also separated by astrocytic processes. Astrocytic filaments (arrow) are commonly seen throughout the nucleus. Part of an astrocytic cell body (AST) is also seen. Bar = 1 μm.

feature that is uncharacteristic of most neural tissue: they often have vacant or incompletely covered postsynaptic densities (Tweedle & Hatton, 1986). The cytoplasm of MNCs is richly endowed with numerous organelles indicative of their peptidergic nature. Most notable are many dense core vesicles that contain the secretory products of these cells. A well-developed Golgi apparatus, abundant free ribosomes, and diffuse endoplasmic reticulum allow these cells to be readily identified.

Magnocellular Projections to the Neural Lobe

MNCs have various projections within the brain stem and spinal cord; the functions and extent of these connections are only now being investigated (Swanson, 1986). Some axons also give off collaterals that terminate in the hypothalamus in areas adjacent to the nuclear borders of both the paraventricular nuclei (Hatton, Cobbett, & Salm, 1985) and the SON (Mason, Ho, & Hatton, 1984). MNCs are best known for their massive projection to the neurohypophysis, which forms the hypothalamo-neurohypophysial tract. This tract is formed by axons from magnocellular neurons that course through the internal zones of the median eminence and terminate in the neurohypophysis (see Figure 15.1). Oxytocin and vasopressin are released by terminals that contact the basal lamina surrounding fenestrated capillaries in the neural lobe of the pituitary. Action potentials that invade terminals cause the calcium-dependent exocytotic release of hormones into the circulation. Several factors may modify hormone release at the level of the neural lobe. A variety of neurotransmitters have been found in the neural lobe, and they may modulate the secretion of these hormones. Pituicytes, the astrocytes of the neurohypophysis (Salm, Hatton, & Nilaver, 1982; Suess & Pliška, 1981), may also modulate release through their association with the basal lamina and axonal terminals. Transmission and scanning electron microscopy have shown that under basal conditions, axonal terminals are often completely enclosed by pituicyte cytoplasm (Hatton, Perlmutter, Salm, & Tweedle, 1984; Tweedle & Hatton, 1980b, 1982). Thin pituicyte processes are also interposed between axonal endings and the basal lamina, through which hormone must pass in order to enter the capillary circulation (Tweedle & Hatton, 1987; Wittkowski, 1986). Evidence that these arrangements between axons, pituicytes, and the vascular system participate in modulating hormone release is presented later in this chapter.

Although the main efferent pathways of axons from the SON to the neurohypophysis are well established, less is known about afferent input to these cells. Intrahypothalamic connections, as well as input from various portions of the brain stem and limbic system, are extensive and complex. Catecholaminergic inputs reach the SON from brain stem nuclei, and many afferent inputs probably influence MNCs via local interneurons located in perinuclear areas. For example, a cholinergic input to the SON from local neurons lying just dorsal and lateral to the SON has recently been described (Hatton, Ho, & Mason, 1983; Mason, Ho, Eckenstein, & Hatton, 1983). Other probable inputs to the SON include γ-aminobutyric acid (GABA) from local neurons, angiotensin from the subfornical organ, serotonin and histamine from more posterior regions of the hypothalamus, and excitatory amino acid inputs from the olfactory system.

PHYSIOLOGICAL FUNCTIONS OF MAGNOCELLULAR NEURONS

A wide variety of central effects for oxytocin and vasopressin have been proposed, including roles in memory consolidation and cardiovascular functioning (reviewed in Meisenberg & Simmons, 1983). These peptides are best known for their peripheral effects: oxytocin, for its role in the contraction of uterine smooth muscle during parturition and the milk-ejection reflex during lactation; and vasopressin, or antidiuretic hormone, as a promoter of water resorption by the kidneys. Typically, experimental manipulations that seek to understand the neurobiology of this system use either dehydrated animals, which causes both peptides to be released, or nursing animals in which there is a rather selective release of oxytocin. It should be recognized, of course, that vasopressin is also of great importance to the lactating animal in maintaining water balance during suckling.

Electrophysiological Studies of Magnocellular Neurons

Electrophysiological studies have determined that oxytocinergic and vasopressin-gergic cells can be distinguished on the basis of their firing characteristics (see Poulain & Wakerley, 1982, for review). Presumed oxytocinergic cells have been characterized in anesthetized lactating rats with suckling pups. In these animals, MNCs may fire in either a "fast continuous" or a "slow irregular" manner; however, presumed oxytocinergic cells occasionally exhibit a synchronized, high-frequency discharge, after which milk ejection occurs. The relationship between these two events is shown in Figure 15.5. Vasopressinergic cells do not generally respond to suckling, and, in osmotically stimulated rats, they display a phasic firing pattern consisting of alternating periods of silence and bursts of action potentials (Cobbett, Smithson, & Hatton, 1986). A variety of mechanisms appear to participate in generating these stereotypic patterns of activity and in producing synchronous firing among oxytocinergic neurons. It may be that some of the many dramatic anatomical changes that occur in this system during activation contribute to the characteristic activity patterns observed. Although our emphasis here is on the alterations that occur in the hypothalamo-neurohypophysial system of female rats during parturition and lactation, many of these changes can also be shown to occur in dehydrated animals of either sex (Perlmutter et al., 1985; Tweedle & Hatton, 1976, 1977, 1980a, 1982, 1984, 1986, 1987).

Oxytocin's Role in Parturition and Lactation

Ever since Dale (1906) discovered that pituitary extracts produce uterine contractions in pregnant cats, experimental work has increased our appreciation for and our knowledge of the role of oxytocin in animal parturition (for review, see Fuchs, 1985). A general conclusion from these experiments is that the hypothalamoneurohypophysial system is activated prior to parturition (i.e., expulsion of the first fetus), with increasing activity measured as parturition is completed. Oxytocinergic cells increase their firing rates during parturition, and bursts of action potentials precede the expulsion of each fetus (Summerlee, 1981). Similarly, the pituitary content of both oxytocin and vasopressin is progressively depleted during parturition (Fuchs & Saito,

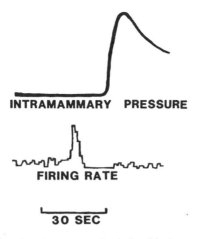

Figure 15.5. Tracings showing the temporal relationship between the firing rate in a magnocellular neuron and the milk-ejection reflex. Upper trace: pressure transducer. Lower trace: rate meter. A significant increase in firing rate precedes the rise in intramammary pressure, indicating a milk ejection. (Modified from Belin et al., 1984)

1971), a process reflected in progressive increases in blood levels of oxytocin (Higuchi, Tadokoro, Honda, & Negoro, 1986).

Oxytocin also plays a pivotal role during lactation, when it serves as the efferent limb of the milk-ejection reflex. Although the complete anatomical pathway for the milk-ejection reflex has not been worked out, it is well established that suckling of pups is synaptically relayed to excite oxytocinergic cells. Despite continuous suckling of pups, milk ejection does not occur continuously, but about every 10–12 min. Calculations of the number of oxytocinergic cells that must be activated in order to release enough hormone for milk ejections to occur suggest that most oxytocinergic cells in the magnocellular nuclei must fire within a very brief time span.

Electrophysiological studies have confirmed that this is indeed the case: just prior to milk ejections, oxytocin neurons fire bursts of action potentials (see Figure 15.5). These bursts occur relatively synchronously both within and between the two nuclei (Belin & Moos, 1986; Belin, Moos, & Richard, 1984). Within seconds after this bursting, a milk ejection occurs. The milk-ejection reflex poses a number of continuing questions for neurobiologists. In addition to the paucity of information about the precise anatomical pathway from the mammary gland to the oxytocinergic cells in the hypothalamus, and the lack of specific information on where and how stimulation from the mammary gland is gated, the mechanism by which oxytocinergic cells can synchronize their firing with such precision remains to the elucidated.

ANATOMICAL TRANSFORMATIONS IN BRAIN AND PITUITARY ASSOCIATED WITH PARTURITION AND LACTATION

Studies using both parturient and lactating females have provided experimental insights into the hypothalamo-neurohypophysial system under physiological conditions of both acute and chronic demand. Given the heightened state of neuronal activity and hormone release during parturition and lactation, it is perhaps not surprising that the organization of the various neural elements of the hypothalamo-neurohypophysial system in parturient and lactating females is dramatically different from that of virgin animals.

Ultrastructural Changes in Glial–Neuronal Relationships

Ultrastructural studies have revealed that a unique transformation in the cellular elements occurs in the SON around the time of parturition and during lactation. These changes are unusual in that they involve both the neurons and the astrocytes of the SON (Hatton & Tweedle, 1982; Theodosis, Poulain, & Vincent, 1981). In prepartum animals (2–24 hr prior to parturition), the glial processes that normally separate adjacent neurons and dendrites of the SON have withdrawn, resulting in a significantly higher proportion of direct neural membrane apposition (Hatton & Tweedle, 1982; Theodosis & Poulain, 1984) (Figure 15.6A). These somatic appositions are maximally elevated after 14 days of lactation, when cellular membrane apposition is increased more than 30-fold compared with that of virgin animals (Hatton & Tweedle, 1982). That direct membrane appositions in lactating animals involve withdrawal of SON glial cell processes has been demonstrated in an immunocytochemical study using an antibody to the astrocytic marker glial fibrillary acidic protein. A redistribution of this marker is seen in the SON of lactating animals compared with virgin SONs (Salm, Smithson, & Hatton, 1985). This redistribution is not seen in control areas in the lateral hypothalamus, indicating that the effect is not a generalized hypothalamic response.

In the dendritic zone of the SON, dendritic bundles (i.e., two or more dendrites with a membrane in direct apposition) (Figure 15.6B) form as a result of the withdrawal of glial processes (Perlmutter et al., 1984). Both the extent of dendritic membrane in apposition and the number of dendrites in a given bundle have been found to be significantly greater in prepartum animals than in virgin animals. In the 24 hr between pre- and post-parturition, further ultrastructural manifestations of activation are detected. Postpartum animals (2–24 hr after parturition) exhibit a level of dendritic bundling that is higher than that of both virgin and prepartum animals. In addition, preliminary evidence suggests that the size of the entire dendritic zone is increased in postpartum animals compared with that of virgins (Taubitz, Smithson, & Hatton, 1987).

Figure 15.6. Ultrastructural appearance of adjacent SON magnocellular neurons and dendrites under conditions of increased hormone demand. (A) Arrows delineate the extent of an individual membrane apposition between two SON cell bodies. Compare with Figure 15.4A, where glial processes are present between adjacent neurons. Bar = 1 μm. (B) Bundles in the dendritic zone of the SON. A small bundle of two dendrites (dendrites numbered with 1's) and a large bundle of six dendrites (dendrites numbered with 2's) are shown. An axon terminal (A) makes synaptic contacts (arrows) with two dendrites, forming a double synapse. Bar = 1 μm.

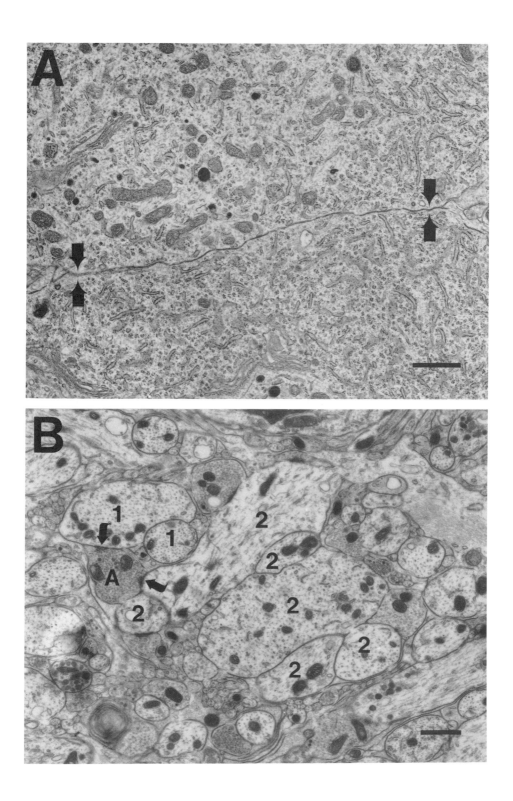

Functionally, the withdrawal of glial processes from between adjacent neural elements may have many different effects. Given that astrocytes act to take up extracellular potassium ($[K^+]_0$) (Orkand, 1977), their relative absence from between neural elements may lead to significant local increases in $[K^+]_0$. These local increases could slightly depolarize those areas of cells in direct apposition, thus increasing the overall excitability in the SON. Another potentially important result of glial withdrawal may be the formation of electrical fields around groups of neurons similar to those that form around pyramidal cells in the hippocampus (Taylor & Dudek, 1982, 1984a, 1984b). Both these effects may serve as mechanisms by which MNCs can modify their electrical activity when increases in hormone demand occur. Local elevations of $[K^+]_0$ probably also aid these peptidergic neurons in the production of their secretory products, since increases in protein synthesis have been shown to occur in brain slices incubated in medium containing even slightly elevated potassium concentrations (Lipton & Heimbach, 1977, 1978).

Synapse Formation

Withdrawal of glial processes may also play a permissive role in what is perhaps the most amazing ultrastructural change that occurs in the hypothalamo-neurohypophysial system. Synapse formation, an uncommon event in the adult animal in response to physiological stimulation, has been shown to occur in both parturient and lactating animals. Evidence for synapse formation is first seen in the dendritic region of the SON in postpartum animals (Perlmutter et al., 1984). These new synapses have a particular arrangement between the dendrites of the SON. They consist of one axon terminal simultaneously forming synaptic contacts with two or more dendrites (see Figure 15.6B). Although these shared axodendritic synapses are found in the dendritic region of the virgin's SON, their frequency is higher in postpartum animals compared with either virgins or prepartum animals. Axodendritic double synapses, which reach their maximum frequency in postpartum animals, decline in frequency with lactation (Perlmutter et al., 1984); however, additional forms of synaptic plasticity have been seen to occur in the time between parturition and day 14 of lactation. Vacant or incompletely covered postsynaptic densities, which are found on about 5% of the dendrites in virgins, are virtually nonexistent in the dendritic zones of lactating animals (Tweedle & Hatton, 1986). A reasonable interpretation of this is that the incompletely occupied or vacant densities have become occupied during stimulation. Synapse formation is also apparent in the somatic region of the SON during lactation and again involves the appearance of new shared synapses similar to those seen in the dendritic region in postpartum animals. New synapse formation in the somatic and dendritic regions of the SON during various functional states is shown in Figure 15.7. In addition, studies of postsynaptic specializations in the paraventricular nucleus using freeze fracture techniques have revealed the presence of unique annular synapses. These annular postsynaptic structures appear to last throughout the mother's life and are not found in virgin females or male rats (Hatton & Ellisman, 1982).

Each of the forms of synaptic modification discussed above may participate in the general activation of the hypothalamo-neurohypophysial system during parturition and lactation. The disappearance of incompletely covered postsynaptic densities (i.e., the complete occupation of the densities by presynaptic terminals) may allow a given

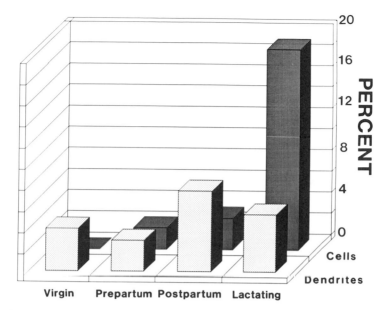

Figure 15.7. Percentages of SON cell bodies and dendrites contacted by double synapses in virgin females around the time of parturition and lactation. Dendritic double synapses were significantly elevated in postpartum animals; somatic double synapses were significantly increased only during lactation. (Data from Hatton & Tweedle, 1982; Perlmutter et al., 1984)

synaptic input to a dendrite to exert a more powerful influence on the dendrite, as postsynaptic densities are presumably the morphological correlate of neurotransmitter receptors and specialized ion channels (for review, see Siekevitz, 1985). Shared synapses may also be an important mechanism that permits oxytocin cells within the SON to fire in such close synchrony. They may accomplish this by either excitatory or inhibitory actions.

Ultrastructural Changes in the Neurohypophysis

In addition to the striking changes in the region of cell bodies and dendrites of MNCs, ultrastructural reorganization in the neurohypophysis is evident in parturient and lactating animals. As mentioned earlier, axonal profiles in the neural lobe of unstimulated animals are often completely surrounded by pituicyte cytoplasm. The number of enclosed axons varies according to the demand for hormone release; thus the incidence of enclosed axons decreases dramatically in the prepartum–postpartum interval. This decrease in the number of axons completely surrounded by pituicyte cytoplasm remains significantly lower than in either virgin or prepartum animals throughout lactation (Tweedle & Hatton, 1982).

Further restructuring occurs in postpartum and lactating animals at the basal lamina, where terminals secrete their hormone into the circulation (Tweedle & Hatton, 1987). Here the percentage of neural coverage of the basal lamina increases significantly, with a corresponding decrease in pituicyte coverage, in postpartum an-

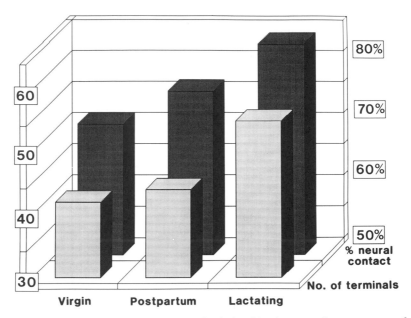

Figure 15.8. Graph illustrating the temporal relationships between the percentage of neural contact at the basal lamina and the number of terminals per 100 μm of basal laminas in virgin, postpartum, and lactating animals. The scale at the left shows the number of terminals; the scale at the right shows percentage of neural contact. (Data from Tweedle & Hatton, 1987)

imals compared with prepartum animals. This increase in neural contact is also maintained throughout lactation. The mechanism by which increases in neural contact occur at the basal lamina appears to differ between parturient and lactating animals. In postpartum animals on the day of parturition, the increased coverage is associated with a dramatic increase in the size of individual terminals, while in lactating animals (10–14 days postpartum), the increased coverage is due to a larger number of terminals in contact with the basal lamina (Figure 15.8).

These changes in the arrangements of pituicytes, axons, and the basal lamina may have profound effects on hormone release during parturition and lactation. Axon terminals engulfed by pituicytes are unable to secrete their hormones into the circulation, since they do not contact the basal lamina. The rapid release of enclosed terminals presumably allows a greater number of terminals to contact the basal lamina and participate in effective hormone release. The withdrawal of pituicyte processes from their positions along the basal lamina permits axon terminals greater access to the basal lamina and thus may serve as an additional mechanism that maximizes hormone release.

Changes in Neuronal Coupling Associated with Lactation

Detecting alterations in the organization of this system has not been left exclusively to the anatomist. Experiments utilizing hypothalamic brain slices have provided further evidence of remodeling in conjunction with increases in hormone demand. The

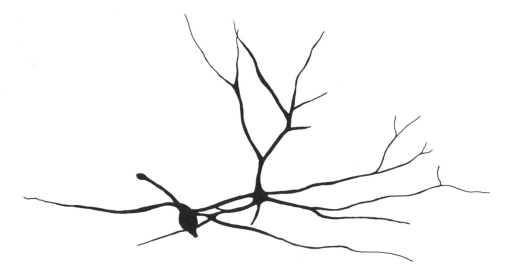

Figure 15.9. Projection drawing of a pair of dendrodendritically dye-coupled neurons in the SON. An intracellular injection of Lucifer Yellow into one cell resulted in dye transfer to the adjacent cell. These cells have at least one point of contact between their dendrites.

transfer from one neuron to another of an intracellularly injected dye of low molecular weight has long been taken as indirect evidence for electrotonic coupling between cells. One such dye is Lucifer Yellow (Stewart, 1978). Interneuronal coupling with this dye was first shown in SON neurons in 1981 (Andrew, MacVicar, Dudek, & Hatton, 1981). An example of dendrodendritic coupling, the predominant type seen in the SON, is shown in a projection drawing in Figure 15.9. Recent studies have found that the incidence of dye coupling varied rather predictably with the physiological state of the animal; an increased incidence of dye coupling was presumed to be related to activation of the system. Thus it has been inferred that electrical coupling among MNCs is another of the plastic properties of this system that is modified according to the functional demands placed on it. Electrical coupling has now been demonstrated directly by recording intracellularly from coupled pairs of SON neurons (Yang & Hatton, 1988).

   In addition to the other ways, reviewed above, that the brains of nursing mothers differ from those of untreated virgins, one would expect the chronically increased demand for oxytocin during lactation to produce observable changes in coupling. In two separate studies, it has been found that the incidence of dye coupling among SON cells in lactating rats is more than double that of virgin female animals (Hatton, Yang, & Cobbett, 1987; Yang & Hatton, 1987a). The observed changes were due to an increased incidence of coupling among oxytocin and vasopressin cells, which was interpreted to reflect both the nursing mother's finely tuned oxytocin response to suckling and her ability to respond readily to any challenges to water balance that may arise. It may be that enhanced coupling among oxytocinergic neurons participates in the synchrony of discharge of these cells and/or in metabolic cooperation among them.

   In support of a role for electrical coupling among oxytocin cells in mother rats are the results of recent experiments demonstrating direct olfactory inputs to SON cells. Anatomical tracing studies (Hollowell, Smithson, & Hatton, 1986; Smithson, Weiss, & Hatton, 1987; Smithson, Zapp, Hatton, & Weiss, 1988) and electrophysiological studies (Dougherty, Yang, & Hatton, 1986) have shown that afferents from both the main and the accessory olfactory bulbs synapse directly on SON neurons. Electrical stimulation of the output pathway from these two structures is exclusively excitatory on oxytocinergic neurons of the SON. Furthermore, such stimulation for brief periods of time in slices of hypothalamus increases the incidence of dye coupling among SON neurons in tissue from lactating rats. The same stimulation in slices from males or untreated virgin females does not affect the amount of dye coupling observed (Yang & Hatton, 1987b).

It may be that this stimulation is ineffective in normal male and virgin females, in part, because the SON cell bodies, and particularly dendrites of those animals, are separated by interposed glial processes, disallowing formation of junctions. In contrast, these processes have long since retracted from between the neural elements of lactating animals, and thus intercellular junction formation would not be disallowed. Our hypothesis is that the olfactory and vomeronasal stimulation that the nursing mother receives from pup licking and retrieval just prior to suckling serves to increase the electrical coupling among oxytocinergic neurons. This would have the effect of making more efficient the cellular coordination necessary for the milk-ejection reflex.

IMPLICATIONS OF CHANGES IN BRAIN MORPHOLOGY

In addition to its well-characterized role in parturition and lactation, oxytocin may also participate in the expression of maternal behavior. Increased bundling among SON dendrites, similar to that seen in lactating nursing mothers, has been observed in virgin females induced by continuous exposure to young pups to engage in full maternal behavior (Salm, Modney, & Hatton, 1988). Intracerebroventricular injections of oxytocin result in similar ultrastructural changes in the SON (Theodosis, Montagnese, Rodriguez, Vincent, & Poulain, 1986) and produce rapid onset of full maternal behavior in virgin animals (Pedersen, Ascher, Monroe, & Prange, 1982). Correspondingly, electrical stimulation of the olfactory pathways in slices of hypothalamus from maternally behaving virgins increases the coupling among SON neurons, just as it does in tissue from nursing mothers (Modney, Yang, & Hatton, 1987). An interaction between olfaction and oxytocin's ability to induce full maternal behavior in virgin animals has also been suggested (Wamboldt & Insel, 1987). Thus it appears that some rather profound brain changes are associated with maternal behavior independently of lactation, some of which have been linked to olfactory stimuli and oxytocin.

These extensive alterations in the organization of the hypothalamo-neurohypophysial system are impressive, since they occur in response to physiological stimuli. Equally impressive is the complete reversibility of most of these changes. That is, when lactation ceases and the requirements for hormone release return to basal levels, the ultrastructural appearance of the SON resembles that seen in virgin animals. Membrane appositions and double synapses (Salm & Hatton, 1986) in the SON of the female disappear 10 days to 1 month postweaning, and vacant postsynaptic densities reappear in the dendritic zone. Axonal profiles are again found to be engulfed by pituicyte cytoplasm, and neural coverage of the basal lamina returns to the levels seen in virgin animals. Of all the changes discussed, the only ones that persist, presumably throughout the life of the animal, are the unique annular postsynaptic specializations that form in the paraventricular nucleus.

SUMMARY

Profound plastic changes in the hypothalamus and posterior pituitary have been found to accompany parturition and subsequent nursing in rats. These changes involve, at specific times, the dendrites, cell bodies, and axon terminals of the oxytocin- and vasopressin-secreting magnocellular neurons that compose the hypothalamo-neurohypophysial system. Also participating in the overall response to the change in status from pregnant to nursing mother rat are the astrocytic glia associated with the different parts of these neurons. By the last day of pregnancy, glial processes have withdrawn from between adjacent magnocellular neurons, resulting in direct soma–somatic apposition and bundling of dendrites. At the same time, the axonal endings in the neurohypophysis are released by the astrocytes that normally enclose them, facilitating hormone secretion during parturition and subsequently during lactation. An increase in axodendritic double synapses occurs around the time of parturition, and the novel appearance of axosomatic double synapses occurs during nursing. Despite glial retraction, expansion of the dendritic zone occurs in postpartum animals in the supraoptic nucleus. Electrical coupling (as indicated by intercellular dye transfer) also increases in nursing mothers. It is likely that many of these changes, at least in the lactating mothers, contribute to the synchrony of firing among oxytocin cells that precedes the milk-ejection reflex. All these modifications appear to be completely reversible. There is evidence from earlier work, however, that certain postsynaptic specializations formed during motherhood remain throughout the life of the animal. Some of these plastic changes, now well documented in mothers, are currently also being found in virgins induced by the presence of pups to behave maternally.

ACKNOWLEDGEMENTS

Thanks to Dr. A. K. Salm for helpful comments on this chapter and to K. Grant, P. Rusch, I. Smithson, and A. Zapp for their technical expertise. Supported by NIH grant NS 09140.

REFERENCES

Andrew, R. D., MacVicar, B. A., Dudek, F. E., & Hatton, G. I. (1981) Dye transfer through gap junctions between neuroendocrine cells of rat hypothalamus. *Science, 211,* 1187–1189.

Armstrong, W. E., Schöler, J., & McNeill, T. H. (1982). Immunocytochemical, Golgi and electron microscopic characterization of putative dendrites in the ventral glial lamina of the rat supraoptic nucleus. *Neuroscience, 7,* 679–694.

Armstrong, W. E., Warach, S., Hatton, G. I., & McNeill, T. H. (1980). Subnuclei in the rat hypothalamic paraventricular nucleus: A cytoarchitectural, horseradish peroxidase and immunocytochemical analysis. *Neuroscience, 5,* 1931–1958.

Belin, V., & Moos, F. (1986). Paired recording from supraoptic and paraventricular oxytocin cells in suckled rats: Recruitment and synchronization. *Journal of Physiology, 377,* 369–390.

Belin, V., Moos, F., & Richard, Ph. (1984). Synchronization of oxytocin cells in the hypothalamic paraventricular and supraoptic nuclei in suckled rats—direct proof with paired extracellular recording. *Experimental Brain Research, 57,* 201–203.

Cobbett, P., Smithson, K. G., & Hatton, G. I. (1986). Immunoreactivity to vasopressin- but not to oxytocin-associated neurophysin antiserum in phasic neurons of rat hypothalamic paraventricular nucleus. *Brain Research, 362,* 7–16.

Dale, H. H. (1906). On some physiological properties of ergot. *Journal of Physiology, 34,* 163–206.

Dougherty, P. J., Yang, Q. Z., & Hatton, G. I. (1986). Synaptically mediated responses recorded intracellularly in rat supraoptic nucleus induced by lateral olfactory tract stimulation. *Society for Neuroscience Abstracts, 12,* 1256.

Fuchs, A.-R. (1985). Oxytocin in animal parturition. In J. A. Amico & A. G. Robinson (Eds.), *Oxytocin: Clinical and laboratory studies* (pp. 207–236). Amsterdam: Elsevier.

Fuchs A.-R., & Saito, S. (1971). Pituitary oxytocin and vasopressin content of pregnant rats before, during and after parturition. *Endocrinology, 88,* 574–578.

Hatton, G. I., Cobbett, P., & Salm, A. K. (1985). Extranuclear axon collaterals of paraventricular neurons in the rat hypothalamus: Intracellular staining, immunocytochemistry and electrophysiology. *Brain Research Bulletin, 14,* 123–132.

Hatton, G. I., Ho, Y. W., & Mason, W. T. (1983). Synaptic activation of phasic bursting in rat supraoptic nucleus neurons recorded in hypothalamic slices. *Journal of Physiology, 345,* 297–317.

Hatton, G. I., Perlmutter, L. S., Salm, A. K., & Tweedle, C. D. (1984). Dynamic neuronal–glial interactions in hypothalamus and pituitary: Implications for control of hormone synthesis and release. *Peptides, 5,* 121–138.

Hatton, G. I., & Tweedle, C. D. (1982). Magnocellular neuropeptidergic neurons in hypothalamus: Increases in membrane apposition and number of specialized synapses from pregnancy to lactation. *Brain Research Bulletin, 8,* 197–204.

Hatton, G. I., Yang, Q. Z., & Cobbett, P. (1987). Dye coupling among immunocytochemically identified neurons in the supraoptic nucleus of lactating rats. *Neuroscience, 21,* 923–930.

Hatton, J. D., & Ellisman, M. H. (1982). A restructuring of hypothalamic synapses is associated with motherhood. *Journal of Neuroscience, 2,* 704–707.

Higuchi, T., Tadokoro, Y., Honda, K., & Negoro, H. (1986). Detailed analysis of blood oxytocin levels during suckling and parturition in the rat. *Journal of Endocrinology, 110,* 251–256.

Hollowell, D. E., Smithson, K. G., & Hatton, G. I. (1986). Anatomical evidence for olfactory input to the supraoptic nucleus of the rat: An anterograde study employing the lectin *Phaseolus vulgaris* leucoagglutinin (PHA-L). *Society for Neuroscience Abstracts, 12,* 1256.

Lipton, P., & Heimbach, C. J. (1977). The effect of extracellular K^+ concentration on protein synthesis in guinea-pig hippocampal slices. *Journal of Neurochemistry, 28,* 1347–1354.

Lipton, P., & Heimbach, C. J. (1978). Mechanism of extracellular potassium stimulation of protein synthesis in the *in vitro* hippocampus. *Journal of Neurochemistry, 31,* 1299–1307.

Mason, W. T., Ho, Y. W., Eckenstein, F., & Hatton, G. I. (1983). Mapping of cholinergic neurons associated with rat supraoptic nucleus: Combined immunocytochemical and histochemical identification. *Brain Research Bulletin, 11,* 617–626.

Mason, W. T., Ho, Y. W., & Hatton, G. I. (1984). Axon collaterals of supraoptic neurons: Anatomical and electrophysiological evidence for their existence in the lateral hypothalamus. *Neuroscience, 11,* 169–182.

Meisenberg, G., & Simmons, W. H. (1983). Centrally mediated effects of neurohypophysial hormones. *Neuroscience and Biobehavioral Reviews, 7,* 263–280.

Modney, B. K., Yang, Q. Z., & Hatton, G. I. (1987). Pup-induced maternal behavior in virgin rats is associated with increased frequency of dye coupling among supraoptic nucleus (SON) neurons. *Society for Neuroscience Abstracts, 13,* 1593.

Pedersen, C. A., Ascher, J. A., Monroe, Y. L., & Prange, A. J., Jr. (1982). Oxytocin induces maternal behavior in virgin female rats. *Science, 216,* 648–650.

Orkand, R. G. (1977). Glial cells. In E. R. Kandel (Ed.), *Handbook of physiology* (Vol. 1, Part 2, pp. 855–875). Baltimore: Waverly Press.

Perlmutter, L. S., Tweedle, C. D., & Hatton, G. I. (1984). Neuronal–glial plasticity in the supraoptic dendritic zone: Dendritic bundling and double synapse formation at parturition. *Neuroscience, 13,* 769–779.

Perlmutter, L. S., Tweedle, C. D., & Hatton, G. I. (1985). Neuronal/glial plasticity in the supraoptic dendritic zone in response to acute and chronic dehydration. *Brain Research, 361,* 225–232.

Peterson, R. P. (1966). Magnocellular neurosecretory centers in the rat hypothalamus. *Journal of Comparative Neurology, 128,* 181–190.

Poulain, D. A., & Wakerley, J. B. (1982). Electrophysiology of hypothalamic magnocellular neurons secreting oxytocin and vasopressin. *Neuroscience, 7,* 773–808.

Salm, A. K., & Hatton, G. I. (1986). Beyond maternity: Electron microscopy of the supraoptic nucleus (SON) of 10-day weaned primiparous rats. *Anatomical Record, 214,* 215–216A.

Salm, A. K., Hatton, G. I., & Nilaver, G. (1982). Immunoreactive glial fibrillary acidic protein in pituicytes of the rat neurohypophysis. *Brain Research, 236,* 471–476.

Salm, A. K., Modney, B. K., & Hatton, G. I. (1988). Alterations in supraoptic nucleus neurons of maternally behaving virgin rats. *Brain Research Bulletin, 21,* 685–691.

Salm, A. K., Smithson, K. G., & Hatton, G. I. (1985). Lactation-associated redistribution of the glial fibrillary acidic protein within the supraoptic nucleus: An immunocytochemical study. *Cell and Tissue Research, 242,* 9–15.

Siekevitz, P. (1985). The postsynaptic density: A possible role in long-lasting effects in the central nervous system. *Proceedings of the National Academy of Sciences USA, 82,* 3494–3498.

Smithson, K. G., Weiss, M., & Hatton, G. I. (1987). Anterograde and retrograde tracing studies of the main olfactory bulb inputs to rat supraoptic neurons. *Society for Neuroscience Abstracts, 13,* 1594.

Smithson, K. G., Zapp, A., Hatton, G. I., & Weiss, M. (1988). Accessory olfactory bulb inputs to the rat supraoptic nucleus: An anterograde and retrograde tracing study. *Anatomical Record, 220,* 91A.

Stewart, W. W. (1978). Functional connections between cells as revealed by dye-coupling with a highly fluorescent naphthalimide tracer. *Cell, 14,* 741–759.

Suess, U., & Pliška, V. (1981). Identification of the pituicytes as astroglial cells by indirect immunofluorescence-staining for the glial fibrillary acidic protein. *Brain Research, 221,* 27–33.

Summerlee, A. J. S. (1981). Extracellular recordings from oxytocin neurons during the expulsive phase of birth in unanesthetized rats. *Journal of Physiology, 321,* 1–9.

Swanson, L. W. (1986). Organization of mammalian neuroendocrine system. In V. B. Mountcastle, F. E. Bloom, & S. R. Geiger (Eds.), *Handbook of physiology* (Vol. 4, Sect. 1, pp. 317–364). Bethesda, MD: Waverly Press.

Taubitz, I., Smithson, K. G., & Hatton, G. I. (1987). Dendritic zone of the supraoptic nucleus (SON): Ultrastructural changes in motherhood. *Society for Neuroscience Abstracts, 13,* 1594.

Taylor, C. P., & Dudek, F. E. (1982). Synchronous neural after-discharges in rat hippocampal slices without active chemical synapses. *Science, 218,* 810–812.

Taylor, C. P., & Dudek, F. E. (1984a). Excitation of hippocampal pyramidal cells by an electrical field effect. *Journal of Neurophysiology, 52,* 126–142.

Taylor, C. P., & Dudek, F. E. (1984b). Synchronization without active chemical synapses during hippocampal after-discharges. *Journal of Neurophysiology, 52,* 143–155.

Theodosis, D. T., Montagnese, C., Rodriguez, F., Vincent, J. C., & Poulain, D. A. (1986). Oxytocin induces morphological changes in the adult hypothalamo-neurohypophysial system. *Nature, 322,* 738–740.

Theodosis, D. T., & Poulain, D. A. (1984). Evidence for structural plasticity in the supraoptic nucleus of the rat hypothalamus in relation to gestation and lactation. *Neuroscience, 11,* 183–193.

Theodosis, D. T., Poulain, D. A., & Vincent, J. C. (1981). Possible morphological bases for synchronization of neuronal firing in the rat supraoptic nucleus during lactation. *Neuroscience, 6,* 919–929.

Tweedle, C. D., & Hatton, G. I. (1976). Ultrastructural comparisons of neurons of supraoptic and circularis nuclei in normal and dehydrated rats. *Brain Research Bulletin, 1,* 103–122.

Tweedle, C. D., & Hatton, G. I. (1977). Ultrastructural changes in rat hypothalamic neurosecretory cells and their associated glia during minimal dehydration and rehydration. *Cell and Tissue Research, 181,* 59–72.

Tweedle, C. D., & Hatton, G. I. (1980a). Evidence for dynamic interactions between pituicytes and neurosecretory axons in the rat. *Neuroscience, 5,* 661–667.

Tweedle, C. D., & Hatton, G. I. (1980b). Glial cell enclosure of neurosecretory endings in the neurohypophysis of the rat. *Brain Reserach, 192,* 555–559.

Tweedle, C. D., & Hatton, G. I. (1982). Magnocellular neuropeptidergic terminals in neurohypophysis: Rapid glial release of enclosed axons during parturition. *Brain Research Bulletin, 8,* 205–209.

Tweedle, C. D., & Hatton, G. I. (1984). Synapse formation and disappearance in adult rat supraoptic nucleus during different hydration states. *Brain Research, 309,* 373–376.

Tweedle, C. D., & Hatton, G. I. (1986). Vacant postsynaptic densities on supraoptic dendrites of adult rats diminish in number with chronic stimuli. *Cell and Tissue Research, 245,* 37–41.

Tweedle, C. D., & Hatton, G. I. (1987). Morphological adaptability at neurosecretory axonal endings on the neurovascular contact zone of the rat neurohypophysis. *Neuroscience, 20,* 241–246.

van den Pol, A. N. (1982). The magnocellular and parvocellular paraventricular nucleus of rat: Intrinsic organization. *Journal of Comparative Neurology, 206,* 317–345.

Wamboldt, M. Z., & Insel, T. R. (1987). The ability of oxytocin to induce short latency maternal behavior is dependent on peripheral anosmia. *Behavioral Neuroscience, 101*, 439–441.

Wittkowski, W. (1986). Pituicytes. In S. Fedoroff & A. Vernadakis (Eds.), *Astrocytes: Development, morphology, and regional specialization of astrocytes* (Vol. 1, pp. 173–200). Orlando, FL: Academic Press.

Yang, Q. Z., & Hatton, G. I. (1987a). Dye coupling among supraoptic nucleus neurons without dendritic damage: Differential incidence in nursing mother and virgin rats. *Brain Research Bulletin, 19*, 559–565.

Yang, Q. Z., & Hatton, G. I. (1987b). Dye coupling incidence among supraoptic nucleus (SON) is increased by lateral olfactory tract (LOT) stimulation in slices from lactating but not virgin or male rats. *Society for Neuroscience Abstracts, 13*, 1593.

Yang, Q. Z., & Hatton, G. I. (1988). Direct evidence for electrical coupling among rat supraoptic nucleus neurons. *Brain Research, 463*, 47–56.

16

Role of Neurogenic Stimuli and Milk Prolactin in the Regulation of Prolactin Secretion During Lactation

CE GROSVENOR, G. V. SHAH,
AND W. R. CROWLEY

In many species, prolactin (PRL) appears to be the major hormone secreted by the anterior pituitary gland that regulates the secretion of milk. This chapter deals, first, with the principal stimuli—suckling and sensory cues from offspring—that affect PRL release in the lactating rat and, second, with the hypothalamic mechanisms by which the neural stimuli are transduced into secretory changes within the pituitary lactotrophe cells that secrete PRL. The concept is presented that PRL within the lactotrophe of the lactating rat exists in two different forms, a prereleasable and a releasable form, that respond differently to transducing hypothalamic factors or hormones. Finally, data are presented that suggest that the secretion of at least one of the transducing factors, dopamine (DA), from its terminal neurons in the median eminence, as well as its action on the pituitary lactotrophe during adulthood, are influenced by the level of PRL present in the mother's milk and subsequently ingested by the offspring during a critical postpartum period. Thus PRL appears to function as a growth/maturation hormone important to the development of certain neuroendocrine and endocrine systems.

AFFERENT CONTROL OF PRL SECRETION DURING LACTATION

In contrast to the "internal" regulation of PRL by ovarian steroids in the nonlactating female, secretion of this hormone during lactation is almost entirely under the control of afferent impulses generated in the mammary glands in response to suckling and by other exteroceptive signals emanating from the litter (Grosvenor & Mena, 1971). PRL is not secreted tonically during lactation but is released phasically in response to such stimulation; very low circulating titers of the hormone are found in the absence of suckling.

There is a lack of detailed information on the specific pathways within the brain that convey the stimulatory influence of suckling to the hypophysiotropic regions of the hypothalamus. It appears that the "suckling-activated" systems follow a diffuse, multisynaptic path throughout the reticular formation and enter the hypothalamus by diverse routes (Dubois-Dauphin, Armstrong, Tribollet, & Dreifus, 1985a, 1985b; Juss & Wakerly, 1981). Some of these fibers terminate in the magnocellular supraoptic and paraventricular nuclei of the hypothalamus to signal release of oxytocin, and others most likely terminate in other hypothalamic regions to stimulate PRL release. For example, in several species, the medial preoptic nucleus also appears to be important, as destruction of this area decreases suckling-induced PRL release and inhibits lactation (Wolinska, Polkowska, & Domanski, 1977). Much remains to be learned about the neuronal networks and, specifically, the afferents to the preoptic area and medial basal hypothalamus that regulate PRL secretion in various physiological conditions. Some of the gaps in our knowledge have been filled by neuropharmacological studies, which suggest the involvement of specific neurotransmitter and hypothalamic hormonal systems.

INFLUENCE OF EXTEROCEPTIVE STIMULATION OF PRL SECRETION DURING LACTATION

Throughout the reproductive life of the animal, the functioning of the PRL-releasing mechanism is under both stimulatory and inhibitory influences of nontactile, exteroceptive stimuli (ECS). Thus stimulation or inhibition of PRL release has been demonstrated in response to various exteroceptive signals in cyclic (Van der Lee & Boot, 1955, 1956) or newly mated female mice (Bruce, 1959; Bruce & Parrot, 1960); in postparturient (Alloiteau, 1962) or lactating rats (Grosvenor, 1965), goats (Bryant, Linzell, & Greenwood, 1970), and cows (Johke, 1969; Tucker, Convey, & Koprowski, 1973); and in pigeons and doves (Patel, 1936; Witschi, 1935). The first experimental evidence demonstrating an influence of ECS provided by the litter on the secretion of PRL by the maternal pituitary gland was obtained in 1965 (Grosvenor, 1965). In this study, placement of rat pups beneath their mother, but physically separated from her by a wire screen, for 30 min following a 10-hr interval of isolation on postpartum day 14 resulted in a fall in maternal pituitary PRL concentration similar in extent to that produced by 30 min of suckling. Self-licking of mammary gland areas, which may indirectly cause PRL release in the pregnant rat (McMurtry & Anderson, 1971; Roth & Rosenblatt, 1968), was ineffective in eliciting the acute fall in pituitary PRL concentration of the lactating rat (Grosvenor & Mena, 1969). In another study (Grosvenor, Mena, Dhariwal, & McCann, 1967), milk reaccumulation was significantly reduced in rats following blockade of the suckling-induced release of PRL. However, placement of the pups under their mother during the milk-reaccumulation period restored milk secretion to normal. These initial studies demonstrated that ECS from the litter exerts a strong stimulatory influence on PRL release and, thus, indirectly on milk secretion. It seems clear from these and other studies (Moltz, Levine, & Leon, 1969; Stern & Siegel, 1978; Zarrow, Johnson, Denenberg, & Bryant, 1973) that PRL may be released in response to ECS from rat pups.

In subsequent studies (Mena & Grosvenor, 1971), the sensory information from the litter responsible for the release of PRL from the maternal pituitary gland was determined to be mainly olfactory. It is noteworthy that these results agree with those found in mice (Whitten, 1966) concerning the essential role of olfaction in the stimulation or inhibition of PRL release by ECS. From the studies in female mice, it was concluded that the effects of ECS on PRL release was mediated by a pheromone produced by the male. Pheromones from the pups may also be instrumental in the ECS-induced release of PRL in the lactating rat, although conclusive information has yet to be obtained.

In the experiments referred to above (Mena & Grosvenor, 1971), olfactory, visual, and auditory stimuli from the pups were blocked, either singly or in combination, during the period of exposure to the mother. In another series of experiments, the stimuli from the pups were not blocked, but there were combinations of surgical eliminations during pregnancy of two of the three sense organs of the mothers. When such mothers were exposed on postpartum day 14 to their pups, a decrease in PRL concentrations occurred not only in those mothers that had only olfaction remaining, but also in those in which only vision was intact (Mena & Grosvenor, 1971). The latter result was interpreted to suggest that in the absence of the primary mediator olfaction—visual stimuli are able to stimulate PRL release. These studies also suggested that visual cues from the pups may influence PRL release in normal rats under specific physiological situations. Auditory cues may also be important for PRL release under certain circumstances (Terkel, Damassa, & Sawyer, 1979).

The efficacy of ECS from the litter as a stimulus for maternal PRL release varies throughout the normal 21-day period of lactation in the rat. In the primiparous rat, the exteroceptive mechanism for PRL release does not function in early lactation that is, through day 7—but it is active in mid- (day 14) and late (day 21) lactation (Grosvenor, Maiweg, & Mena, 1970). Apparently, this function develops somewhere between early and mid-lactation, although the precise time and mechanism involved are not yet known. Whatever the mechanism of this development, it is clear that by postpartum day 14, the configuration of the exteroceptive cues from the pups is well established as a stimulus, inasmuch as on day 14 the mothers can discriminate between those cues emanating from much older or from younger pups placed under them (Grosvenor et al., 1970). Moreover, these rats do not respond to exteroceptive cues either from their own or from variously aged pups of other lactating rats housed in adjacent or nearby cages in the same room (Grosvenor et al., 1970). This observation suggests that the spatial relationship of mothers and pups during suckling may be important for the exteroceptive release of PRL.

The changing characteristics of the pups, and/or the changing behavioral interrelationships between pups and mother as the pups develop, may also contribute to the onset of the exteroceptive mechanism in the primiparous rat. This is suggested by the finding that PRL release occurs in primiparous rats in response to ECS on day 7, provided that the mother is given 7- to 8-day-old foster pups on day 2 (Grosvenor et al., 1970). The finding that primiparous rats on day 7 did not show a reduction in PRL when exposed for the first time to 14-day-old pups further indicates that the exteroceptive mechanism develops gradually. Once the configuration of stimuli is established, it is apparently retained through a second pregnancy following a normal first lactation, since exposure of the mother to the pups on day 7 of a second lactation elicits PRL release to the same extent as that following suckling (Grosvenor et al., 1970).

Between mid- and late lactation, the exteroceptive mechanism for PRL release undergoes two important modifications. First, there is an increased sensitivity of the mother to exteroceptive signals. In contrast to the 14-day postpartum rat, in which the exteroceptive mechanism is activated only when the pups are underneath the mother, the 21-day postpartum rat releases PRL in response to exteroceptive signals emanating from other lactating rats and/or their litters (Mena & Grosvenor, 1972). Thus in order to demonstrate the effect of a stimulus such as suckling on PRL release on day 21, it is necessary first to isolate the mothers from the animal-room environment (Mena & Grosvenor, 1972). The changing characteristics of the pups from 14 to 21 days of age are also involved in some manner in increasing the responsiveness of the exteroceptive mechanism for PRL release. For example, release of PRL occurs in response to animal-room environmental stimuli in the day 14 primiparous rat, provided that 13- and 14-day-old foster pups are substituted for the mother's own pups on day 7 (Mena & Grosvenor, 1972).

The second modification is the generation by the pups of a peripheral inhibitory influence on the action of PRL in stimulating milk secretion. In contrast to day 14, when exposure of the mothers to their litters leads to depletion of pituitary PRL, which then stimulates milk secretion, the same procedure performed on day 21 on isolated postpartum rats results in a normal depletion of PRL, but accompanied by an inhibition, rather than a stimulation, of milk secretion (Mena & Grosvenor, 1972). These observations suggest that simultaneous with the stimulatory influence of ECS from the pups on PRL release, as evidenced by the depletion of the hormone from the pituitary gland, a blocking influence is exerted by stimuli emanating from the pups on the action of the hormone, perhaps at the level of the mammary gland, which prevents it from stimulating milk secretion.

In late lactation, the normal spatial relationship of the mother to her pups at the time of suckling—that is, the pups under the mother—is required before the stimuli from the pups are inhibitory for milk secretion. No inhibition of milk secretion occurs, for example, if the pups are placed alongside the mother in another cage. These observations suggest a mechanism whereby galactopoiesis is reduced at the time when the litter is less dependent on the milk of the mother for survival, even though the pituitary and mammary glands are still quite functional. The mechanism of the inhibitory effect is unknown, but it may be related to sympathetic adrenal activation.

It is apparent, therefore, that as with the stimulus of suckling, ECS from the litter plays a decisive role in PRL secretion during lactation in the rat. The importance of each type of stimulus, however, varies throughout the lactation period. Thus the finding that the exteroceptive mechanism is not apparent in the primiparous rat during early lactation suggests that suckling constitutes the major mechanism for release of PRL, which is necessary for the maintenance of milk secretion during this period. During the later stages of lactation, suckling maintains its effectiveness, as in the previous stages, but the increased responsiveness of the exteroceptive mechanism may shift the major role to this mechanism as the principal regulator of PRL release.

THE INTRACELLULAR TRANSFORMATION OF PRL
IN RESPONSE TO SUCKLING

It has been demonstrated, initially with bioassay and subsequently with radioimmu-noassay and polyacrylamide gel electrophoresis (Grosvenor & Mena, 1971, 1982), that the concentration of PRL in the lactating rat pituitary gland is significantly re-duced by suckling. However, the amounts of PRL (in micrograms) that appeared to be "lost" from the pituitary during the first few minutes of suckling could not be accounted for simply as hormone released into the circulation at the same time. Only when suckling was sufficiently prolonged did the amount of PRL released into the circulation approximate the amount reduced in the pituitary. A series of observations (Grosvenor & Mena, 1982) has led to the hypothesis that PRL is rapidly transformed from a prereleasable to a releasable form within the pituitary of the lactating rat during the first few minutes of suckling. Continued suckling then promotes the steady release into the circulation of small amounts (400–800 ng/min) of transformed PRL. Recent evidence suggests that the transformation and start of release occur 1–2 hr after synthesis of the hormone has been completed (Mena, Martinez-Escalera, Clapp, & Grosvenor, 1984).

The nature of the transformation process is not known, although it is inhibited by the PRL-inhibiting hormone dopamine (DA) (Grosvenor, Mena, & Whitworth, 1980) and is provoked by the DA antagonists haloperidol and domperidone (Gros-venor, Goodman, & Mena, 1984). Evidence suggests that suckling causes PRL trans-formation by transiently suppressing the release of DA from the hypothalamus (Chiocchio, Cannata, Cordero Funes, & Tramezzani, 1979; Mena, Enjalbert, Car-bonell, Priam, & Kordon, 1976; Plotsky & Neill, 1982a). A striking concomitance to the brief removal of the dopaminergic tone is that the release of transformed PRL is rendered much more sensitive to putative PRL-releasing factors, such as thyrotro-pin-releasing hormone (TRH) (deGreef & Visser, 1981; Fagin & Neill, 1981; Gros-venor & Mena, 1980; Plotsky & Neill, 1982b). Releasable PRL within the anterior pituitary gland has a lower threshold for activation (Grosvenor, Mena, & Whitworth, 1979; Grosvenor, Whitworth, & Mena, 1981) and is less soluble than PRL that has not undergone transformation (Mena, Martinez-Escalera, Clapp, Aguayo, Forray, & Grosvenor, 1982). It was this difference in the solubility of transformed PRL that, in the early studies, made it seem as though it had disappeared from the pituitary gland with the onset of suckling. (See Table 16.1 for a comparison of characteristics of the two forms of PRL.)

At the same time that transformed PRL is released, its concentration in the prereleasable pool is restored. This process takes 1.5–2 hr. The concentration of the prereleasable form of PRL is not reduced by further suckling during this time; that is, the mechanism that functions to transform PRL has become refractory (for review, see Grosvenor & Mena, 1982). The histidyl-proline metabolite of TRH (cyclo-his-pro) inhibits the transformation phase after suckling (Shyr & Grosvenor, 1985). This substance may interact with DA or may sensitize the inhibitory effectiveness of DA on the transformation process.

Table 16.1 Comparison of the characteristics of secretion of prereleasable and transformed PRL in the lactating rat

Characteristics of suckling-induced transformation of prereleasable to releasable PRL in rat lactotrophe	Characteristics of suckling-induced release of transformed PRL into the circulation
1. Occurs rapidly (within the first 5 min of suckling).	1. Occurs in a steady, minute-by-minute manner.
2. 15–60 µg of PRL are transformed.	2. 400–800 ng/min (RP-1 Standard) is released.
3. Occurs periodically (every 1.5–2 hr).	3. PRL is released only when suckling is applied; i.e., is stimulus dependent.
4. Has a high threshold of activation.	4. Has a lower threshold of activation than PRL transformation.
5. Is inhibited by DA.	5. Is stimulated by TRH and possibly other prolactin-releasing factors
6. Prereleasable PRL is soluble at pH 7.0–7.2	6. Releasable PRL is insoluble at pH 7.0–7.2
7. PRL synthesized either <2 hr or >12 hr beforehand is *not* transformed; that synthesized >12 hr is destroyed by lysosomes.	7. PRL transformed >12 hr earlier is not released but is lysed.

PRL-INHIBITING HORMONES: DOPAMINE

It is now well established that DA, which is secreted into the hypophysial portal blood from the nerve endings of cells located in the arcuate nucleus, acts on membrane receptors on the lactotrophe cells in the pituitary gland to inhibit the secretion of PRL (Ben-Jonathan, 1985; Grosvenor & Mena, 1982; MacLeod, 1976; Weiner & Ganong, 1978). Thus DA serves in this region as a hypothalamic hormone, and several unique features of these tuberoinfundibular dopamine (TIDA) neurons distinguish them from the dopaminergic neurotransmitter systems in other areas of the brain (Moore & Demarest, 1982). Numerous studies have shown that DA exerts a tonic inhibition of basal PRL release in animals and humans, and that PRL release is markedly elevated following the administration of drugs that disrupt DA neurotransmission (Weiner & Ganong, 1978). DA or DA agonists, such as bromocriptine, also depress the physiological increases of PRL that occur in lactation (deGreef & Visser, 1981; Plotsky & Neill, 1982b; Smalstig, Sawyer, & Clemens, 1974). TIDA neurons therefore appear to account for a major portion of the inhibitory hypothalamic tone governing PRL secretion.

Physiological stimuli for PRL release, such as suckling, estrogen, and mating, appear to decrease the activity of TIDA neurons, as evidenced by their ability to reduce DA concentrations (Chiocchio et al., 1979; deGreef & Neill, 1979; Mena et al., 1976) and to reduce the turnover of DA in the median eminence (Crowley, 1982; Demarest, McKay, Riegle, & Moore, 1983; Selmanoff & Gregerson, 1985; Selmanoff & Wise, 1981). In addition, within several minutes after the onset of suckling or stimulation of the mammary nerve, there is a transient (3–5 min) decrease in the secretion of DA into portal blood (deGreef & Visser, 1981; Plotsky & Neill, 1982a). Recent data suggest that the level of functioning of TIDA neurons in the adult rat is influenced by exposure of these neurons to maternally derived PRL during a critical postpartum period (see below).

IMPORTANCE OF NEONATAL HORMONES FOR ADULT NEUROENDOCRINE REGULATION

Exposure to or deprivation of specific hormones during brief perinatal critical periods has profound effects on adult neuroendocrine regulation. This is clearly exemplified by the perinatal actions of androgen and androgen metabolites in the organization of sexually dimorphic patterns of gonadotrophin and PRL secretion and mating behavior in rodents (Harris, 1964; MacLusky & Naftolin, 1981; McEwen, 1981; Neill, 1972; see chapter 5). In addition to the endocrine glands of the offspring, a potentially important source of hormones for the neonatal animal is the mother, which can supply these substances to the offspring via the milk during lactation (Baram, Koch, Hazum, & Fridkin, 1977; Kacsoh, Terry, Crowley, & Grosvenor, unpublished data; Smith & Ojeda, 1984). While some differences exist among species, rat milk contains neuropeptides (luteinizing hormone releasing hormone [LHRH], growth hormone releasing hormone [GHRH], TRH) and pituitary (PRL, growth hormone [GH]) and other protein and peptide hormones (e.g., insulin, epidermal growth factor, and erythropoietin), (Grosvenor, Shyr, & Crowley, 1986; Koldovsky, 1980; Pope & Swinburne, 1980). It is important to note that the immature gastrointestinal tract of neonatal animals is permeable to large molecules (Clark & Hardy, 1969; Morris & Morris, 1974), which can be absorbed in biologically active form to exert effects in the developing organism. For example, there is evidence for an action of milk-derived LHRH in the development of ovarian function (Smith & Ojeda, 1984) and for milk-borne GHRH in the stimulation of GH release during the neonatal period (Kacsoh, Crowley, & Grosvenor, unpublished data). Until recently, however, there has been no evidence that a neonatal *deficiency in milk-derived hormones* or other factors can alter subsequent endocrine regulation.

Neuroendocrine Effects of a Neonatal Deficiency in Milk-Derived PRL

Earlier studies from this laboratory (Grosvenor & Whitworth, 1976; Whitworth & Grosvenor, 1978) and others (Koldovsky, 1980; McMurtry & Malven, 1974a, 1974b; McMurtry, Malven, Arave, Erb, & Harrington, 1975) have established the presence of PRL, in concentrations of 200–400 ng/ml, in the milk of lactating rats and other species. It has been possible to trace the accumulation of radiolabeled PRL from the maternal circulation into the milk within the mammary gland and its passage from the stomach of the suckling offspring into their systemic circulation (Grosvenor & Whitworth, 1976; Whitworth & Grosvenor, 1978). Through the episodic feeding and absorption from the gastrointestinal tract, neonatal rats may be exposed to a total of 300 ng of PRL reaching the circulation per 24 hr (Whitworth & Grosvenor, 1978), and it is interesting to note that this occurs during a period when the neonatal pituitary gland secretes relatively little PRL (Döhler, von zur Mühlen, & Döhler, 1977; Döhler & Wuttke, 1975; Ojeda & McCann, 1974).

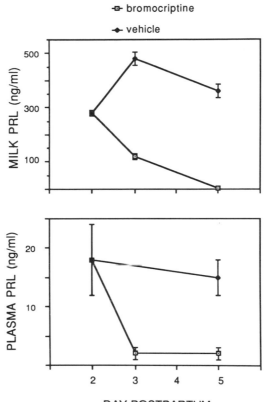

Figure 16.1. Plasma (right) and milk (left) concentrations of PRL in lactating rats injected sc twice daily with saline or 0.25 mg bromocriptine (CB-154), commencing on postpartum day 2, and killed on either postpartum day 3 or 5. Each data point is the mean ± SEM of five or six rats.

An important role for milk-derived PRL has been suggested by recent results from this laboratory (Shyr, Crowley, & Grosvenor, 1986). Lactating mothers were treated on each of days 2–5 postpartum with the DA agonist bromocriptine mesylate (250 μg/300-g rat per day in divided doses) or an ethanol/saline vehicle. For most lactating rats, the bromocriptine treatment markedly reduced the concentration of PRL in milk (Figure 16.1), with little or no effect on lactation per se. Except for daily weighings to monitor weight gain, the offspring received no direct treatments, and were weaned at 21 days of age. When the offspring of the two groups of mothers were 33–35 days of age, measurements of catecholamine turnover and serum PRL were made. Compared with the offspring of vehicle-treated mothers, the male and female offspring of bromocriptine-treated mothers—that is, those consuming PRL-deficient milk on days 2–5 postpartum—exhibited the following: (1) significantly decreased steady-state concentrations of DA in the median eminence (Table 16.2); (2) significantly reduced turnover rates of DA in the median eminence (assessed from the rate of decline after synthesis inhibition) (see Table 16.2); and, consistent with

Table 16.2 Effects of maternal treatment with bromocriptine (CB-154), with or without ovine(o)PRL, from postpartum days 2–5 on the concentrations and turnover of DA in the TIDA and tuberohypophyseal dopamine (THDA) neurons in 33- to 35-day-old offspring

Treatment of mothers (no. of offspring)	TIDA neurons		
	Initial concentration (pg/μg protein)	k	Turnover rate (pg/μg protein·hr)
Exp I (postpartum days 2–5)			
Saline (22)	120 ± 12	0.838 ± 0.133	101 ± 16
CB-154 (32)	86 ± 9^{a}	0.203 ± 0.062^{b}	19 ± 5^{b}
oPRL + CB-154 (26)	124 ± 3	0.671 ± 0.110	65 ± 7^{a}
Exp II (postpartum days 9–12)			
Saline (31)	84.5 ± 8.9	0.685 ± 0.120	66 ± 12
CB-154 (22)	96.2 ± 9.4	0.848 ± 0.119	82 ± 11
Treatment of mothers (no. of offspring)	THDA neurons		
	Initial concentration (pg/μg protein)	k	Turnover rate (pg/μg protein·hr)
Exp I (postpartum days 2–5)			
Saline (22)	12.1 ± 0.5	0.639 ± 0.080	6.9 ± 1.1
CB-154 (32)	10.8 ± 0.5	0.853 ± 0.108	9.2 ± 1.2
oPRL + CB-154 (26)	9.6 ± 0.7	0.537 ± 0.080	5.2 ± 0.8
Exp II (postpartum days 9–12)			
Saline (31)	9.0 ± 0.5	0.576 ± 0.075	5.2 ± 0.7
CB-154 (22)	11.0 ± 0.8	0.810 ± 0.090	8.9 ± 0.9

Notes: Values are the mean ± SEM. k, rate constant for decline. Significance was determined by single-factor analysis of variance and Newman-Keuls tests.
[a] $p < .05$ vs. all others.
[b] $p < .01$ vs. all others.

the above observations, (3) significantly elevated serum PRL concentrations (Figure 16.2). These effects were completely abolished by concomitant administration of ovine PRL to bromocriptine-treated mothers (see Figure 16.2, Table 16.2). Neonatal PRL deficiency did not affect the tuberohypophyseal DA system innervating the neu-rointermediate lobe of the pituitary (see Table 16.2) and did not affect noradrenergic activity in either of these regions. That a critical period may be involved for these effects of neonatal PRL deficiency is suggested by the further finding that neither the reduction in median eminence DA turnover nor the elevation of serum PRL occurred if the maternal bromocriptine treatments were delayed until days 9–12 of lactation (see Table 16.2). The hyperprolactinemia resulting from neonatal PRL deficiency persists beyond the onset of puberty until at least 90–100 days of age (Shah, Shyr, Grosvenor, & Crowley, 1988) (Figure 16.3). The onset of puberty is not altered by

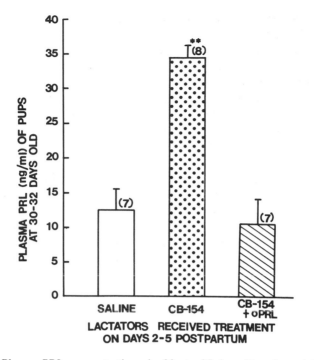

Figure 16.2. Plasma PRL concentrations in 30- to 35-day old male and female pups of mothers treated daily from postpartum days 2 to 5 with saline or bromocriptine (CB-154), with or without oPRL. Values are the mean ± SEM. **the value is significantly ($p < .01$) elevated (by Student's t-test) compared with the value in the saline group. The number of litters is given in parentheses.

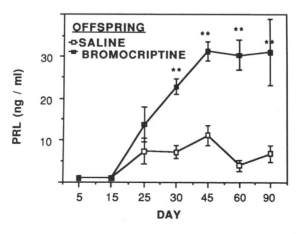

Figure 16.3. Serum PRL concentrations (±SEM) at various postnatal ages of control (saline) and neonatal PRL-deficient (bromocriptine) female offspring. **$p < .01$ vs. respective saline.

this treatment, but neonatal PRL-deficient offspring show estrous acyclicity for 20–30 days during the immediate postpubertal period.

The results of a number of control experiments (Shyr et al., 1986) suggest strongly that these effects can be attributed to a reduction of milk-derived PRL rather than to nonspecific effects of the bromocriptine treatment. First, the neonatal rates of weight gain were no different for the offspring of control, bromocriptine-, or bromocriptine-plus-PRL-treated mothers, implying that nutrition is not adversely affected with the doses of bromocriptine used. Second, concomitant infusion of ovine (o) PRL via osmotic minipump at a dose (12 μg/hr) comparable with the amount of rat PRL released during suckling completely reversed both the decrease in DA turnover (see Table 16.2) and the hyperprolactinemia (see Figure 16.2) in the offspring at 35 days of age, suggesting that the effects of maternal bromocriptine treatment are due specifically to reduction of PRL. Moreover, neither chronic nor acute treatment of lactating rats with bromocriptine reduced the basal levels of PRL in the pups (Shyr et al., 1986); this implies that bromocriptine itself does not enter the neonate via the milk in amounts sufficient to suppress endogenous PRL secretion. This conclusion is buttressed by studies in which radiolabeled bromocriptine was administered acutely and chronically to lactating rats. Levels of radioactivity in maternal milk were quite low, and in the serum of the offspring were approximately 0.001% of the administered dose; unmetabolized bromocriptine, as measured by radioimmunoassay, was essentially undetectable in milk and offspring.

These results suggest that one function of milk-derived PRL in the neonate is to influence the maturation of the inhibitory TIDA dopaminergic controls over PRL secretion, and that a deficiency in milk-derived PRL during a brief postnatal critical period leads to a reduction of the dopaminergic inhibition of PRL secretion that persists into adulthood.

Development of the TIDA System Controlling PRL Secretion

The functional development of the TIDA system occurs postnatally in rats. Although present on the day of birth, neurons in the arcuate nucleus do not begin to synthesize or release DA until approximately postnatal day 5 (Hyyppa, 1969; Loizou, 1971; Smith & Simpson, 1970) as the processes of these neurons grow into the median eminence to meet the vessels of the primary capillary plexus, which also form at this time (Glydon, 1957). The dopaminergic inhibition of PRL secretion gradually develops during the first and second postnatal weeks (Ojeda & McCann, 1974), even though DA receptors are demonstrable in the pituitary gland on postnatal day 1 (Becú de Villalobos, Yacos, Cardinali, & Libertun, 1984). Thus it is conceivable that milk-derived PRL, perhaps acting in concert with endogenously secreted PRL, stimulates the synthesis and/or release of DA from developing TIDA neurons. In the mature animal, elevations of circulating PRL increase TIDA activity (Ben-Jonathan, 1985; Demarest, Riegle, & Moore, 1984, 1986; Moore, 1987; Moore & Demarest, 1982; Selmanoff, 1981) by enhancing DA synthesis (Arita & Kimura, 1986; Demarest & Moore, 1980; Perkins, Westfall, Paul, MacLeod, & Rogol, 1979) and release (Foreman & Porter, 1981; Perkins & Westfall, 1978). This action has been observed as early as postnatal day 15 (Honma, Hohn, & Wuttke, 1979), but it is not known whether it occurs earlier.

Hypoprolactinemia decreases TIDA activity in adulthood (Demarest, Riegle, &

Moore, 1985). Because PRL deficiency on days 2–5 postpartum results in decreased DA turnover some time later (Shah et al., 1988; Shyr et al., 1986), PRL may also exert a stimulatory "organizational" effect during the immediate postnatal period. A precedent for the hormonal regulation of DA neuronal development comes from work showing that the postnatal onset of dopaminergic inhibition of alpha-melanocyte stimulating hormone (α-MSH) secretion from the intermediate lobe is stimulated by α-MSH itself (Lichtensteiger, Schlumpf, & Davis, 1983). Thus the two neuroendocrine DA systems may be similar in that their postnatal functional development may, in part, be regulated by the hormones they are destined to inhibit in the mature animal.

Neonatal PRL Deficiency and Pituitary Regulation of PRL Secretion

Additional results from our laboratory show that neonatal PRL deficiency alters certain aspects of the hypothalamic control over PRL secretion in adulthood. First, a

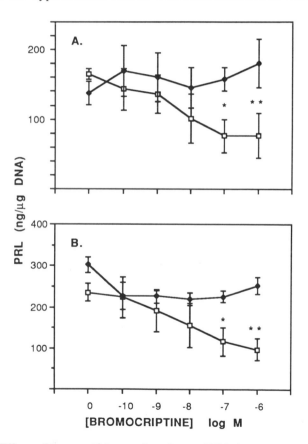

Figure 16.4.   Effects of bromocriptine on the release of PRL from cultured anterior pituitary cells obtained from 100-day-old control (white symbols) or neonatal PRL-deficient (black symbols) female rats during (A) 30-min or (B) 4-hr incubations. $*p < .05$; $**p < .01$ vs. respective control (no bromocriptine) and vs. same dose level in bromocriptine offspring. Bars indicate SEM.

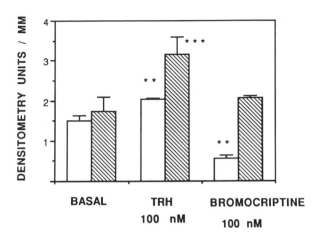

Figure 16.5. Relative levels of cytoplasmic PRL mRNA (±SEM) from dot-blot hybridizations on cultured anterior pituitary cells of control (open bars) and neonatal PRL-deficient (hatched bars) female offspring. **$p < .01$ vs. respective basal level; ***$p < .01$ vs. control with TRH.

degree of pituitary hyperplasia may be present in view of the approximately 20% increase in cells per pituitary obtained during enzymatic dispersal (unpublished observations). Second, in contrast to normal anterior pituitary cells maintained in culture, pituitary cells obtained from 100-day-old offspring of bromocriptine-treated mothers are almost completely unresponsive to the effects of a DA agonist in vitro in decreasing the release of PRL (Figure 16.4) and reducing the cytoplasmic levels of PRL messenger RNA (mRNA) (Figure 16.5). In addition, these cells are somewhat more responsive to TRH on these parameters (Figures 16.5 and 16.6) (Shah et al., 1988). Thus much like neonatal deprivation of androgen (Harris, 1964; Mac-

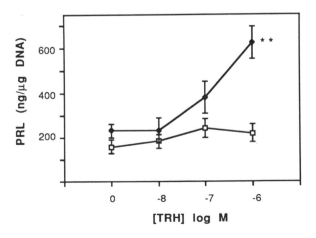

Figure 16.6. Effects of TRH on the release of PRL from cultured anterior pituitary cells during a 4-hr incubation obtained from 100-day-old control (white symbols) or neonatal PRL-deficient (black symbols) female rats. **$p < .01$ vs. respective control (no bromocriptine) and vs. same dose level of control offspring. Bars indicate SEM.

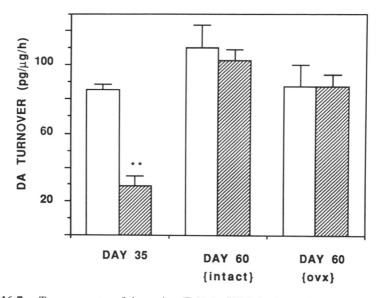

Figure 16.7. Turnover rates of dopamine (DA) (±SEM) in the median eminence of control (open bars) and neonatal PRL-deficient (hatched bars) female offspring at postnatal days 35 and 60. **$p < .01$ vs. respective saline.

Lusky & Naftolin, 1981; McEwen, 1981; Neill, 1972), neonatal deprivation of PRL alters sensitivity to hormones in adulthood. Because TIDA activity in neonatal PRL-deficient offspring eventually returns to normal (Figure 16.7), the persistently elevated serum PRL may, in part, be maintained by the depressed responses of the pituitary lactotrophe to DA (Shah et al., 1988).

SUMMARY

A great deal of research has been devoted to the effects of suckling and ECS from neonates on the secretion of PRL during lactation and to the mechanisms involved in

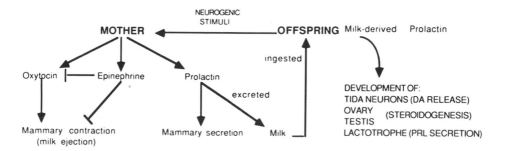

Figure 16.8. Interrelationships between PRL secretion in response to neurogenic stimulation and the influence of milk-derived PRL on the neonatal development of elements involved in PRL secretion during adulthood.

this neuroendocrine reflex. Recent work suggests that maternal PRL, which is secreted into the circulation in response to suckling and to ECS, and which is concentrated in the milk and ingested by the neonate, is absorbed into the circulation of the neonate. Milk-borne PRL may modulate the development and/or maintenance of the functional status of DA-secreting TIDA neurons, and gonads, the pituitary lactotrophe cells, and possibly, although unexplored, maternal behavior elements that are intimately and interactively involved in the regulation of PRL secretion in the adult. These interrelationships are shown in Figure 16.8. Thus neonatal animals may stimulate the development of their own neuroendocrine regulatory mechanisms through this complex interanimal loop, in which suckling and ECS evoke the release of maternal PRL, which is then ingested and subsequently exerts its important developmental effects.

REFERENCES

Alloiteau, J. J. (1962). Le contrôle hypothalamique de l'adenohypophyse: III. Regulation de la fonction gonadotrophe femelle. Activité LTH. *Biological Medicine, 51*, 250–259.

Arita, J., & Kimura, F. (1986). Characterization of *in vitro* dopamine synthesis in the median eminence of rats with haloperidol-induced hyperprolactinemia. *Endocrinology, 119*, 1666–1673.

Baram, T., Koch, Y., Hazum, E., & Fridkin, M. (1977). Gonadotropin-releasing hormone in milk. *Science, 198*, 300–301.

Becú de Villalobos, D., Yacos, M. I., Cardinali, D. P., & Libertun, C. (1984). A developmental study of adenohypophyseal dopaminergic receptors and of haloperidol-induced prolactin. *Developmental Brain Research, 12*, 167–171.

Ben-Jonathan, N. (1985). Dopamine: A prolactin-inhibiting hormone. *Endocrinology Review, 6*, 564–589.

Bruce, H. M. (1959). An exteroceptive block to pregnancy in the mouse. *Nature, 184*, 105.

Bruce, H. M., & Parrot, D. M. V. (1960). Role of olfactory sense in pregnancy block by strange males. *Science, 131*, 1526.

Bryant, G. D., Linzell, J. L., & Greenwood, F. C. (1970). Plasma prolactin in goats measured by radioimmunoassay: The effects of teat-stimulating mating behavior, stress, fasting and oxytocin, insulin and glucose injections. *Hormones, 1*, 26–35.

Chiocchio, S. R., Cannata, M. A., Cordero Funes, J. R., & Tramezzani, J. H. (1979). Involvement of adenohypophysial dopamine in the regulation of prolactin release during suckling. *Endocrinology, 105*, 544–547.

Clark, R. M., & Hardy, R. N. (1969). The use of [^{125}I] polyvinyl pyrrolidone K. 60 in the quantitative assessment of the uptake of macromolecular substances by the intestine of the young rat. *Journal of Physiology, 204*, 113–125.

Crowley, W. R. (1982). Effects of ovarian hormones on norepinephrine and dopamine turnover in individual hypothalamic and extrahypothalamic nuclei. *Neuroendocrinology, 34*, 381–386.

deGreef, W. J., & Neill, J. D. (1979). Dopamine levels in hypophysial stalk plasma of the rat during surges of prolactin secretion induced by cervical stimulation. *Endocrinology, 105*, 1093–1099.

deGreef, W. J., & Visser, T. J. (1981). Evidence for the involvement of hypothalamic dopamine and thyrotropin-releasing hormone in suckling-induced release of prolactin. *Journal of Endocrinology, 91*, 213–223.

Demarest, K. T., McKay, D. W., Riegle, G. D., & Moore, K. E. (1983). Biochemical indices of tuberoinfundibular dopaminergic neuronal activity during lactation: A lack of response to prolactin. *Neuroendocrinology, 36*, 130–137.

Demarest, K. T., & Moore, K. E. (1980). Accumulation of L-dopamine in the median eminence: An index of tuberoinfundibular dopaminergic nerve activity. *Endocrinology, 106*, 463–468.

Demarest, K. T., Riegle, G. D., & Moore, K. E. (1984). Prolactin-induced activation of tuberoinfundibular dopaminergic neurons: Evidence for both a rapid tonic and a delayed induction component. *Neuroendocrinology, 38*, 467–475.

Demarest, K. T., Riegle, G. D., & Moore, K. E. (1985). Hypoprolactinemia induced by hypophysectomy and long-term bromocriptine treatment decreases tuberoinfundibular dopaminergic neuronal activity and the responsiveness of these neurons to prolactin. *Neuroendocrinology, 40*, 369–376.

Demarest, K. T., Riegle, G. D., & Moore, K. E. (1986). The rapid "tonic" and the delayed "induction" components of the prolactin-induced activation of tuberoinfundibular dopaminergic neurons following the systemic administration of prolactin. *Neuroendocrinology, 43*, 291–299.

Döhler, K -D , von zur Mühlen, A , & Döhler, U (1977) Pituitary luteinizing hormone (LH), follicle stimulating hormone (FSH) and prolactin from birth to puberty in female and male rats. *Acta Endocrinologica, 85*, 718–728.

Döhler, K.-D., & Wuttke, W. (1975). Changes with age in levels of serum gonadotropins, prolactin and gonadal steroids in prepubertal male and female rats. *Endocrinology, 97*, 898–907.

Dubois-Dauphin, M., Armstrong, W. E., Tribollet, E., & Dreifus, J. J. (1985a). Somatosensory systems and the milk-ejection reflex in the rat: I. Lesions of the mesencephalic lateral tegmentum disrupt the reflex and damage mesencephalic somatosensory connections. *Neuroscience, 15*, 1111–1129.

Dubois-Dauphin, M., Armstrong, W. E., Tribollet, E., & Dreifus, J. J. (1985b). Somatosensory systems and the milk-ejection reflex in the rat: II. The effects of lesions in the ventroposterior thalamic complex, dorsal columns and lateral cervical nucleus-dorsolateral funiculus. *Neuroscience, 15*, 1131–1140.

Fagin, K. D., & Neill, J. D. (1981). The effect of dopamine in thyrotropin-releasing hormone–induced prolactin secretion in vitro. *Endocrinology, 109*, 1835–1840.

Foreman, M. M., & Porter, J. C. (1981). Prolactin augmentation of dopamine and norepinephrine release from superfused medial-basal hypothalamus fragments. *Endocrinology, 108*, 800–804.

Glydon, R. S. (1957). The development of the blood supply of the pituitary in the albino rat, with special reference to the portal vessels. *Journal of Anatomy, 91*, 237–244.

Grosvenor, C. E. (1965). Evidence that exteroceptive stimuli can release prolactin from the pituitary gland of the lactating rat. *Endocrinology, 76*, 340–342.

Grosvenor, C. E., Goodman, G. T., & Mena, F. (1984). Control of the multiphase secretion of prolactin in the lactating rat. In F. Mena & C. Valverde (Eds), *Prolactin secretion: A multidisciplinary approach* (pp. 275–284). Orlando, FL: Academic Press.

Grosvenor, C. E., Maiweg, H., & Mena, F. (1970). A study of factors involved in the development of the exteroceptive release of prolactin in the lactating rat. *Hormones and Behavior, 1*, 111–120.

Grosvenor, C. E., & Mena, F. (1969). Failure of self-licking of nipples to alter pituitary prolactin concentration in lactating rats. *Hormones and Behavior, 1*, 85–91.

Grosvenor, C. E., & Mena, F. (1971). Effect of suckling upon the secretion and release of prolactin from the pituitary gland of the lactating rat. *Journal of Animal Science, 32* (Suppl. 1), 115–136.

Grosvenor, C. E., & Mena, F. (1980). Evidence that thyrotropin-releasing hormone and a hypothalamic prolactin-releasing factor may function in the release of prolactin in the lactating rat. *Endocrinology, 107*, 863–868.

Grosvenor, C. E., & Mena, F. (1982). Regulating mechanisms for oxytocin and prolactin secretion during lactation. *Neuroendocrine Perspectives, 1*, 69–102.

Grosvenor, C. E., Mena, F., Dhariwal, A. P. S., & McCann. S. M. (1967). Reduction of milk secretion by prolactin-inhibiting factor: Further evidence that exteroceptive stimuli can release pituitary prolactin in rats. *Endocrinology, 81*, 1021–1028.

Grosvenor, C. E., Mena, F., & Whitworth, N. S. (1979). Ether releases large amounts of prolactin from rat pituitaries previously depleted by short-term suckling. *Endocrinology, 105*, 884–887.

Grosvenor, C. E., Mena, F., & Whitworth, N. S. (1980). Evidence that the dopaminergic-PIF mechanism regulates the depletion-transformation phase and not the release phase of prolactin secretion during suckling in the rat. *Endocrinology, 106*, 481–485.

Grosvenor, C. E., Shyr, S. W., & Crowley, W. R. (1986). Effect of neonatal prolactin deficiency on prepubertal tuberoinfundibular and tuberohypophyseal dopaminergic neuronal activity. *Endocrinologica Experimentalis, 20*, 223–228.

Grosvenor, C. E., & Whitworth, N. S. (1976). Incorporation of rat prolactin into rat milk *in vivo* and *in vitro*. *Journal of Endocrinology, 70*, 1–9.

Grosvenor, C. E., Whitworth, N. S., & Mena, F. (1981). Evidence that the depletion and release phases of prolactin secretion in the lactating rat have different thresholds in response to exteroceptive stimulation from rat pups. *Endocrinology, 108*, 820–824.

Harris, G. W. (1964). Sex hormones, brain development and brain function. *Endocrinology, 75*, 627–647.

Honma, K., Höhn, K. -G., & Wuttke, W. (1979). Involvement of catecholamines in eliciting LH peaks in 15-day-old female rats: Effects of treatment with prolactin. *Brain Research, 179*, 271–279.

Hyyppa, M. (1969). A histochemical study of the primary catecholamines in the hypothalamic neurons of the rat in relation to the ontogenetic and sexual differentiation. *Cell and Tissue Research* (Z. Zellforsch.), *98*, 550–560.

Johke, T. (1969). Prolactin release in response to milking stimulus in the cow and goat estimated by radioimmunoassay. *Endocrinologia Japonica, 16*, 179–185.

Juss, T. S., & Wakerly, J. B. (1981). Mesencephalic areas controlling pulsatile oxytocin release in the suckled rat. *Journal of Endocrinology, 91*, 233–244.

Koldovsky, O. (1980). Minireview. Hormones in milk. *Life Sciences, 26*, 1833–1836.

Lichtensteiger, W., Schlumpf, M., & Davis, M. D. (1983). *Monographs on Neurology and Science, 9*, 213–224.

Loizou, L. A. (1971). The postnatal development of monoamine-containing structures in the hypothalamo-hypophyseal system of the albino rat. *Cell and Tissue Research, 114*, 234–252.

Loizou, L. A. (1972). The postnatal ontogeny of monoamine-containing neurons in the central nervous system of the albino rat. *Brain Research, 40*, 395–418.

MacLeod, R. M. (1976). Regulation of prolactin secretion. In L. Martini & W. F. Ganong (Eds.), *Frontiers in neuroendocrinology* (pp. 169–194). New York: Raven Press.

MacLusky, N. J. & Naftolin, F. (1981). Sexual differentiation of the central nervous system. *Science, 211*, 1294–1303.

McEwen, B. S. (1981). Neural gonadal steroid actions. *Science, 211*, 1303–1311.

McMurtry, J. P., & Anderson, R. R. (1971). Prevention of self-licking on mammary gland development in pregnant rats. *Proceedings of the Society for Experimental Biology and Medicine, 137*, 354–356.

McMurtry, J. P., & Malven, P. V. (1974a). Experimental alterations of protein levels in goat milk and blood plasma. *Endocrinology, 95*, 559–564.

McMurtry, J. P., & Malven, P. V. (1974b). Radioimmunoassay for endogenous and exoge-nous prolactin in milk of rats. *Journal of Endocrinology, 61,* 211–217.
McMurtry, J. P., Malven, P. V., Arave, C. W., Erb, R. E., & Harrington, R. B. (1975). Environmental and lactation variables affecting prolactin concentrations in bovine milk. *Journal of Dairy Science, 58,* 181–189.
Mena, F., Enjalbert, A., Carbonell, L., Priam, M., & Kordon, C. (1976). Effect of suckling on plasma prolactin and hypothalamic monoamine levels in the rat. *Endocrinology, 99,* 445–451.
Mena, F., & Grosvenor C. E. (1971). Release of prolactin by exteroceptive stimulation: Sen-sory stimuli involved. *Hormones and Behavior, 2,* 107–116.
Mena, F., & Grosvenor C. E. (1972). Effect of suckling and exteroceptive stimulation upon prolactin release in the rat during late lactation. *Journal of Endocrinology, 52,* 11–22.
Mena, F., Martinez-Escalera, G., Clapp, C., Aguayo, D., Forray, C., & Grosvenor C. E. (1982). A solubility shift occurs during depletion-transformation of prolactin within the lactating rat pituitary. *Endocrinology, 111,* 1086–1091.
Mena, C., Martinez-Escalera, G., Clapp, C., & Grosvenor C. E. (1984). In vivo and in vitro secretion of prolactin by lactating rat adenohypophyses as a function of intracellular age. *Journal of Endocrinology, 101,* 27–32.
Moltz, H., Levine, R., & Leon, M. (1969). Prolactin in the postpartum rat: Synthesis and release in the absence of suckling stimulation. *Science, 163,* 1083–1084.
Moore, K. E. (1987). Interactions between prolactin and dopaminergic neurons. *Biology of Reproduction, 36,* 47–58.
Moore, K. E., & Demarest, K. T. (1982). Tuberoinfundibular and tuberohypophyseal dopa-minergic neurons. In W. F. Ganong & L. Martini (Eds.), *Frontiers in neuroendo-crinology* (Vol. 7, pp. 161–189). New York: Raven Press.
Morris, B., & Morris, R. (1974). The absorption of ^{125}I-labeled immunoglobulin G by differ-ent regions of the gut in young rats. *Journal of Physiology, 241,* 761–770.
Neill, J. D. (1972). Sexual differences in the hypothalamic regulation of prolactin secretion. *Endocrinology, 90,* 1154–1159.
Ojeda, S. R., & McCann, S. M. (1974). Development of dopaminergic and estrogenic control of prolactin release in the female rat. *Endocrinology, 95,* 1499–1505.
Patel, M. D. (1936). The physiology of the formation of pigeons' milk. *Physiological Zool-ogy, 9,* 129–152.
Perkins, N. A., & Westfall, T. C. (1978). The effect of prolactin on dopamine release from rat striatum and medial basal hypothalamus. *Neuroscience, 3,* 59–63.
Perkins, N. A., Westfall, T. C., Paul, C. V., MacLeod, R. M., & Rogol, A. D. (1979). Effect of prolactin on dopamine synthesis in medial basal hypothalamus: Evidence for a short loop feedback. *Brain Research, 160,* 431–444.
Plotsky, P. M., & Neill, J. D. (1982a). The decrease in hypothalamic dopamine secretion induced by suckling: Comparison of voltammetric and radioisotopic methods of mea-surement. *Endocrinology, 110,* 691–697.
Plotsky, P. M., & Neill, J. D. (1982b). Interactions of dopamine and thyrotropin-releasing hormone in the regulation of prolactin release in lactating rats. *Endocrinology, 111,* 168–173.
Pope, G. S., & Swinburne, J. K. (1980). Reviews in the progress of diary science: Hormones in milk: Their physiological significance and value as diagnostic aids. *Journal of Dairy Research, 47,* 427–449.
Roth, L. L., & Rosenblatt, J. S. (1968). Self-licking and mammary development during preg-nancy in the rat. *Journal of Endocrinology, 42,* 363–378.
Selmanoff, M. (1981). The lateral and medial median eminence: Distribution of dopamine,

norepinephrine, and luteinizing hormone-releasing hormone and the effect of prolactin on catecholamine turnover. *Endocrinology, 108,* 1716–1721.

Selmanoff, M., & Gregerson, K. A. (1985). Suckling decreases dopamine turnover in both medial and lateral aspects of the median eminence in the rat. *Neuroscience Letters, 57,* 25–30.

Selmanoff, M., & Wise, P. M. (1981). Decreased dopamine turnover in the median eminence in response to suckling in the lactating rat. *Brain Research, 212,* 101–115.

Shah, G. V., Shyr, S. W., Grosvenor C. E., & Crowley, W. R. (1988). Hyperprolactinemia after neonatal prolactin (PRL) deficiency in rats: Evidence for altered anterior pituitary regulation of PRL secretion. *Endocrinology, 122,* 1883–1890.

Shyr, S. W. Crowley, W. R., & Grosvenor C. E., (1986). Effect of neonatal prolactin deficiency in prepubertal tuberoinfundibular and tuberohypophyseal dopaminergic neuronal activity. *Endocrinology, 119,* 1217–1221.

Shyr, S. W., & Grosvenor C. E. (1985). Histidyl-proline-diketopiperazine (cyclo-his-pro) inhibits the suckling and domperidone-induced transformation of prolactin in the lactating rat. *Endocrinology, 117,* 2170–2175.

Smalstig, E. B., Sawyer, B. D., & Clemens, J. A. (1974). Inhibition of rat prolactin release by apomorphine in vivo and in vitro. *Endocrinology, 95,* 123–130.

Smith, G. C., & Simpson, W. R. (1970). Monoamine fluorescence in the median eminence of foetal, neonatal and adult rats. *Cell and Tissue Research, 104,* 541–556.

Smith (White), S. S., & Ojeda, S. R. (1984). Maternal modulation of infantile ovarian development and available ovarian luteinizing hormone-releasing hormone (LHRH) receptors via milk LHRH. *Endocrinology, 115,* 1973–1983.

Stern, J. M., & Siegel, H. I. (1978). Prolactin release in lactating primiparous and multiparous thelectomized and maternal virgin rats exposed to pup stimuli. *Biology of Reproduction, 19,* 177–182.

Terkel, J., Damassa, D. A., & Sawyer, C. H. (1979). Ultrasonic vocalization of infant rats stimulate prolactin release in lactating females. *Hormones and Behavior, 12,* 95–102.

Tucker, H. A., Convey, E. M., & Koprowski, J. A. Milking-induced release of endogenous prolactin in cows infused with exogenous prolactin. *Proceedings of the Society for Experimental Biology and Medicine, 142,* 72–75.

Van der Lee, S., & Boot, L. M. (1955). Spontaneous pseudopregnancy in mice. *Acta Physiologica et Pharmacologica Neerlandia, 4,* 442–444.

Van der Lee, S., & Boot, L. M. (1956). Spontaneous pseudopregnancy in mice: II. *Acta Physiologica et Pharmacologica Neerlandia, 5,* 213–215.

Weiner, R. I., and Ganong, W. F. (1978). Role of brain monoamines and histamine in regulation of anterior pituitary secretion. *Physiological Reviews, 58,* 905–976.

Whitten, W. K. (1966). Pheromones and mammalian reproduction. *Advances in Reproductive Physiology, 1,* 155–177.

Whitworth, N. S., & Grosvenor C. E. (1978). Transfer of milk prolactin to the plasma of neonatal rats by intestinal absorption. *Journal of Endocrinology, 79,* 191–199.

Witschi, E. (1935). Seasonal sex characteristics in birds and their hormonal control. *Wilson Bulletin, 47,* 177–188.

Wolinska, E., Polkowska, J., & Domanski, E. (1977). The hypothalamic centers involved in the control of production and release of prolactin in sheep. *Journal of Endocrinology, 73,* 21–29.

Zarrow, M. X., Johnson, N. P., Denenberg, V. H., & Bryant, L. P. (1973). Maintenance of lactational diestrum in the postpartum rat through tactile stimulation in the absence of suckling. *Neuroendocrinology, 11,* 150–155.

IV

BIOBEHAVIORAL CORRELATES OF PARENTING

The six chapters of this part focus on the experiential and sensory factors that influence and help regulate parenting behavior in the mammals that have been studied over the organism's life span. Among the issues highlighted are the role of males in parenting, the biobehavioral variables that impringe on organisms prenatally and during adolescence to affect the expression of parenting, and the bases for parental recognition of offspring.

Craig Kinsley's chapter reviews the literature on the influences of gonadal hormones on parenting in both male and female rodents. The data on the infanticide (killing of young) observed in various species of mice, and the interpretation of these data, are summarized. Kinsley provides an overview of prenatal factors that may influence parental behavior when rodents become adults. Some of the variability, he asserts, can be attributed to the intrauterine position of the fetus and the resultant hormonal milieu in which development during gestation occurs. The chapter also describes how stress (a regimen of heat and restraint) on the pregnant female can affect the subsequent parenting behavior of male and female rats and mice born to dams so treated. The findings provide a complex picture of the endogenous and exogenous environmental factors that can significantly alter the biological and behavioral mechanisms thought to underlie parenting behavior in rodents.

In the next chapter, Susan Brunelli and Myron Hofer examine the ontogeny of parental behavior in rats. The authors review data that suggest that exposure to pups can induce parental behaviors in juvenile rats. Such behaviors differ from the mature patterns observed in dams that have given birth and are sexually dimorphic between juvenile males and females. Also stressed is the strong effect of the prenatal hormonal milieu (see chapter 17) on hormonal and stimulus factors that may regulate the expression of maternal behavior. The authors' own studies indicate that the expression of parental behaviors in female rats begins to differentiate as early as the prepubertal phase. Brunelli and Hofer raise the interesting hypothesis that parental behaviors are linked, temporally and perhaps functionally, to play behaviors. They suggest that the development of maternal behavior patterns is linked to and shaped by other behavioral patterns, which are themselves dependent on the age, sex, hormonal status, and experience of rodents.

The following chapter, by Michael Leon, Robert Coopersmith, Laura Beasley, and Regina Sullivan, reviews the literature on the dependence of mother–infant interactions on thermal factors that may serve to regulate parental behavior of the female Norway rat. As the authors demonstrate, Norway rat mothers are susceptible to acute hyperthermia during contact with their young because of their chronically high heat load. The data on endocrine control of this maternal heat load are summarized. The authors also present evidence on other factors, such as diet, that influence the heat load and the way it varies in conjunction with diurnal rhythms. In addition, recent experiments are described that demonstrate how contact bouts between mother and pups are initiated, maintained, and curtailed in relation to thermal factors under the control of the dam and the litter, respectively. These experiments strongly suggest that one aspect of maternal behavior (covering the litter during feeding) is regulated by biobehavioral factors that alter maternal brain temperature.

In the next chapter, by Jeffrey Alberts and David Gubernick, parental systems of the Norway rat *(Rattus norvegicus)* and the California mouse *(Peromyscus californicus)* are compared. The rat's system is characterized by the mother–pup dyad, while that of the mouse involves a triadic relationship (mother–pup–father). The dynamics of parental care are examined in terms of the cyclic and temporal characteristics of the animals' interactions. This comparison of the two rodent systems reveals that the male mouse spends as much time as the female in contact with the young, and this pattern is maintained throughout the rearing period. The authors describe the symbiotic exchange of resources (e.g., water, elecrolytes, and thermal energy) that benefits both parents and offspring. They also describe findings that indicate that in the triadic system illustrated by *P. californicus,* males as well as females show a readiness for parenting just before parturition; males exhibit parental behavior some 10 days prior to the birth of their offspring compared with their nonreproductive conspecifics. Findings on the hormonal bases of parental behavior, as well as on neuronal mechanisms that subsume this behavior, are summarized. The authors show that *P. californicus* exhibits a sexual dimorphism in the volume of the MPOA, with that of virgin males being larger than that of virgin females (see chapters 5 and 12). When females become mothers, the volume of their MPOA approximates that of virgin females. In contrast, the MPOA volume of males, when they become fathers, decreases, approximating that of females. These findings await functional studies to determine the role of the MPOA in the parenting behavior of *Peromyscus* males.

The next chapter, by Warren Holmes, focuses on the question of why adults most often provide care and sustenance to their own offspring, or those closely related, rather than to unrelated young. The author provides a comparative overview of the mechanisms by which kin recognition between parent and offspring takes place. There are three general requirements for behavioral recognition to occur: (1) unique phenotypic traits, (2) specialized sensory apparatus, and (3) decision rules (often based on social interaction experience) that allow kin-differential behavior. Holmes provides an evolutionary framework (see chapter 3) for parent–offspring recognition and summarizes the literature on kin signatures and the errors made in recognizing such markers. The empirical findings on sex differences in parent–offspring recognition are outlined. The chapter concludes with a review of the findings from the author's extensive research on the Beldings ground squirrel that illustrates the learning and sensory mechanisms employed by this species to accomplish kin recognition.

The last chapter in this part, by Michael Yogman, focuses on research suggesting that the father's role is determined more by contextual than by biological factors. Specifically, Yogman examines (1) cross-species and ethnographic comparisons of data on males designed to demonstrate the range and environmental influences on their parenting; (2) evidence on the male's competence in caring for human infants; (3) differences in male and female parenting styles to uncover unique parent–infant patterns exhibited by males; and (4) antecedents and consequences of male involvement with infants. From this analysis of the cross-species and cross-cultural literature on the topic, Yogman concludes that fathers play a unique and necessary role in raising offspring. Moreover, he asserts that during the prenatal, perinatal, and infancy periods, fathers both are skilled at and have the capacity for establishing effective and meaningful relationships with their children. In many respects, their interaction pattern with their offspring closely resembles the mother–child relationship. The author provides an overview of cross-species evidence concerning differences between paternal and maternal interaction styles with offspring. He concludes that the father is predominantly responsible for initiating and engaging in play with the baby. Furthermore, Yogman asserts that the father's relative distancing and approach–withdrawal relationship to his infant may complement the maternal pattern of continual availability. These two patterns could, he suggests, expose the infant to behaviors that, respectively, arouse and soothe the infant.

17

Prenatal and Postnatal Influences on Parental Behavior in Rodents

CRAIG H. KINSLEY

The developmental events that shape the nascent organism can affect its expression of parental behavior as an adult. This chapter discusses these factors and how they contribute to the display of parental behavior in male and female rodents during postnatal and prenatal development.

The chapter is divided in three sections. In the first section, work on sex differences in parental care, including infanticide, is discussed. In the second section, data on the perinatal effects of hormones and the modification of parental care arising from intrauterine effects are reviewed. In the third section, an overview of the effects of prenatal stress on parental care is provided. It should be emphasized that parental care is not an all-or-none phenomenon. Rather, it is exhibited to varying degrees by both males and females and it can, depending on the perinatal hormonal experiences of the animal, be modified.

SEX DIFFERENCES IN PARENTAL BEHAVIOR: AN OVERVIEW OF NORMATIVE RESPONSES

Sex differences in parental behavior exist in a number of rodent species. Females of most species respond to young more promptly than males and exhibit a higher incidence of the various components of pup-directed behaviors. Among rodents, male rats are capable of exhibiting parental behavior following prolonged exposure to young, with latencies similar to those of females (Rosenblatt, 1967). Males, however, are more likely to commit infanticide, injure the pups, or fail to care for them on initial exposure (Jakubowski & Terkel, 1985a, 1985b; Rosenberg, Denenberg, Zarrow, & Frank, 1971). Furthermore, there exist sex differences in the various components of parental responding. Males do not build nests, lick, crouch over, or otherwise provide for the care of young as effectively as females (McCullough, Quadagno, & Goldman, 1974; Quadagno, McCullough, Ho, & Spevak, 1973; Rosenberg & Herrenkohl, 1976). The different patterns of parental responding shown by the sexes are due in part to sex differences in hormone exposure during early development. Again, the differences are not absolute, as variation is apparent in the responses of the two sexes.

During the past decade, the phenomenon of infanticide, the killing of young, has been studied more extensively in the context of parental behavior. Once believed to be an aberrant response, the killing of young under certain conditions is now recognized as part of some animals' behavioral repertoire. Infanticide contributes to reproductive fitness and is considered by some to be the complement of parental care (Hausfater & Hrdy, 1984). Infanticidal behavior is subject to endocrinological and social influences at various stages of an animal's life span. The influence of factors functioning during the perinatal period that can affect an animal's tendency to commit infanticide is especially profound. More recently, the examination of infanticide has provided fruitful hypotheses and extensive debate about the sociobiological role of kin selection in population dynamics. In rodents, infanticide is displayed primarily by the male. In contrast, virgin females rarely kill young; instead, they either are indifferent to or care for them. As an example of the normality of infanticide at particular times, newly parturient female hamsters engage in infanticide by selectively culling their litters to achieve a more manageable litter size (Day & Galef, 1977). Work with rats indicates that males are more likely to commit infanticide than females. This difference appears to arise in part from the presence of circulating testosterone (T) (Rosenberg et al., 1971).

PERINATAL FACTORS AFFECTING PARENTAL BEHAVIOR

The Role of Exposure to Gonadal Hormones

The developing organism is a type of tabula rasa on which both endocrinological and other chemical forces may exert strong influences on the development of parental responsiveness. The extent to which parental behavior differences are influenced by gonadal hormones during postnatal life was first examined by Quadagno and Rockwell (1972). Neonatally castrated male rats were compared with intact males, and normal with neonatally androgenized females, for parental behavior. Neonatally gonadectomized males and intact females exhibited similar patterns of rapid onset of maternal behavior. Intact males and perinatally androgenized females, on the contrary, exhibited reduced levels of maternal care. Therefore, it was concluded that postnatal exposure to androgens reduces parental responsiveness. Bridges, Zarrow, and Denenberg (1973) extended these findings by examining parental responsiveness following the administration of a 21-day-duration female-hormone regimen in adulthood (concurrent estradiol [E_2] and progesterone [P] treatment). The presence of postnatally administered androgens inhibited the parental responsiveness of neonatally intact males and androgenized females as measured by pup retrieval in a novel environment (a T-maze), not in the home cage. The data of Bridges et al. suggest that the effects of postnatal T on the expression of parental behavior may become apparent only when tested in situations that depart from simple assessments of parental responsiveness (such as home-cage retrieval behavior). In general, then, early exposure to androgens appears to suppress the display of parental behavior in rodents.

As mentioned above, male rats are more likely than females to neglect or kill pups on initial exposure. This predisposition to not exhibit parental responsiveness is also due to perinatal exposure to androgens. Since the male is endocrinologically active and sexual differentiation begins prenatally (see chapter 5), to what extent do prenatal androgens contribute to this behavior? Quadagno, Ho, and Spevak (1973) treated female rats with 2 mg testosterone propionate (TP) in late pregnancy (days 16–20). Although the genitalia of these rats' female pups were masculinized, there was no observable difference in parental responsiveness of these females as adults. In another study, Rosenberg and Sherman (1975) investigated the extent to which pup killing could be modified by prenatal exposure to androgens. Pregnant female rats were injected with TP on gestation days 16–20. On postnatal days 3, 5, 7, and 9, the female offspring were injected with TP. At 40 days of age, the females were ovariectomized and implanted with T-containing capsules. A greater percentage of the females exposed to T prenatally and postnatally were infanticidal compared with vehicle-treated controls. However, the percentage of killers did not differ between females treated prenatally and postnatally and those treated postnatally only (93% and 80%, respectively). While it is unclear to what extent prenatal treatment alone may influence pup killing, it is clear that exposure to androgens neonatally facilitates the expression of infanticide in adult rats.

More recently, Ichikawa and Fujii (1982) investigated the extent to which prenatally administered androgen affected the parental responsiveness of female rats. In this study, the fetuses were directly injected with 0.25 mg TP on day 19 or 21 of gestation. Testing for parental behavior occurred in adulthood. Reductions in the percentage of animals retrieving, nursing and crouching in the nest, and licking were found in the prenatally TP-treated females relative to the controls and most postnatally T-treated groups. Moreover, the latencies to exhibit these parental behaviors were longer in androgen-exposed animals. The authors also reported unpublished observations that male rats castrated at birth and tested as adults exhibit enhanced parental responsiveness to newborns. Based on these findings, Ichikawa and Fujii concluded that prenatal treatment with or exposure to androgens "influences the sexual differentiation of the CNS controlling maternal behavior . . . and that the critical androgen-sensitive period of the CNS for maternal behavior exists in the fetal stage of late pregnancy" (p. 230). However, since injected controls were not run for the prenatal treatment by Ichikawa and Fujii, further work in this area is necessary before these conclusions about "androgen-sensitive" maternal-behavior structures can be made.

Anderson, Zarrow, and Denenberg (1970) investigated the prenatal organizing effects of androgens on maternal nest-building behavior in rabbits. Pregnant rabbits were treated with TP during gestation, and in adulthood their female offspring were ovariectomized and treated with estradiol and progesterone to stimulate nest-building behavior. Only 11% of the TP-treated offspring responded with nest building, compared with 90% of the controls. Since none of the male offspring of control mothers engaged in maternal nest building, the authors concluded that androgens masculinize this form of parental behavior by inducing insensitivity to the nest-building-promoting qualities of estradiol and progesterone.

In outbred mice (Rockland-Swiss [R-S] or CF-1), approximately 35–50% of males commit infanticide, compared with 5–10% of females (Svare, Broida, Kinsley, & Mann, 1984; Svare, Kinsley, Mann, & Broida, 1984; vom Saal, 1984). The sex difference apparent in this behavior is also mediated by exposure to androgens during both adulthood and the perinatal period. Infanticide in males becomes apparent at 30–40 days of age (Gandelman, 1973), coincident with pubertal surges of androgens (Svare, Bartke, & Macrides, 1978). Gonadectomized adult males show low levels of infanticide; treatment with TP reinstates the behavior (Gandelman & vom Saal, 1975).

It appears that exposure to androgens during early development in mice results in a pattern of parental behavior different from that expected. For example, evidence presented over the past 10 years demonstrates that early exposure to androgens in mice suppresses infanticide, a response that departs from the conventional behavioral effects of perinatal exposure to androgens, which would be expected to facilitate the expression of masculine behaviors. For example, administration of androgens to neonatal males attenuates, whereas the mere act of gonadectomy enhances, pup killing. If the 50–65% of male mice that do not commit infanticide are treated with androgens in adulthood, no increase in infanticide follows (Gandelman & vom Saal, 1975). Yet, 100% of females given chronic injections of T in adulthood can be induced to commit infanticide (Svare, 1979). Because females, unlike males, are not exposed to high or moderate levels of androgens perinatally, they are not subject to the suppressive effects of androgens on infanticide. This may account for their greater responsivity to the infanticide-promoting qualities of exogenously administered androgens in adulthood. It therefore is believed that in mice, perinatal exposure to androgens affords a type of "immunization" against the infanticide-promoting effects of androgen exposure in adulthood. Exactly how this effect is brought about remains unclear.

The exhibition of infanticide is also under the influence of genotypic factors. In the outbred R-S mouse, as stated above, approximately 35–50% of males commit infanticide, compared with 5–10% of females. The percentage of infanticidal versus noninfanticidal animals changes dramatically in the case of two inbred mouse strains: C57BL/6J (C57) and DBA/2J (DBA). In the former, 75–85% of males will kill newborns, compared with 20–25% of DBA males, percentages that persist in the face of removal of gonadal steroids (Svare, Broida, et al., 1984; Svare, Kinsley, et al., 1984). Thirty to 40% of C57 females, compared with 0–5% of DBA females, will kill young (Svare, Broida, et al., 1984; Svare, Kinsley, et al., 1984). Interestingly, prior to 40 days of age, a significantly greater proportion of C57 females will kill pups than after that time (Mann, Kinsley, Broida, & Svare, 1983). This is presumably due to some suppressive effect on infanticide provided by the ovaries, as gonadectomy will facilitate killing in older C57 females (Mann et al., 1983).

Another component of parental behavior, postpartum aggression (PPA), displayed by females is influenced by prenatal and postnatal factors as well as by strain or genetic factors. Whereas female C57 mice show low levels of PPA, DBA females are highly aggressive (Broida & Svare, 1982). Initially, the findings for both the males and the females of these two strains were believed to result from differential androgen exposure prenatally or postnatally (i.e., low levels of androgen exposure in C57 mice vs. high levels in DBA mice). The discrepancy between the outbred and the inbred mice, as well as between C57 and DBA mice, is now believed to be due to a combination of androgen insensitivity and/or intrauterine hormonal factors. Furthermore, recent work suggests that that the increased sensitivity to adrenal androgens of neonatally gonadectomized C57 males may account for their high incidence of infanticide as adults (Kinsley & Svare, 1987b).

The findings discussed in the first section of this chapter relate to sex differences in the display of parental behavior. In general, perinatal exposure to androgen suppresses parental responding in rats. In mice, however, the reverse appears to be true; that is, early exposure to androgens appears to facilitate parental responding (i.e., reduce infanticide and promote retrieval, crouching, and PPA). The perinatal actions of hormones, then, exert strong, long-lasting effects on this behavior. As will be shown, these effects of hormones can be extended to the early prenatal period. For instance, fetal position within the uterus relative to same- andopposite-sex fetuses affects the display and intensity of various parental behaviors. The next section describes this phenomenon as it relates to pup-directed and maternal aggressive behaviors.

Intrauterine Position

Not all female rodents are immediately responsive to young on initial exposure, nor are all males unresponsive to young or infanticidal. There is substantial variation in the exhibition of parental behavior (as pointed out earlier for male mice, upward of 50% of outbred males fail to show pup killing, whereas 5–10% of females kill pups). Since the display of parental behavior can be influenced by early exposure to steroid hormones, could there exist some contribution to this behavior, other than that provided by the individual itself, that exerts a similar effect on the development of parental behavior? In polytocous (i.e., having many offspring at any one time) species, male and female fetuses are randomly juxtaposed in the uterus. Because of this random event in development, male and female fetuses can reside adjacent to the same- or opposite-sex fetuses. Hormonal secretions leaching across from one fetus to the next, coincident with the onset and timing of ontogeny of sensitive neural structures, can affect the organization of parental behavior during this period of sexual differentiation. An unresolved issue is whether contiguity or position in the uterus relative to either the vagina or the ovary accounts for the uterine influences on behavior (Meisel & Ward, 1981). In this chapter, the intrauterine classification system of vom Saal (1981) and of Clemens, Gladue, and Coniglio (1978) is used, with the male as the referent: that is, animals flanked by two males are referred to as "2M"; those between two females are referred to as "0M"; and those flanked by one male and one female are referred to as "1M." For example, a 2M female is one that was between two males in utero. This phenomenon, referred to as "intrauterine position" (IUP), has provided some novel ways of looking at sexual differentiation of parental and other behaviors.

To generate IUP animals, timed-mated females are caesarean sectioned 12–24 hr prior to delivery, and their litters are removed, cleaned and sexed, stimulated, and palpated by hand (pup survival rates are enhanced by these procedures). IUP is noted, and the animals are cross-fostered to untreated, lactating dams. IUP is, as stated above, a random event, with the percentages of each of 0M, 2M, and 1M animals produced approximating 30–55%, 15–23%, and 50%, respectively, of the total number of young produced, depending on litter size (ranging from 12–20 per litter; vom Saal, 1981).

Intrauterine Position and Parental Behavior in Mice

IUP has been identified as a major source of the variability inherent in a number of sexually dimorphic behaviors (for reviews, see vom Saal, 1981, 1983b, 1983c, 1984), including parental behavior. Studies of females derived from the three IUPs have been directed at the elucidation of the early hormonal actions on a variety of female-typical behaviors, morphology, and physiological responses. Included among these responses are interfemale aggression, sexual attractiveness, anogenital distance (AGD), body weight, and locomotor activity (Gandelman, vom Saal, & Reinisch, 1977; Kinsley, Miele, Konen, Ghiraldi, & Svare, 1986; Kinsley, Miele, Konen-Wagner, et al., 1986; vom Saal & Bronson, 1980). More recently, attention has been directed at the influence of IUP on the parental responses exhibited by female mice.

In these studies, adult R-S female mice from the 0M and 2M IUPs were compared on five aspects of parental behavior and physiology (Kinsley, Konen, Miele, Ghiraldi, & Svare, 1986). Given the endocrine profiles reported for the two groups (high amniotic T levels in the 2M females, high amniotic E_2 levels in the 0M females [vom Saal & Bronson, 1980; vom Saal, Grant, McMullen, & Laves, 1983]), one would expect that parental behavior (i.e., infanticidal behavior) in 0M and 2M females would parallel that of 0M and 2M males, respectively (2M males responsive to young, 0M males infanticidal; see below). However, when exposed to newborn pups and examined for spontaneous parental behavior, 0M and 2M females showed no differences in responses (Kinsley, Konen, et al., 1986).

When separate groups of 0M and 2M females were assessed for aggressive behavior during pregnancy (referred to as "pregnancy-induced aggression" [PIA]), 2M females were found to be much more aggressive than 0M females, engaging in significantly greater numbers of lunges (rapidly thrusting toward an intruder male, failing contact) and attacks (biting and grappling with the male) (Kinsley, Konen, et al., 1986). Furthermore, 2M females fought on a greater number of test days than did 0M females. Intrauterine contiguity, then, apparently influences this female-typical aggressive behavior.

The 0M and 2M females used in this study of PIA were allowed to deliver their young, whereupon the litters were culled to five pups. On postpartum days 6–8, these females were tested for PPA. The 2M females again displayed higher levels of aggression relative to 0M females, showing a larger number of both lunges and attacks. Thus 2M females exhibit significantly higher levels of aggression than do 0M females during pregnancy and lactation.

The reason 2M females are much more aggressive than 0M females during pregnancy and lactation may be due to the higher titers of androgens to which 2M females are exposed prenatally. Vom Saal and Bronson (1980) have reported that 2M female mice possess higher levels of fetal and amniotic T than do 0M females. Experimental support for these correlative findings comes from the work of Mann and Svare (1983), who injected pregnant mice with low, nonsterilizing doses of TP and raised the female offspring to adulthood, when they were timed-mated and tested for PPA in early lactation. Prenatal TP was found to elevate aggression significantly when these mice were lactating. The mechanism underlying IUP effects in the 2M female is thought to involve the interamniotic transfer of steroid hormones of fetal origin from adjacent male fetuses (notwithstanding the tenable demonstration of Meisel & Ward, 1981).

These effects of IUP on maternal aggression may therefore indicate that female mice are sensitive to the early organizing effects of fetal androgens. Furthermore, the work indicates that females are sensitive to a treatment known to affect male aggression. In total, it appears that aggression may possess homologous elements in both sexes that are dependent on the sex-typical expression of the aggressive behavior.

IUP also affects parental behavior in male mice (vom Saal, 1983a). 2M male mice exposed to newborn pups are more parental than are 0M males, a significant proportion of the latter engaging in pup killing. Males from the 1M category are intermediate in response. Because there is no difference between 0M and 2M males in prenatal T (vom Saal et al., 1983), it has been suggested that the exposure of 0M males to elevated levels of prenatal E_2 may contribute to the decrease in parental responsiveness observed in them.

It should also be pointed out that IUP effects are modulated by genotype. As mentioned earlier, two inbred mouse strains (C57 and DBA) differ markedly in their infanticidal tendencies, with C57 mice exhibiting high levels and DBA mice low levels of pup killing, and their PPA, with DBA females exhibiting higher levels of aggression than C57 females. Over several years of experimentation and numerous manipulations with these two strains, consistently similar percentages of between- and within-group variability in infanticidal behavior and PPA have been observed: for the C57 males, 70–75% commit infanticide, whereas for DBA males, 10–20% do; 50–70% of female DBAs show PPA, whereas 10–20% of C57 females do (Svare, Broida, et al., 1984; see chapter 7). From where do these consistent percentages come? Since inbred mouse strains possess nearly 100% of their genes in common, the variability most likely is of environmental origin; IUP may account for the variability observed within the inbreds. In order to address this question, 0M and 2M males and females from the C57 and DBA inbred strains were compared for parental and postpartum aggressive behavior. In addition, separate groups of 0M and 2M inbred males and females were weighed, and AGDs were measured as indicators of perinatal androgenization. Finally, in order to compare the efficacy of the various classification systems (i.e., contiguity or the "upstream–downstream" of Meisel and Ward, 1981), all the tested animals were categorized in several ways (e.g., contiguity, number of males above or below in a particular uterine horn, number of females above or below, absolute position within a uterine horn, adjacent to either the ovary or the vagina, and position relative to the number of fetuses within a uterine horn) (Kinsley et al., 1984).

In contrast to what was expected, there was no relationship of contiguity or any of the other classification criteria to the exhibition of infanticide in the males, PPA in the females, or AGD in either sex. Variation in the behavioral measures was similar to that previously observed. Male C57 mice were infanticidal at the rate of 70–80%, whereas DBA males remained low, at about 10–20%. Female C57 mice exhibited low levels of PPA, and DBA females showed much higher aggression. In terms of morphology, DBA animals, on average, had longer AGDs than C57 animals, irrespective of sex. The origin of the variability in the expression of infanticide and PPA in the C57 and DBA is, therefore, still open to question. Of interest is a report by Robinson, Fox, and Sidman (1985) stating that C57 and DBA animals differ with regard to the presence of a sexually dimorphic nucleus in the preoptic area (POA). C57 animals do not appear to have this nucleus, whereas in DBA animals, the sex difference (normally larger in males) is reversed, with females having the larger volume. The POA is a behaviorally important area for numerous sexually dimorphic behaviors and physiological responses (Simerly & Swanson, 1987). Differences in the manner in which this region responds to either hormones or other stimulation, either prenatally or in adulthood, could subsequently affect parental behavior.

Intrauterine Position and Parental Behavior in Rats

In rats, IUP effects on pup-induced parental behavior have recently been investigated (Kinsley & Bridges, unpublished observations). Groups of 0M and 2M male rats were generated as described above and in adulthood were tested for parental behavior toward foster young. 2M male rats responded significantly faster to young with 2 consecutive days of full parental behavior (FPB)—retrieving, grouping, and crouching over neonates—compared with 0M males (median = 2.0 vs. 9.0 days, respectively). No differences were observed with regard to infanticide on first exposure to pups. At present, no data exist on prenatal levels of steroids in rats from different IUPs. It is difficult, therefore, to draw firm conclusions about how differences in steroid levels could account for the findings. They suggest, however, that exposure to elevated prenatal E_2 (in the case of the 0M male rat, following from data in the mouse) suppresses parental responding. It is interesting that the mean of the latencies to respond for 0M and 2M males is similar to the male rat latency reported in other work (where both 0M and 2M males, in addition to 1M males, would be expected to contribute randomly to the overall effect; Bridges, 1984; Kinsley & Bridges, 1988b; Samuels & Bridges, 1983). Prenatal exposure to varying levels of gonadal steroids influences the display of parental behavior in the male rat, and in some respects the effects are similar to those reported for the male mouse (vom Saal, 1983a)—that is, low parental responsiveness in the 0M male and high responsiveness in the 2M male.

In female rats, the exhibition of both parental behavior and neuroendocrinological responses is under the influence of organizing effects of gonadal steroids (Harlan & Gorski, 1977a, 1977b; Ichikawa & Fujii, 1982). There are IUP influences on both of these responses as well. When 0M and 2M female rats were examined for parental behavior in adulthood (Kinsley & Bridges, unpublished observations), no significant difference was observed between them in their latencies to respond to young, although the direction of the difference was the same as it was in the males (median = 7.0 days for 0M females vs. 5.0 days for 2M females). In a second study, a neuroendocrinological response, prolactin (PRL) release to young, was measured in 0M and 2M females. 0M and 2M female rats were timed-mated. At parturition, the litters were culled to 10 pups, and on day 3 of lactation the females were fitted with intra-atrial cannulas. Sequential blood samples were initiated on the following day. Following a 180-min period of pup removal, a baseline blood sample (0.3 ml) was taken (0 min time point), and the pups were returned to the mothers. Subsequent samples in the presence of pups were taken at 10, 30, and 60 min. Blood samples were collected for 4 days. The results are presented in Table 17.1. Generally, 0M females secreted significantly less PRL in response to young than did 2M females. Both groups exhibited increases in PRL with advancing days. Because there were no differences between the two groups at the baseline time point, the results must be

Table 17.1   Mean (±SEM) levels of suckling-induced PRL release in 2M and 0M females

Time	OM female (N = 5)				2M female (N = 6)			
	Day 1	Day 2	Day 3	Day 4	Day 1	Day 2	Day 3	Day 4
0 min	4.64	12.29	3.13	4.58	5.25	5.29	7.24	19.36
	±1.23	±3.44	±1.11	±1.32	±2.24	±1.10	±2.60	±3.50
10	5.17	17.04	50.70*	31.80*	12.12	57.43	229.28	192.52
	±0.87	±3.78	±8.66	±5.76	±3.79	±5.88	±20.09	±18.65
30	11.90	161.76	281.85*	317.71	68.17	199.23	362.46	362.05
	±2.20	±11.11	±17.70	±20.86	±12.22	±25.54	±15.87	±13.99
60	45.56	97.09	119.06	150.79*	73.09	127.74	147.97	206.00
	±8.90	±7.78	±14.07	±11.38	±7.91	±15.00	±11.06	±10.01

Note: 0′ represents the sample following 3 hr of pup deprivation. Values are expressed as nanograms PRL/mL plasma (rat RP-3).

*$p < .05$, vs. 2M female at same time point, same day.

interpreted in light of differences in the manner in which the animals responded to suckling and/or pup stimulation. Thus these results support the idea that prior intra-uterine contiguity influences neuroendocrine responsivity to suckling stimuli as well as parental behavior in the female rodent (Kinsley, Konen, et al., 1986).

Overall, IUP has provided some interesting ways of looking at and accounting for the variability inherent in sexually differentiated behavior. In regard to parental responsiveness, IUP has been shown to play a significant role in the exhibition of pup killing by male mice and pup-induced parental behavior by male rats. More significant, however, are the effects of IUP on female parental behaviors—in particular, such maternal aggressive behaviors as PIA and PPA. Moreover, pup-induced physiological responses like those described earlier indicate that at another level—neuroendocrine responsivity—IUP may contribute to the intensity of parental behavior displayed by the female through actions of steroid hormones during early development. In this regard, then, the IUP phenomenon may represent differences in a type of natural "masculinization" of neuroendocrine function and behavior in the female, and may raise the question of what constitutes a "true" female. The work presented so far demonstrates that parental behavior can be influenced by early exposure to gonadal steroids as well as IUP. Both exogenous and endogenous hormonal stimulation may affect the behavior. If it were possible to eliminate exposure to superfluous gonadal steroids, to what extent would parental behavior be affected?

The "Singleton" Preparation

The findings just discussed, in combination with earlier work on the IUP phenomenon, suggest that contiguity to male littermates in utero and the associated exposure to steroid substances produces a form of endogenous "contamination" that affects physiology, morphology, and behavior. There is a corollary to this idea, particularly in the female, which is believed to develop passively in the absence of the steroids that affect the male. In a polytocous species, there is no such thing as an "uncontaminated'' animal—that is, one that develops without the influence of inadvertent steroid stimulation. If it were possible to produce a female in which such steroid exposure was precluded, would this "pure" animal exhibit any trace of masculine characteristics? Gandelman and Graham (1986a, 1986b) reported just such a preparation in the mouse. Time-mated females had their litters reduced in utero to a single embryo on day 8 of pregnancy, prior to the onset of fetal gonadal steroidogenesis. Control pregnant females were given sham operations. One day before the expected delivery of young, the single male or female fetus (referred to as a "singleton") was removed by caesarean section. Using this preparation, Gandelman and Graham found that singleton females were unresponsive to T-induced aggression. Thus masculinization of the female, as measured by latency to exhibit T-induced aggression, cannot occur in the absence of steroid stimulation by adjacent fetuses in utero.

Recently, the singleton phenomenon has been used in rats to examine the contribution of adjacent fetuses to parental behavior (Kinsley & Bridges, unpublished observations). Since T exposure perinatally is thought to suppress parental responding in the rat, it was asked whether a female that had no prenatal hormonal stimulation would respond to pups more quickly than a female from a normal-size litter. Similarly, would singleton males exhibit alterations in male-typical parental behavioral responses due to the reduction in prenatal hormonal stimulation from adjacent fetuses? In this study, timed-mated rats had their uteri exteriorized on day 8 of gestation. All but one embryo, randomly determined, were removed from the uteri by means of a small diameter suction pipette. Control females were sham operated. On day 21 of pregnancy, 24 hr prior to delivery, the singleton and control fetuses were removed by caesarean section, cleaned, weighed, sexed, and placed with a normal, lactating foster mother. All groups were weaned on day 25 and placed with same-sex, nonexperimental animals until the time of testing for parental behavior at 80–100 days of age. Both male and female singletons and their respective controls were tested for parental behavior using foster rat young. Singleton males responded more quickly to the young than did control males (median latency in days = 2.0 vs. 6.5, respectively). Singleton females, however, took longer to respond to the young than did control females (median = 9.5 vs. 4.0 days, respectively).

If singleton males are not exposed to the steroid secretions of in utero littermates (which follows from the work of Gandelman and Graham, 1986a, in mice), the finding of increased responsiveness in singleton males relative to control males would be expected. 2M males, however, like the previously described mice, when exposed to lower levels of E_2 than 0M males, might be expected to resemble singleton males not because of any similarities regarding prenatal T exposure, but because of alterations in prenatal E_2 exposure, provided that both 2M and singleton males are exposed to low levels of E_2 relative to 0M and control males, respectively. Therefore, it may be prenatal E_2 that contributes to retarding pup-induced parental responsiveness, a possibility also suggested by Ichikawa and Fujii (1982). (Moreover, the next section describes work by vom Saal, 1983a, that suggests that a treatment that shifts parental responding away from the infanticidal phenotype of the 0M male mouse, prenatal stress, may do so by changing prenatal E_2 titers.)

In singleton female rats, however, the situation appears to be different. Recall that although in the same direction as the males, no difference was observed between

0M and 2M female rats in pup-induced parental responsiveness. Singleton females, presumably exposed to little or no steroids from other fetuses, resembled 0M females in their parental behavior, more so than 2M females. The singleton female more closely resembles an unexposed/uncontaminated female but, in terms of parental behavior, responds as though she was exposed to steroids (e.g., E_2 or T) prenatally. The absence of an effect on parental behavior in the nulliparous, pup-induced parental, singleton female rat, however, may also be due to the reproductive state of the test animal. Since the mechanisms underlying the normal display of parental behavior in the female (including the requisite physiological responses to ovarian-steroid stimulation) may be apparent only following pregnancy and parturition, the failure to observe differences in the nulliparous condition in the above study (Kinsley & Bridges, unpublished observations) does not preclude differences under these other conditions.

Exposure to gonadal steroids early in development influences parental behavior in both male and female rats and mice. Under normal development, males are exposed to higher levels of androgens than are females. Also, males or females positioned between two males in utero are exposed to elevated levels of androgens compared with animals flanked by two females in utero. By eliminating such exposure, it is possible to modify the parental behavioral responses of both sexes. In the next section, evidence is presented demonstrating alterations in parental behavior associated with the disruption of prenatal exposure to hormones in males and females.

PRENATAL STRESS: A FACTOR ALTERING PARENTING CAPACITY

The exhibition of sexually dimorphic behaviors is generally determined by the presence or relative absence of hormones at discrete developmental stages. If males, for instance, are deprived of the necessary hormonal stimulation during the pre- and postnatal organizational periods, their subsequent behavior is severely affected, relative to the male mean. One treatment that affects behavioral as well as physiological aspects of sexual differentiation is prenatal stress (PS). When exposed to a stressful regimen of heat and restraint during gestation, male rat offspring are demasculinized and feminized with respect to copulatory behavior as adults (Ward, 1972), and display a variety of reductions in male-typical copulatory behavior (Meisel, Dohanich, & Ward, 1979; Ward, 1972), intermale aggression (Harvey & Chevins, 1984; Kinsley & Svare, 1986), and open field behavior (Thompson, 1957). In the female, PS is associated with reductions in fertility and fecundity (Herrenkohl, 1979), disruptions of the estrous cycle, and effects on sexual behavior in rats as well as mice (Allen & Haggett, 1977; Herrenkohl & Politch, 1978; Herrenkohl & Scott, 1984; Politch & Herrenkohl, 1984; but see Beckhardt & Ward, 1983).

Although examinations of PS effects in the male have tended to focus on disruption of and/or variation in the expression of male-typical sexual behavior, several reports have examined male parental responsiveness. Vom Saal (1983a), in mice, and Miley, Frank, and Hoxter (1981), in rats, found that PS reduced the infanticidal tendencies of males. McLeod and Brown (1986) reported that in male rats, PS facilitated retrieval, crouching, and nest building.

Prenatal Stress and Parental Behavior in Mice

In the mouse, vom Saal (1983a) investigated PS effects on the incidence of infanticidal behavior in 0M and 2M males. As previously described, a greater percentage of 0M compared with 2M male mice are infanticidal toward newborns. Given the disruptive actions of PS on perinatal hormone titers and timing (Ward & & Weisz, 1980, 1984) and the role of perinatal hormones in "organizing" infanticide, in what manner would IUP influence PS effects in the male? Vom Saal (1983a) stressed pregnant mice and caesarean-delivered 0M and 2M males. In adulthood, the males were exposed to pups. PS eliminated any effect of IUP; that is, all the males resembled highly parental 2M males. Whereas 70% of control 0M males were infanticidal, only 25% of prenatally-stressed (P-S) 0M males killed. PS not only decreased the proportion of males that were infanticidal, but also increased the proportion that were parental toward young. Vom Saal suggests that PS somehow disrupts the secretion of E_2 in the female and, thereby, levels of that steroid in the 0M male fetus. Eliminating the influences of prenatal E_2 apparently could shift the infanticidally predisposed 0M male toward the phenotype of the parentally behaving 2M male mouse.

Kinsley and Svare (1988) have also examined parental behaviors in P-S female mice. PS had no effect on females' spontaneous parental behavior as adults. Most of the females from both groups (P-S and controls) acted parentally toward the young, although there was a nonsignificant tendency for P-S females to exhibit slightly less infanticidal behavior relative to controls (0% vs. 11%, respectively). In another study, the effects of PS on nest building were measured in separate groups of P-S and control female mice. Beginning 24 hr after mating, nest building was assessed in all females, using the protocol previously described for 0M and 2M female mice (Kinsley, Konen, et al., 1986). PS reduced nest building, especially during the first to early-second trimester of pregnancy (Kinsley & Svare, 1988). However, in keeping with previously published observations (Broida & Svare, 1982; Kinsley, Konen, et al., 1986), both groups tended to build larger and more elaborate nests with advancing gestation.

PS also affects aggression shown by females during pregnancy and lactation. When tested for PIA on days 16, 17, and 18 of gestation, P-S females exhibited significantly less aggression toward an adult intruder male than did control females. This shift in the level of PIA was reflected primarily in a reduction in the number of attacks directed at the male. These females were allowed to deliver their young, and the litters were culled to five pups. On lactation days 6, 7, and 8, the P-S and control females were assessed for PPA against males with which the females had no previous experience. Interestingly, in an apparent reversal of the level of aggressive behavior displayed during pregnancy, P-S females exhibited more intense aggressive behavior compared with control females (Kinsley & Svare, 1988). P-S females exhibited aggression on more test days and engaged in a greater number of lunges compared with controls.

These results on nest building, PIA, and PPA in mice indicate that PS in the female may affect the degree of either androgenization or virilization by androgen-mimetic substances. Observations of the effects of PS on ovarian cyclicity and sexual behavior in the female rat also indicate that P-S females are exposed to higher than normal levels of androgens during development (Herrenkohl, 1983; Herrenkohl &

Politch, 1978; Herrenkohl & Scott, 1984). Behavioral evidence also indicates that the P-S female mouse may be exposed to increased prenatal levels of androgen. For example, the effects of exposure to exogenous perinatal androgens on nest building (Lisk, Russell, Kahler, & Hanks, 1973) and PPA (Kinsley, Konen, et al., 1986; Mann & Svare, 1983) often mimic the sequelae of PS, as does prenatal treatment with P (Konen, Kinsley, & Svare, 1986). Furthermore, the effects may be due to alterations of levels of and/or sensitivity to P and E_2 in the P-S female, since both nest building (Anderson et al., 1970; Broida & Svare, 1982; Lisk, Pretlow, & Fried-man, 1969) and PIA (Mann, Konen, & Svare, 1984; Svare, Miele, & Kinsley, 1986) are dependent to some degree on these hormones.

Politch and Herrenkohl (1979) examined PPA in P-S female mice and found that PS *reduced* PPA. Although their findings are at variance with Kinsley and Svare's (1988), there are a number of reasons that may account for the findings. First, there were differences in the intensity and timing of the stress procedure administered to the pregnant female. Furthermore, Kinsley and Svare tested females on 3 consecutive days, whereas Politch and Herrenkohl did so on only 1 day. Next, R-S albino mice were used by Kinsley and Svare, whereas Politch and Herrenkohl used the CD-1 strain. As will be discussed later, genetic differences play a very important role in determining the direction and intensity of the behavioral as well as the physiological effects of PS.

Prenatal Stress and Parental Behavior in Rats

Kinsley and Bridges (1988b) examined a number of parental behaviors in P-S virgin male and female rats. (The procedure for producing P-S animals is described in Kinsley and Bridges, 1987.) In an initial study, P-S males and females were observed as adults for parental behavior and nest building in the presence of foster young. P-S males exhibited a significantly shorter latency to initiate FPB than did control males (median = 5.0 vs. 8.0 days, respectively) (Figure 17.1). P-S males also displayed shorter latencies to crouch over and build nests in response to neonates. P-S females, on the contrary, exhibited a significantly longer latency than control females to show FPB (7.0 vs. 3.0 days, respectively), as well as longer latencies to retrieve one, two, or three pups, to begin to crouch over pups, and to build nests in response to young.

Table 17.2 compares simple sex differences in the components of parental re-sponsiveness (control male vs. control female). Females were, in general, more re-sponsive to young than were males, and they more rapidly built nests and began to crouch over young. P-S males and control females exhibited similar latencies to show 2 consecutive days of FPB (3–5 days) and differed in latencies to retrieve pups and engage in 1 day of FPB, with control females exhibiting more rapid onset of the behavioral components. When the responses of P-S females and control males were compared, no apparent behavioral differences were observed in the overall display of parental behavior (latencies for 2 consecutive days of FPB, 7–9 days), save for la-tency to retrieve one pup, in which the P-S female latency was longer. These data demonstrate that PS eliminates the sex difference normally observed in pup-induced maternal behavior and, moreover, that PS renders more "female-like" the male's responsiveness to young, while conversely rendering the response of the female more

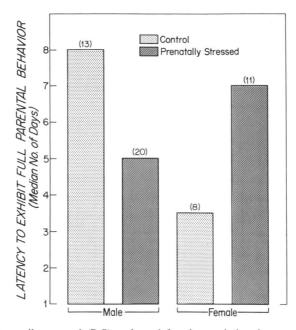

Figure 17.1.   Prenatally stressed (P-S) male and female pup-induced parental behavior. Adult P-S males and females were exposed to foster young continuously over the course of 10 days until 2 consecutive days of full parental behavior were observed. (Reprinted by permission of *Hormones and Behavior* and Academic Press, adapted from Kinsley & Bridges, 1988b)

"male-like." These effects are thought to be due to alterations in the levels and timing of exposure to gonadal hormones during prenatal "sensitive" periods.

These PS effects support the idea that PS is a potent disruptor of sexual differentiation in both sexes. These effects on behavior, however, were detected in adulthood. How would PS affect the parental responses of juvenile animals? Juvenile (18–25 days of age) male and female rats display a rapid onset (2- to 3-day latency) of parental-like behavior (Bridges, Zarrow, Goldman, & Denenberg, 1974; Brunelli, Shindledecker, & Hofer, 1985; Gray & Chesley, 1984; Mayer & Rosenblatt, 1979). The rapid onset of parental behavior at this age stands in marked contrast to the latencies shown by virgin and nulliparous adult rats. Furthermore, in juveniles there is an apparent elimination of the sex difference normally accompanying pup-induced parental behavior in the adult animal. Whereas adult males take longer to respond to pups than do adult females, juvenile males tend to respond more rapidly to foster young than do juvenile female rats (Bridges et al., 1974; Gray & Chesley, 1984; Kinsley & Bridges, 1988a). In order to evaluate the possible effect of PS on the development of parental behavior, P-S males and females were tested for onset of parental responsiveness at 25 days of age with three 1- to 7-day old neonates (Kinsley & Bridges, unpublished observations) (Figure 17.2). P-S males tested as juveniles took significantly longer to begin responding to pups compared with juvenile controls (median = 8.0 vs. 2.0 days, respectively). Female P-S juvenile rats had latencies similar to those of the P-S males and significantly longer than those of female con-

Table 17.2 Median latencies (in days) for P-S and control male and female rats to exhibit various components of parental behavior

Parental behavior	Males		Females	
	P-S	Control	P-S	Control
Two consecutive days of FPB	5.0**	8.0[c]	7.0***	3.0
One day of FPB	4.0	6.0[c]	6.0+	3.0
Retrieve one pup	4.0	4.0	5.0+	2.5
Retrieve two pups	4.0	4.0	6.0‡	2.5
Retrieve three pups	4.0	5.0[b]	6.0‡	3.0
Latency to crouch (over at least one pup)	4.0**	5.0[a]	5.0+	3.0
Latency to build a good to excellent nest	2.0+	3.0[b]	5.0*	2.0

P-S vs. controls:
*$p < .05$
**$p < .04$
***$p < .025$
†$p < .01$
‡$p < .001$
Males vs. females:
[a]$p < .05$
[b]$p < .025$
[c]$p < .002$
Source: adapted with permission from Kinsley & Bridges, 1988b.

trols (8.0 vs. 4.0, respectively). PS, therefore, lengthens the latency to respond to young in both male and female juvenile rats. To date, the possible physiological factors that account for the rapid onset of FPB in juvenile rats have not been completely characterized. Recently, however, a possible contribution of PRL to this behavior was demonstrated in juvenile male rats (Kinsley & Bridges, 1988a). Given the role of PRL in the exhibition of parental behavior in adult female rats (Bridges, DiBiase, Loundes, & Doherty, 1985), and the fact that P-S male rats secrete less PRL than control males in response to estrogen stimulation (Kinsley & Bridges, 1987), it would be of interest to determine whether the P-S juvenile animal may exhibit reductions in either levels of or sensitivity to the parental behavior–promoting properties of PRL.

In addition to the effects on parental behavior just described, some alterations in the pup-directed parental behavior of postpartum females have been reported, although not all are in accordance with the present findings. Herrenkohl and Gala (1979) reported that P-S females were able to care for young behaviorally but were deficient in PRL secretion (measured by lactation performance and levels of systemic PRL). Fride, Dan, Gavish, and Weinstock (1985) examined maternal behavior (pup retrieval) in postpartum P-S female rats. The animals were required to enter an alley, at the end of which their pups were placed. There was no difference between P-S and control females in retrieval latencies or in the percentage of females retrieving

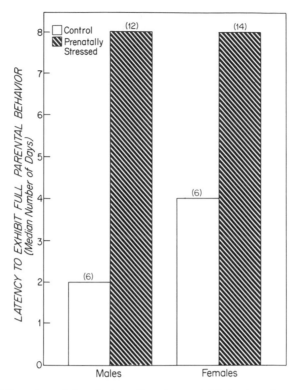

Figure 17.2. P-S male and female juvenile rat parental behavior. Juvenile P-S male and female rats (2᠄ days of age) were exposed to foster young over the course of 10 days or until 2 consecutive days of full parental behavior were observed.

the pups. In a subsequent test, in which the females were required to traverse the alley in the presence of a conflict situation (an air stream was introduced into the alley), all but one control female retrieved their pups, whereas nearly half the P-S females failed to retrieve (96% vs. 52%, respectively). Therefore, differences became apparent only after the activation of stress- or anxiety-producing mechanisms in the animal (in a novel test environment similar to that of the Bridges et al., 1973, study, which utilized the T-maze). Fride et al. (1985) also found alterations in hippocampal benzodiazepine receptors in P-S females, alterations that might underlie the behavioral disruptions in their P-S females.

Prenatal Stress and Genotypic Interactions in Parental Behavior

Sex differences as well as genetic differences are evident that help to "color" the eventual exhibition of parental behavior, as was described with regard to IUP. It has recently been demonstrated that genetic differences interact with PS (Kinsley & Svare, 1987a). C57 and DBA females were timed-mated and exposed to the stress session during the last trimester of pregnancy. Their P-S male and female offspring were tested as adults for a variety of aggressive and parental behaviors. Normally, C57 males are much more infanticidal than DBA males (Svare, Broida, et al., 1984;

Svare, Kinsley, et al., 1984), and vom Saal (1983a) demonstrated that PS eliminates infanticidal behavior in 0M males, which are usually highly infanticidal. PS failed to have any effect on the percentage of C57 and DBA males that committed infanticide: the percentages were nearly identical to control values; that is, C57s were highly infanticidal, whereas DBAs showed very little of the behavior. When tested for intermale aggression, P-S C57 males were much more aggressive than controls, whereas DBA males were unaffected. Harvey and Chevins (1984) and Kinsley and Svare (1986a) reported that PS reduced intermale aggression in two different outbred mouse strains. The same evidently does not hold for inbred mice.

In the case of females, PS increased the incidence and intensity of PPA in C57s, but reduced PPA in DBAs. In measures of reproductive and lactation performance, differential effects were also observed. For example, P-S DBA females produced smaller litters relative to control DBAs, whereas there was no effect in C57s. P-S C57 females, though, weighed significantly less than control C57s on day 6 postpartum, but no difference was evident between P-S and control DBA females. No other measures were significantly different between P-S and control mice of either strain. As discussed earlier, PS reduced PIA and increased PPA in outbred R-S mice (Kinsley & Svare, 1988), whereas P-S CD-1 mice showed reductions in PPA (Politch & Herrenkohl, 1979). Moreover, measures of AGD were a poor predictor of subsequent parental and maternal aggressive behavior in the C57 and DBA inbred mice, at least in terms of what has previously been referred to as "masculinization." These morphological findings, then, in combination with the behavioral work in the inbred strains and other work in outbred strains, demonstrate that genotype must be considered as an important source of variation influencing the effects of PS on parental as well as other forms of behavior.

Prenatal-Stress Effects on Brain Chemistry and Anatomy

PS exerts many effects on behavior and neuroendocrine physiology in male and female rodents. What variety of actions on the central nervous system can account for this variety of effects? Some of the effects of PS may be due to alterations of the ontogeny of neurotransmitter/neurochemical systems—in particular, the catecholaminergic systems. PS alters dopamine and norepinepherine levels within the areas of the adult brain (specifically, the medial preoptic area [MPOA]; Moyer, Herrenkohl, & Jacobowitz, 1978) associated with maternal and sexual behavior (Numan, 1983) and with neuroendocrine regulation of pituitary PRL (Pan & Gala, 1985a, 1985b). Furthermore, P-S males and females are significantly less sensitive to the PRL-releasing capacity of E_2 (Kinsley & Bridges, 1987). Thus normal neurochemical regulation of PRL may be affected by exposure to PS.

In addition to a possible effect of PS on catecholaminergic systems, other data suggest that PS may disrupt the development of the opioid system(s), which normally play a role in the physiology and behavior of the adult animal. The development of the mu (μ)-type opioid receptor system is sexually dimorphic (Hammer, 1985). Endogenous opioid systems modulate a wide variety of physiological and behavioral variables in both the males and females, including PRL secretion, gonadotropin release, and sexual and parental behaviors (Allen, Renner, & Luine, 1985; Benton, 1985; Bridges & Grimm, 1982; Kalra & Simpkins, 1981; Kinsley & Bridges, 1986).

Furthermore, Ward, Monaghan, and Ward (1986) reported that treatment with the opiate antagonist naltrexone concurrent with the PS regimen reduced the disruptive effects of PS on male sexual behavior in the rat. The perturbation of sexual differentiation that underlies the effects of PS may concurrently affect the normal development of the opioid receptor system. In adulthood, therefore, the normal modulatory role performed by the opioids in the aforementioned behavioral and physiological systems may be modified.

If opioid systems are affected in P-S animals, such differences might be reflected in traditional assessments of opiate responsivity. Changes in sensitivity to the analgesic properties of morphine, as measured by tail-flick latencies, were recently reported in P-S rats (Kinsley, Mann, & Bridges, 1988). P-S male rats were significantly less sensitive to a dose of 5.0 mg/kg morphine (i.e., lower pain thresholds) at each time point following morphine administration. PS *increased* the sensitivity to morphine in female rats (i.e., significantly longer tail-flick latencies). Therefore, effects on opiate systems may underlie the effects of PS on other sexually dimorphic behaviors—in particular, those, such as parental and aggressive behaviors, that are mediated by opiates.

Other actions of PS on the central nervous system of the rodent may contribute to the range of effects described thus far. Both the E_2-induced PRL response (Pan & Gala, 1985a, 1985b) and the expression of maternal behavior (Numan, 1983) are under the regulation of the MPOA. It has been shown that the sexually dimorphic nucleus of the preoptic area (SDN-POA) is reduced in volume in P-S males, and thus more closely resembles the normal female SDN (Anderson, Fleming, Rhees, & Kinghorn, 1986; Anderson, Rhees, & Fleming, 1985). A preliminary report on the possible alteration of the structure of the SDN-POA in the P-S female rat (Herrenkohl, 1983) stated that the nucleus is actually *increased* in volume; that is, it is masculinized. Together, these data provide evidence that the range of effects of PS may originate in alterations in the central nervous system of the adult, alterations that were induced by the significant disruptions of prenatal hormone activity accompanying the stress.

SUMMARY AND CONCLUSIONS

That early, particularly prenatal, events exert strong, long-lasting effects on the exhibition of parent-related behaviors in rodents is well documented. Figures 17.3 and 17.4 provide a summary and generalization of the data presented in this chapter, without reference to what may be considered an improvement in behavior, as qualitative changes are sometimes difficult to interpret. For instance, does the the fact that IUP influences female aggressive behavior and male infanticidal behavior in a particular direction represent an improvement in those behaviors? Without information about the ramifications of such changes, labels such as ''good'' and ''bad'' may be premature. As reviewed in this chapter, pre- and postnatal treatment with steroid hormones of both endogenous and exogenous means, or alterations in either the timing or the level of exposure to such hormones, may shift the threshold of responsiveness for a wide variety of parental behaviors of both mice and rats away from that nor-

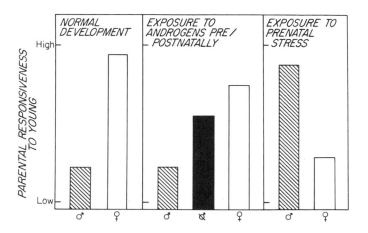

Figure 17.3. Schematic summary of perinatal influences on relative levels of parental behavior in male and female rodents.

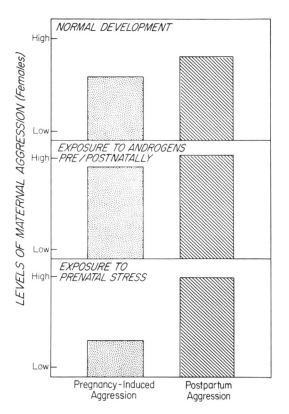

Figure 17.4. Summary of perinatal influences on maternal aggressive behaviors in female rodents.

mally displayed in the absence of such effects. Conversely, by preventing or altering the quality of such exposure through means like artificial litter reduction in utero (in the case of males) or PS, the exhibition of parental behaviors can be substantially affected. The degree to which early influences have ramifications for understanding the substrates associated with parental responsiveness, and the implications of such influences for population dynamics in naturally occurring feral populations of rodents, await further examination.

ACKNOWLEDGEMENTS

I wish to acknowledge and thank my co-workers, colleagues, and mentors who participated in and contributed to much of the research presented here: Drs. Robert S. Bridges, John P. Broida, and Loraina Ghiraldi; Christine Konen-Wagner; Drs. Martha A. Mann, Phyllis E. Mann, and Joseph L. Miele; Paul Ronsheim; and Drs. Beverly S. Rubin and Bruce B. Svare.

REFERENCES

Allen, D. L., Renner, K. J., & Luine, V. N. (1985). Naltrexone facilitation of sexual receptivity in the rat. *Hormones and Behavior, 19*, 98–103.

Allen, T. O., & Haggett, B. N. (1977). Group housing of pregnant mice reduces copulatory receptivity of female progeny. *Physiology and Behavior, 19*, 61–68.

Anderson, C. O., Zarrow, M. X., & Denenberg, V. H. (1970). Maternal behavior in the rabbit: Effects of androgen during gestation upon the nest-building behavior of the mother and her offspring. *Hormones and Behavior, 1*, 337–345.

Anderson, D. K., Rhees, R. W. & Fleming, D. E. (1985). Effects of prenatal stress on differentiation of the sexually dimorphic nucleus of the preoptic area (SDN-POA) of the rat brain. *Brain Research, 332*, 113–118.

Anderson, R. H., Fleming, D. E., Rhees, R. W., & Kinghorn, E. (1986). Relationships between sexual activity, plasma testosterone, and the volume of the sexually dimorphic nucleus of the preoptic area in prenatally stressed and non-stressed rats. *Brain Research, 370*, 1–10.

Beckhardt, S., & Ward, I. L. (1983). Reproductive functioning in the prenatally stressed female rat. *Developmental Psychobiology, 16*, 111–118.

Benton, D. (1985). Mu and kappa opiate receptor involvement in agonistic behaviour in mice. *Pharmacology, Biochemistry and Behavior, 23*, 871–875.

Bridges, R. S. (1984). A quantitative analysis of the roles of dosage, sequence, and duration of estradiol and progesterone exposure in the regulation of maternal behavior in the rat. *Endocrinology, 114*, 930–940.

Bridges, R. S., DiBiase, R., Loundes, D. D., & Doherty, P. C. (1985). Prolactin stimulation of maternal behavior in female rats. *Science, 227*, 782–784.

Bridges, R. S., & Grimm, C. T. (1982). Reversal of morphine disruption of maternal behavior by concurrent treatment with the opiate antagonist naloxone. *Science, 218*, 166–168.

Bridges, R. S., Zarrow, M. X. & Denenberg, V. H. (1973). The role of neonatal androgen in the expression of hormonally induced maternal responsiveness. *Hormones and Behavior, 4*, 315–322.

Bridges, R. S., Zarrow, M. X., Goldman, B. D., & Denenberg, V. H. (1974). A developmental study of maternal responsiveness in the rat. *Physiology and Behavior, 12*, 149–151.

Broida, J. P., & Svare, B. B. (1982). Postpartum aggression in C57BL/6J and DBA/2J mice: Experiential and environmental influences. *Behavioral and Neural Biology, 35,* 76–83.

Brunelli, S. A., Shindledecker, R. D., & Hofer, M. A. (1985). Development of maternal behaviors in prepubertal rats at three ages: Age-characteristic patterns of responses. *Developmental Psychobiology, 18,* 309–326.

Clemens, L. G., Gladue, B., & Coniglio, L. (1978). Prenatal endogenous androgenic influences on masculine sexual behavior and genital morphology in male and female rats. *Hormones and Behavior, 10,* 40–53.

Day, C. S. D., & Galef, B. G. (1977). Pup cannibalism: One aspect of maternal behavior in golden hamsters. *Journal of Comparative and Physiological Psychology, 91,* 1179–1189.

Fride, E., Dan, Y., Gavish, M., & Weinstock, M. (1985). Prenatal stress impairs maternal behavior in a conflict situation and reduces benzodiazepine receptors. *Life Sciences, 36,* 2103–2109.

Gandelman, R. (1973). The development of cannibalism in male Rockland-Swiss albino mice and the influence of olfactory bulb removal. *Developmental Psychobiology, 6,* 159–164.

Gandelman, R. (1983). Hormones and infanticide. In B. B. Svare (Ed.), *Hormones and aggressive behavior* (pp. 105–118). New York: Plenum Press.

Gandelman, R., & Graham, S. (1986a). Development of the surgically produced singleton mouse fetus. *Developmental Psychobiology, 19,* 343–350.

Gandelman, R., & Graham, S. (1986b). Singleton mouse fetuses are subsequently unresponsive to the aggression-activating property of testosterone. *Physiology and Behavior, 37,* 465–467.

Gandelman, R., & vom Saal, F. S. (1975). Exposure to early androgen attenuates androgen-induced pup-killing in male and female mice. *Behavioral Biology, 20,* 252–260.

Gandelman, R., vom Saal, F. S., & Reinisch, J. M. (1977). Contiguity to male foetuses affects morphology and behaviour of female mice. *Nature, 266,* 722–724.

Gray, P., & Chesley, S. (1984). Development of maternal behavior in nulliparous rats *(Rattus norvegicus):* Effects of sex and early maternal experience. *Journal of Comparative Psychology, 98,* 91–99.

Hammer, R. P., Jr. (1985). The sex hormone-dependent development of opiate receptors in the rat medial preoptic area. *Brain Research, 360,* 65–72.

Harlan, R. W., & Gorski, R. A. (1977a). Steroid regulation of luteinizing hormone secretion in normal and androgenized rats at different ages. *Endocrinology, 101,* 741–749.

Harlan, R. W., & Gorski, R. A. (1977b). Correlations between ovarian sensitivity, vaginal cyclicity and luteinizing hormone and prolactin secretion in lightly androgenized rats. *Endocrinology, 101,* 750–759.

Harvey, P. W., & Chevins, P. F. D. (1984). Crowding pregnant mice affects attack and threat behavior of male offspring. *Hormones and Behavior, 18,* 101–110.

Hausfater, G, & Hrdy, S. B. (Eds.). (1984). *Infanticide: Comparative and evolutionary perspectives.* New York: Aldine.

Herrenkohl, L. R. (1979). Prenatal stress reduces fertility and fecundity in female offspring. *Science, 206,* 1097–1099.

Herrenkohl, L. R. (1983). Prenatal stress may alter sexual differentiation in male and female offspring. *Monographs in Neural Science, 9,* 176–183.

Herrenkohl, L. R., & Gala, R. R. (1979). Serum prolactin levels and maintenance of progeny by prenatally stressed female offsping. *Experientia, 35,* 702–704.

Herrenkohl, L. R., & Politch, J. A. (1978). Effects of prenatal stress on the estrous cycle of offspring as adults. *Experientia, 23,* 1240–1241.

Herrenkohl, L. R., & Scott, S. (1984). Prenatal stress and postnatal androgen: Effects on reproduction in female rats. *Experientia, 40*, 101–103.

Ichikawa, S., & Fujii, Y. (1982). Effect of prenatal androgen treatment on maternal behavior in the female rat. *Hormones and Behavior, 16*, 224–233.

Jakubowski, M., & Terkel, J. (1985a). Incidence of pup killing and parental behavior in virgin female and male rats *(Rattus norvegicus):* Differences between Wistar and Sprague-Dawley stocks. *Journal of Comparative Psychology, 99*, 93–97.

Jakubowski, M., & Terkel, J. (1985b). Transition from pup killing to parental behavior in male and female albino rats. *Physiology and Behavior, 34*, 683–686.

Kalra, S. P., & Simpkins, J. W. (1981). Evidence for noradrenergic mediation of opioid effects on luteinizing hormone secretion. *Endocrinology, 109*, 776–782.

Kinsley, C. H., & Bridges, R. S. (1986). Opiate involvement in postpartum aggression in rats. *Pharmacology, Biochemistry and Behavior, 25*, 1007–1011.

Kinsley, C. H., & Bridges, R. S. (1987). Prenatal stress reduces estradiol-induced prolactin release in male and female rats. *Physiology and Behavior, 40*, 647–653.

Kinsley, C. H., & Bridges, R. S. (1988a). Prolactin modulation of the maternal-like behavior exhibited by juvenile rats. *Hormones and Behavior, 22*, 49–65.

Kinsley, C. H., & Bridges, R. S. (1988b). Prenatal stress and maternal behavior in intact virgin rats: Response latencies are decreased in males and increased in females. *Hormones and Behavior, 22*, 76–89.

Kinsley, C. H., Broida, J. P., Konen, C. M., Miele, J. L., Ghiraldi, L., & Svare, B. B. (1984). Intrauterine position fails to account for variation in the morphology and behavior of inbred mice. *Society for Neuroscience Abstracts, 10*, 825.

Kinsley, C. H., Konen, C. M., Miele, J. L., Ghiraldi, L., & Svare, B. B. (1986). Intrauterine position modulates maternal behaviors in female mice. *Physiology and Behavior, 36*, 793–799.

Kinsley, C. H., Mann, P. E., & Bridges, R. S. (1988). Prenatal stress alters morphine and stress-induced analgesia in male and female rats. *Pharmacology, Biochemistry, and Behavior, 30*, 123–128.

Kinsley, C. H., Miele, J. L., Konen, C. M., Ghiraldi, L., & Svare, B. B. (1986). Intrauterine contiguity influences regulatory activity in adult female and male mice. *Hormones and Behavior, 20*, 7–12.

Kinsley, C., Miele, J., Konen-Wagner, C., Ghiraldi, L., Broida, J., Svare, B. (1986). Prior intrauterine position influences body weight in male and female mice. *Hormones and Behavior 20*, 201–211.

Kinsley, C. H., & Svare, B. B. (1986). Prenatal stress reduces intermale aggression in mice. *Physiology and Behavior, 36*, 783–786.

Kinsley, C. H., & Svare, B. B. (1987a). Genotype modulates prenatal stress effects on aggression in male and female mice. *Behavioral and Neural Biology, 47*, 138–150.

Kinsley, C. H., & Svare, B. B. (1987b). Adrenal testosterone maintains infanticidal behavior in neonatally-gonadectomized C57BL/6J male mice. *Physiology and Behavior, 41*, 519–521.

Kinsley, C. H., & Svare, B. B. (1988). Prenatal stress alters maternal aggression in mice. *Physiology and Behavior, 42*, 7–13.

Konen, C. M., Kinsley, C. H., & Svare, B. B. (1986). Mice: Postpartum aggression is elevated following prenatal progesterone exposure. *Hormones and Behavior, 20*, 212–221.

Lisk, R., Pretlow, R., & Friedman, S. (1969). Hormonal stimulation necessary for elicitation of maternal nestbuilding in the mouse. *Animal Behaviour, 17*, 730–737.

Lisk, R., Russell, J. A., Kahler, S. G., & Hanks, J. B. (1973). Regulation of hormonally mediated maternal nest structure in the mouse *(Mus musculus)* as a function of neonatal hormone manipulation. *Animal Behaviour, 21*, 296–301.

Mann, M. A., Kinsley, C. H., Broida, J. P., & Svare, B. B. (1983). Infanticide exhibited by female mice: Genetic, developmental, and hormonal influences. *Physiology and Behavior, 30,* 697–702.

Mann, M. A., Konen, C. M., & Svare, B. B. (1984). The role of progesterone in pregnancy-induced aggression in mice. *Hormones and Behavior, 18,* 140–160.

Mann, M. A., & Svare, B. B. (1983). Prenatal testosterone exposure elevates maternal aggression in mice. *Physiology and Behavior, 30,* 503–507.

Mayer, A. D., & Rosenblatt, J. S. (1979). Ontogeny of maternal behavior in the laboratory rat: Early origins in 18–27-day-old young. *Developmental Psychobiology, 12,* 407–424.

McCullough, J., Quadagno, D. M., & Goldman, B. D. (1974). Neonatal gonadal hormones: Effect on maternal and sexual behavior in the male rat. *Physiology and Behavior, 12,* 183–188.

McLeod, P. J., & Brown, R. E. (1986). *The effects of prenatal stress and post-weaning housing conditions on parental behavior of male Long-Evans rats.* Paper presented at the 18th annual conference on Reproductive Behavior, Montreal, Canada.

Meisel, R. L., Dohanich, G. P., & Ward, I. L. (1979). Effects of prenatal stress on avoidance acquisition, open-field performance and lordotic behavior in male rats. *Physiology and Behavior, 22,* 527–530.

Meisel, R., & Ward, I. L. (1981). Fetal female rats are masculinized by male littermates located caudally in the uterus. *Science, 213,* 239–242.

Miley, W. M., Frank, M., & Hoxter, A. L. (1981). Rat-pup killing and maternal behavior in male Long-Evans rats: Prenatal stimulation and postnatal testosterone. *Bulletin of the Psychonomic Society, 17,* 119–122.

Moyer, J. A., Herrenkohl, L. R., & Jacobowitz, D. M. (1978). Stress during pregnancy: Effect on catecholamines in discrete brain regions of offspring as adults. *Brain Research, 144,* 173–178.

Numan, M. (1983). Brain mechanisms of maternal behaviour in the rat. In J. Balthazart, E. Prove, & R. Gilles (Eds.), *Hormones and behaviour in higher vertebrates* (pp. 69–85). New York: Springer-Verlag.

Pan, J. T., & Gala, R. R. (1985a). Central nervous system regions involved in the estrogen-induced afternoon prolactin surge: 1. Lesion studies. *Endocrinology, 117,* 382–387.

Pan, J. T., & Gala, R. R. (1985b). Central nervous system regions involved in the estrogen-induced afternoon prolactin surge: 2. Implantation studies. *Endocrinology, 117,* 388–395.

Politch, J. A., & Herrenkohl, L. R. (1979). Prenatal stress reduces maternal aggression in mice. *Physiology and Behavior, 23,* 415–418.

Politch, J. A., & Herrenkohl, L. R. (1984). Effects of prenatal stress on reproduction in male and female mice. *Physiology and Behavior, 32,* 95–99.

Quadagno, D. M., McCullough, J., Ho, G.K.-H., & Spevak, A. M. (1973). Neonatal gonadal hormones: Effects on maternal and sexual behavior in the female rat. *Physiology and Behavior, 11,* 251–254.

Quadagno, D. M., & Rockwell, J. (1972). The effect of gonadal hormones in infancy on maternal behavior in the adult rat. *Hormones and Behavior, 3,* 55–62.

Robinson, S. M., Fox, T. O., & Sidman, R. L. (1985). A genetic variant in the morphology of the medial preoptic area in mice. *Journal of Neurogenetics, 2,* 381–388.

Rosenberg, K. M., Denenberg, V. H., Zarrow, M. X., & Frank, B. L. (1971). Effects of neonatal castration and testosterone on the rat's pup-killing behavior. *Physiology and Behavior, 7,* 363–368.

Rosenberg, K. M., & Sherman, G. F. (1975). The role of testosterone in the organization, maintenance and activation of pup killing in the male rat. *Hormones and Behavior, 6,* 173–179.

Rosenberg, P. A., & Herrenkohl, L. R. (1976). Maternal behavior in the male rat: Critical times for the suppressive actions of androgens. *Physiology and Behavior, 16,* 293–297.

Rosenblatt, J. S. (1967). Nonhormonal basis of maternal behavior in the rat. *Science, 156,* 1512–1514.

Samuels, M. H., & Bridges, R. S. (1983). Plasma prolactin cncentrations in parental male and female rats: Effects of exposure to rat young. *Endocrinology, 113,* 1647–1654.

Simerly, R. B., & Swanson, L. W. (1987). The organization of neural inputs to the medial preoptic nucleus of the rat. *Journal of Comparative Neurology, 246,* 312–342.

Svare, B. B. (1979). Steroidal influences on pup-killing behavior in mice. *Hormones and Behavior, 13,* 153–164.

Svare, B. B., Bartke, A., & Macrides, F. (1978). Juvenile male mice: An attempt to accelerate testis function by exposure to female stimuli. *Physiology and Behavior, 21,* 1009–1013.

Svare, B. B., Broida, J. P., Kinsley, C. H., & Mann, M. A. (1984). Psychobiological mechanisms underlying infanticide in small mammals. In G. Hausfater & S. Hrdy (Eds.), *Infanticide: Comparative and evolutionary perspectives* (pp. 387–400). New York: Aldine.

Svare, B. B., Kinsley, C. H., Mann, M. A., & Broida, J. P. (1984). Infanticide: Accounting for genetic variation. *Physiology and Behavior, 33,* 137–152.

Svare, B. B., Miele, J. L., & Kinsley, C. H. (1986). Progesterone stimulates aggression in pregnancy-terminated mice. *Hormones and Behavior, 20,* 194–200.

Thompson, W. R. (1957). Influences of prenatal maternal anxiety on emotionality in young rats. *Science, 125,* 698–699.

vom Saal, F. S. (1981). Variations in phenotype due to random intrauterine positioning of male and female fetuses in rodents. *Journal of Reproduction and Fertility, 62,* 633–650.

vom Saal, F. S. (1983a). Variation in infanticide and parental behavior in male mice due to prior intrauterine proximity to female fetuses: Elimination by prenatal stress. *Physiology and Behavior, 30,* 675–681.

vom Saal, F. S. (1983b). The interaction of circulating oestrogens and androgens in regulating mammalian sexual differentiation. In J. Balthazart, E. Prove, & R. Gilles (Eds.), *Hormones and behaviour in higher vertebrates* (pp. 159–177). New York: Springer-Verlag.

vom Saal, F. S. (1983c). Models of early hormonal effects on intrasex aggression in mice. In B. B. Svare (Ed.), *Hormones and aggressive behavior* (pp. 197–222). New York: Plenum Press.

vom Saal, F. S. (1984). Proximate and ultimate causes of infanticide and parental behavior in male house mice. In G. Hausfater & S. Blaffer-Hrdy (Eds.), *Infanticide: A comparative analysis* (pp. 401–424). New York: Aldine.

vom Saal, F. S., & Bronson, F. H. (1980). Sexual characteristics of adult female mice are correlated with their blood testosterone levels during prenatal development. *Science, 208,* 597–599.

vom Saal, F. S., Grant, W., McMullen, C., & Laves, K. (1983). High fetal estrogen concentrations: Correlation with increased adult sexual performance and decreased aggression in male mice. *Science, 220,* 1306–1308.

Ward, I. L. (1972). Prenatal stress feminizes and demaculinizes the behavior of males. *Science, 175,* 82–84.

Ward, I. L., & Weisz, J. (1980). Maternal stress alters plasma testosterone in fetal males. *Science, 207,* 328–329.

Ward, I. L., & Weisz, J. (1984). Differential effects of maternal stress on circulating levels of corticosterone, progesterone, and testosterone in male and female rat fetuses and their mothers. *Endocrinology, 114,* 1635–1644.

Ward, O. B., Monaghan, E. P., & Ward, I. L. (1986). Naltrexone blocks the effects of prenatal stress on sexual differentiation in male rats. *Pharmacology, Biochemistry and Behavior, 25,* 573–576.

Weisz, J., & Ward, I. L. (1980). Plasma testosterone and progesterone titers of pregnant rats, their male and female fetuses, and neonatal offspring. *Endocrinology, 106,* 306–316.

18

Parental Behavior in Juvenile Rats: Environmental and Biological Determinants

SUSAN A. BRUNELLI AND MYRON A. HOFER

In pyschobiology, there is a growing literature (as evidenced by the chapters in this volume) on the experimental study of parental behavior: the neural, sensory, neuroendocrine, and neurochemical events surrounding the *adult* expression of maternal behavior in mammalian species. Despite this burgeoning interest in parental behavior at all levels of analysis, until recently there has been little systematic study of the *development* of parental behavior—that is, parental behavior shown by the young of various species. As expressed by Mayer (1983), "an interest in ontogeny was almost nonexistent; perhaps the belief was that maternal behavior is an adult pattern not to be found amoung young animals" (p. 1). And yet, young animals from a wide range of species show parental or parenting-like behavior long before they reach reproductive maturity. Recent field research has shown that in animals as diverse as cichlid fish, passerine and corvid bird species, mongooses, coyotes, wolves and other carnivores, some rodent species, and a large number of nonhuman primate species, as well as humans, food provision, defense, retrieval and transport, babysitting, and socialization of young can be seen in adolescents or juveniles (for reviews and examples of this extensive literature, see Bekoff, 1981; Bekoff & Wells, 1982; Brown, 1985; Moehlman, 1987; Ostermeyer & Elwood, 1984; Rood, 1978; Taborsky, 1985).

Given the extent of parenting behavior by young animals across species and orders, what is the adaptive value of such behavior? The selective pressures contributing to the evolution of helping behavior are still a matter of controversy (Bekoff, 1981; Brown, 1985), although early models favored a modification of Hamilton's (1964) "kin selection" theory (cited in Brown, 1975), in which a helper increases the reproductive success of siblings or parents, and thus indirectly its own. Brown (1985) speculates that for juveniles or immatures, "helping" behavior, defined as "parentlike behavior toward young that are not offspring of the helper" (p. 137), very likely confers a variety of both present and future benefits on the helpers. He states that in the case of juveniles and immatures, the adaptive value of engaging in parenting behavior may lie in postponing parenthood (delayed breeding) until foraging efficiency has increased or until animals rise high enough in the dominance hierarchy to establish optimal breeding conditions, either territorial or social. In species in which consecutive litters or broods temporally overlap, first-born litters can benefit by prolonging their association with the mother—for example, by extending the nursing period—while providing such parental care as nest attendance (Calhoun, 1949; Gilbert, Burgoon, Sullivan, & Adler, 1983). Young may also derive substantial gains by "inheriting" portions of parental territory in time, as well as by reaping the benefits of being familiar with and recognized by offspring that have been helped and that will help in their turn (Brown, 1985). Another obvious possibility is that young animals learn parenting skills by engaging in parenting behavior early on. For example, nest building in Mexican jays requires skill and practice, and while 2-year-old jays attempt nest building (with little success), 1-year-olds do not (Brown, 1985). A variation of this theme comes from experimental studies in rats, which suggest that familiarity with young at an early age may predispose older animals to behave maternally later on (Gray & Chesley, 1984; Moretto, Paclik, & Fleming, 1986), although there is little evidence for the early learning of parental skills in rodents (see Mayer, 1983, for a review of juvenile parenting in rodent species).

Thus although juvenile parental behaviors have been described, and there is much speculation about their adaptive value to both juveniles and their recipients, little attention has been given to the mechanisms involved in their role in the ontogeny of adult parental behavior. Nevertheless, the sheer number of descriptions of parental behavior in young animals suggests that this period of life may represent a universally important time for the development of parental behavior.

CHARACTERISTICS OF JUVENILE PARENTAL BEHAVIOR

Characterizations of parental behavior in young animals include such terms as "clumsy" (Owens, 1975), "piecemeal" (Mayer, 1983), "inept" (Trollope & Blurton-Jones, 1975), and "playful" (MacDonald, 1979; Redican & Mitchell, 1973). In monkeys and apes, numerous accounts exist of parenting behavior on the part of young animals (usually females ranging from juvenile to subadult ages)—carrying, cradling, grooming, sniffing, and retrieving infants (Epple, 1975; Estrada & Sandoval, 1977; Horvich, 1974; Lancaster, 1971; Owens, 1975; Trollope & Blurton-Jones, 1975). In some species, parenting behavior shown by young nonhuman primates has been described as occurring within the context of play or having a playful aspect (Owens, 1975; Trollope & Blurton-Jones, 1975), and Lancaster (1971) and others have described these behaviors as "play parenting," since the behaviors in many respects conform to definitions of play. In that sense, instances of parenting shown by juveniles often superficially resemble adult parental behavior sequences, but (1) behavioral sequences are reordered; (2) elements of other behaviors may be intermingled; (3) components in a sequence may be omitted; and (4) movements may be exaggerated or inhibited during play (Loizos, 1966). As an example, Owen (1975) described infant-carrying behavior in *Papio anubis:*

> Characteristically, one juvenile female would chase another who carried an infant ventrally, and rough-and-tumble play sometimes followed . . . this often gave the other juvenile a chance to seize it and carry it away herself. Between play bouts, and often during rough-and-tumble play, the infant would be groomed intensely by the female in possession. (p. 397)

In experimental studies of rodents, juvenile caretaking activities have been similarly described (Mayer, 1983). In an early study of maternal behavior in juvenile hamsters, Rowell (1961) described the behavior as follows:

> The "maternal" behavior of the pups, though composed of the same behavioural elements as that of the lactating female . . . has many of the attributes of play activity: the patterns are often not complete but are broken off suddenly . . . the young animal gives an impression of "not knowing what to do next." (p. 13)

Parental behavior, when it occurs in young children, often occurs during play, and studies of children's "play parenting" abound, conducted in the context of either development of gender role (e.g., Connolly, Doyle, & Ceschin, 1983; Pitcher & Schultz, 1983) or development of prosocial behavior, often in naturalistic studies of play (e.g., Radke-Yarrow, Zahn-Waxler, & Chapman, 1983). Several lines of research with humans suggest that play parenting occurs most often in girls and, to a lesser extent, in boys (e.g., Fagot & Leinbach, 1983; Grief, 1976); the same appears to be true of nurturant behaviors (Maccoby & Jacklin, 1974), although some studies have found that nurturant, caretaking behaviors are exhibited equally by both sexes (Pelletier-Stiefel et al., 1986; Radke-Yarrow et al., 1983). In such studies, parental and play behaviors are frequently intermingled, such as caretaking and agonistic behaviors (Pelletier-Stiefel et al., 1986), and parental and other types of roles (e.g., fire fighter) (Pitcher & Schultz, 1983). Berman (1986), for example, found that under experimental conditions when young children were asked to care for infants, attempts at nurturance, "although . . . strikingly adultlike, . . . often seemed poorly directed. At times the . . . interactions were prolonged, but at other times the behavior seemed inappropriate or badly timed" (p. 41).

Thus parental behavior expressed by young of many species has a "juvenile" quality in that the *patterns* of parental behaviors expressed are incomplete and show the intrusion of other behaviors. In particular, sequences of play are interwoven with parental behaviors, so that definitional lines become blurred (is this really parental behavior, or are they just "playing around"?). This may represent mere inexperience on the part of juveniles, but, of even more interest to developmental investigators, it may also show a unique concurrence of behavioral systems during ontogeny. It has been suggested that play is the "cradle of adult behavior" in that "it provides the first expression of behaviors we characterize as exploratory, aggressive, sexual, affiliative, and parental" (Hofer, 1981, p. 272). It is important to trace the origins of parental behavior in these first fragments enacted in play, as well as the forces that shape them into cohesive, integrated sequences of behavior. So, for example, it is known that children are capable of displaying nurturant behavior from a very early age under eliciting conditions (Rheingold, Hay, & West, 1976; Zahn-Waxler, Friedman, & Cummings, 1983). The question that is being asked here is, with respect to the development of parental behavior, are these inherent nurturant tendencies organized and expressed through play? And if so, does this process require certain kinds of experiences to optimally organize particular patterns of parental behavior? At a time when attitudes toward child rearing are changing with regard to the roles of the sexes, it is especially important to define these processes early in development.

This chapter describes parental behavior in preadolescent (juvenile) rats *(Rattus norvegicus),* as examined in a series of studies conducted in our laboratory and by others. These studies considered aspects of juvenile parental behavior and caregiving that have not been addressed in previous rodent studies—that is, the playful quality of parental behavior observed in juveniles across species. The focus was on the occurrence of social behaviors shown by juvenile rats in response to neonates: play and exploratory behaviors, other social behaviors, and parental behaviors. A major interest was in determining the relationships of these behaviors to one another in order to elucidate their role in the development of parental behavior during this period.

PARENTAL BEHAVIOR AND SENSITIZATION IN JUVENILE RATS

In the normal course of events, a postparturitional rat engages in an organized, integrated pattern of functionally related behaviors that promote the survival, growth, and development of the young. Four of them have been operationally defined as central, or "index," maternal behaviors by those who observe them experimentally: (1) feeding pups by adopting the highly characteristic nursing posture of *crouching* over pups with the back arched and the legs splayed, exposing the ventrum and nipple line; (2) *nest building* of a maternal nest designed to enclose, protect, and provide heat for the young, which are incapable of independent thermoregulation; (3) *retrieving* stray young to the nest by the nape of the neck; and (4) licking pups to clean them—in particular, *anogenital licking* to stimulate bladder and rectal emptying (Rosenblatt & Lehrman, 1963; Wiesner & Sheard, 1933; see also Rosenblatt, Siegel, & Mayer, 1979).

Parental behaviors can also be elicited in adult male and female rats through a process called "sensitization" (Rosenblatt, 1967), an experimental procedure in which individuals housed singly with pups for long periods eventually come to show the four index behaviors seen in mother rats (see, e.g., chapters 4 and 6 for discussions of sensitization in adults). Under these circumstances, sensitized adult females show an organized pattern of maternal behaviors similar to that displayed by lactating females (Fleming & Rosenblatt, 1974a).

Juvenile rats, like their older conspecifics, can be sensitized to show parental behaviors when in continuous contact with pups. In fact, juvenile rats (18–40 days of age) of both sexes can be induced to show parental behaviors with extreme rapidity, 24-day-olds showing an average latency of about 2 days (Bridges, Zarrow, Goldman, & Denenberg, 1974). This is in contrast to adult latencies, which average 4 to 6 days in females and vary between 6 and 14 days in males, depending on the strain of rats used or the conditions of testing (Brown, 1986; Jakubowski & Terkel, 1985; Lubin, Leon, Moltz, & Numan, 1972; Rosenblatt, 1967; Rosenblatt et al., 1979). Unlike adults, juveniles younger than about 24–25 days tend to exhibit little or no fear of neonates on initial exposure; on the contrary, they appear to be highly attracted to pups and in their presence show behaviors that may be described as "social," which take the form of sniffing, licking, pawing, and huddling in contact with pups (Mayer & Rosenblatt, 1979a, 1979b; Moretto et al., 1986). After this time, juveniles begin to show more adult-like response patterns characterized by increasing initial avoidance of neonates, which has been shown to be olfactory mediated (Fleming & Luebke, 1981; Fleming & Rosenblatt, 1974b; Mayer & Rosenblatt, 1979b).

Table 18.1   Description of behavioral variables

Retrieve	Juvenile picks up pup and carries it to nest or sleeping area.
Nest build	Juvenile picks up nesting material and places it in nest or arranges placement of nest material.
Anogenital lick	Juvenile engages in licking pup's anogenital area for a 2- to 5-sec period.
Crouch	Juvenile crouches over pup with back raised in an arch and legs splayed, exposing its ventral area (nipple line).
Sniff-lick-paw	Juvenile sniffs, licks, or paws pup, exclusive of anogenital area (after Mayer & Rosenblatt, 1979b).
Charge	Juvenile moves rapidly toward or away from pup with bounding, leaping, or hopping motions. Behavior normally seen in context of sibling play bouts (after Poole & Fish, 1975).
Pounce	Juvenile hops in an arc, landing with front paws and muzzle on pup. Motion usually directed at nape of neck in juvenile play bouts (after Poole & Fish, 1975).
Pup carry	Juvenile attempts to pick up or move pup, but not any appreciable distance.
Contact	Juvenile engages in any vegetative activity in body contact with pup (e.g., groom, lie, sit, eat, drink).
Activity	Any locomotor act (walking, running, climbing, etc.), exclusive of those listed above.
Other	Any vegetative activity *not* in body contact with pup (e.g., groom, lie, sit, eat, drink).

Juvenile latencies to show maternal behaviors rise to about 6 days in 30-day-olds (Mayer, Freeman, & Rosenblatt, 1979), although in some strains they remain somewhat lower (Stern, 1987). Also, unlike adults, juvenile (24-day-old) males of some strains show more retrieving behavior than females, and with shorter latencies (Barron & Riley, 1985; Bridges et al., 1974; Gray & Chesley, 1984; Kinsley & Bridges, 1988; Stern, 1987). This sex difference begins to reverse itself by about 30 days (Bridges et al., 1974; Mayer et al., 1979; Stern, 1987). In other strains, this difference is not reliable and requires special conditions to elicit it (Brunelli, Lloyd, & Hofer, 1986; Mayer et al., 1979).

Given these findings, a series of questions was asked about the nature and organization of parental behaviors shown by juvenile rats after sensitization. The first question was, do juveniles show *organized* parental behavior as a result of exposure to young, or are behavioral components fragmented and unconnected to one another? Next, is the organization of components of parental behavior affected by antecedent events? Finally, is there any relationship of parental behaviors to play in juvenile rats?

Brunelli, Shindledecker, and Hofer (1985) observed basic behavior patterns shown by male and female juvenile rats at 18, 24, and 30 days of age. In this and subsequent studies, animals were tested daily over 5 days; singly housed juveniles were tested for 15 min each day with freshly fed 3- to 8-day-old pups; the pups then remained with them for 24 hr. In order to determine the frequency of occurrence of these behaviors in each age group, each animal was observed once a minute for 15 min, for a total of 16 observations per day per animal over 5 days. The behaviors

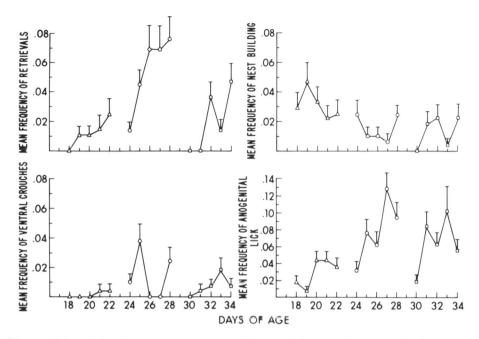

Figure 18.1.   Daily changes in group means (frequencies) over 5 days of testing for retrieving, nest building, crouching, and anogenital licking. Days 0–4 of testing for 18-day-olds denoted by triangles; for 24-day-olds, denoted by circles; for 30-day-olds, denoted by boxes. (Copyright © 1985, John Wiley & Sons. Reprinted by permission of the publisher)

observed can roughly be divided into four categories: maternal behaviors, play and exploratory behaviors, other (noncontact) behaviors, and passive contact (Table 18.1).

In this initial study, it was found that of the maternal behaviors tested, retrieving and anogenital licking increased daily over the 5 days of testing in all three groups, as would be expected if they were being "induced" by exposure to pups (Figure 18.1). This is in accord with what other reearchers investigating juvenile parental behavior have found (Barron & Riley, 1984; Bridges et al., 1974; Gray & Chesley, 1984; Kinsley & Bridges, 1988; Mayer & Rosenblatt, 1979). Crouching and nest-building behaviors apparently were not induced by pup exposure, as evidenced by the fact that there as no increase in these behaviors over time. In this study, no sex differences were found for any of the behaviors measured, although in a subsequent study others' findings were replicated showing that males retrieve more than females in the 24-day-old group (Brunelli et al., 1986).

A survey of studies investigating juvenile maternal behavior shows that most report some form of crouching behavior on the part of juvenile rats, but the form and frequency of the behavior appear to depend on the strain of rats used, age of juveniles, conditions of testing, and possibly criteria for crouching. Mayer and Rosenblatt (1979) informally observed juveniles in their Sprague-Dawley colony "spread-eagled" over pups (Mayer, personal communication). Gray and Chesley (1984) and Moretto, Paclik, and Fleming (1987) noted that juveniles rarely crouched in the lactating posture but were frequently seen lying on top of pups. Kinsley and Bridges

(1988) observed juveniles crouching over pups and included it as a criterion for maternal behavior, as did Stern (1987). On the basis of observations in adult lactating females with their litters, one would predict that juveniles tend to crouch more during the light portion of the day–night cycle (Grota & Ader, 1969; Myers, Brunelli, Squire, Shindedecker, & Hofer, 1989), but reports of juvenile crouching are not consistent with this finding. They varied from study to study, regardless of the light cycle or the length of the test period. Such inconsistencies suggest that crouching, while it does occur, may not be a reliable phenomenon in juveniles, especially given the proclivity of juveniles to huddle with pups (Mayer & Rosenblatt, 1979). Further systematic work should document the parameters of its occurrence in juveniles, and the sufficient and necessary conditions under which it occurs.

Of the other social behaviors observed, social sniffing-licking-pawing decreased over time in all three age groups, indicating that this behavior occurred in response to some aspect of novelty in the pups, as suggested by Mayer and Rosenblatt (1979b). Pouncing, the play-initiatory behavior, like retrieving and anogenital licking, also increased with exposure to pups in the same pattern as retrieving and anogenital licking. While this suggests that pouncing may be induced in the same manner as retrieving or anogenital licking, it is also true that pouncing behavior increases over time in animals isolated from peers (Einon & Morgan, 1976; Panksepp & Beatty, 1980; Stevens & Alberts, 1981), as the juveniles were in this study, their only companions being neonates for the 5 days of sensitization. Pouncing behavior also increases with age over the age range tested (Meaney & Stewart, 1981). As in the Mayer and Rosenblatt study (1979b), it was found that whereas younger animals spent most of their time in contact with pups, older animals tended to spend time away from pups, suggesting the inhibition of responsiveness characteristic of adults (Fleming & Luebke, 1981).

Due to an interest in patterns of relationships among the various classes of behavior, the data were subjected to principal component (factor) analyses. This is a method of organizing intercorrelations among variables into new sets of composite variables (principal components) that are orthogonal (uncorrelated) with other sets (components) (Kim & Mueller, 1978).* If parental behaviors were well organized— that is, occurring in close contiguity as a *patterned* behavioral output—one would expect to see them intercorrelated across observation periods, and thus clustered together on one factor in juveniles.

Table 18.2 shows that juvenile parental behaviors were not well correlated in any of the three groups studied. Instead, the data reveal a pattern of parental behaviors (principal component) in which one behavior, retrieving, is more highly correlated with the play behaviors of charging and pouncing than with other parental behaviors. This was true at all three ages. There was some indication (as a third

---

*It is important to note that, as used in these studies, principal component analyses are descriptive in nature; therefore, no inferences can be made about differences among groups underlying populations. Nor are assumptions of a causal relationship made between some common underlying factor structures and patterns of play and maternal behaviors (as is done in common factor analysis). Play or maternal behavior is not an entity (or set of them), but consists of integrated patterns that are subject to change over time and across conditions. The objective of this work is to look at changes in interrelationships with manipulations of development, and at patterns that are stable at a given age but change over time. Although based on multiple observations of animals, these analyses have shown remarkable stability across many sets of data, yielding an informal test of reliability of the data set as a whole.

Table 18.2   Principal component analyses: relations among behavioral variables

Age group (days)	Principal components	
18	1 (14.2)[a]	2 (10.8)
	Retrieve	Contact
	(0.38)[b]	(0.82)
	Pup carry	
	(0.80)	
	Charge	
	(0.73)	
	Pounce	
	(0.84)	
24	1 (13.5)	2 (9.6)
	Retrieve	Contact
	(0.35)	(0.77)
	Pup carry	
	(0.74)	
	Charge	
	(0.66)	
	Pounce	
	(0.85)	
30	1 (13.2)	2 (9.9)
	Retrieve	Contact
	(0.56)	(0.74)
	Pup carry	
	(0.59)	
	Charge	
	(0.74)	
	Pounce	
	(0.85)	

[a] The percentage of total variance accounted for by each component.

[b] Numbers in parentheses below variables are coefficients called "loadings." They express the association of the variable to the component.

Source: Copyright © 1985, John Wiley & Sons. Reprinted by permission of the publisher.

component, not shown) of a developmental progression in older groups, with crouching and anogenital licking becoming increasingly associated. Nest building was orthogonal (unrelated) to any other parental behaviors.

These data for juvenile patterns of parental behaviors are different from the patterns shown by adult females. Under similar conditions of sensitization, naive, virgin adult females at 100 days of age show statistical evidence of a high degree of integration of most maternal index behaviors by principal component analyses (Brunelli, Shindledecker, & Hofer, 1989), confirming Fleming and Rosenblatt's (1974a) findings. That is, retrieving, anogenital licking, crouching, and contact appear on the first principal component with very high loadings. It is of note that nest building is not correlated with other maternal behaviors in sensitized adult females as well as in juveniles. Slotnick (1967) also found that in newly parturient mothers, measures of pup licking and nursing correlated with each other and with time in the nest, but showed no relationship to retrieving or nest-building measures. The independence of retrieving in this situation is probably due to the fact that under natural nesting conditions, retrieving is much less likely to occur than during sensitization. These data suggest that while maternal behaviors are also independent in inexperienced adult females, they are organized after parturition and during lactation. Moreover, the adult female is *prepared* to behave in an organized fashion prior to parturition, as evidenced by the behavior of virgins during sensitization.

Thus juvenile parental behavior is different from the adult form in that its components are distinct from one another in their expression, are not elicited in an integrated pattern after sensitization, and are intermixed with juvenile play behaviors in one parental behavior, retrieving.

HORMONAL FACILITATION OF BEHAVIOR

Given these findings, the next attempt was to manipulate juveniles in such a way as to exert an organizing effect on their parental behavior. In an early study, Koranyi, Lissak, Tamasy, and Kamaras (1976) were able to facilitate maternal behavior with startlingly short latencies (average, 12 hr) in 18- to 24-day-old juvenile rats by injecting blood plasma from newly parturient rats. Earlier, similar procedures by Terkel and Rosenblatt (1968) had facilitated maternal behavior in adult virgin females with 48-hr latencies, indicating that hormones present in maternal blood plasma were acting on the neural substrate to produce these short latencies.

Brunelli, Shindledecker, and Hofer (1987) incorporated methods from these two studies to examine the patterns of maternal and play behaviors expressed by juveniles under the facilitative influence of hormones in postparturient blood plasma. Two questions were posed: (1) can parental behaviors shown by juveniles become organized in an adult-like fashion in response to hormone action? and (2) would the relations first observed between play and maternal behaviors be altered as a consequence of hormonal manipulations?

On the day before testing (day −1), 18- and 30-day-old male and female subjects were implanted with chronic jugular cannulas under ether anesthesia. On the day of testing (day 0), blood was collected from newly parturient donor dams, centrifuged to collect plasma, and infused slowly through the chronic catheters into the subjects (plasma [PL] group), which were gently restrained in a plastic chamber. A second group of siblings (infused controls [IC]) received an equal volume of 5% dextrose solution to control for manipulations received as a result of infusion—for example, anesthesia, tissue damage, blood volume expansion, restraint during infusion—which were called "handling." A third sibling group (nonhandled controls [NHC]) received no handling prior to testing.

It is noteworthy that almost all effects occurred in the older group of 30-day-olds, and these are reported here. The clearest effect of hormone facilitation was to enhance retrieving (Figure 18.2A) in 30-day-old females relative to either handled or nonhandled females. The latency for this plasma effect was similar to the 48 hr reported for adult virgin females by Terkel and Rosenblatt (1968). In this respect, maternal responsiveness in older juvenile females appears to share a hormonally responsive neural substrate. However, major effects were due to handling, in which

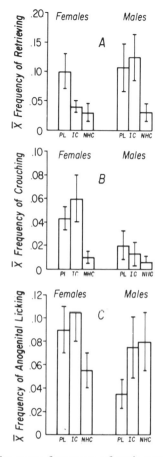

Figure 18.2. Thirty-day-olds: mean frequency of each maternal behavior over 5 days for males and females in each treatment group. PL = subjects infused with maternal blood plasma; IC = infusion control group, infused with 5% dextrose in water; NHC = nonhandled control group, receiving no catheter implantation or infusion. (Copyright © 1987 by the American Psychological Association. Reprinted by permission of the publisher)

the effects of plasma (PL) could not be distinguished from the nonspecific effects of the infusion procedure itself (IC group). This was true for retrieving in males (Figure 18.2A), in which PL and IC groups showed higher levels of retrieving relative to NHCs. Crouching (Figure 18.2B) and anogenital licking (Figure 18.2C) were also increased in handled (PL, IC) females over nonhandled (NHC) levels. Thus there were different effects of handling on parenting behaviors for the two sexes, suggesting that the mechanisms mediating hormonal and handling effects are different.

Interestingly, handling had a similar effect on pouncing in both sexes, increasing levels of pouncing relative to NHCs. This indicates that pouncing and retrieving may be under the control of separate systems, since retrieving appears to be hormonally mediated, at least in part, whereas pouncing is not. For males, there was no evidence of hormonal facilitation of any behaviors.

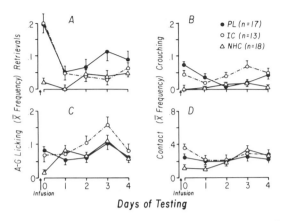

Figure 18.3. Daily changes in 30-day-old group means over 5 days of testing for (A) retrieving, (B) crouching, (C) anogenital licking, and (D) contact. Days 0–4 of testing for plasma subjects (PL), infusion controls (IC), and nonhandled controls (NHC). (Copyright © 1987 by the American Psychological Association. Reprinted by permission of the publisher)

The differences between hormone and handling effects in the sexes may be explained, in part, by an examination of the immediate and delayed effects of the two manipulations. As Figure 18.3 shows, changes that occurred in retrieving appeared to be hormonally mediated (Figure 18.3A) and were delayed by several days after PL infusion. But in addition to these changes, there is an immediate effect of handling on retrieving (Figure 18.3A), as evidenced by the relatively high frequencies shown on day 0 by the PL and IC groups. This was true as well for crouching (Figure 18.3B), anogenital licking (Figure 18.3C), and contact (Figure 18.3D) on the first observation. There was a similar short-term effect on pouncing behavior.

Unpublished data from our laboratory indicate that this acute effect of handling occurs for both sexes and thus appears to be a basic, undifferentiated juvenile pattern. In contrast, the facilitation that occurs with hormone manipulation appears to be confined to retrieving in older females. This effect is consistent with reports that for juvenile females, adult-like responses are established with the advent of puberty (Mayer et al., 1979; Mayer & Rosenblatt, 1979a) and with hormonal events occurring at puberty that would heighten responsiveness to hormonal facilitation (Ojeda, Andrews, Advis, & Smith-White, 1980; Ramirez & Sawyer, 1965). These events are the maturation of the neuroendocrine reproductive axis, increased uterine weight, increasing serum estradiol levels, increasing sensitivity of central components of steriod positive feedback, and the growth of estrogen receptors in the hypothalamus, pituitary, and amygdala (Dudley, 1981; McEwen, 1983; Ojeda et al., 1980; Parker & Mahesh, 1976).

In contrast, the *decline* of responsiveness (increased latency) shown by juveniles appears to be unaffected by hormonal manipulations. For example, Mayer and Rosenblatt (1979a) found that pregnant mare's serum given to juvenile females at 21 days, which induced premature cycling, failed to alter characteristic juvenile responses at 30 days. In a recent study by Stern (1987), neither gonadectomy nor hypophysectomy in females at 21 days altered the increase in latencies to maternal responsiveness at 42 days, implying that affected gonadal and pituitary hormones had no influence on the decline of maternal responsiveness with age. Thus while our data and those of Koranyi et al. (1976) suggest that at least retrieving behavior may be *facilitated* hormonally in juvenile females, to date little can be said about the endocrine basis underlying parental responsiveness under more natural conditions prior to puberty.

In males, the picture is also incomplete, but it suggests little gonadal basis for juvenile parental behavior. For example, testosterone administered to males at 21 days produced no change in latency (0.4–2 days) or in the number of males showing maternal behavior (64–88%) when tested at 24 days of age; this is in contrast to latencies shown by adult male rats (6 to >15 days; Bridges, Zarrow, & Denenberg, 1973; Fleischer et al., 1981; LeRoy & Krehbiel, 1978). As in females, gonadectomy of males at 21 days did not alter the increase in latencies to maternal responsiveness occurring at 42 days (Stern, 1987), again suggesting the lack of testosterone involvement in the pubertal decline in juvenile parental behavior. Recent work by Kinsley and Bridges (1988), however, suggests that prolactin levels may contribute to the greater *responsiveness* that has been found in younger (24-day-old) juvenile males. These authors speculate that juvenile females may be susceptible to prolactin influence somewhat earlier in development, which may account for the differences in retrieving at this age. They also suggest that prolactin in blood plasma may account for some of its facilitative effects. However, as the authors point out, bromocriptine used to alter prolactin levels in their experiments may have had direct dopaminergic effects on behavior, as well as other endocrine effects (see also Martin & Bateson, 1982). Given the fact that catecholamine systems have been implicated in the mediation of handling effects (Weiner & Bethea, 1981), these possibilities have to be further disentangled before definitive statements about hormonal action in males can be made.

How did the changes in individual behaviors brought about by these manipulations affect the structure of relationships among parental and other behaviors? The results of principal component analyses (Table 18.3) showed that in 18-day-olds and nonhandled 30-day-olds, the structure of relationships in the first two factors remained unchanged, replicating the findings of the first study. That is, in older, unmanipulated animals and in younger animals regardless of manipulation, retrieving remained associated with the play behaviors charging and pouncing. In contrast, in older, handled animals that received an infusion of plasma or a control substance, an interesting shift in the structure of behaviors occurred. In these groups, retrieving was no longer associated with the play behaviors charging and pouncing, but with two other maternal behaviors, crouching and anogenital licking. Nest building was uncorrelated with maternal or play behaviors. Pup carrying, which by definition represented incomplete retrieving, remained associated with play behaviors, even though facilitated by handling in the short term. These findings were true for both sexes.

Thus the procedural manipulations associated with intravenous infusion had the unpredicted effect of integrating three maternal behaviors—retrieving, crouching, and licking—in older but not in younger juveniles. These results parallel the finding by Mayer, Freeman, and Rosenblatt (1979) that handling facilitated the appearance of maternal behavior in 30-day-old juveniles of both sexes. On the basis of this result, Mayer et al. suggested that the growth of pup avoidance occurring during the juvenile

Table 18.3 Principal component analyses: relations among
behavioral variables

Treatment group	Principal components	
PL	1 (18.1)[a]	2 (11.4)
	Pup carry	Retrieve
	(0.57)[b]	(0.54)
	Charge	Crouch
	(0.86)	(0.45)
	Pounce	Anogenital lick
	(0.90)	(0.74)
IC	1 (17.2)	2 (11.6)
	Pup carry	Retrieve
	(0.48)	(0.39)
	Charge	Crouch
	(0.87)	(0.48)
	Pounce	Anogenital lick
	(0.90)	(0.69)
NHC	1 (16.4)	2 (10.9)
	Retrieve	Contact
	(0.44)	(0.70)
	Pup carry	
	(0.61)	
	Charge	
	(0.74)	
	Pounce	
	(0.85)	

[a]The percentage of total variance accounted for by each component.
[b]Numbers in parentheses below variables are coefficients called "loadings." They express the association of the variable to the component.
PL = plasma subjects; IC = infusion control; NHC = nonhandled controls.
Source: Copyright © 1987 by the American Psychological Association. Reprinted by permission of the publisher.

period is an expression of a generalized development of fear responses characteristic of adult rats that is not present in younger animals (Bronstein & Hirsch, 1976). Stern (1987) has suggested that in the normal course of events, weanling rats respond socially to pups prior to the advent of adult neophobia; at puberty, this responsiveness is suppressed by maturing inhibitory pathways, noted below. In our study, the fact that handling facilitated maternal responsiveness in 30-day-old but not in younger juveniles is consistent with this notion, since handling is known to influence "fear-mediated" behaviors (Archer, 1973).

A great deal of evidence indicates that in rats, avoidance responses to pups are mediated by olfaction (Fleming & Rosenblatt, 1974b; Mayer & Rosenblatt, 1975, 1977, 1979a) and associated limbic structures that are thought to inhibit maternal behavior (Fleming, Micelli, & Moretto, 1983; Fleming, Vaccarino, & Luebke, 1980; Numan, 1987; see chapter 12). Investigating the role of olfaction in juvenile and postadolescent rats, Mayer et al. (1979) found some interesting ontogenetic changes in olfactory-mediated responses to pups in relation to handling. Using zinc sulfate to produce anosmia, which in adult females facilitates maternal responsiveness (e.g.,

Mayer & Rosenblatt, 1975), Mayer et al. found that in females, zinc sulfate was more effective in facilitating sensitization with increasing age between 30 and 60 days. Similar results were obtained with males. However, handling facilitated maternal responsiveness in both sexes at 30 days; its importance declined in females with age, but not in males, up to about 90 days. In a separate experiment, Mayer et al. found that males required more handling (from day 21 to day 30) than females to maximize maternal behavior at day 30. They concluded that while olfactory-mediated avoidance appears not to be gender specific, there is another dimension of avoidance of pups, corresponding to generalized fear of novelty exhibited by rats, that appears to have a sexually dimorphic component (for discussion, see Mayer, 1983).

Findings on the organization of maternal behaviors suggest that the effects of handling may go beyond simple fear reduction in ways that cannot yet be specified. However, mechanisms that mediate the widespread effects of handling may include neuroendocrine or neurochemical modulation in response to handling, which may substitute at an earlier age for the normal role played by maternal or ovarian hormones. One such candidate is prolactin, which has been implicated in the stimulation of maternal care in both adult and juvenile rats (Bridges, DiBiase, Loundes, & Doherty, 1985; Jakubowski & Terkel, 1986; Kinsley & Bridges, 1988; Loundes & Bridges, 1986; Stern & Siegel, 1978) and in the response to stressors encountered during handling (Weiner & Bethea, 1981). Prolactin has been thought to act in conjunction with elevated levels of estrogen (which was likely to be the case in parturient blood plasma), along with reductions in progesterone levels (Bridges et al., 1985; Rosenblatt et al., 1979), to stimulate maternal behavior. Concomitant changes in neurotransmitter systems accompanying stress (e.g., serotonin, endogenous opiates, norepinephrine, catecholamines; Weiner & Bethea, 1981) may act to mediate hormone effects or may act independently. Changes in adrenal hormones accompanying handling (e.g., Denenberg, 1964) could also be involved, possibly by altering olfactory sensitivity. Clearly, though, whatever the mechanisms are, handling older juveniles promotes a pattern of behavior characteristic of adult females. Whether these mechanisms are the same in juvenile males and females, and at what age they diverge, also remains to be determined.

SOCIAL FACTORS CONTRIBUTING TO PARENTAL BEHAVIOR IN JUVENILE RATS

Having tried to integrate patterns of parental behaviors hormonally, the next attempt was to shift the relationships between parental and play behaviors, using manipulations of early social experience known to affect behavior organization (Brunelli et al., 1989; Einon & Morgan, 1977; Hofer, 1975a; Panksepp & Beatty, 1980; Stevens & Alberts, 1981). Observations were made of how early weaning in sibling groups (EG) or early weaning plus social isolation (EI) from 14 to 24 days of age would selectively alter the frequencies of maternal, filial, and play behaviors shown by 24-day-old juvenile rats in the presence of neonates. Also of interest were the possible effects produced by these early experiences on the integration of these behaviors. Our purpose was to test the degree to which retrieving and play were linked in the

Table 18.4 Principal component analyses: relations among behavioral variables

Treatment group	Principal components	
NW	1 (19.1)[a]	2 (15.4)
	Retrieve	Activity
	(0.29)[b]	(0.70)
	Pup carry	Sniff-lick-paw
	(0.72)	(0.61)
	Charge	Rear
	(0.72)	(0.58)
	Pounce	
	(0.82)	
EG	1 (18.6)	2 (16.4)
	Retrieve	Sniff-lick-paw
	(0.35)	(0.77)
	Pup carry	Rear
	(0.74)	(0.62)
	Charge	Activity
	(0.53)	(0.62)
	Pounce	
	(0.86)	
EI	1 (23.8)	2 (14.3)
	Pup carry	Sniff-lick-paw
	(0.80)	(0.71)
	Charge	Activity
	(0.77)	(0.70)
	Pounce	Rear
	(0.88)	(0.50)

[a]The percentage of total variance accounted for by each component.

[b]Numbers in parentheses below variables are coefficients called "loadings." They express the association of the variable to the component.

NW = normally weaned subjects; EG = early weaned group; EI = early weaned isolate.

Source: Copyright © 1989, John Wiley & Sons. Reprinted by permission of the publisher.

organization of juvenile parental behavior. If they were only weakly or fortuitously associated, they could become dissociated by the developmental effects of these long periods of early social deprivation.

Previous studies demonstrated that retrieving behavior in juveniles shares a statistical commonality with the play behaviors of charging, pouncing, and pup carrying, and the results of this study, shown in Table 18.4, showed that normally weaned 24-day-olds showed the same relationship. But surprisingly, in this study retrieving could be disengaged from these other behaviors by a 10-day prior period of social isolation. As illustrated by Figure 18.4a, early weaned and isolated (EI) animals retrieved less than their siblings early weaned in groups (EG) or their normally weaned siblings (NW), while the other behaviors changed in the opposite direction with isolation.

Pup carrying, an incomplete form of retrieval, increased with isolation (Figure 18.4b), as did the behaviors pouncing (Figure 18.4c) and charging (Figure 18.4d).

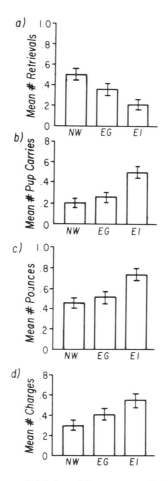

Figure 18.4. (a) Mean scores of 24-day-old groups retrieving over 5 days of testing. (b) Mean scores for each group pup carrying over 5 days of testing. NW = normally weaned; EG = early weaned—group; EI = early weaned—isolate. (c) Mean scores of 24-day-old groups pouncing over 5 days of testing. (d) Mean scores for each group charging. (Copyright © 1989 by John Wiley & Sons. Reprinted by permission of the publisher).

Thus a 10-day period of prior isolation appeared to produce a reduction in retrieving by substituting other behaviors in the repertoire of these juveniles. Principal component analyses indicated this as well: while normally weaned and early weaned group animals showed retrieving, pup carrying, charging, and pouncing together on the first component, for the early-weaned isolates, retrieving was not associated with this group of behaviors (see Table 18.4).

Several hypotheses could be advanced to explain how social isolation disrupts retrieving behavior (Brunelli et al., 1989). One favored hypothesis is that rats subjected to prolonged early social isolation have difficulty in inhibiting responding or switching between different patterns of behavior (Einon & Morgan, 1976). One hour of social play a day alleviates such deficits in isolates (Einon & Morgan, 1978), leading to the suggestion that the characteristics of social exposure that facilitate the ability to switch between patterns of behavior may arise from the rapid alternation of roles and behavior patterns encountered in play. By this hypothesis, socially isolated rats in this study were unable to change from a playful pattern of behavior to an appropriate caretaking behavior, such as retrieving, on the basis of cues encountered in association with week-old pups.

What is the nature of the connections between play and maternal behaviors in normal juveniles? If one considers principal components as weighted sets of correlations (Kim & Mueller, 1978), one can extrapolate the conclusion that if individual animals charge, pounce, pup carry, and retrieve, then those behaviors will appear together on a component. The main hypothesis was that the temporal coupling of behaviors accounts for these correlations (Brunelli et al., 1985); that is, individual animals charge, then pounce, then attempt to pick up pups, and finally retrieve pups in a behavior chain (Figure 18.5a). This interpretation is modeled on Poole and Fish's (1975) study of the temporal relationships among the elements of social play in juvenile rats, in which the behaviors charging, pouncing, aggressive grooming, and wrestling follow one another with high probabilities (Figure 18.5b), with charging and pouncing serving to initiate play bouts (see also Pellis & Pellis, 1983).

This hypothesis of temporal coupling was examined by Brunelli, Lloyd, and Hofer (1986). The reasoning was that the sensitization procedure itself constitutes social deprivation, based on the fact that juveniles will display increased play solicitation following a period of social isolation of as little as 2 hr (Panksepp & Beatty, 1980; Stevens & Alberts, 1981). Therefore, instead of isolating all juveniles from social interactions with peers throughout sensitization, as is usually done, one group was provided with a 1-hr period of play interaction with three littermates after the removal of pups from the previous day and 30 min before the presentation of fresh pups. The idea was that juveniles would play vigorously during this period and, as a result, would show less charging and pouncing when confronted later with neonates. If play and retrieving were linked in a behavior chain, retrieving would be reduced, along with the antecedent behaviors charging and pouncing.

Following the presentation of fresh pups, the first 5 min of the 15-min test on each of the 5 days of testing were videotaped to allow for later sequence analysis in both groups.* The collapsed data were subjected to first-order transitional (Markov chain) analyses (Fagen & Young, 1979) in order to determine the temporal relationships among behaviors in the first 5 min of testing in each of the groups.†

Analyses of frequencies of behaviors showed that during the 15-min test, animals allowed the 1-hr interaction with agemates (Play group) retrieved significantly less than animals without this experience (Isolate group) (Figure 18.6). In fact, they exhibited fewer of all maternal and play behaviors but did not differ in general activity. Observation of the 1-hr interaction with agemates in the Play group showed high levels of rough-and-tumble play, as expected.

Figure 18.7 shows the temporal relationships among behaviors. The analysis indicates that in animals having the usual sensitization experience (Isolate), charging leads to pouncing, which leads to pup carrying and retrieving in a probabilistic be-

---

*This period was chosen on the basis of the observation that the behaviors of interest—retrieving, pouncing, charging, and pup carrying—occurred primarily during this time.
†Because observations were made over 5 days, assumptions of stationarity are probably violated (Fagen & Young, 1979); therefore, these results must be considered preliminary to more rigorous analyses.

Hypothesized Play–Maternal Behavior Sequence

(a) Juvenile–Pup: Charge ⟶ Pounce ⟶ Pupcarry (?) ⟶ Retrieve
(b) Juvenile–Sibling: Charge ⟶ Pounce ⟶ Aggressive Groom ⟶ Wrestle

Figure 18.5.   (a) Hypothesized temporal relationships between juvenile play behaviors and retrieving behavior during sensitization. (b) Juvenile–sibling sequence. (Based on Poole & Fish, 1975)

havior chain, accounting for the factor structure seen in juveniles. Other relationships also exist, in which sniffing-licking-pawing leads to anogenital licking, and both lead to contact (lying) with pups, in what appears to be another subset of behavior, and conforming to the picture presented by the second juvenile principal component factor structure (see Tables 18.2, 18.3, and 18.4).

In contrast, animals having 1 hr of play prior to testing (Play) showed remarkably altered patterns of relationships (Figure 18.8). First, it was found that their patterns were less complicated, reflecting fewer temporal pathways between behaviors. More striking was the finding that retrieving was disconnected from the usual path: pouncing seemed to occur randomly, with no predictable connection between pouncing, pup carrying, or retrieving, and the relationship between charging and pouncing was reversed. Thus the behavior sequence leading to retrieving appeared to be eliminated by the reversal of play deprivation inherent in the normal sensitization procedure. Some of the pathways seen in normally sensitized juveniles still existed, particularly among the licking behaviors and contact (lying) with pups; these seemed

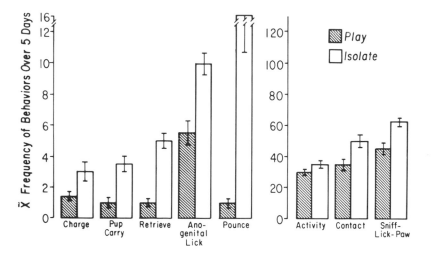

Figure 18.6.   Mean frequencies of behaviors shown during the first 5 min of testing by juveniles allowed to play for 1 hr prior to testing each day for 5 days (Play group) vs. animals not given play experience (Isolate group). Panel at right shows higher frequencies of behaviors than does panel at left.

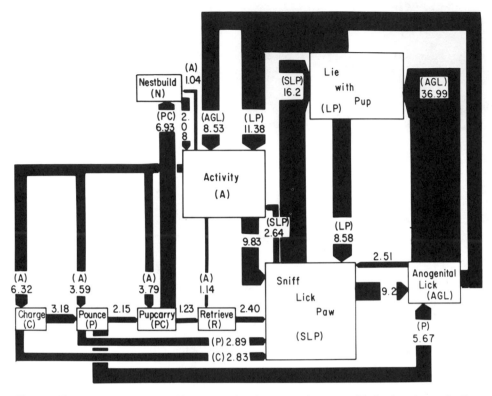

Figure 18.7. Significant transitions ($p < .05$) between elements of behavior shown by Isolate group during the first 5 min of daily testing during sensitization. The Z-scores associated with the transitions are printed on the arrows.

not to have been affected by the play manipulation, indicating that play behaviors were therefore separable from contact and licking, as suggested by the presence of two separate and orthogonal factor structures in the previous experiments.

The tentative conclusion from this is that play satiation appears to affect not only the frequency of behaviors shown, but also the organization of behaviors leading to retrieving. Follow-up studies should analyze the 1-hr interaction experience for those factors responsible for its effects—for example, sensory cues associated with agemate companions (thermal, tactile, olfactory), as well as behavioral interactions (huddling contact, play-initiatory behaviors, wrestling, following, grooming, and so on). Another factor to be considered is the possibility that the play experience produced generalized fatigue, but the lack of change in general activity levels during subsequent testing argues against this. In addition, stimulus properties of pups may be important in that different-age pups may elicit different classes of behaviors in juveniles, as suggested by studies by Mayer and Rosenblatt (1979b). For example, older, livelier pups may elicit more play behaviors than younger pups, but the same amount of retrieving. These kinds of studies should allow one to analyze the interaction of stimulus factors with the organization of juvenile parental behaviors in interpreting the present results.

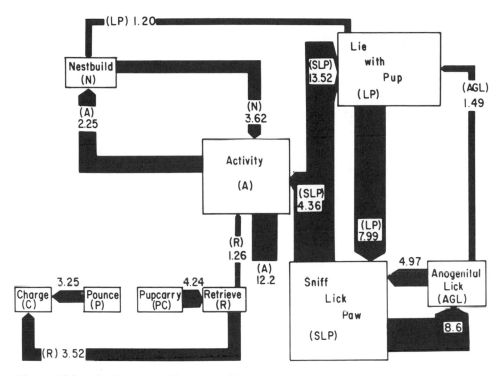

Figure 18.8. Significant transitions ($p < .05$) between elements of behavior for the Play group.

DISCUSSION AND CONCLUSIONS

The data reviewed in these experiments—ours and those of others—are the beginning of the study of the ontogeny of parental behavior in rats, from which some preliminary conclusions can be drawn. First, the data presented support research showing that in juvenile rats of various ages, parental behavior can be induced by continuous exposure to pups. Such findings suggest a degree of similarity between early and later maternal behaviors. In contrast, the behavior of juvenile rats shows qualitative differences from that of adult females under the same conditions. In juveniles, maternal behaviors are neither consistent nor integrated in their expression, and some maternal behaviors are facilitated more easily than others. Moreover, there are age-related differences in the expression of maternal behaviors. These data indicate the gradual maturation of adult-type functioning in a number of domains.

Second, evidence is accumulating that differences in parental behavior exist between males and females during the juvenile period along a number of dimensions. The most obvious difference is in retrieving, noted by many investigators and in different stocks (Brunelli et al., 1986). Other gender differences in the parental behavior of juveniles are more subtle, and seem to be related to developing differences in traits that are not specifically related to parental behavior but that are likely to affect its expression. For example, fear of novelty and related traits appear to be differentially affected by handling in males, and they presumably affect maternal responsiveness (for discussion, see Mayer, 1983). The expression of these behaviors, and of maternal behavior as well, is influenced by the presence of perinatal androgens in males (e.g., Beatty, 1979; Gorski, 1971; Goy & McEwen, 1978; Quadagno, Briscoe, & Quadagno, 1977). To the extent that maternal behavior is more likely to occur in adult males that have been neonatally castrated than in males that have not, exposure to androgens (specifically testosterone) perinatally appears to suppress parental behavior (McCullough, Quadagno, & Goldman, 1974; Quadagno & Rockwell, 1972; Rosenberg & Herrenkohl. 1976). This conclusion also appears to be true of females treated perinatally with androgen (Ichikawa & Fujii, 1982). It is unclear whether the critical androgen-sensitive period for suppression of maternal behavior is the same for females as for males (Bridges et a, 1973; Guerra & Hancke, 1982; Quadagno, DeBold, Gorzalka, & Whalen, 1974). Nor is it clear what level of androgen treatment is necessary for suppression (or enhancement) of maternal behavior in neonatally treated females (see Gonzalez & Deis, 1986, for a discussion of the possible mechanisms of androgenization of maternal behavior in females). Nevertheless, the likelihood is strong that perinatal hormones affect later sensitivity to hormonal or stimulus factors that act to facilitate maternal behavior, although they may not be responsible for the basic capacity to behave maternally (see chapter 17).

Two pieces of evidence bear on these findings. One is the result of recent factor analyses of 24-day-old females compared with males during sensitization. The data indicate that for females, retrieving, crouching, licking, and contact behaviors cluster together on one factor, as is typical of adult females, while play behaviors cluster with retrieving on another. In 24-day-old males, in contrast, the typical juvenile pattern is seen, in which maternal behaviors are *not* associated and in which play behaviors occur on the same factor with retrieving (Brunelli, unpublished data). A second, related finding comes from factor analysis of adult maternal behavior during sensitization. In the same sensitization context in which adult females show the organized pattern of parental behavior discussed earlier, adult *male* patterns more closely resemble those of juveniles. In this pattern, there appears to be no organization of maternal behaviors, retrieving is dissociated from other maternal behaviors, and retrieving can be associated with play behaviors (Brunelli et al., 1989).

These data suggest that the basic behavioral response to pups may be a juvenile pattern, which begins to differentiate in females as early as the prepubertal period. During the pubertal period and later, under the influence of ovarian hormones, maternal behaviors become organized and integrated into the functional adult female maternal pattern. In the absence of ovarian hormones, or as a consequence of the maturation of reduced sensitivity to pup stimulus factors in males, maternal behavior in the male retains its juvenile quality into adulthood. Hypothetically, the male developmental pattern outlined earlier could occur in neonatally androgen-treated, masculinized females. Conversely, males castrated neonatally (and treated with an appropriate female hormone regimen; e.g., Bridges et al, 1973) could be expected to show maternal behavior in a more functionally integrated fashion, both as juveniles and as adults.

These findings also demonstrate that juveniles of both sexes show a unique pattern in which parental behaviors are apparently coupled, temporally and perhaps even functionally, with play behaviors. One of the hypothesized functions of play is that it serves to acquaint young animals with the stimulus–response contingencies involved in social interactions (Bekoff & Byers, 1981). Another is that play acts to promote flexibility, to make behavioral adjustments in adaptation to changing properties and contexts in the physical and social environments (Fagen, 1984). For rodents specifically, pouncing initiates play-fighting, bringing young animals in contact with one another in social interaction (Pellis & Pellis, 1983). Thus in the context of sensitization, under stimulus conditions not usually encountered by juveniles, pouncing may serve to bring juveniles into contact with pups, acquainting them with their specific stimulus properties, and acting as a mediator through which retrieving, and thus parental behavior, is established. In play-satiated animals that have experienced play just prior to sensitization, the stimulus cues emitted by pups may be insufficient, in comparison with those just encountered with agemates, to elicit pounces and other play behavior that leads to retrieving. Another process is at work in juveniles socially isolated since late infancy. For them, the contingencies between play behavior and retrieving are missed, through either an inability to discriminate the stimulus charactertistics of pups or an inability to alternate behavioral modes in response to the stimulus cues emitted by pups. Whether this behavioral inflexibility continues into adulthood as a consequence of early social deprivation very likely depends on subsequent life events, which may ameliorate its effects (Ackerman, Hofer, & Weiner, 1975; Hofer, 1975b).

The plasticity of behavior inherent in the responsiveness of juveniles may stem, in part, from the fact that the components of parental behavior appear to occur as unconnected or loosely associated motor acts. However, it is evident that juvenile parental behaviors are not simply disorganized, but can be organized and reorganized, depending on the context in which they are expressed. The same components can be combined in various ways to subserve functionally different activities during development, including play. This hypothesis may account for other phenomena in development—for example, the development of the lordosis reflex (Beach, 1966; Williams & Lorang, 1987; see also Komisaruk, 1978, for a discussion of hormone-sensitive behavior patterns). Thus for juveniles, parental behaviors can be coupled with other behavioral acts or integrated into a maternal pattern, depending on the age, sex, hormonal status, and experience of the animal. As suggested at the beginning of the chapter, this malleability is likely to be adaptive in the face of a variety of circumstances early in life, and ultimately in parenting behavior later in life.

ACKNOWLEDGEMENTS

This research was supported by a National Institute of Mental Health project grant and Research Scientist Award to M.A.H. and by National Institute of Mental Health research training grants (predoctoral and postdoctoral) to S.A.B.

We wish to thank Augustus Romano and Harriet Lloyd for their help in data collection.

REFERENCES

Ackerman, S. H., Hofer, M. A., & Weiner, H. (1975). Age at maternal separation and gastric erosion susceptibility in the rat. *Psychosomatic Medicine, 37,* 180–184.

Archer, J. (1973). Tests for emotionality in rats and mice: A review. *Animal Behavior, 21*, 205–235.

Barron, B. S., & Riley, E. P. (1985). Pup-induced maternal behavior in adult and juvenile rats exposed to alcohol prenatally. *Alcoholism: Clinical and Experimental Research, 9*, 360–365.

Beach, F. A. (1966). Ontogeny of "coitus related" reflexes in the female guinea pig. *Proceedings of the National Academy of Sciences, USA, 56*, 526–533.

Beatty, W. W. (1979). Gonadal hormones and sex differences in nonreproductive behaviors in rodents: Organizational and activational influences. *Hormones and Behavior, 12*, 112–163.

Bekoff, M. (1981). Mammalian sibling interactions: Genes, facilitative environments and the coefficient of familiarity. In D. J. Gubernick & P. Klopfer (Eds.), *Parental care in mammals* (pp. 307–346). New York: Plenum Press.

Bekoff, M., & Byers, J. A. (1981). A critical reanalysis of the ontogeny and phylogeny of mammalian social and locomotor play: An ethological hornet's nest. In K. Immelman, G. W. Barlow, L. Petrinovich, & M. Mann (Eds.), *Behavioral development: The Bielefeld Interdisciplinary Project* (pp. 296–337). Cambridge: Cambridge University Press.

Bekoff, M., & Wells, M. C. (1982). Behavioral ecology of coyotes: Social organization, rearing patterns, space use, and resource defense. *Zeitschrift für Tierpsychology, 60*, 281–305.

Berman, P. (1986). Young childen's responses to babies: Do they foreshadow differences between maternal and paternal styles? In A. Fogel & G. F. Melson (Eds.), *Origins of nurturance: Developmental, biological and cultural perspectives on caregiving* (pp. 25–51). Hillsdale, NJ: Erlbaum.

Bridges, R. S., DiBiase, R., Loundes, D. D., & Doherty, P. (1985). Prolactin stimulation of maternal behavior in female rats. *Science, 227*, 782–784.

Bridges, R. S., Zarrow, M. X., & Denenberg, V. H. (1973). The role of neonatal androgen in the expression of hormonally induced maternal responsiveness in the adult rat. *Hormones and Behavior, 4*, 315–322.

Bridges, R. S., Zarrow, M. X., Goldman, B. D., & Denenberg, V. H. (1974). A developmental study of maternal responsiveness in the rat. *Physiology and Behavior, 12*, 149–151.

Bronstein, P. M., & Hirsch, S. M. (1976). Ontogeny of behavioral reactions in Norway rats. *Journal of Comparative and Physiological Psychology, 90*, 620–629.

Brown, J. L. (1975). *The evolution of behavior.* New York: Norton.

Brown, J. L. (1985). The evolution of helping behavior—an ontogenetic and comparative perspective. In E. S. Gollin (Ed.), *The comparative development of adaptive skills: Evolutionary implications* (pp. 137–171). Orlando, FL: Academic Press.

Brown, R. E. (1986). Paternal behavior in the male Long-Evans rat *(Rattus norvegicus). Journal of Comparative Psychology, 100*, 162–172.

Brunelli, S. A., Lloyd, H. O., & Hofer, M. A. (1986). *Play interactions and juvenile maternal behavior in the rat.* Paper presented at the Conference on Reproductive Behavior, Montreal.

Brunelli, S. A., Shindledecker, R. D., & Hofer, M. A. (1985). Development of maternal behaviors in prepubertal rats at three ages: Age-characteristic patterns of responses. *Developmental Psychobiology, 18*, 309–326.

Brunelli, S. A., Shindledecker, R. D., & Hofer, M. A. (1987). Behavioral responses of juvenile rats *(Rattus norvegicus)* to neonates after infusion of maternal blood plasma. *Journal of Comparative Psychology, 101*, 47–59.

Brunelli, S. A., Shindledecker, R. D., & Hofer, M. A. (1989). Early experience and maternal behavior in rats. *Developmental Psychobiology, 22,* 295–314.

Calhoun, J. B. (1949). *The ecology and sociology of the Norway rat.* Bethesda, MD: Department of Health, Education, and Welfare.

Connolly, J., Doyle, A-B, & Ceschin, F. (1983). Forms and functions of social fantasy play in preschoolers. In M. B. Liss (Ed.), *Social and cognitive skills: Sex roles and children's play* (pp.71–92). New York: Academic Press.

Denenberg, V. H. (1964). Critical periods, stimulus input and emotional reactivity: A theory of infant stimulation. *Psychological Review, 71,* 335–351.

Dudley, S. (1981). Prepubertal ontogeny of responsiveness to estradiol in the female rat central nervous system. *Neuroscience and Behavioral Reviews, 5,* 421–435.

Einon, D. F., & Morgan, M. J. (1976). Habituation of object contact in socially-reared and isolated rats. *Animal Behaviour, 24,* 415–420.

Einon, D. F., & Morgan, M. J. (1977). A critical period for social isolation in the rat. *Developmental Psychobiology, 10,* 123–132.

Einon, D. F., & Morgan, M. J. (1978). Early isolation produces enduring hyperactivity in the rat, but no effect on spontaneous alternation. *Quarterly Journal of Experimental Psychology, 30,* 151–156.

Epple, G. (1975). Parental behavior in *Saguinus fuscicollis* (sp. *callithricidae*). *Folia Primatologica, 24,* 221–238.

Estrada, A., & Sandoval, J. M. (1977). Social relations in a free-ranging troop of stumptail macaques *(Macaca arctoides)*: Male-care behavior 1. *Primates, 18,* 793–813.

Fagen, R. (1984). Play and behavioral flexibility. In P. K. Smith (Ed.), *Play in animals and humans* (pp. 159–173). New York: Basil Blackwell.

Fagen, M. F., & Young, D. Y. (1979). Temporal patterns of behaviors: Durations, intervals, latencies, and sequences. In P. W. Colgan (Ed.), *Quantitative ethology* (pp. 79–114). New York: Wiley.

Fagot, B. I., & Leinbach, M. D. (1983). Play styles in early childhood: Social consequences for boys and girls. In M. B. Liss (Ed.), *Social and cognitive skills: Sex roles and children's play* (pp. 93–116). New York: Academic Press.

Fleischer, S., Kordower, J. H., Kaplan, B., Dicker, R., Smerling, R., & Ilgner, J. (1981). Olfactory bulbectomy and gender differences in maternal behaviors of rats. *Physiology and Behavior, 26,* 957–959.

Fleming, A. S., & Luebke, C. (1981). Timidity prevents the virgin female rat from being a good mother: Emotionality differences between nulliparous and parturient females. *Physiology and Behavior, 27,* 863–838.

Fleming, A. S., Micelli, M., & Moretto, D. (1983). Lesions of the medial preoptic area prevent the facilitation of maternal behavior produced by amygdala lesions, *Physiology and Behavior, 31,* 503–510.

Fleming, A. S., & Rosenblatt, J. S. (1974a). Maternal behavior in the virgin and lactating rat. *Journal of Comparative and Physiological Psychology, 86,* 957–972.

Fleming, A. S., & Rosenblatt, J. S. (1974b). Olfactory regulation of maternal behavior in rats: I. Effects of olfactory bulb removal in experienced and inexperienced lactating and cycling females. *Journal of Comparative and Physiological Psychology, 86,* 221–232.

Fleming, A. S., Vaccarino, F., & Luebke, C. (1980). Amygdaloid inhibition of maternal behavior in the nulliparous female rat. *Physiology and Behavior, 25,* 731–743.

Gilbert, A. V., Burgoon, D. A., Sullivan, K. A., & Adler, N. T. (1983). Mother–weanling interactions in Norway rats in the presence of a successive litter produced by a postpartum mating. *Physiology and Behavior, 30,* 267–271.

Gonzalez, D. E., & Deis, R. P. (1986). Maternal behavior in cyclic and androgenized female rats: Role of ovarian hormones. *Physiology and Behavior, 38*, 789–793.

Gorski, R. A. (1971). Gonadal hormones and the perinatal development of neuroendocrine function. In L. Martini & W. F. Ganong (Eds.), *Frontiers in neuroendocrinology* (pp. 237–290). New York: Oxford University Press.

Goy, R. W., & McEwen, B. S. (1978). *Sexual differentiation of the brain.* Cambridge, MA: MIT Press.

Goy, R. W., & McEwen, B. S. (1980). *Sexual differentiation of the brain.* Cambridge, MA: MIT Press.

Gray, P., & Chesley, S. (1984). Development of maternal behavior in nulliparous rats *(Rattus norvegicus):* Effects of sex and early maternal experience. *Journal of Comparative Psychology, 98*, 91–99.

Grief, E. B. (1976). Sex role playing in preschool children. In J. S. Bruner, A. Jolly, & K. Sylva (Eds.), *Play, its role in development and evolution.* (pp. 385–391). New York: Basic Books.

Grota, L. G., & Ader, R. (1969). Continuous recording of maternal behaviour in *Rattus norvegicus. Animal behavior, 17*, 722–729.

Guerra, F., & Hancke, J. L. (1982). Neonatal masculinization affects maternal behavior sensitivity in female rats. *Experientia, 38*, 868–869.

Hofer, M. A. (1975a). Studies on how maternal separation produces behavioral change in young rats. *Psychosomatic Medicine, 37*, 245–264.

Hofer, M. A. (1975b). Survival and recovery of physiologic function after early maternal separation in rats. *Physiology and Behavior, 15*, 475–480.

Hofer, M. A. (1981). *The roots of human behavior: An introduction to the psychobiology of early development.* San Francisco: Freeman.

Horvich, R. H. (1974). Development of behaviors in a male spectacled langur *(Presbytis obscurus). Primates, 15*, 151–178.

Ichikawa, S., & Fujii, Y. (1982). Effect of prenatal androgen treatment on maternal behavior in the female rat. *Hormones and Behavior, 16*, 224–233.

Jakubowski, M., & Terkel, J. (1985). Incidence of pup killing and parental behavior in virgin female and male rats *(Rattus norvegicus):* Differences between Wistar and Sprague-Dawley stocks. *Journal of Comparative Psychology, 99*, 93–97.

Jakubowski, M., & Terkel, J. (1986). Nocturnal surges and reflexive release of prolactin in parentally behaving virgin female and male rats. *Hormones and Behavior, 20*, 270–286.

Kim, J. O., & Mueller, C. W. (1978). *Factor analysis: Statistical methods and practical issues.* Beverly Hills, CA: Sage.

Kinsley, C. H., & Bridges, R. S. (1988). Prolactin modulation of maternal-like behavior displayed by juvenile rats. *Hormones and Behavior, 221*, 49–65.

Komisaruk, B. K. (1978). The nature of the neural substrate of female sexual behavior in mammals and its hormone sensitivity: Review and speculation. In J. B. Hutchinson (Ed.), *Biological determinants of sexual behavior* (pp. 349–393). New York: Wiley.

Koranyi, L., Lissak, K., Tamasy, V., & Kamaras, L. (1976). Behavioral and electrophysiological attempts to elucidate central nervous system mechanisms responsible for maternal behavior. *Archives of Sexual Behavior, 11*, 232–247.

Lancaster, J. B. (1971). Play-mothering: The relations between juvenile females and young infants among free-ranging vervet monkeys. In J. S. Bruner, A. Jolly, & K. Sylva (eds.), *Play: Its role in development and evolution* (pp. 368–382). New York: Basic Books.

LeRoy, L. M., & Krehbiel, D. A. (1978). Variations in maternal behavior in the rat as a function of sex and gonadal state. *Hormones and Behavior, 11*, 232–247.

Loizos, C. (1966). Play in mammals. In P. A. Jewell & C. Loizos (Eds.), *Play, exploration and territory in mammals* (pp. 1–9). New York: Academic Press.

Loundes, D. D., & Bridges, R. S. (1986). Length of prolactin priming differentially affects maternal behavior in female rats. *Hormones and Behavior, 11,* 232–247.

Lubin, L. M., Leon, M., Moltz, H., & Numan M. (1972) Hormones and maternal behavior in the male rat. *Hormones and Behavior, 3,* 369–374.

Maccoby, E. E., & Jacklin, C. N. (1974). *The Psychology of Sex Differences.* Stanford, CA: Stanford University Press.

MacDonald, D. W. (1979). "Helpers" in fox society. *Nature, 282,* 69–71.

Martin, P., & Bateson, P. (1982). The lactation-blocking drug bromocriptine and its application to studies of weaning and behavioral development. *Developmental Psychobiology, 15,* 139–157.

Mayer, A. D. (1983). The ontogeny of maternal behavior in rodents. In R. W. Elwood (Ed.), *Parental behavior of rodents* (pp. 1–20). Chichester: Wiley.

Mayer, A. D., Freeman, N. G., & Rosenblatt, J. S. (1979). Ontogeny of maternal behavior in the laboratory rat: Factors underlying changes in responsiveness from 30 to 90 days. *Developmental Psychobiology, 121,* 425–439.

Mayer, A. D., & Rosenblatt, J. S. (1975). Olfactory basis for the delayed onset of maternal behavior in virgin female rats: Experiential effects. *Journal of Comparative and Physiological Psychology, 89,* 701–710.

Mayer, A. D., & Rosenblatt, J. S. (1977). Effects of intranasal zinc sulfate on open field and maternal behavior in female rats. *Physiology and Behavior, 18,* 101–109.

Mayer, A. D. & Rosenblatt, J. S. (1979a). Hormonal influences during ontogeny of maternal behavior in female rats. *Journal of Comparative and Physiological Psychology, 93,* 879–898.

Mayer, A. D., & Rosenblatt, J. S. (1979b). Ontogeny of maternal behavior in the laboratory rat: Early origins in 18–27-day-old young. *Developmental Psychobiology, 12,* 407–424.

McCullough J., Quadagno, D. M., & Goldman, B. D. (1974). Neonatal gonadal hormones: Effect on maternal and sexual behavior in the male rat. *Physiology and Behavior, 12,* 183–188.

McEwen, B. S. (1983). Neural gonadal steroid actions. *Science, 211,* 1303–1311.

Meaney, M. J., & Stewart, S. J. (1981). A descriptive study of social development in the rat *(Rattus norvegicus). Animal Behaviour, 29,* 34–45.

Moehlman, P. D. (1987). Social organization in jackals. *American Scientist, 75,* 366–375.

Moretto, D., Paclik, L., & Fleming, A. S. (1986). The effects of early rearing environments on maternal behavior in adult female rats. *Developmental Psychobiology, 19,* 581–591.

Myers, M. M., Brunelli, S. A., Squire, J. M., Shindledecker, R. D., & Hofer, M. A. (1989). Maternal behavior of SHR rats and its relationship to offspring blood pressures. *Developmental Psychobiology, 22,* 29–53.

Numan, M. (1987). Preoptic area neural circuitry relevant to maternal behavior in the rat. In N. A. Krasnegor, E. M. Blass, M. A. Hofer, & W. P. Smotherman (Eds.), *Perinatal development: A psychobiological persepective.* (pp. 275–298). Orlando, FL: Academic Press.

Ojeda, S. R., Andrews, W. W., Advis, J. P., & Smith-White, S. (1980). Recent advances in the endocrinology of puberty. *Endocrinology Reviews, 1,* 228–257.

Ostermeyer, M. C., & Elwood, R. W. (1984). Helpers(?) at the nest in the Mongolian gerbil, *Meriones unguiculatus. Behaviour, 91,* 61–77.

Owens, N. W. (1975). Social play behaviour in free-living baboons, *Papio anubis. Animal Behaviour, 23,* 387–408.

Panksepp, J., & Beatty, W. W. (1980). Social deprivation and play in rats. *Behavioral and Neural Biology, 30,* 197–206.

Parker, C. R., & Mahesh, V. B. (1976). Hormonal events surrounding the natural onset of puberty in female rats. *Biology of Reproduction, 14,* 347–353.

Pelletier-Stiefel, J., Pepler, D., Crozier, K., Stanhope, L., Corter, C., and Abramovitch, R. (1986). Nurturance in the home: A longitudinal study of sibling interaction. In A. Fogel & G. F. Melson (Eds.), *Origins of nurturance: Developmental, biological and cultural perspectives on caregiving* (pp. 3–23). Hillsdale, NJ: Erlbaum.

Pellis, S. M., & Pellis, V. C. (1983). Locomotor-rotational movements in the ontogeny of play of the laboratory rat *Rattus norvegicus. Developmental Psychobiology, 16,* 269–286.

Pitcher, S. L., & Schultz, M. S. (1983). *Boys and girls at play: The development of sex roles.* New York: Bergin & Garvey.

Poole, T. B., & Fish, J. (1975). An investigation of playful behaviour in *Rattus norvegicus* and *Mus musculus* (Mammalia). *Journal of Zoology, 175,* 61–71.

Quadagno, D. M., Briscoe, R., & Quadagno, J. S. (1977). Effect of perinatal gonadal hormones on selected nonsexual behavior patterns: A critical assessment of the nonhuman and human literature. *Psychological Bulletin, 84,* 62–80.

Quadagno, D. M., DeBold, J. F., Gorzalka, B. B., & Whalen, R. E. (1974). Maternal behavior in the rat: Aspects of concaveation and neonatal treatment. *Physiology and Behavior, 12,* 1071–1074.

Quadagno, D. M., & Rockwell, J. (1972). The effect of gonadal hormones in infancy on maternal behavior in the adult rat. *Hormones and Behavior, 3,* 55–62.

Radke-Yarrow, M., Zahn-Waxler, C., & Chapman, M. (1983). Children's prosocial dispositions and behavior. In E. M. Heatherington (Ed.), *Socialization, personality and social development* (Vol. 4) P. H. Mussen (Ed.), *Handbook of Child Psychology* (pp. 469–545). New York: Wiley.

Ramirez, V. D., & Sawyer, C. H. (1965). Advancement of puberty in the female rat by estrogen. *Endocrinology, 76,* 1158–1168.

Redican, W. K., & Mitchell, G. (1974). Play between adult male and infant rhesus monkeys. *American Zoologist, 14,* 295–302.

Rheingold, H. L., Hay, D. F., & West, M. J. (1976). Sharing in the second year of life. *Child Development, 47,* 1148–1158.

Rood, J. P. (1978). Dwarf mongoose helpers at the den. *Zeitschrift für Tierpsychology, 48,* 277–287.

Rosenberg, P. A., & Herrenkohl, L. R. (1976). Maternal behavior in male rats: Critical times for the suppressive action of androgens. *Physiology and Behavior, 16,* 293–297.

Rosenblatt, J. S. (1967). Nonhormonal basis of maternal behavior in the rat. *Science, 156,* 1512–1514.

Rosenblatt, J. S., & Lehrman, D. S. (1963). Maternal behavior in the laboratory rat. In H. L. Rheingold (Ed.), *Maternal behavior in mammals* (pp. 8–57). New York: Wiley.

Rosenblatt, J. S., Siegel, H. I., & Mayer, A. D. (1979). Progress in the study of maternal behavior in the rat: Hormonal, nonhormonal, sensory, and developmental aspects. In J. S. Rosenblatt, R. A. Hinde, C. Beer, & M. C. Busnel (Eds.), *Advances in the study of behavior* (Vol. 10, pp. 225–311). New York: Academic Press.

Rowell, T. E. (1961). Maternal behavior in nonmaternal golden hamsters. *(Mesocricetus auratus). Animal Behaviour, 9,* 11–15.

Slotnick, B. M. (1967). Intercorrelations of maternal activities in the rat. *Animal Behaviour, 15,* 267–269.

Stern, J. S. (1987). Pubertal decline in maternal responsiveness in Long-Evans rats: Maturational influences. *Physiology and Behavior, 41,* 93–98.

Stern, J. S., & Siegel, H. I. (1978). Prolactin release in lactating, primiparous and multiparous thelectomized, and maternal virgin rats exposed to pup stimuli. *Biology of Reproduction, 19,* 177–182.

Stevens, S. S., & Alberts, J. R. (1981). *An experimental analysis of play deprivation in Rattus norvegicus.* Paper presented at the meeting of the International Society for Developmental Psychobiology, New Orleans.

Taborsky, M. (1985). Breeder–helper conflict in a cichlid fish with broodcare helpers: An experimental analysis. *Behaviour, 85,* 45–75.

Terkel, J., & Rosenblatt, J. S. (1968). Maternal behavior induced by maternal blood plasma injected into virgin rats. *Journal of Comparative and Physiological Psychology, 65,* 479–482.

Trollope, J., & Blurton-Jones, N. G. (1975). Aspects of reproduction and reproductive behaviour in *Macaca arctoides. Primates, 16,* 191–205.

Weiner, R. I., & Bethea, C. L. (1981). Hypothalamic control of prolactin secretion. In R. B. Jaffe (Ed.), *Prolactin* (pp. 19–55). New York: Elsevier.

Weisner, B. P., & Sheard, N. M. (1933). *Maternal behavior in the rat.* London: Oliver & Boyd.

Williams, C. L. & Lorange, D. (1987). Brain transections differentially alter lordosis and ear wiggling of 6-day-old rats. *Behavioral Neuroscience, 101,* 819–826.

Zahn-Waxler, C., Friedman, S. L., & Cummings, E. M. (1983). Children's emotions and behaviors in response to infants' cries. *Child Development, 54,* 1522–1528.

19

Thermal Aspects of Parenting

MICHAEL LEON, ROBERT COOPERSMITH,
LAURA J. BEASLEY, AND REGINA M. SULLIVAN

PARENTING

The care of offspring is a critical aspect of mammalian reproduction. It is important for the early survival of the young and is the basis for their future interactions with the world. Therefore, the potential importance of research on animal models for the care of human infants is enormous. The importance of even subtle aspects of the physiological–psychological relationship for survival and development in humans is just beginning to be seriously considered in the scientific and medical communities.

There is a recent example of the ability of research on animal models to generate important advances in human neonatal care. Tactile stimulation of neonatal rats initiates physiological events that are required for normal growth and development (Schanberg, Evoniuk, & Kuhn, 1984). Similar tactile stimulation of premature human neonates living in an incubator has been shown to promote their growth and development (Scafidi et al., 1986). Such tactile stimulation also has thermal consequences in rat neonates (Sullivan, Shokrai & Leon, 1988; Sullivan, Wilson, & Leon, 1988), raising the possibility that the thermal aspects of mother–young interactions discussed in this chapter may play a role in producing the beneficial effects of tactile stimulation in both rat and human neonates.

In the review that follows, we focus on the thermal aspects of mother–young interactions in the Norway rat. Some comparative issues will then be addressed in a discussion of thermal aspects of maternal care in pallid bats.

THERMAL LIMITATION OF MOTHER–YOUNG CONTACT

During periodic contact bouts, Norway rat mothers deliver milk and warmth to their young; between bouts, the young are kept warm and protected in insulated nests (Leon, 1985b). Since all mother–young interactions must occur during these contact bouts, the mechanism controlling their duration should be an important aspect of parental behavior in this species. Indeed, the pattern of contact bout duration is interesting and perhaps counterintuitive. Rat mothers spend a progressively decreasing amount of time in contact with their pups during the first 2 weeks postpartum due to a decrease in the length of individual bouts (Leon, Croskerry, & Smith, 1978). The curtailment in the duration of bouts occurs despite the fact that the growing pups require more and more milk during this period (Grota & Ader, 1969; Leon et al., 1978).

Mother rats appear to control the duration of contact bouts, since their young maintain nipple contact unless forced to disengage (Hall, Cramer, & Blass, 1975). The question, then, is, what stimulates the dams to terminate contact with the litters? Mothers do not termiante bouts when they have delivered a specific amount of milk to the litter (Adels & Leon, 1986; Leon et al., 1978). Rather, mothers appear to terminate bouts when they experience acute hyperthemia. When acute maternal hyperthemia is prevented, either by placing the mothers in a cool environment or by continuously cooling the pups, mothers have very long contact bouts (Leon et al., 1978) (Figure 19.1). If the ambient temperature is high, mothers rapidly experience hyperthermia and quickly terminate contact bouts. Mothers kept with artificially warmed pups also experience rapid hyperthermia, and they rapidly terminate bouts. Since the pups were always warm, these data suggest that mothers do not monitor pup temperature and remain on the pups long enough to warm them to a particular temperature.

Reducing maternal heat load by removing the dam's fur prevents acute maternal hyperthermia and prolongs contact bouts. Increasing the maternal heat load by removing the tails of the dams (the tails are normally a major avenue of heat loss) provokes rapid hyperthermia and short contact bouts (Leon et al., 1978). Subcutaneous or core heating curtails contact bouts, but only after a long latency, while heating of the preoptic area induces rapid bout termination (Woodside, Pelchat, & Leon, 1980). Since the brain has neurons that respond to changes in local temperature (Boulant & Bignall, 1973), these data suggest that mothers may endure peripheral and core hyperthermia but terminate contact once their brain temperature rises.

Homeotherms normally guard their brain temperature from retained body heat (Baker, 1982), thereby avoiding the hazards associated with brain hyperthermia (Burger & Furman, 1964). Mammals normally keep their brain temperature lower than their core temperature by cooling the warm blood that enters the brain (Baker, 1982). It therefore seemed unlikely that mothers would allow their brain temperature to increase during maternal care. However, 79% of the 279 bouts that were analyzed had a stereotyped brain temperature rise that began 2–5 min prior to bout termination (Leon, Adels, & Coopersmith, 1985).

The thermal stimulus that prompts the dams to leave the nest is not a change in the rate of brain temperature rise, nor is it a response to exceeding a critical brain temperature. Rather, mothers appear to become more and more vulnerable to a prolonged brain temperature rise over the course of the contact bout that forces them to interrupt pup contact. This increase in vulnerability to an acute brain temperature rise may be due to a progressive increase in the amount of heat retained by the dam, as reflected in an increase in the number of transitory peaks in brain temperature over the course of a bout. While these transitory rises do not drive the dam off the pups, they may reflect increased difficulty in maintaining brain thermal homeostasis.

Chronic Elevation of Maternal Heat Load

Mother rats are vulnerable to acute hyperthermia during contact bouts because their heat load is chronically high, placing them at their physiological limit for heat dissi-

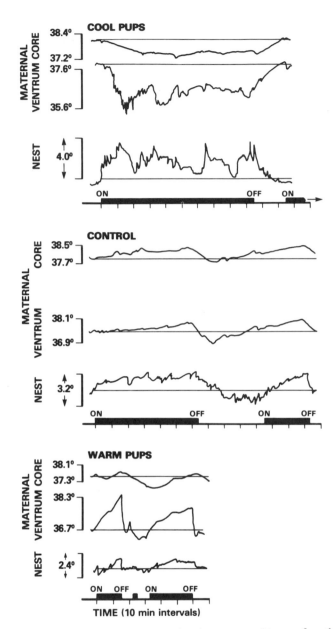

Figure 19.1. Top set of records shows ventral and core temperatures of mother Norway rats, as well as pup skin temperature (nest temperature) when pups are cooled continuously by a coil beneath the litter. The temperatures are shown in relation to the pattern of mother–young contact; "On" indicates the time that the dam began a contact bout, and "Off" indicates bout termination. The middle set of records shows the same recordings when pup temperature is unmanipulated. The bottom set shows the recordings when the temperature of the pups is kept artificially high. Note the decrease in contact bout duration with the increase in maternal heat load. (From Leon, et al., 1978)

pation. The additional acute loss of surface area available for heat dissipation during pup contact eventually increases retained heat to the point where the dams are forced to terminate pup contact.

Jans and Leon (1983b) found that mother rats have a chronically elevated core temperature. Dams are also far more vulnerable to hyperthermia than are nonlactating females. In addition, mother rats in a cold environment (4°C) are far more able to prevent their body temperature from decreasing than are nonlactating females (Jans & Leon, 1983b). Using more direct measures of maternal heat production, Adels and Leon (1986) found that mother rats have a 17.3% chronic increase in oxygen consumption relative to that of nonlactating females. In a cold environment (4°C), lactating and nonlactating rats increase their heat production to similar levels, but while the mothers maintain their high core temperature, the core temperature of nonlactating females decreases (Adels & Leon, 1986). Additionally, mothers maintain their body temperature in the cold with heat production no greater than that of nonlactating females, indicating an increased ability of mother rats to retain heat.

Endocrine Control

An obvious candidate for the mechanism underlying the chronically increased maternal heat load is the metabolic process involved in milk production. However, removal of the mammary tissue does not affect maternal hypermetabolism (Adels & Leon, 1986; Denckla & Bilder, 1975). Similarly, inhibiting milk delivery by sealing the mother's nipples does not decrease maternal core temperature or disrupt maternal contact patterns (Leon et al., 1978). Pups are rotated between milk-producing and non-milk-producing dams in these experiments to keep the young healthy and to give both experimental and control mothers equivalent pup stimulation.

Another candidate for the increase in heat production is brown adipose tissue, the only tissue that is specialized for heat production in rats (Nicholls & Locke, 1984). This tissue is normally activated by noradrenergic stimulation in a cold environment or after eating (Nicholls & Loche, 1984). Heat production by brown adipose tissue, however, is suppressed during lactation, probably to conserve energy for milk production (Isler, Trayhurn, & Lunn, 1984; Trayhurn, Douglas, & McGluckin, 1982). Neither removal of the brown adipose tissue nor suppression of noradrenergic activity affects maternal temperature (Adels & Leon, 1986).

The possibility that the pups induce their mothers to have a chronically elevated heat load by provoking the release of maternal hormones has also been considered. It seemed possible that these secretions could mediate a chronic increase in maternal heat load by increasing heat production and/or heat retention in the mothers. Rat pups induce their dams to secrete a variety of hormones, and these secretions were eliminated selectively to determine their role in the chronic increase in maternal heat production. Again, litters were rotated among dams to maintain healthy pups despite any impairments in lactation.

Removal of the adrenal glands along with the ovaries of mother rats chronically reduces their core temperature to the level of nonlactating females and sharply increases pup contact time (Leon et al., 1978). Removal of the ovaries or the adrenal medulla alone does not affect either maternal core temperature or contact time, while removal of the entire adrenal gland (with both medulla and cortex) decreases mater-

Chronic Component

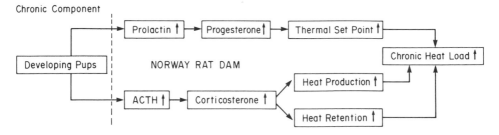

Figure 19.2. A model of the endocrine mechanisms underlying the chronic increase in maternal temperature. The developing pups induce their mothers to release both prolactin and ACTH, which provoke the release of progesterone and corticosterone, respectively. Progesterone then elevates the maternal thermal set point, and corticosterone then increases maternal heat production and possibly heat retention. The resulting chronic increase in maternal heat load makes the mothers vulnerable to a further acute increase in their heat load, eventually driving up maternal brain temperature and forcing the interruption of pup contact. (From Adels & Leon, 1986)

nal temperature and increases maternal contact time to the same extent as the double operation (Leon et al., 1978). These data suggest that adrenocortical glucocorticoids, which are normally elevated during lactation (Simpson, Simpson, Sinha, & Schmidt, 1973), induce the chronic increase in maternal temperature. Replacement therapy with corticosterone and cortisol, but not aldosterone (the dominant adrenal mineralocorticoid) or progesterone (the precursor of both adrenal mineralo- and glucocorticoids), restores the normally elevated maternal temperature and the normal contact pattern (Woodside & Leon, 1980). The frequency of contact bouts remains unchanged by endocrine manipulation. Chronic heat load replacement for adrenalectomized-ovariectomized dams by housing them in a warm environment decreases the duration of their contact bouts (Woodside & Leon, 1980).

Corticosterone increases the maternal heat load, at least in part, by increasing maternal heat production (Adels & Leon, 1986). Prolactin is indirectly involved in chronically increasing the maternal heat load by stimulating the production of progesterone, which, in turn, elevates the maternal thermal set point (Adels & Leon, 1986; Woodside et al, 1980). Progesterone itself, however, does not increase maternal heat production (Adels & Leon, 1986). Nor do thyroid hormones appear to play a role in increasing maternal heat production. Indeed, these secretions are actually depressed during lactation (Adels, Leon, Wiener, & Smith, 1986; Lorscheider & Reineke, 1972). This pattern of secretion may be a compensatory response to the chronically high maternal temperature, since dams can increase their thyroid hormone production in a cold environment (Adels et al., 1986). Oxytocin is released by suckling and provokes milk ejections in mother rats (Lincoln & Paisley, 1982). This hormone can raise body temperature when infused into the brain (Fahrbach, Morrell, & Pfaff, 1984), raising the possiblity that oxytocin contributes to the chronic, or even the acute, increase in maternal temperature. Figure 19.2 summarizes the proposed endocrine model for the mediation of the chronically elevated heat load of mother rats.

Diet

Factors other than hormones chronically alter maternal heat load, thereby changing the vulnerability of mother rats to acute hyperthermia. Restriction of maternal food intake depresses the maternal temperature chronically, allowing the mothers to increase the amount of time spent with their litters (Leon, Chee, Fischette, & Woodside, 1983). The increase in maternal contact time occurs even when pup body weight and temperature do not differ between diet-restricted and well-fed mothers. Diet restriction does not suppress circulating levels of either progesterone or corticosterone, suggesting that decreased food availability may decrease maternal temperature by limiting the available fuel for heat production (Leon et al., 1983).

Diurnal Rhythmicity

The temperature of lactating rats has a diurnal cycle, with the peak occurring during the night and the trough occurring during the day (Figure 19.3) (Leon, Adels, Coopersmith, & Woodside, 1984). Mothers also spend much more time with their offspring during the day than during the night (Ader & Grota, 1970; Croskerry, Smith, Leon, & Mitchell, 1976; Lee & Williams, 1977), due to a decreased duration of dark-phase contact bouts (Leon et al, 1984). There is a high inverse correlation between maternal brain temperature at bout onset and the duration of individual bouts ($r = -.67$; $p < .001$), indicating that mothers may be more vulnerable to acute brain hyperthermia during the night than during the day. Adrenalectomy-ovariectomy also supresses maternal temperature during both the day and the night. The bouts in both day and night are prolonged under these conditions, again suggesting that the diurnal cycle of maternal temperature mediates the daily cycle in maternal contact time.

In contrast to nonlactating rats, mothers kept in constant light do not have disrupted food- or water-intake cycles (Leon, 1985a). Mothers in constant light, however, gain less weight than their counterparts in a light–dark cycle. Although mothers in constant light do not differ in their total contact time or daily bout frequency from controls, the distribution of contact time is shifted in constant light. These data suggests that mothers maintain a free-running temperature and contact bout rhythm even under constant conditions.

Reproductive Strategy of Mother Rats

Since the duration of contact bouts is limited by hyperthemia, mothers can prevent an increase in their temperature by choosing to care for their young in a cool environment. However, mother rats, with or without their pups, choose a relatively warm area on a thermal gradient (Jans & Leon, 1983a). While dams with access to nest material move to a somewhat cooler area on a thermal gradient, the temperature within the nest remains high enough to increase maternal temperature during pup contact. Therefore, mother rats either seek out or construct a warm area in which to care for pups, thereby limiting their pup contact time. In fact, when dams can choose the temperature at which to care for their pups, they spend half as much contact time as they do in the cooler laboratory (Jans & Leon, 1983a).

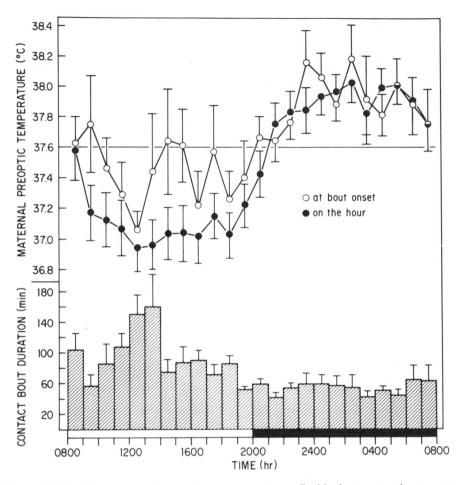

Figure 19.3. Mean maternal preoptic temperature transcribed both at contact bout onset and on the hour. A guideline is provided to facilitate light–dark comparisons. The dark bar indicates the dark phase of the light–dark cycle. The lower portion of the figure depicts the mean contact duration of bouts initiated during each hour. SEMs are shown. (From Leon et al., 1984)

What determines the duration of the interbout interval? Mother rats must dissipate the increased heat load incurred during the bout before they return to their pups (Jans & Leon, 1983a). When mothers are kept warm during the interbout interval on a warmed floor, the interbout interval is greatly prolonged. Mothers prefer to remain in a warm area during the interbout interval, thereby prolonging the duration of the interval. Therefore, mother rats seek neither to minimize their time away from their pups nor to maximize their pup contact time.

Contact Bout Initiation and Maintenance

In addition to those contact bouts which can last for 1 hr or longer, there are brief bouts that last for less than 10 min. While the long bouts are typically accompanied by a rise in maternal temperature and appear to be thermally limited, the short bouts are not accompanied by a rise in maternal temperature. What determines whether such short bouts are maintained? Perhaps dams periodically initiate contact to monitor the state of the litter within the first few minutes. Contact with the pups would be maintained only if the young were cool and/or if they attached to their dam's nipples. In fact, mothers returning to warm pups have significantly longer intervals between long bouts (> 10 min) than do mothers returning to cool pups or pups whose temperature is unmanipulated. These intervals, however, are punctuated by many short visits to the nest (Jans & Leon, 1983a). These data suggest that mothers monitor their pups during periodic brief visits and maintain contact if the pups are cool.

The roles of pup temperature, nipple attachment, and ultrasound emission in determining either the latency of the mother to return to the nest or the maintenance of the contact bout were then assessed. When mothers terminated a contact bout, a new litter was slid into the nest area with a specially designed double-nest tray (Jans & Leon, 1983a). The pups in the nest tray were either warm and awake, warm and anesthetized, cool and awake, or cool and anesthetized.

Although the temperature of the pups does not influence the initial latency of the mother to return to the nest, it does influence the probability of bout maintenance. All dams returning to cool pups had long contact bouts, regardless of the pups' ability to attach to their mother's nipples. Dams were much less likely to remain with warm pups, particularly if they were incapable of attaching to the dam's nipples. Of those bouts in which mothers remained with warm pups, 80% of the litters were attached to the nipples within 10 min (Jans & Leon, 1983a). These data suggest that if the pups are warm, nipple attachment may be a factor in maintaining mother–young contact.

A model summarizing the findings on the control of mother–young contact is presented in Figure 19.4. Specifically, a contact bout is likely to be maintained if the pups are cool and if they attach to their mother's nipples. If the bout is terminated before about 10 min, the mother returns to the young periodically, apparently without a direct signal from the pups. Eventually, when the pups present the appropriate stimulus to the dam following bout initiation, pup contact is maintained. The mother then begins to deliver milk to the young.

At the same time, the maternal heat load begins to rise acutely due to the decrease in the surface area available for heat loss when the mother contacts the pups with her ventrum. Any factor that changes the heat load of mother rats will increase the likelihood of a brain temperature rise that cannot be controlled by physiological means. The dams then use behavioral means to thermoregulate by interrupting their contact with the pups. The factors that affect the acute rise in maternal brain temperature include ambient temperature, pup temperature, the conductivity of the mother, and the surface area/mass ratio of the mother–young unit. The choice or the construction of a warm place in which to care for the young ensures that the mothers will be subjected to periodic hyperthermia while in contact with the pups.

The vulnerability of mothers to this acute hyperthermia is a result of their chronically elevated heat load, induced by the dual action of progesterone and corticosterone. Progesterone increases the thermal set point of the dams, while corticosterone increases maternal heat production and possibly heat retention. The fuel provided by

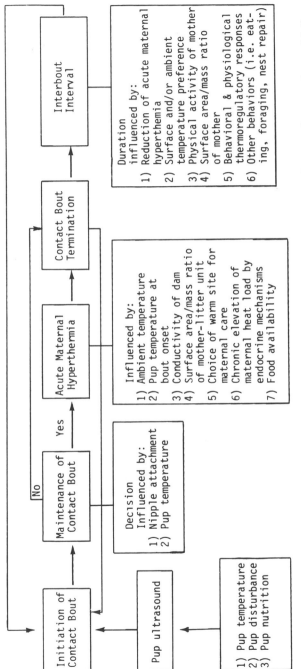

Figure 19.4. A model of the mechanisms underlying the initiation, maintenance, and duration of maternal contact bouts. Mothers monitor the pups during brief visits, which are maintained if the pups are cold or hungry. If the mother stays, her brain temperature eventually rises, depending on a variety of factors that affect her heat load, thereby limiting the duration of the bout. The mother must dissipate the heat gained during pup contact before she will again visit the litter, and the rate of heat loss during this time is also affected by a variety of factors that influence her heat load. Ultrasonic distress calls may bring the mother to the pups in an emergency situation. (From Jans & Leon, 1983a)

the increased food intake of lactation is also necessary to maintain the elevated heat production during that period (Leon & Woodside, 1983).

During the interbout interval, dams must dissipate sufficient heat to be able to return to their pups. The duration of the interbout interval may also be due, in part, to those factors that influence maternal heat load during that period. One such factor is the temperature of the preferred area in which mothers spend the interbout interval. They prefer to spend the interval in an area warm enough to prevent their rapid return to the pups (Jans & Leon, 1983a). Another factor that could change the mother's heat load, and thereby the duration of the interbout interval, is the physical activity patterns and the postures assumed by the mother during that time.

Maternal Compensatory Responses

Pups maintain their growth rate under a variety of circumstances that drastically reduce the duration of maternal contact bouts (Jans, de Villiers, & Woodside, 1984; Jans & Leon, 1983a). One mechanism by which mothers can maintain the pup growth rate with curtailed contact bouts is by decreasing the latency for milk delivery. Milk is normally ejected periodically into the nursing offspring (Lincoln & Paisley, 1982) only when the dam is in slow wave sleep (Lincoln et al., 1980). A periodic increase in paraventricular and supraoptic area neural activity precedes a release of oxytocin from the posterior pituitary gland (Lincoln & Paisley, 1982; Summerlee & Lincoln, 1981), and the oxytocin causes the mammary gland to expel the milk that has collected in alveolar structures.

Pups have a stereotyped behavioral response as the milk is received, allowing the time of milk delivery to be recorded (Wakerley & Drewett, 1975). Jans and Woodside (1987) observed these responses and found that milk delivery begins sooner for mothers nursing older rats and mothers caring for pups in a warm environment, situations in which contact bouts are brief. In addition, most of the milk delivered to the litters is provided in the first few milk ejections (Drewett & Trew, 1978; Grosvenor & Mena, 1983). Mothers with short contact bouts have prolonged interbout intervals (Jans & Leon, 1983a) and increase their available milk stores during such periods (Friedman, 1975). Together, these data suggest that mothers are able to deliver milk to their offspring in the shortened bouts necessitated by the thermal consequences of mother–young contact.

Pup Compensatory Responses

While this discussion has focused on the vulnerability of mother rats to hyperthermia during mother–young contact, the young also share that vulnerability. During contact bouts, mothers prevent pup heat loss and serve as a heat source for the pups, thereby increasing pup temperature (Leon et al., 1978). While pups are capable of moving away from heat sources to reduce their heat loads (Alberts, 1978a, 1978b; Fowler & Kellogg, 1975; Kleitman & Satinoff, 1982; Leon, 1985b), there is a large cost to moving away from the mother, since the young cannot obtain milk while away from her.

Since neonates are vulnerable to heat-induced seizures that can cause permanent neural damage (Gilbert & Cain, 1985; Holtzman, Obana, & Olsen, 1981) what de-

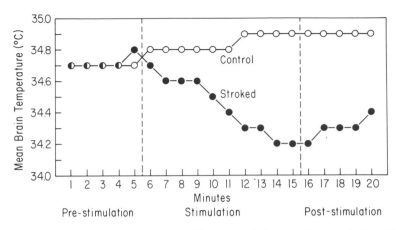

Figure 19.5. Brain temperature of 5-day-old rat pups before, during, and after 10 min of tactile stimulation. The decrease during stimulation is significant ($p < .05$). (From Sullivan et al., 1988)

termines whether pups can decrease their heat load during maternal contact bouts? Pups are quite vulnerable to hyperthermia in a warm environment in the absence of their mother (Sullivan, Shokrai, et al., 1988; Sullivan, Wilson, et al., 1988). When maternal contact is mimicked by stroking 5-day-old pups with a small brush, the elevated pup temperature observed in a warm environment is ameliorated. The mechanism for this action appears to involve a remarkable response by the pups to tactile stimulation. Both their body and their brain temperatures decrease reliably in response to such stimulation (Figure 19.5), suggesting that maternal contact provokes an anticipatory, compensatory thermoregulatory response by the pups.

Maternal tactile stimulation reliably precedes maternal heat exchange with the pups. The young have apparently evolved to use the tactile stimulus, rather than an elevated heat load, to initiate a compensatory thermoregulatory response in anticipation of the warm mother. Tactile stimulation produces a decrease in pup temperature by increasing pup respiration, thereby drawing cool, dry air across the respiratory membranes and the nasal venous plexus (Sullivan, Shokrai, et al., 1988; Sullivan, Wilson, et al., 1988). The convective and evaporative heat loss that occurs thereby decreases pup temperature; tactile stimulation in humid air prevents the decrease in pup temperature. Tactile stimulation decreases pup body temperature through postnatal day 20 but fails to affect brain temperature starting between postnatal days 5 and 10, when brain temperature becomes buffered from core thermal excursions (Sullivan, Wilson, et al., 1988).

The insulation provided by the maternal nest affects not only maternal heat load and contact pattern (Jans & Leon, 1983a), but also the development of the offspring. Pups reared by mothers with or without a nest have equal thermoregulatory abilities, but pups reared in nests normally have lower body temperatures, smaller adrenal glands, and less brown adipose tissue. Females are more affected by rearing conditions than are males (Jans et al., 1984).

LACTATION IN A HETEROTHERM

Given the importance of thermal considerations to homeothermic Norway rat mothers and their young, an analysis of temperature in a heterothermic lactating species is of interest. Pallid bats *(Antrozous pallidus)* were selected for this comparison. These bats forage at dusk and dawn, leaving their twin batlets with the other offspring in a crèche located in a cave or barn attic. Since bat mothers leave the young during the coolest part of the night, hyperthermia is unlikely to limit mother–young interactions in this species. Nevertheless, thermal considerations are critical for an understanding of the reproduction of this species.

Pallid bats are insectivores found in western North America. The females are inseminated in October and store sperm in their reproductive tracts until ovulation occurs in April (Orr, 1954). They are thought to hibernate during the winter; in June, all 20–200 females in the roost give birth in synchrony. (Brown, 1976).

Bats are particularly interesting with respect to thermoregulation and reproduction because they are normally heterotherms, allowing their body temperature to rise and fall with the ambient temperatures (Trune & Slobodchikoff, 1976). While this energetic strategy is normally an effective means of conserving energy, homeothermy (maintenance of body temperature above or below ambience) may be required for the success of both pregnancy and lactation (Racey, 1983). Alternatively, pregnancy and lactation may preclude homeothermy because of the high metabolic cost of reproduction; mothers may require their limited energetic resources to support lactation rather than a high maternal body temperature (Studier, Lysengen, & O'Farrell, 1973).

In fact, pregnant pallid bats shift to homeothermy as early as 3 weeks before giving birth (Beasley & Leon, 1986). Homeothermy is maintained both day and night throughout lactation, as measured by core temperature and heat production. Mothers bats raise their heat production by 200% in a cool environment, indicating their ability to maintain homeothermy under a thermal challenge. Unsuccessful mothers, who are no longer nursing their young, continue to maintain homeothermy while the other mothers are lactating. These data suggest that there may be a circannual rhythm in which female bats become homeothermic in anticipation of reproduction. Indeed, there is evidence for such a rhythm, with the shift back to heterothermy facilitated by an autumnal rise in melatonin (Beasely & Leon, 1986).

In the wild, mother bats are found in the warmest parts of the attics in which they care for their young (Licht & Leitner, 1967). In the laboratory, mother pallid bats always prefer to go to a warm area on a thermal gradient, a behavioral pattern that conserves energy (Beasley & Leon, 1986). This preference may reflect the tropical ancestry of this species (Jepson, 1970).

Two-week-old batlets are not able to thermoregulate effectively alone, even in a relatively warm environment (Beasely & Leon, 1986). Their ability to thermoregulate increases dramatically when they are allowed to huddle with either their mother or, in their mother's absence, a member of the crèche. Since the mother must be absent to acquire the food needed to maintain lactation, the crèche appears to serve a critical role in reproduction. Birth synchrony may be necessary for the reliable production of crèches, and such an adaptation may have been a prerequisite for the successful invasion of the temperate zone by bats. The delayed ovulation observed in such species may facilitate synchronous production of young, even in the absence of synchronous mating behavior.

CONCLUSIONS AND SUMMARY

To return to the Norway rats, it is clear that thermal interactions between mother and young are involved in the control of moment-to-moment changes in maternal behavior. Norway rats appear to be geared to care for their young unless another motivational state, such as thermoregulation, takes precedence. Mothers seem to monitor their own internal state, rather than that of their young, to modulate maternal behavior. By taking care of their own needs, however, the pups are given adequate, although not necessarily maximal, nurturance. For example, although rat pups grow faster with increased amounts of milk (Friedman, 1975), mothers defend their own energetic state rather than deliver maximal amounts of milk to their litters (Leon & Woodside, 1983). This energetic strategy is similar to that used in limiting contact bout duration; maternal rather than pup hyperthermia is the critical factor. Similarly, mothers transport their young to a place of safety in response to maternal rather than pup disturbance (Brewster & Leon, 1980). Mothers certainly do not ignore the young, but they are only indirectly sensitive to pup needs. They are thereby able to provide the young with adequate care without debilitating themselves.

In summary, the duration of contact bouts between Norway rat mothers and young is limited by an acute increase in maternal brain temperature. The vulnerability of mothers to this acute hypethermia during pup contact is due to their chronic hyperthermia, which is induced by the action of several hormones. Nightly hyperthermia further limits the duration of maternal contact bouts during that period. Mothers either seek out or build a warm area to care for their litters, ensuring that their contact time will be thermally limited. The hyperthermia experienced by pups during maternal contact bouts is ameliorated by a compensatory decrease in both brain and body temperature that is produced by maternal contact. These mechanisms act in concert to provide the transfer of heat to the young, while ensuring that the mothers will be driven away from the pups long enough to forage for the increased amounts of food required to support lactation.

ACKNOWLEDGEMENTS

The research for this chapter was supported, in part, by grant BNS 8616815 from the NSF to M.L., who holds Research Scientist Development Award MH 00371 from the NIMH. R.C. was supported by postdoctoral training grant AG 00096 from the NIA. L.J.B was supported by postdoctoral fellowship HD 06551 from the NICHD. R.M.S. was supported by postdoctoral fellowship HD 06818 from the NICHD, grant NS 26100 from the NINCDS, and a grant from the Pew Foundation.

REFERENCES

Adels, L. E., & Leon, M. (1986). Thermal control of mother–young contact in Norway rats: Factors mediating the chronic elevation of maternal temperature. *Physiology and Behavior, 36*, 183–196.

Adels, L. E., Leon, M., Wiener, S. G., & Smith, M. S. (1986). Endocrine response to acute cold exposure by lactating and non-lactating Norway rats. *Physiology and Behavior, 36*, 179–181.

Ader, R., & Grota, L. J. (1970). Rhythmicity in the maternal behaviour of *Rattus norvegicus*. *Animal Behaviour, 18*, 144–150.

Alberts, J. R. (1978a). Huddling by rat pups: Multisensory control of contact behavior. *Journal of Comparative and Physiological Psychology, 92*, 220–230.

Alberts, J. R. (1978b). Huddling by rat pups: Group behavioral mechanisms oftemperature regulation and energy conservation. *Journal of Comparative and Physiological Psychology, 92*, 231–245.

Alberts, J. R., & May, B. (1984). Nonnutritive, thermotactile induction of filial huddling in rat pups. *Developmental Psychobiology, 17*, 161–181.

Baker, M. A. (1982). Brain cooling in endotherms in heat and exercise. *Annual Review of Physiology, 44*, 85–96.

Beasley, L. J., & Leon, M. (1986). Metabolic strategies of pallid bats, *Antrozous pallidus*, during reproduction. *Physiology and Behavior, 36*, 159–166.

Boulant, J. A., & Bignall, K. W. (1973). Determinants of hypothalamic neuronal thermosensitivity in ground squirrels and rats. *American Journal of Physiology, 225*, 306–310.

Brewster, J., & Leon, M. (1980). Relocation of the site of mother–young contact: Maternal transport behavior in Norway rats. *Journal of Comparative and Physiological Psychology, 94*, 69–79.

Brown, P. (1976). Vocal communications in the pallid bat, *Antrozous pallidus*. *Zeitschrift für Tierpsychologie, 41*, 34–54.

Burger, F. J., & Furman, F. A. (1964). Evidence of heat injury in mammalian tissues. *American Journal of Physiology, 206*, 1057–1061.

Croskerry, P. G., Smith, G. K., Leon, L. N., & Mitchell, E. A. (1976). An inexpensive system for continuously recording maternal behavior in the laboratory rat. *Physiology and Behavior, 16*, 403–414.

Denckla, W. D., & Bilder, G. E. (1975). Investigations into the hypermetabolism of pregnancy and lactation. *Life Sciences, 16*, 404–414.

Drewett, R. F., & Trew, A. M. (1978). The milk ejection of the rat as a stimulus and a response to the litter. *Animal Behaviour, 26*, 982–987.

Fahrbach, S. E., Morrell, J. I. & Pfaff, D. W. (1984). Oxytocin induction of short-latency maternal behavior in nulliparous, estrogen primed female rats. *Hormones and Behavior, 18*, 267–286.

Fowler, S. J., & Kellogg, C. (1975). Ontogeny of thermoregulatory mechanisms in the rat. *Journal of Comparative and Physiological Psychology, 89*, 738–746.

Friedman, M. I. (1975). Some determinants of milk ingestion of suckling rats. *Journal of Comparative and Physiological Psychology, 89*, 636–647.

Gilbert, M. E., & Cain, D. P. (1985). A single neonatal pentylenetetrazol or hyperthermia convulsion increases kindling susceptibility in the adult rat. *Developmental Brain Research, 22*, 169–180.

Grosvenor, C. E., & Mena, F. (1983). Effect of underfeeding upon the rate of milk ejection in the lactating rat. *Journal of Endocrinology, 96*, 215–222.

Grota, L., & Ader, R. (1969). Continuous recording of maternal behaviour in *Rattus norvegicus*. *Animal Behaviour, 17*, 722–729.

Hall, W. G., Cramer, C. P., & Blass, E. M. (1975). Developmental changes in suckling of rat pups. *Nature, 258*, 318–320.

Holtzman, D. K., Obana, K., & Olson, J. (1981). Hyperthermia-induced seizures in the rat pup: A model for febrile convulsions in children. *Science, 213*, 1034–1036.

Isler, D., Trayhurn, P., & Lunn, P. G. (1984). Brown adipose tissue metabolism in lactating rats: The effects of litter size. *Annals of Nutrition and Metabolism, 28,* 101–109.

Jans, J., de Villiers, S., & Woodside, B. (1984). The effect of rearing environment on pup development. *Developmental Psychobiology, 18,* 341–347.

Jans, J., & Leon, M. (1983a). Determinants of mother–young contact in Norway rats. *Physiology and Behavior, 30,* 919–935.

Jans, J., & Leon, M. (1983b). The effects of lactation and ambient temperature on the body temperature of female Norway rats. *Physiology and Behavior, 30,* 959–961.

Jans, J., & Woodside, B. (1987). Effects of litter age, litter size and ambient temperature on the milk ejection reflex in lactating rats. *Developmental Psychobiology, 20,* 333–344.

Jepson, G. L. (1970). Bat origins and evolution. In W. A. Wimsatt (Ed.), *Biology of bats (Vol. 1,* pp. 1–64). New York: Academic Press.

Kleitman, N., & Satinoff, E. (1982). Thermoregulatory behavior in rat pups from birth to weaning. *Physiology and Behavior, 29,* 537–541.

Lee, M. H. S., & Williams, D. I. (1977). A longitudinal study of mother–young interaction in the rat: The effects of infantile stimulation, diurnal rhythms and pup maturation. *Behaviour, 63,* 241–261.

Leon, M. (1985a). The effect of constant light on maternal physiology and behavior. *Physiology and Behavior, 34,* 631–633.

Leon, M. (1985b). Development of thermoregulation. In E. Blass (Ed.), *Handbook of behavioral neurobiology* (Vol. 8, pp. 297–322). New York: Plenum Press.

Leon, M., Adels, L., & Coopersmith, R. (1985). Thermal limitation of mother–young contact in Norway rats. *Developmental Psychobiology, 18,* 85–105.

Leon, M., Adels, L., Coopersmith, R., & Woodside, B. (1984). Diurnal cycle of mother–young contact in Norway rats. *Physiology and Behavior, 32,* 999–1003.

Leon, M., Chee, P., Fischette, C., & Woodside, B. (1983). Energetic limits on reproduction: Interaction of thermal and dietary factors. *Physiology and Behavior, 30,* 939–943.

Leon, M., Croskerry, P. G., & Smith, G. K. (1978). Thermal control of mother–young contact in rats. *Physiology and Behavior, 21,* 793–811.

Leon, M., & Woodside, B. (1983). Energetic limits on reproduction: Maternal food intake. *Physiology and Behavior, 30,* 945–957.

Licht, P., & Leitner, P. (1967). Behavioral responses to high temperatures in three species of microchiropteran bats. *Comparative Biochemistry and Physiology, 22,* 371–387.

Lincoln, D. W., Hentzen, K., Hin, T., Van der Schoot, P., Clarke, G., & Summerlee, A. J. S. (1980). Sleep: A prerequisite for reflex milk ejection in the rat. *Experimental Brain Research, 38,* 151–162.

Lincoln, D. W., & Paisley, A. C. (1982). Neuroendocrine control of milk ejection. *Journal of Reproduction and Fertility, 65,* 571–586.

Lorscheider, F. L., & Reineke, E. P. (1972). Thyroid hormone secretion rate in the lactating rat. *Journal of Reproduction and Fertility, 30,* 269–279.

Nicholls, D. G., & Locke, R. M. (1984). Thermogenic mechanisms in brown fat. *Physiological Reviews, 64,* 1–64.

Orr, R. T. (1954). Natural history of the pallid bat, *Antrozous pallidus. Proceedings of the California Academy of Sciences, 28,* 165–246.

Racey, P. A. (1983). Ecology of bat reproduction. In T. H. Kunz (Ed.), *Ecology of bats* (pp. 57–104). New York: Plenum Press.

Scafidi, F. A., Field, T. M., Schanberg, S. M., Bauer, C. R., Vega-Lahr, N., Garcia, R., Poirier, J., Nystrom, G., & Kuhn, C. M. (1986). Effects of tactile/kinesthetic stimulation on the clinical course and sleep/wake behavior of preterm neonates. *Infant Behavior and Development, 9,* 91–105.

Schanberg, S. M., Evoniuk, G., & Kuhn, C. M. (1984). Tactile and nutritional aspects of maternal care: Specific regulators of neuroendocrine function and cellular development. *Proceedings of the Society for Experimental Biology and Medicine, 175,* 135–146.

Simpson, A. A., Simpson, M. H. W., Sinha, Y. N., & Schmidt, G. H. (1973). Changes in concentration of prolactin and adrenal corticosteroids in rat plasma during pregnancy and lactation. *Journal of Endocrinology, 58,* 675–676.

Studier, E. H., Lysengen, V. L., & O'Farrell, M. J. (1973). Biology of *Myotis thysanodes* and *M. lucifungis* (Chiroptera: Vespertilionidae): II. Bioenergetics of pregnancy and lactation. *Comparative Biochemistry and Physiology, 44,* 467–471.

Sullivan, R. M., Shokrai, N., & Leon, M. (1988). Physical stimulation reduces body temperature of infant rats. *Developmental Psychobiology, 21,* 225–235.

Sullivan, R. M., Wilson, D. A., & Leon, M. (1988). Physical stimulation reduces brain temperature of infant rats. *Developmental Psychobiology, 21,* 237–250.

Summerlee, A. J. S., & Lincoln, D. W. (1981). Electrophysiological recordings from oxytocinergic neurons during milk ejection in the unanesthetized lactating rat. *Journal of Endocrinology, 90,* 255–265.

Trayhurn, P., Douglas, J. B., & McGluckin, M. (1982). Brown adipose tissue thermogenesis is "suppressed" during lactation in mice. *Nature, 298,* 59–60.

Trunc, D. R., & Slobodchikoff, C. N. (1976). Social effects on the metabolism of the pallid bat *(Antrozous pallidus). Journal of Mammalogy, 57,* 656–663.

Wakerley, J. B., & Drewett, R. F. (1975). The pattern of suckling in the infant rat during spontaneous milk ejection. *Physiology and Behavior, 15,* 277–281.

Woodside, B., & Leon, M. (1980). Thermoendocrine influences on maternal nesting behavior in rats. *Journal of Comparative and Physiological Psychology, 94,* 41–60.

Woodside, B., Leon, M., Attard, M., Feder, H. H., Siegel, H. I., & Fischette, C. (1981). Prolactin-steroid influences on the thermal basis for mother–young contact in Norway rats. *Journal of Comparative and Physiological Psychology, 95,* 771–780.

Woodside, B., Pelchat, R., & Leon, M. (1980). Acute elevation of the heat load of mother rats curtails maternal nest bouts. *Journal of Comparative and Physiological Psychology, 94,* 61–68.

20

Functional Organization of Dyadic and Triadic Parent–Offspring Systems

JEFFREY R. ALBERTS AND DAVID J. GUBERNICK

This is an era of reductionistic and mechanistic fervor. Advances in biological and biomedical technique have made feasible measurements and manipulations that were virtually inconceivable only a decade ago. Researchers who study biological aspects of behavior have productively applied many of these new tools. Such technical advances are amply reflected in research progress concerning mechanisms, particularly in the molecular (reductionistic) understanding of mechanisms. In contrast, progress has been relatively modest in the more molar realms of biological research, including behavioral analyses.

Disproportionate attention to mechanisms can lead to a sanguine attitude toward the nature of phenomena that originally stimulated the search for mechanisms. This is a potential pitfall in regard to behavioral phenomena, for we remain at a relatively basic, even primitive, level of sophistication in our ability to describe and analyze behavior in a manner that is readily integrated with other levels of biological organization and function.

A goal of this chapter is to view parental behavior on levels that are sufficiently molar to include the offspring and to make it apparent that parent–offspring interactions constitute a unified *system*. To highlight this view of parental behavior, two mammalian systems are described and compared: one involves a species in which the mother is the sole source of parental care; the other, a system in which the male participates vigorously in rearing the offspring. In concert with their offspring, parents in these two species provide examples of dyadic and triadic systems.

The chapter begins with the assertion that parental behavior is more than a static behavioral category. To appreciate its organization, one must adopt perspectives that (1) capture the temporal characteristics of behavioral organization, notably the parental behavior *cycle,* and (2) include the functional, regulatory contributions of the offspring. To illustrate this approach, it is helpful to introduce the subjects of our presentation: the Norway rat *(Rattus norvegicus)* and the California mouse *(Peromyscus californicus).* These rodent species exhibit maternal and biparental care, respectively.

Next, parent–offspring relationships are discussed as systems, and this is exemplified by a ''symbiotic model.'' Emphasis is on physiological exchanges and behavioral interactions. Included in this analysis are some of the non-physiological, more psychological exchanges as well.

Following this depiction of behavioral and physiological interactions between parents and infants, the focus shifts to the concept of "parental responsiveness," the hypothetical internal state that modulates the expression of parental behavior (see chapter 4). This concept is used as a common dimension on which to compare the different determinants of the onset and maintenance of parental behavior in rat mothers, mouse mothers, and mouse fathers. The chapter concludes with additional comparisons of hormonal and neural mechanisms that mediate parental behavior in these two familial systems.

THE PARENT–OFFSPRING RELATIONSHIP AS A SYSTEM

Parental Behavior

Parental behavior can be described in terms of a constituent set of behavioral activities, such as nursing, licking, and carrying the young. Other types of activities that are not directed at offspring, such as nest building, can also be part of the cluster of activities we refer to as "parental behavior." Singular terms do not necessarily imply singular activities or common underlying mechanisms (Slotnick, 1967), but the term "parental behavior" denotes a functional category of behavior.

Each of the activities composing parental behavior can be operationally defined—that is, described so precisely and objectively that careful, trained researchers can repeat observations and measure the same behavioral units in different experiments. Operational definitions of behavior and the quantification of behavioral categories are important empirical tools. Nevertheless, there are at least two deficiencies in any approach that is based solely on such static, compartmentalized views of parental behavior.

The first deficiency of a purely categorical analysis of parental behavior is that it lacks attention to the *dynamics* of behavioral organization. The expression of parental behavior changes over time, creating a dynamic profile of behavior whose components change in a predictable, orderly sequence that forms a *cycle* of activities termed the "parental behavior cycle" (Rosenblatt & Lehrman, 1963; see also Alberts, 1986). The profile of parental behavior over time is as much a part of the behavioral organization as its hormonal activation or neural mechanisms.

The second deficiency is that categorical analyses of parental activities are prone to overlook active regulation of the parent by the offspring. Even the most "helpless" altricial infant can profoundly influence the activities of its parent(s). In rodents, as in humans, the infant emits sounds and other stimuli that alter parental hormone secretions and activate or direct parental behavior. In addition to such moment-to-moment controls, the offspring can also exert long-term modulation of parental physiology and behavior. Such offspring effects are closely linked to parental stimulation and, indeed, can form a "reciprocal dance" of action and interaction (Alberts & Gubernick, 1983). Any view of parental behavior that overshadows such linkages or effects is a limited perspective.

The Relationship as a System

One solution to these potential deficiencies is to treat the parent–offspring relationship as a unitary *system*. From a systems viewpoint, the parent–offspring relationship becomes a functional unit, not the simple juxtaposition of separate entities (the parent[s] and the young). A systems view of the parent–offspring relationship provides access to an integrated level of organization and can also contribute to fresh perspectives on both parents and offspring. Thus the functional significance of both parental and offspring behavior is found, in part, in its relationship to the other components of the system. Parental behavior can be better understood in relation to the offspring, and the offsprings' characteristics are clarified by knowledge of the parental context.

Many mammalian parent–offspring systems involve multiple offspring. It is often expedient and valid to view the offspring aggregate as a singular unit (Alberts, 1978) in the relationship. In this way, the parent and a litter of 10 offspring can be approached as a dyadic system.

When studying such functional systems, organizing principles and mechanisms can be found on several levels. Some of the most basic ways of understanding and analyzing these systems include (1) manifestations of stereotyped motor patterns and sequences, (2) the sensory controls of the behavior, (3) the neural and endocrine controls, (4) the fulfillment of the participants' needs, and (5) the interrelationships between one player's actions and the other player's reactions (on both the behavioral and the physiological levels). Many of the questions asked during traditional analyses of behavior can be repeated, except that the approach is applied to the system, rather than to an individual. Naturally, it is important to retain an awareness of and interest in the individuals that make up the system, but explicit attention is paid to the functional integrity of the system as a unit.

A DYADIC RAT, A TRIADIC MOUSE

One benefit of a comparative approach in biology is that it allows the researcher to exploit the opportunities offered by naturally occurring variety. Although this book presents a rich sample of the kinds of research and analysis that constitute the empirical and conceptual foundations of contemporary views of parental behavior, it will be noted that most of this work has been conducted on a couple of species of laboratory rodents, usually the rat *(Rattus norvegicus)* and the mouse *(Mus musculus)*. The Norway rat is an excellent subject, but it displays a particular form of parental behavior—similar to that of many species but different from that of many others.

A Dyadic Family System: *Rattus norvegicus*

The Norway rat belongs to the family Muridae (genus, *Rattus;* species, *norvegicus*) and exists in both a wild form and a variety of domesticated or laboratory strains. Rats tend to live in large social groups or colonies. Mating is "promiscuous," meaning that females mate with more than one male and males mate with more than one female. Indeed, as McClintock (1987) has argued persuasively, the copulatory behavior of Norway rats appears specifically adapted to *group mating,* with multiple

Figure 20.1. Drawings of three maternal activities of the rat dam: nursing, licking of the young, and transport of the young.

females and males mating in different combinations. Paternity is thereby indeterminant; in addition, single litters can be multiply sired (McClintock, 1984).

Gestation in *R. norvegicus* usually lasts for 21–22 days. Litter sizes range from about 6–12 altricial pups, each weighing about 7 g at birth. The pups are born glabrous, with sealed eyes and ears. They are pink and so thin skinned that it is often possible to see the milk in their stomachs.

Infant rats require extensive care and maintenance, but they develop rapidly. The ears usually open on day 12, and the eyes unseal about 3 days later. For the first 15 days or so, the pups are obligate sucklers. Onset of food intake is gradual but appears to be a growth requirement by day 19 (Alberts & Gubernick, 1983). Weaning is usually complete by day 35 (Thiels & Alberts, 1985; Thiels, Cramer, & Alberts, 1988).

The activities that constitute parental care in the rat, like those of many mammals, are nursing, licking of the young, nest building, and transport. Some of these behavioral categories are illustrated in Figure 20.1. Each behavior (and many others that can be cast as parental activities) can be operationally defined and quantified in terms such as frequency, duration, probability of occurrence, and so on. When such parameters are plotted over an appropriate developmental time frame, they form a characteristic cyclic organization.

The cycle of maternal behavior in the rat has been described in quantitative detail (e.g., Rosenblatt, 1965), and its temporal form and expression have been related beautifully to the age-typical characteristics of the developing offspring (Rosenblatt & Lehrman, 1963). For instance, Figure 20.2 illustrates the time spent licking offspring each day during the rat's lactation cycle. The form of the curve is charac-

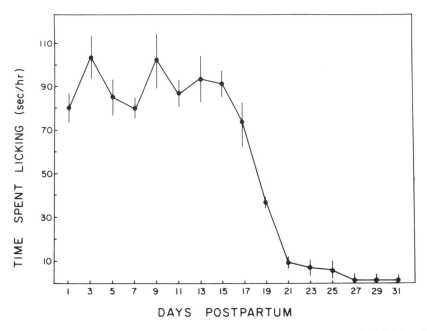

Figure 20.2. The amount of time that rat mothers engaged in pup anogenital licking. Each point represents the mean (±SE) for six dams and their litters that were videotaped continuously for 12 hr/day on alternate days from day 1 to day 31 postpartum. (From Gubernick & Alberts, 1983)

teristic of many measures of maternal behavior and represents the general form of the postpartum cycle. The development of spontaneous urination by rats is synchronized with the decline in licking (Gubernick & Alberts, unpublished data).

Observations in the field (Calhoun, 1962), in large outdoor enclosures, and in the laboratory indicate that only adult females reliably display the full profile of parental activities. From a sociobiological perspective, this type of postpartum parental investment might be predicted in a species in which paternity is indeterminate, although known paternity is no guarantee of postpartum paternal care (Kleiman, 1977; Trivers, 1972).

Interestingly, male rats can be induced to display the full range of maternal activities (excluding lactation; however, males will crouch over the pups in a stereotyped nursing posture). Induction of "maternal" behavior in male rats can be accomplished simply by exposing intact males to infant rats for 5–7 days (Rosenblatt, 1967). Other, more invasive methods, such as cannulation of hormones into brain nuclei, have also been used, but there is little evidence that the expression of maternal behavior by male rats is hormone dependent (Brown & Moger, 1983; Jakubowski & Terkel, 1986). Under conditions thought to be typical for the Norway rat, the burden of postnatal parental care is borne exclusively by the mother. The ecology of the Norway rat is so broad that it is impossible to specify the typical ecological, territorial, or social conditions of the species. The rat is one of the great "generalists" of the world, is commensal with humanity, and has accompanied human settle-

Figure 20.3. A female California mouse, *P. californicus,* huddled over her 5-day-old young.

ment throughout the world. In contrast, most other rodent species are more selective in the niches they inhabit, and they show contrasting reproductive strategies.

A Triadic Family System: *Peromyscus californicus*

The California mouse (*Peromyscus californicus*) is a member of the family Cricetidae (genus, *Peromyscus;* subgenus, *Haplomylomys*) and is found in chaparral, sage scrub, and oak-woodland regions of coastal California, from San Francisco to Baja California (Cranford, 1982; McCabe & Blanchard, 1950; McClosky, 1972). Figure 20.3 is a photograph of these relatively unfamiliar but delightful rodents. *P. californicus* is the largest *Peromyscus* in the United States: adults of both sexes weight 40–70 g (Gubernick & Alberts, unpublished data). Their natural diet consists of seeds, vegetation, and invertebrates (Merritt, 1978).

Adult *Peromyscus* are fairly sedentary, with low but stable population sizes and low reproductive potential (McCabe & Blanchard, 1950). Life expectancy in the field is 9–18 months (Chandler, 1979; McCabe & Blanchard, 1950), which is relatively long for a rodent. Breeding occurs throughout the year in both the laboratory and the field (Drickamer & Vestal, 1973; McCabe & Blanchard, 1950). Limited available field and laboratory data suggest that persistent, if not monogamous, associations are formed between males and females (Dewsbury, 1981).

After a relatively long gestation period of 31–33 days (Gubernick, 1988), females produce one to four altricial young; the typical litter size is two (Drickamer & Vestal, 1973). The pups are relatively heavy (3–5 g), pigmented (black dorsum and pink ventrum), and more developed at birth than other *Peromyscus* neonates (Layne, 1968). By day 10, pups can thermoregulate (unpublished observations). Their eyes and ears open on about day 15 (range, 13–17 days). Pups first leave the next on day 16 (range, 13–19 days) and eat solid food on about day 20 (Gubernick, unpublished observations).

P. californicus exhibit extensive biparental care of their young. Both mothers and fathers are intimately involved, and participate actively in rearing their offspring (Dudley, 1974; Gubernick & Alberts, 1987b). Fathers exhibit all the components of parental behavior displayed by mothers except lactation. The new father typically sniffs and licks the pup at the moment of birth or immediately thereafter. Fathers often retrieve a newborn pup, lick it, and huddle over it while the female gives birth to another young. After parturition is completed, the father, the mother, and their young remain in physical contact in the nest. Fathers and mothers build nests and carry the young.

Figures 20.4 and 20.5 are from a recent quantitative study of parental behavior by *P. californicus:* Continuous 12-hr-long time-lapse video recordings were taken of six family groups on alternate days from postpartum day 1 to day 31. Thus well over 1000 hr of observational data were accumulated. The vigorous involvement of the father is clearly reflected in Figure 20.4, which shows the average number of hours of pup contact provided by each of the two parents across the full postpartum rearing period. From this graph, it can be seen that the male spends as much time as his mate in contact with the offspring and that this commitment is maintained throughout the rearing period. Thus both parents initially show high levels of attendance, which decline gradually as the offspring mature. It is noteworthy that these values (4–10 hr/day of contact) include the female's nursing time. Thus even without the contact involved in milk transfer, the father provides abundant contact, comfort, and thermal protection to the offspring. As a consequence, *Peromyscus* pups are rarely left unattended.

The father's active role in nurturance is further demonstrated by Figure 20.5, which quantifies the licking component of mouse parental behavior. Again, both parents display similar profiles of activity over the cycle of parental care, but the father is more active than the mother in this parameter of parental behavior. The triadic family system of the California mouse provides a useful model for exploration

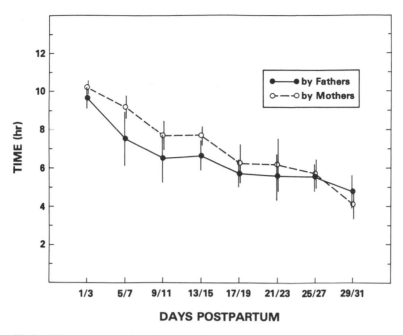

Figure 20.4. The amount of time during a 12-hr observation period that mothers and fathers were in physical contact with a pup. Pup contact for mothers included nursing and nonursing contact. Each data point represents the mean (±SE) for six fathers and six mothers averaged over 2 days. Parents were videotaped continuously for 12 hr/day on alternate days from day 1 to day 31 postpartum. (From Gubernick & Alberts, 1983b)

of mother–father–offspring interactions with other family systems, notably humans (Parke, Power, & Gottman, 1979).

RESOURCE EXCHANGE IN DYADIC AND TRIADIC SYSTEMS

Thus far, the phenomena of dyadic and triadic family systems have been documented in two rodent species, and some of the temporal features of their parental behavior cycles have been described. Additional components of temporal significance, the *onset* and *maintenance* of the parental cycle, will be described later. At this point, however, attention is turned to a novel set of the events and processes that provide an even fuller functional account of parental behavior and the parent–offspring relationship. The perspective has been cast metaphorically as a "symbiosis model" of parent–offspring relationships, because it treats the parent–offspring aggregate as a unified, organized system in which the participants function in a manner *analogous* to that of members of a symbiotic community (Alberts, 1986; Alberts & Gubernick, 1983). Like more traditional symbiotic relationships, the parent–offspring system is characterized as one in which (1) the participants live in close association with each other, (2) their activities are integrated, (3) there is interdependence, often involving

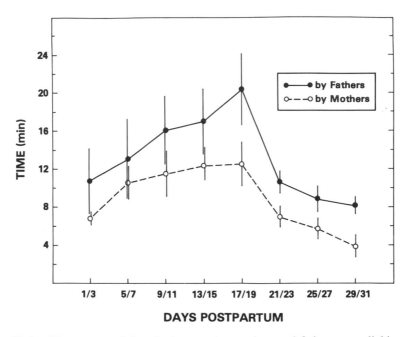

Figure 20.5. The amount of time in the nest that mothers and fathers spent linking a pup. Pup licking included anogenital and nonanogenital licking. See Figure 20.4 for details. (From Gubernick & Alberts, 1983b)

shared metabolic resourcces, and (4) there is a degree of mutual benefit. (See Alberts & Gubernick, 1983), for further discussion of the historical roots and conceptual applications of the symbiosis concept.)

The metaphor of symbiosis has heuristic value; it helps to illuminate levels of interaction that otherwise might go unappreciated. The more common traditional view of mammalian parent–offspring relations emphasizes the manner in which parents provide for the needs of the offspring. In this model, there is an emphasis on the basic unidirectional flow of goods, time, attention, and resources from parent to offspring, usually at some cost to the parent. (Alberts, 1986, discusses a variety of such perspectives.) In contrast, the symbiotic perspective emphasizes *bidirectional* exchanges of various physiological resources and other stimuli between parent and young that may be of mutual, although not necessarily equal, benefit.

Water and Electrolytes

The most complete example of bidirectional resource exchange involves the physiological commodities of body water and some electrolytes. Consider the *licking* behavior of the rodent parent. By licking their offspring, many parents, including those in most rodent species, cleanse their offspring. Licking can modulate infant arousal and facilitate kidney function. In addition, licking—especially of the anogenital region—stimulates the infants' micturition reflex and helps them to void. These are

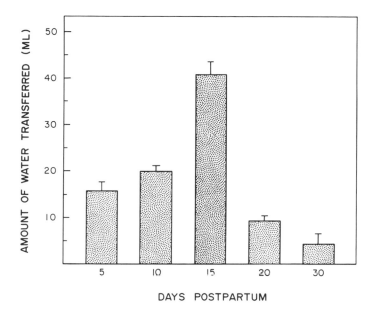

DAYS POSTPARTUM

Figure 20.6. Amount of water transferred from a litter of eight pups to their dams through-out lactation. Each bar represents the mean ± SE for six mothers. Water transfer was estimated by injecting some of the pups with tritium and measuring the amount of radioactivity recovered in the mother 24 hr later. (From Gubernick & Alberts, 1981)

typical examples of the myriad ways in which parental behavior is functionally or-ganized around the needs of the offspring.

But there is more to it. In many species, the mother ingests offspring urine (e.g., Baverstock & Green, 1975; Friedman & Bruno, 1976). Most quantitative stud-ies of offspring urine ingestion by parents have employed tracer techniques. The method used in many of the studies described below involves injecting some of the pups in a litter with radioactively labeled water (tritium). The next day, the tritium label can be found in the mother rat's blood plasma, but not in that of the uninjected pups. If urination by injected pups is blocked, the tritium label does not transfer. By measuring the levels of tritium in the bladders of infected offspring and determining the total body water in the mother rat, the amount of labeled urine the dam consumed can be determined.

The maternal rat consumes physiologically significant quantities of the pups' salty but hypotonic urine. Figure 20.6 illustrates the quantities of water reclaimed by dams. Each day, the mother ingests about two-thirds of the lactational water that she transferred to her litter the day before (Friedman, Bruno, & Alberts, 1981; Gubernick & Alberts, 1983). The litter not only receives essential water and nutrients from the dam, but also serves as a source of water used to keep the dam in fluid balance (Friedman et al., 1981).

Rat dams display vigorous licking of their pups, including anogenital licking, in the presence of ad libitum water. Pup urine is preferred to water. It appears that the

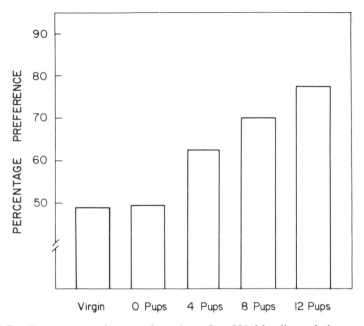

Figure 20.7. Percentage preference of rat dams for .001 M saline solution over water in daily two-choice tests as function of litter size. Litter size was manipulated on the day of birth.

attraction is the sodium salt in pup urine. Lactating rats show an enhanced appetite for NaCl (but not KCl) (Alberts & Gubernick, 1983; Richter & Barelare, 1938), and pup anogenital licking is decreased when mothers have access to extra sodium solution (Gubernick & Alberts, 1983).

Interestingly, the rat dam's appetite for sodium is proportional to the amount of suckling she receives. Figure 20.7 shows the results of a study in which the number of pups that were nursed by dams was manipulated experimentally at birth. The histograms show the dams' salt preference for dilute (0.001 M NaCl) saline, averaged over a week of continuous testing. There is an orderly relationship between the magnitude of the dams' appetite for dilute saline and the amount of suckling they received (Alberts & Gubernick, unpublished results).

Comparison of Parental Licking in Dyadic and Triadic Family Systems

In the rat, there is a simple and understandable relationship among the processes of lactation, suckling, maternal licking, micturition, and urine ingestion. The mother's licking behavior can be understood in the metaphorical economy of mutualistic water balance. In this regard, the vigorous pup licking exhibited by *Peromyscus* fathers raises some interesting questions: Does the mouse mother provide lactational water to the litter and reclaim proportionately less than female rats, because the mouse father is also licking pups and ingesting urine? Is paternal behavior in the male California mouse "bought" by the urine resource? Is mouse parental behavior regulated by a salt appetite such as that exhibited by the rat dams?

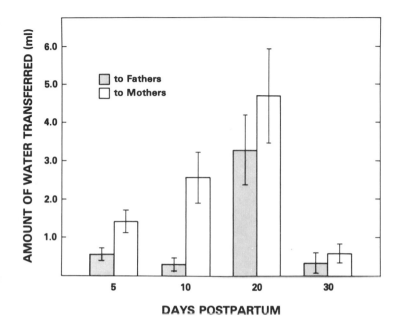

Figure 20.8. Amount of water transferred from a litter of two pups to their father and mother throughout lactation. Each bar represents the mean ± SE for six fathers and six mothers. Water transfer was estimated by injecting of the some pups with tritium and measuring the amount of radioactivity recovered in the father and mother 24 hr later. (From Gubernick & Alberts, 1987a)

There are answers to some of these questions (Gubernick & Alberts, 1987a). First, the amount of water transferred from pups to mothers and fathers in the biparental *Peromyscus* was determined by injecting 5-, 10-, 20-, and 30-day-old pups with tritiated water and measuring the label in the mothers' and fathers' plasma after 24 hr of interaction with their litter. Figure 20.8 illustrates that on days 5 and 10, mothers obtained more pup urine than did fathers. Parents consumed equivalent amounts of pup urine on days 20 and 30.

Although the mothers consumed as much urine as the father, or even more, the fathers ingested some pup urine. There are several possible explanations for this arrangement, whereby the female provides all the water to the litter and the male intervenes in the recycling of the valuable water resource. One explanation is that the mother is imperfectly tuned to the time or amount of urine availability, thereby making it possible for the male to gain access to some remaining urine. The mother may also indirectly influence the father to continue caring for the young by allowing him to consume pup urine, which implies that urine consumption may have some functional significance for the father. Perhaps males urine-mark territories more frequently while parental, and reclaim lost water and electrolytes from ingestion of pup urine. Because the father remains with the pups for substantial periods while the mother is out of the nest (Gubernick & Alberts, 1987b), urine consumption may compensate him for the reduction in foraging. These questions have not yet been studied experimentally.

The possibility that pup licking by *Peromyscus* mothers and/or fathers is regulated by an enhanced sodium appetite, as it is in rat mothers, has been examined. Similar preference tests indicated no enhanced salt preference in *Peromyscus* fathers or mothers. Thus, the same behavior is under different sets of controls in the two species.

Thermal-Energy Exchanges in the Dyad

Heat is another physiological commodity that can be transferred between members of the dyad. Measurements of the thermal gradient between various body regions of mother rats and their litters at postnatal ages 5 to 25 days revealed that the litter's surface temperatures are consistently cooler than that of the dam, thereby establishing a consistent pattern of thermal flux from mother to litter throughout the nursing period (Alberts, unpublished observations). Thermal energy appears to be a resource that consistently moves from the mother to the pups. Leon and his associates (Leon, Croskerry, & Smith, 1978) have studied some of the ways that thermal interactions can modulate bout duration or nest visitation by rat mothers.

These exchanges between rat dams and the young can have profound consequences. One outcome involves the age-related emergence of the pups' filial preference for other rats. A rat's filial preference can be measured in terms of its selection of the animate or inanimate stimuli with which it *huddles*. Huddling is a species-typical behavior displayed readily by rat pups and adults in both the laboratory and the field.

Developmental studies of pups' huddling preferences reveal the dramatic emergence of species affiliation. Rat pups 15 days of age display an overwhelming preference for huddling with conspecifics in relation to other acceptable huddling partners. This preference is not seen in younger age groups (Alberts & Brunjes, 1978). The pups' filial preference is guided by olfactory cues, but the emergence of the species' filial preference is not determined simiply by the development of specific olfactory capabilities or a perceptual bias for rat-typical olfactory cues (Alberts & Brunjes, 1978; Alberts & May, 1980). *Experiences* in the nest with the mother and littermates are the source of the pups' olfactory-filial preference (Brunjes & Alberts, 1979).

A series of experiments was used to identity the experiences in the nest that are sufficient to induce a filial huddling preference. One of the prime candidates was the nursing interactions, for this is a known source of reward for infants (Amsel, Burdette, & Letz, 1976; Kenny & Blass, 1977). Contrary to predictions, no evidence was found for either nutritive (receipt of milk) or nonnutritive rewards related to suckling (the nipple and associated oral stimuli of suckling). The effective events proved to be *thermotactile* stimulation, most notably that involving conductive heat exchange from the mother's body to the litter (Alberts & May, 1984).

The results of these experiments on the source of rat filial preference also point to a class of nonphysiological exchanges that can occur between parent and offspring. In the rat, for example, odors and sounds can come to exert profound behavioral and physiological effects on each member of the dyad. Odors and sounds are forms of psychological rather than purely physiological stimulation, yet they can activate or attenuate behavior, regulate the intensity of activity, and either raise or lower hormone titers. In the framework of a family system as we have been discussing it, this adds such nonphysiological resources to the list of active ingredients involved in its structure and function. The symbiosis model, with its emphasis on bidirectional exchanges of resources, helps attune us to the range of stimuli that can enter into the form and function of the mother–litter dyad (Alberts & Gubernick, 1983).

ONSET AND MAINTENANCE OF PARENTAL BEHAVIOR

The Concept of Parental Responsivity

There is an important analogue to the act or expression of parental behavior. The analogue is *parental responsivity*. This is a label for a hypothetical internal condition that reflects the probability, frequency, duration, and/or intensity of an individual's parental performance. Although parental responsivity is an intervening variable and cannot be directly observed, it can be operationalized and measured with objective, standardized, quantitative tests. Such tests with rodents usually involve presentation of infant pups to an adult under controlled environmental conditions. Variations in parental responsivity are inferred by the adults' reactions to the alien infants: the adult can ignore, attack, or exhibit parental behavior (such as collecting the stimulus pups, licking them, and adopting a nursing posture over them). The nursing posture is readily displayed by nonlactating rats (Rosenblatt, 1967). Such tests of maternal responsivity are a powerful analytical tool.

It can thus be empirically shown that rats and mice are not continuously or spontaneously parental. That is, a randomly chosen, nonreproductive rodent is more likely to ignore or even attack an alien pup than it is to direct nurturant behavior toward it. Parental responsivity is thus normally presumed to exist at very low levels. Nevertheless, the parental repertoires of rat mothers and mouse parents are ready to be activated when the young arrive. It has been shown that the probability that a pregnant female rat or mouse will act parental toward an infant increases dramatically shortly before the time of parturition.

Prenatal Onset: The "Pregnancy Effect"

It is reasonable to consider that the internal physiological conditions of late pregnancy cause an increase in maternal readiness shortly before parturition (Rosenblatt, 1987). With this explanation in mind, it was especially stunning to discover that male *Peromyscus* also show a "pregnancy effect": they begin to display parental behavior toward stimulus pups when their mates are within 10 days of parturition, whereas nonreproductive males are unlikely to be spontaneously parental. Under identical testing conditions, the onset of parental responsiveness of the male *Peromyscus* precedes that of his pregnant partner!

Further experimentation showed that the onset of male parental behavior is linked to historical as well as proximate stimuli associated with his mate. That is, the onset of parental responsivity in the male *P. californicus* depends on (1) having copulated with a female mate and (2) cohabitation with that pregnant partner. Mere cohabitation with a pregnant conspecific is not sufficient to induce parental behavior in the male, nor is the experience of having copulated about 4 weeks earlier. Both of these forms of stimulation/experience are apparently required (see also Elwood, 1985; vom Saal, 1984).

Without probing the internal states of responsivity in pregnant rodents and their partners, it can be clearly observed that when the young arrive in the natal world, the mother's behavior and physiology are set for the challenge of sustaining the lives of a multitude of offspring. The male California mouse, as we have seen, is similary prepared.

Postnatal Maintenance of Parental Behavior and Responsiveness

Early lactation in the female rodent is considered to be a transitional stage between a hormonally primed initiation of maternal behavior and a subsequent nonhormonal *maintenance* phase in which the sustaining force is stimulation from the offspring. The active role of pup stimulation in the maintenance of postpartum maternal behavior can be demonstrated by removal of the pups. If her pups are taken away at birth and a rat dam is tested for maternal responsivity 4 days later, the results show a precipitous drop in her maternal responsiveness (Bridges, 1975; Rosenblatt, 1965; Rosenblatt & Lehrman, 1963; chapter 6). Interestingly, similar removal of pup stimulation 4 days later is far less devastating, suggesting some form of cumulative effect of pup stimulation that can assist in postpartum maintenance of maternal behavior later in the cycle. Similar results have been obtained in the Golden hamster (Siegel & Greenwald, 1978) and the Mongolian gerbil (Elwood, 1981).

Pup stimulation is indispensable for the postpartum maintenance of maternal behavior in the California mouse (Gubernick & Alberts, 1989). Mice were tested individually for 10 min with a 1- to 3-day-old alien pup. Animals were considered parental if they spent 1 min or more licking or crouching over the pup. As in the Norway rat, removal of pup stimulation immediately following parturition reduced maternal responsiveness in *Peromyscus*. The left panel of Figure 20.9 shows the 45% decline in the number of mothers responding maternally following 3 days of pup removal.

In contrast, pup stimulation is not essential for maintenance of paternal responding in the California mouse. The right panel of Figure 20.9 shows the immunity of the male to the effects of pup removal. Three days after the removal of his litter, the father *Peromyscus* continues to display paternal behavior at levels comparable with those of fathers living with their mate and offspring (Gubernick & Alberts, in press).

Whereas maintenance of maternal responsivity in the California mouse mother is dependent on stimulation arising from the offspring, parental responsivity in the mouse father is dependent on stimulation arising from the mother. Removal of infant stimulation has little effect on the father, but separation from his partner has the same devastating effect on his paternal behavior as that seen with maternal behavior following removal of the offspring. Figure 20.10 illustrates the contrasting effects of pup and mate removal on female and male *P. californicus*. In contrast, the fathers' presence had no effect on the maintenance of maternal responsiveness. In the absence of pup stimulation, the mother was not parental regardless of the father's presence.

Thus there has been progress toward identifying a critical stimulus that is sufficient to maintain the male's parental care for the first 3 days postpartum. Shortly

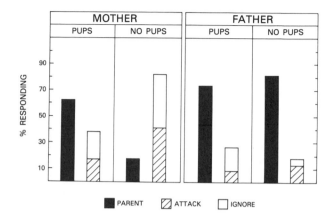

Figure 20.9. The percentage of mothers and fathers exhibiting parental or nonparental behavior toward a test pup on day 3 postpartum. Parents either remained together with their own
pups until testing, or their pups were removed with 4 hr after birth. Each parent was given a
10-min test with a 1- to 3-day-old alien pup. A mouse was considered parental if it spent 1
min or more licking or huddling over the test pup. Animals were tested only once. (Adapted
from Gubernick & Alberts, 1989)

after the birth of the litter, *Peromyscus* pairs were separated. The female was housed
in a specially designed cage above the father's cage. This "town-house" arrangement was rigged such that the mother's excreta rained down into the male's com-
partment, thus exposing him to his partner's urine and feces but not to nonexcretory
cues associated with her. Other fathers were housed without the female. Fathers
exposed to maternal excreta were parental compared with unexposed males (Gubernick & Alberts, 1989).

Further tests indicated that a chemical signal in the mother's urine is sufficient

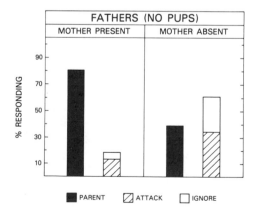

Figure 20.10. The percentage of fathers exhibiting parental or nonparental behavior toward
a test pup on day 3 postpartum. The fathers' own pups were removed on the day of birth, and
the mother either remained with the father (Mother Present) or was removed (Mother Absent).
See Figure 20.9 for other details. (Adapted from Gubernick & Alberts, 1989)

to maintain male parental care for the first 3 days postpartum. It was demonstrated that maternal urine carries the effective cue(s), because fathers painted on the nose with 100 μl of their mates' urine twice each day were parental (80%) compared with fathers painted with distilled water (20%). The urinary cue that maintains paternal responsiveness is apparently specific to the male's mate. Fathers exposed to excreta from a virgin female or another lactating female in the town-house arrangement were less likely to act parental than were fathers exposed to the excreta of their partner.

Perhaps in species such as the California mouse, in which males and females form persistent pair bonds, fathers are more responsive to maternal cues than are males in polygynous species. Disruption of the pair bond reduces paternal responsiveness, whereas maternal urine reinstates paternal solicitude. Interestingly, in the biparental southern grasshopper mouse, *Onychomys torridus,* maternal behavior is affected by a paternal chemosignal. Mothers exposed postpartum to their mates' urine exhibit elevated pup licking (Duvall, Scudder, Southwick, Schultz, 1982).

HORMONAL BASIS OF PARENTAL BEHAVIOR

The hormonal basis of maternal behavior has been extensively investigated (Rosenblatt & Siegel, 1981; see chapters 4 and 6), particularly in rodents. Maternal behavior in the Norway rat is established by the endocrine processes of pregnancy (Rosenblatt & Siegel, 1981). The initiation of maternal behavior occurs prepartum and is due mainly to an increase in estrogen and a decrease in progesterone that occurs as parturition draws near (Rosenblatt, 1987; see chapters 4 and 6). Progesterone withdrawal appears to enhance the effects of estrogen in initiating maternal behavior. Estradiol benzoate implanted into the medial preoptic area (MPOA) facilitated maternal behavior in rats that were ovariectomized and hysterectomized at midpregnancy (Numan, Rosenblatt, & Komisaruk, 1977). Oxytocin (Pedersen & Prange, 1987) and prolactin (Bridges, DiBiase, Loundes, & Doherty, 1985) may also play a role in the onset of maternal behavior associated with ovarian endocrine changes at the end of pregnancy.

Whereas the hormonal basis of maternal behavior has been well documented, parallel aspects of male parental care in mammals remain relatively unexplored. The few studies of hormonal correlates of paternal behavior in rodents used species in which males do not normally display parental care but are induced to act parental either through repeated exposure to infants or through hormonal manipulations (Bridges, 1983; Lubin, Leon, Moltz, & Numan, 1972; Tate-Ostroff & Bridges, 1985). Under such conditions, males are hormonally less responsive than females (Samuels & Bridges, 1983; Sodersten & Eneroth, 1984); in one case, hormonal changes were correlated with the male's behavior (Brown & Monger, 1983; but see Jakubowski & Terkel, 1986, for contradictory findings). The *Peromyscus* male's naturally robust paternal activities were used to determine whether fathers differed behaviorally and hormonally from expectant fathers and from virgin males. As predicted from earlier work, more fathers and expectant fathers were parental than were virgin males. To determine whether there were concomitant hormonal differences between fathers and nonfathers, blood samples were obtained from fathers 2 days postpartum, from expectant

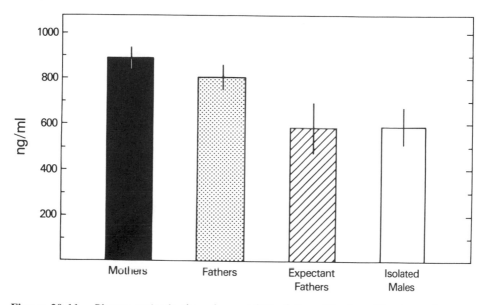

Figure 20.11.   Plasma prolactin titers (mean ± SE) of *P. californicus*. Blood samples were obtained from mothers and fathers 2 days postpartum, from expectant fathers within 10 days before parturition, and from isolated, adult virgin males. (From Gubernick & Nelson, 1989)

fathers within 10 days before parturition, and from isolated adult virgin males. For comparison, blood samples were taken from mothers 2 days postpartum. Plasma prolactin and testosterone were measured by radioimmunoassay (Gubernick & Nelson, 1989).

Plasma prolactin levels of fathers were significantly elevated compared with those of expectant fathers and virgin males, as can be seen in Figure 20.11. Prolactin titers of fathers resembled those of mothers. Plasma testosterone levels, in contrast, did not differ, although testosterone tended to decrease among the groups with increased paternal care. In all cases, androgen levels were low, which may be typical of this species. Testosterone appears not to be associated with spontaneous male parental care in adult virgin laboratory mice (Svare, Bartke, & Gandelman, 1977).

In earlier work with rats, investigators sensitized males to infants and induced male parental behavior. Although actively paternal, these rats showed either no change in prolactin (Grosvenor, 1965; Samuels & Bridges, 1983) or a decrease in the hormone (Brown & Moger, 1983).

Suckling stimulation and olfactory cues from pups enhance prolactin secretion in female rats (Grosvenor, 1965; Stern & Siegel, 1978). Tactile stimulation and olfactory cues from pups might similarly affect *Peromyscus* males, whose prolonged contact with the litter provides ample opportunity for such effects. Indeed, *Peromyscus* pups can be seen to actively probe the ventrum of fathers as they crouch over the litter, although the adult males lack nipples. Another possibility is that the mother affects paternal prolactin secretion. Whether increased prolactin is a cause of paternal behavior or a consequence of caring for young remains to be determined.

To date, only one other study has examined hormonal correlates of paternal behavior in a mammal that normally displays male care (Dixson & George, 1982). Like the California mouse, parental male marmosets (a New World primate) exhibited elevated prolactin levels compared with those of males living with either a pregnant or a nonpregnant female. Testosterone levels did not differ among the groups. Elevated prolactin levels in fathers may be a pattern characteristic of mammalian species in which males normally care for their young.

NEURAL MECHANISMS OF PARENTAL BEHAVIOR

The neuroanatomical basis of maternal behavior has been extensively investigated in rats (see chapter 12). A selected portion of this literature is briefly reviewed mainly to set the stage for some recently collected, contrasting data from *Peromyscus*.

In rats, the MPOA of the basal telencephalon is crucial for the regulation of maternal behavior (Numan, 1987; see chapter 12). Lesions of the MPOA disrupt maternal behavior, and the lateral efferent projections from the MPOA are the most crucial.

Relatively little information is available on the brain mechanisms underlying parental behavior in other species (Carlson & Thomas, 1968; Cruz & Beyer, 1972; Slotnick & Nigrosh, 1975; Voci & Carlson, 1973), and nothing is known about the neural circuitry involved in male parental care. Many species of rodents do not exhibit paternal behavior. In these species, there may be structural sex differences in the brain underlying the behavioral differences between males and females in the care of the young. This hypothesis is based on findings that there are large sex differences in brain regions controlling vocal behavior in zebra finches and canaries. In these species only males sing, and the vocal nuclei of the brain are significantly larger in males than in females (Nottebohm & Arnold, 1976). Interestingly, the preoptic area (POA) of the rat hypothalamus is important in the regulation of several sexually dimorphic behaviors (Gorski, see chapter 5) and contains a sexually dimorphic nucleus (SDN-POA) that is larger in males than in females (Gorski, Gordon, Shryne, & Southam, 1978; see chapter 5).

The California mouse is a particularly promising species in which to explore the brain systems underlying parental care. Fathers are behaviorally similar to mothers in the quantity and quality of care provided the young, and virgin males and females are infrequently parental. Do similar neural regions (e.g., the MPOA) subserve maternal and paternal behavior in *P. californicus*? Are there sexual dimorphisms in brain mechanisms underlying parental behavior? Does the brain change when males and females become parental?

To explore these questions, the MPOA of mothers and fathers 5 days postpartum was compared with the MPOA of nonparental virgin males and females (Gubernick, Sengelaub, & Kurz, in preparation). Outlines of the whole MPOA and whole brain sections containing MPOA were traced and computer digitized for analysis. The ratio of total MPOA to brain size was used to correct for possible differences between groups in brain size.

There was a sexual dimorphism in the volume of the MPOA. The nucleus was significantly larger in virgin males than in virgin females (Figure 20.12). When fe-

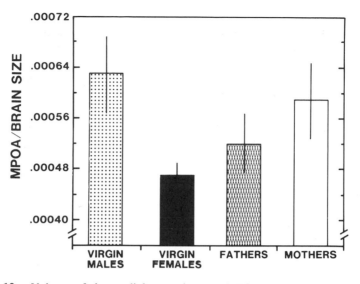

Figure 20.12. Volume of the medial preoptic area (MPOA) (mean ± SE) of mothers and fathers 5 days postpartum and of adult virgin males and females. (Virgin males and females were nonparental in a 10-min test with 1- to 3-day-old pup.)

males became mothers, the MPOA remained relatively the same size, suggesting that the brain of females is ready for parenting and perhaps simply has to be activated. The most striking finding is that the MPOS of males gets smaller when they become fathers, thus becoming equal in size to that of mothers. Mothers and fathers display similar parental behavior and equivalent MPOA volumes. Interestingly, in species of duetting birds, in which both males and females sing, the size of the brain nuclei controlling vocal behavior is the same in both sexes. (Brenowitz, Arnold, & Levin, 1985). We do not yet know whether the MPOA is actively involved in parental behavior in *P. californicus*.

The MPOA may integrate hormonal and sensory inputs that regulate maternal behavior. In the rat, hormonal influences on the MPOA may facilitate maternal behavior by affecting the female's response to olfactory input from pups (Numan, 1987). Olfactory stimulation of accessory olfactory bulbs via the vomeronasal organ inhibits maternal behavior in virgin females (Fleming, Vaccarino, Tambosso, & Chee, 1979). The accessory olfactory bulb (AOB) connects eventually with the MPOA (Scalia & Winans, 1975). A similar mechanism may operate in the control of paternal behavior in *P. californicus*. Fathers are responsive to chemosignals in maternal urine and differ hormonally from nonparental males, as discussed earlier.

CONCLUDING REMARKS

Psychobiological analyses of parenting and our ability to comprehend parent–offspring relationships as integrated biological systems are in an exciting phase of advancement. To predict specific scientific achievements is imprudent and belies the caprices of basic research. Nevertheless, it is likely that exciting progress will continue in endeavors that involve both proximate and ultimate (evolutionary) analyses. Both of these lines of research can and should proceed in parallel and converge periodically, at which points integrative leaps are likely to be most apparent.

Whereas reductionistic and molecular approaches are revealing underlying physiological and neural mechanisms, more molar approaches are providing insights into the interindividual structure of systems that constitute the behavioral world. Parenting is a universal feature of mammalian behavioral life, essential to all individuals and species, including our own. Mammalian parenting is thus an ideal phenomenon for empirical focus. It is universally represented but richly varied, thus affording the advantages of the comparative approaches that have historically served biological analyses. This chapter, although focused on only two rodent species, represents part of this comparative tradition. The continuities emphasized by comparative perspectives presage the inclusion of human behavior in the psychobiological understanding of mammalian parenting.

ACKNOWLEDGEMENTS

The original research reported here was supported in part by grants from the National Institute of Mental Health (MH-28355) to J.R.A. and the National Institute of Child Health and Human Development (HD-21233) to D.J.G. and J.R.A. The authors thank N. Krasnegor and R. Bridges for their editorial guidance, particularly in encouraging a broader, comparative approach for this chapter.

REFERENCES

Alberts, J. R. (1978). Huddling by rat pups: Multisensory control of contact behavior of *Rattus norvegicus*. *Journal of Comparative and Physiological Psychology, 92,* 220–230.

Alberts, J. R. (1986). New views of parent–offspring relationships. In W. T. Greenough & J. Juraska (Eds.), *Developmental Neuropsychobiology* (pp. 449–478). Orlando, FL: Academic Press.

Alberts, J. R., & Brunjes, P. (1978). Ontogeny of thermal and olfactory determinants of huddling in the rat. *Journal of Comparative and Physiological Psychology, 92,* 897–906.

Alberts, J. R., & Gubernick, D. J. (1983). Reciprocity and resource exchange: A symbiotic model of parent–offspring relations. In L. A. Rosenblum & H. Moltz (Eds.), *Symbiosis in parent–offspring Interactions* (pp. 7–45). New York: Plenum Press.

Alberts, J. R., & May, B. (1980). Ontogeny of olfaction: Development of the rats' sensitivity to urine and amyl acetate. *Physiology and Behavior, 24,* 965–970.

Alberts, J. R., & May, B. (1984). Non-nutritive, thermotactile induction of olfactory preferences for huddling in rat pups. *Developmental Psychobiology, 17,* 161–181.

Amsel, A., Burdette, D. R., & Letz, R. (1976). Appetitive learning, patterned alternation and extinction in 10-day-old rats with non-lactating suckling as reward. *Nature, 262,* 816–818.

Baverstock, P., & Green, B. (1975). Water recycling in lactation. *Science, 187,* 657–658.

Brenowitz, E. A., Arnold, A., & Levin, R. N. (1985). Neural correlates of female song in tropical duetting birds. *Brain Research, 334,* 104–112.

Bridges, R. S. (1975). Long-term effects of pregnancy and parturition upon maternal responsiveness in the rat. *Physiology and Behavior, 14,* 245–249.

Bridges, R. S. (1983). Sex differences in prolactin secretion in parental male and female rats. *Psychoneuroendocrinology, 8,* 109–116.

Bridges, R. S., DiBiase, R., Loundes, D. D., & Doherty, P. C. (1985). Prolactin stimulation of maternal behavior in female rats. *Science, 227,* 782–784.

Brown, R. E. (1985). Hormones and paternal behavior in vertebrates. *American Zoologist, 25,* 895–910.

Brown, R. E., & Moger, W. H. (1983). Hormonal correlates of parental behavior in male rats. *Hormones and Behavior, 17,* 356–365.

Brunjes, P. C., & Alberts, J. R. (1979). Olfactory stimulation induces filial huddling preferences in pups. *Journal of Comparative and Physiological Psychology, 93,* 548–555.

Calhoun, J. B. (1962). *The ecology and sociology of the Norway rat* (Public Health Service Document No. 1008). Washington, DC: Government Printing Office.

Carlson, N. R., & Thomas, G. J. (1968). Maternal behavior of mice with limbic lesions. *Journal of Comparative and Physiological Psychology, 66,* 731–737.

Chandler, T. (1979). *Population biology of coastal chaparral rodents.* Unpublished doctoral dissertation, University of California, Los Angeles.

Cranford, J. A. (1982). The effect of woodrat houses on population density of *Peromyscus. Journal of Mammalogy, 63,* 663–666.

Cruz, M. L., & Beyer, C. (1972). Effects of septal lesions on maternal behavior and lactation in the rabbit. *Physiology and Behavior, 9,* 361–365.

Dewsbury, D. A. (1981). An exercise in the prediction of monogamy in the field from laboratory data on 42 species of Muroid rodents. *Biologist, 63,* 138–162.

Dixson, A. F., & George, L. (1982). Prolactin and parental behavior in a male New World primate. *Nature, 299,* 551–553.

Drickamer, L. C., & Vestal, B. M. (1973). Patterns of reproduction in a laboratory colony of *Peromyscus. Journal of Mammalogy, 54,* 523–528.

Dudley, D. (1974). Paternal behavior in the California mouse, *Peromyscus californicus. Behavioral Biology, 11,* 247–252.

Duvall, D., Scudder, K. M., Southwick, C. H., & Schultz, N. J. (1982). Paternal urine elicits increased maternal care in grasshopper mice. *Behavioral and Neural Biology, 34,* 221–225.

Elwood, R. W. (1977). Changes in the responses of male and female gerbils *(Meriones unguiculatus)* towards test pups during the pregnancy of the female. *Animal Behavior, 25,* 46–51.

Elwood, R. W. (1981). Postparturitional reestablishment of pup cannibalism in female gerbils. *Developmental Psychobiology, 14,* 209–212.

Elwood, R. W. (1985). Inhibition of infanticide and onset of paternal care in male mice *(Mus musculus). Journal of Comparative Psychology, 99,* 457–467.

Fleming, A. F., Vaccarino, F., Tambosso, L., & Chee, P. (1979). Vomeronasal and olfactory system modulation of maternal behavior in rats. *Science, 203,* 372–373.

Friedman, M. I., & Bruno, J. P. (1976). Exchange of water during lactation. *Science, 197,* 409–410.

Friedman, M. I., Bruno, J. P., & Alberts, J. R. (1981). Physiological and behavioral consequences in rats of water recycling during lactation. *Journal of Comparative and Physiological Psychology, 95,* 26–35.

Gorski, R. A. (1984). Critical role for the medial preoptic area in the sexual differentiation of the brain. In G. J. De Vries, J. P. C. De Bruin, H. B. M. Uylings, & M. A. Corner (Eds.), *Sex differences in the brain: The relation between structure and function* (Vol. 61, pp. 129–145). Amsterdam: Elsevier.

Gorski, R. A., Gordon, J. H., Shryne, J. E., & Southam, A. M. (1978). Evidence for a morphological sex difference within the medial preoptic area of the rat brain. *Brain Research, 148*, 333–346.

Grosvenor, C. E. (1965). Evidence that exteroceptive stimuli can release prolactin from the pituitary gland of the lactating rat. *Endocrinology, 76*, 340–342.

Gubernick, D. J. (in press). Reproduction in the California mouse, *Peromyscus californicus. Journal of Mammalogy*.

Gubernick, D. J., & Alberts, J. R. (1983). Maternal licking of young: Resource exchange and proximate controls. *Physiology and Behavior, 31*, 593–601.

Gubernick, D. J., & Alberts, J. R. (1987a). "Resource" exchange in the biparental California mouse, *Peromyscus californicus:* Water transfer from pups to parents. *Journal of Comparative Psychology, 101*, 328–334.

Gubernick, D. J., & Alberts, J. R. (1987b). The biparental care system of the California mouse, *Peromyscus californicus. Journal of Comparative Psychology, 101*, 169–177.

Gubernick, D. J., & Alberts, J. R. (1989). Postpartum maintenance of paternal behaviour in the biparental California mouse, *Peromyscus californicus. Animal Behaviour, 37*, 656–664.

Gubernick, D. J., & Nelson, R. (1989). Prolactin and paternal behavior in the biparental California mouse, *Peromyscus californicus. Hormones and Behavior, 23*, 203–210.

Gubernick, D. J., Sengelaub, D., & Kurz, L. (in preparation).

Jakubowski, M., & Terkel, J. (1986). Nocturnal surges and reflexive release of prolactin in parentally behaving virgin female and male rats. *Hormones and Behavior, 20*, 270–286.

Kenny, J. T., & Blass, E. M. (1977). Suckling as incentive to instrumental learning in pre-weanling rats. *Science, 196*, 898–899.

Kleiman, D. G. (1977). Monogamy in mammals. *Quarterly Review of Biology, 52*, 39–69.

Layne, J. M. (1968). Ontology. In J. A. King (Ed.), *Biology of Peromyscus (Rodentia)* (pp. 37–53). Special publication of the American Society of Mammalogists (Vol. 22).

Leon, M., Croskerry, P. G., & Smith, G. K. (1978). Thermal control of mother–infant contact in rats. *Physiology and Behavior, 21*, 793–811.

Lubin, M., Leon, M., Moltz, H., & Numan, M. (1972). Hormones and maternal behavior in the male rat. *Hormones and Behavior, 3*, 369–374.

McCabe, T. T., & Blanchard, B. D. (1950). *Three species of Peromyscus.* Santa Barbara, CA: Rood Associates.

McClintock, M. K. (1984). Group mating in the domestic rat as a context for sexual selection: Consequences for the analysis of sexual behavior and neuroendocrine responses. *Advances in the Study of Behavior, 14*, 1–15.

McClintock, M. K. (1987). A functional approach to the behavioral endocrinology of rodents. In D. Crews (Ed.), *Psychobiology of reproductive behavior: An evolutionary perspective* (pp. 176–203). Englewood Cliffs, NJ: Prentice-Hall.

McCloskey, R. T. (1972). Temporal changes in populations and species diversity in a California rodent community. *Journal of Mammalogy, 53*, 657–676.

Merritt, J. F. (1978). *Peromyscus californicus. Mammalian Species, 85*, 1–6.

Nottebohm, F., & Arnold, A. P. (1976). Sexual dimorphism in vocal control areas of the songbird brain. *Science, 194*, 211–213.

Numan, M. (1987). Preoptic area neural circuitry relevant to maternal behavior in the rat. In N. A. Krasnegor, E. M. Blass, M. A. Hofer, & W. P. Smotherman (Eds.), *Perinatal development: A psychobiological perspective* (pp. 275–298). Orlando FL: Academic Press.

Numan, M., Rosenblatt, J. S., & Komisaruk, B. R. (1977). Medial preoptic area and onset of maternal behavior in the rat. *Journal of Comparative and Physiological Psychology, 91,* 146–164.

Parke, R. D., Power, T. G., & Gottman, J. M. (1979). Conceptualizing and quantifying influence patterns in the family triad. In M. E. Lamb, S. J. Suomi, & S. R. Stephenson (Eds.), *Social interaction analysis.* Madison: University of Wisconsin Press.

Pedersen, C. A., & Prange, A. J., Jr. (1987). Evidence that central oxytocin plays a role in the activation of maternal behavior. In N. A. Krasnegor, E. M. Blass, M. A. Hofer, & W. P. Smotherman (Eds.), *Perinatal development: A psychobiological perspective* (pp. 299–320). Orlando, FL: Academic Press.

Richter, C, P., & Barelare, B. (1938). Nutritional requirements of pregnant and lactating rats studied by the self-selection method. *Endocrinology, 23,* 15–24.

Rosenblatt, J. S. (1965). The basis of synchrony in the behavioral interaction between the mother and her offspring in the laboratory rat. In B. M. Foss (Ed.), *Determinants of infant Behaviour* (Vol 3, pp. 3–41). London: Methuen

Rosenblatt, J. S. (1967). Nonhormonal basis of maternal behavior in the rat. *Science, 156,* 1512–1514.

Rosenblatt, J. S. (1987). Biologic and behavioral factors underlying the onset and maintenance of maternal behavior in the rat. In N. A. Krasnegor, E. M. Blass, M. A. Hofer, & W. P. Smotherman (Eds.), *Perinatal development: A psychobiological perspective* (pp. 321–341). Orlando, FL: Academic Press.

Rosenblatt, J. S., & Lehrman, D. (1963). Maternal behavior of the laboratory rat. In H. L. Rheingold (Ed.), *Maternal behavior in mammals* (pp. 8–57). New York: Wiley.

Rosenblatt, J. S., & Siegel, H. I. (1981). Factors governing the onset and maintenance of maternal behavior among nonhuman mammals. In D. J. Gubernick & P. H. Klopfer (Eds.), *Parental care in mammals* (pp. 13–76). New York: Plenum Press.

Samuels, M. H., & Bridges, R. S. (1983). Plasma prolactin concentrations in parental male and female rats: Effects of exposure to rat young. *Endocrinology, 113,* 1647–1654.

Scalia, F., & Winans, S. S. (1975). The differential projections of the olfactory bulb and accessory olfactory bulb in mammals. *Journal of Comparative Neurology, 161,* 31–55.

Siegel, H. I., & Greenwald, G. S. (1978). Effects of mother–litter separation on later maternal responsiveness in the hamster. *Physiology and Behavior, 21,* 147–149.

Slotnick, B. M. (1967). Intercorrelations of maternal activities in the rat. *Animal Behaviour, 15,* 267–269.

Slotnick, B. M., & Nigrosh, B. J. (1975). Maternal behavior of mice with cingulate cortical, amygdala, or septal lesions. *Journal of Comparative and Physiological Psychology, 88,* 118–127.

Sodersten, P., & Eneroth, P. (1984). Effects of exposure to pups on maternal behaviour, sexual behaviour and serum prolactin concentrations in male rats. *Journal of Endocrinology, 102,* 115–119.

Stern, J. M., & Siegel, H. I. (1978). Prolactin release in lactating, primiparous and multiparous thelectomized and maternal virgin rats exposed to pup stimuli. *Biology of Reproduction, 19,* 177–182.

Svare, B., Bartke, A., & Gandelman, R. (1977). Individual differences in the maternal behavior of male mice: No evidence for a relationship to circulating testosterone levels. *Hormones and Behavior, 8,* 372–376.

Tate-Ostroff, B., & Bridges, R. S. (1985). Plasma prolactin levels in parental male rats: Effects of increased pup stimuli. *Hormones and Behavior, 19,* 220–226.

Thiels, E., & Alberts, J. R. (1985). Role of milk availability in weaning by the Norway rat. *Journal of Comparative Psychology, 99,* 447–456.

Thiels, E., Cramer, C. P., & Alberts, J. R. (1988). Behavioral interactions rather than milk availability determine decline in milk intake of weanling rats. *Physiology and Behavior, 42,* 507–515.

Trivers, R. L. (1972). Parental investment and sexual selection. In B. Campbell (Ed.), *Sexual selection and the descent of man, 1871–1971* (pp. 136–179). Chicago: Aldine.

Voci, V. E., & Carlson, N. R. (1973). Enhancement of maternal behavior and nest building following systemic and diencephalic administration of prolactin and progesterone in the mouse. *Journal of Comparative and Physiological Psychology, 83,* 388–393.

vom Saal, F. S. (1984). Proximate and ultimate causes of infanticide in male house mice. In G. Hausfater & S. B. Hrdy (Eds.), *Infanticide: Comparative and evolutionary perspectives* (pp. 401–424). New York: Aldine.

21

Parent–Offspring Recognition in Mammals: A Proximate and Ultimate Perspective

WARREN G. HOLMES

Indiscriminate care of young by adults is rare among animals, including humans. Why is it that adults most often care for their own biological offspring or those of closely related kin rather than unrelated young? The purpose of this chapter is to examine the question of discriminate care by considering parent–offspring relations in a comparative framework and to seek both proximate and ultimate answers to the question. Tinbergen (1963) explained that proximate answers to behavioral questions emphasize the immediate causes and development of a trait (e.g., the physiology of behavior, the role of experience in behavior), whereas ultimate answers emphasize the long-term reproductive consequences of behavior and behavioral evolution across generations. Thus to understand fully why parental care is not provided indiscriminately and why related young are cared for preferentially over unrelated young, one must incorporate both proximate and ultimate thinking.

Preferential treatment of an individual's biological offspring is frequently based on a parent's ability to discriminate between its own and alien offspring (i.e., those produced by other adults). Discrimination of parents, offspring, and other kin occurs in several animal groups and has recently been a topic of intense investigation (reviews in Fletcher & Michener, 1987; Sherman & Holmes, 1985; Waldman, 1988). Here, parent–offspring recognition and, more generally, kin recognition refer to the differential treatment of conspecifics that correlates with genetic relatedness. At the organismic level, recognition is inferred from differential responsivity or treatment and implies nothing specific about underlying neural processes or cognition (contra Byers & Bekoff, 1986).

Waldman, Frumhoff, and Sherman (1988) distinguish between kin discrimination—the differential treatment of kin—and kin recognition—the internal, unobservable, physiological events that can result in kin-differential treatment. This distinction is valuable because the absence of differential treatment may mean that (1) the organism cannot make the discrimination or (2) conditions are inappropriate to reveal a discrimination that the organism can actually make; that is, recognition is possible, but discrimination is not shown. The ability to recognize kin is often inferred from the preferential treatment of relatives, which occurs, for example, in many Old World monkey species that form dominance hierarchies in which rank is based largely on agonistic support from matrilineal kin (reviewed in Gouzoules & Gouzoules, 1987). However, recognition of kin and preferential treatment of them are not synonymous because kin that are recognized are not always treated favorably (Hoogland, 1985, 1986).

Three general requirements must be met for the behavioral discrimination of kin (Waldman et al., 1988). First, individuals must possess perceivable phenotypic traits that differ from those of other individuals or groups of them. Beecher (1982) introduced the term "kin signatures" to refer to a complex of phenotypic traits that is distinct and allows discrimination of one relative (or group of them) from another. Second, individuals must possess the sensory apparatus to perceive kin signatures. Finally, individuals must acquire a kin-recognition mechanism or decision-making rule, often based on social experience, that results in kin-differential behavior. In this chapter, "recognition mechanism" does not refer to the physiological substrates that support discrimination (although such substrates must exist), but to the role played by environmental experience in recognition behavior at the organismic level.

The meaning of a "kin-recognition mechanism" can be exemplified by defining the association mechanism, one such decision-making rule: "When relatives predictably interact in unambiguous social contexts where kinship is not likely to be confounded by the mixing of unequally related individuals, recognition may be based on the timing, rate, frequency, or duration of such interactions" (Holmes & Sherman, 1983, p. 47). For example, in a laboratory study of dam–young recognition in Belding's ground squirrels *(Spermophilus beldingi)*, pups were cross-fostered within 24 hr of birth to produce four types of dam–young pairs: related-familiar, unrelated-familiar, related-unfamiliar, and unrelated-unfamiliar (Holmes, 1984). Related pairs were composed of genetic relatives, whereas unrelated pairs were not. Familiar pairs included indivuduals housed together for about 35 days in rearing cages after pup exchange, whereas unfamiliar pairs included individuals housed in different cages. During paired-encounter tests of dams and 35-day-old young, familiar pairs were less agonistic than unfamiliar pairs, and agonism did not differ as a function of genetic relatedness (Figure 21.1). Dam–young familiarity established by social interactions sometime during rearing most likely explains the differences in the behavior of familiar and unfamiliar pairs (see the general discussion in Bekoff, 1981).

Two points emerge from this study (Holmes, 1984). First, a kin-recognition mechanism based on social learning does not allow assessment of kinship or genetic relatedness per se, but allows an individual to behave as though it can assess relatedness in species-typical circumstances. Thus "kin recognition" could more precisely be called "kin-correlate recognition," where the factor(s) correlated with kinship (rearing association in the preceding example) vary with the ecology and social organization of the species being considered (Holmes & Sherman, 1983). Indeed, one analytic approach used to investigate a recognition mechanism is to manipulate putative correlates of relatedness, such as rearing association in the case of parent–offspring recognition. If nonkin are treated like kin following the manipulation, it is likely that an important correlate of relatedness in species-typical environments has been identified.

Second, recognition mechanisms are best considered "rules of thumb" or "behavioral heuristics" that typically, but not always, result in kin-differential behavior. Association during rearing mediates dam–offspring recognition in Belding's ground squirrels (Figure 21.1) (Holmes, 1984), and because in nature females rear their litters alone for several weeks in underground burrows, a "familiarity rule" (learn

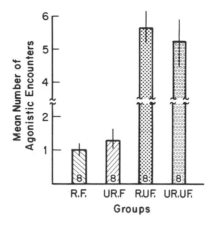

Figure 21.1. Mean (±SE) number of agonistic interactions between Belding's ground squirrel dams and young observed during paired-encounter tests. Numbers of pairs tested are shown inside bars. Cross-fostering pups at birth meant that dams and young tested together as a pair were either familiar (F) or unfamiliar (UF) with each other and were also either related (R) or unrelated (UR) to each other (see text). During 5-min tests, agonistic interactions were much less frequent within familiar pairs—those in which dams and young shared a rearing cage during the pups' development—than they were within unfamiliar pairs, regardless of genetic relatedness.

the kin signatures of conspecifics in your natal burrow) reliably produces differential treatment of a female's own versus alien young (details below). In nature, alien young sometimes enter a dam's burrow before her own litter has emerged and may be treated by the resident dam as her own offspring (Sherman, 1980), which demonstrates that the association mechanism, like other kin-recognition rules, does not produce error-free discrimination.

PARENT–OFFSPRING RECOGNITION IN AN EVOLUTIONARY PERSPECTIVE

Recognition Mechanisms

Because parental care is common in humans and other mammals, it is easy to overlook its absence in many vertebrate species and to forget that the form and complexity of parental care vary considerably across species that manifest it (Gubernick & Klopfer, 1981; Rosenblum & Moltz, 1983). In an evolutionary framework, parental care or, more generally, parental effort is one of two major forms of reproductive effort (the other form is mating effort) (Low, 1978; see chapter 3). Because an individual's time, energy, and resources are limited, parental effort is something an individual *expends,* and in so doing reduces the availability of such effort for the future. Thus natural selection acts on individuals so that parental care will evolve only if the reproductive benefits to the parent outweigh the reproductive costs (see the discussion in chapter 3). (Williams, 1966, presented a general argument to explain why the process of natural selection operates at the individual level rather than at the group or species level and why the individual's rather than the group's reproductive success is the target of selection. Traits that benefit the group, but not the individual, are unlikely to evolve. Modern evolutionists believe that behaviors like parental care and parent–offspring recognition evolved in response to selection operating at no higher than the individual level [see the discussions in Daly & Wilson, 1983; Trivers, 1985].)

Caregiving adults that rear their offspring in a social group may be solicited by alien young (those produced by other adults) to provide care. Such care would usually reduce rather than enhance adults' reproductive success (exceptions are considered below), which suggests that natural selection would favor adults that channeled assistance to their own rather than alien young and used whatever cues were available to dispense care discriminately. For instance, discriminitive care could occur by assisting young confined to specific locations, such as nests or burrows occupied by the parent (e.g., Shugart, 1987; recognition by the spatial distribution mechanism [Holmes & Sherman, 1983]. One might argue that because parents responded to locational cues rather than traits of young, "true" parent–offspring recognition did not occur. Indeed, use of the term "recognition" could be restricted to cases in which animals' phenotypic traits are discriminated, but in an evolutionary analysis the *consequences* of acts rather than their proximate control are paramount. (Waldman et al., 1988, distinguish between *direct* recognition, wherein relatives' phenotypic traits are critical cues, and *indirect* recognition, wherein contextual features associated with kin are critical cues.) There are several proximate mechanisms that can mediate parent–offspring recognition, which are considered below. Here, however, it is emphasized that natural selection would favor adults and young that used whatever correlates of relatedness were available to them to respond discriminately to kin. Such correlates (traits) may or may not have evolved specifically to facilitate parent–offspring recognition (Beecher, Medvin, Stoddard, & Loesche, 1986), but regardless of their historical genesis, such correlates may currently underlie discriminative care.

Should parent–offspring recognition be asymmetrical? From the viewpoint of the young, the identity of an adult caregiver should not matter so long as care is provided adequately. The problem with this scenario is that it implies that young would gain no benefit from being able to recognize their parent(s). As Beecher (1981) explained, the scenario is flawed because parental recognition of young is unlikely to be perfect and is subject to two kinds of errors: (1) providing care to alien young and (2) failing to provide care to related young. The likelihood of the latter error would be reduced if selection favored the ability of young to recognize their own parents, which they do in some species (discussions in Swartz & Rosenblum, 1981; Beecher, Stoddard, & Loesche, 1985). In this chapter, the ability of parents to recognize young and of young to recognize parents is considered, and terms like "parent–offspring recognition" are not meant to specify who makes the discrimination.

The importance to young of recognizing their parents was also implied by Trivers (1974), who pointed out that because the genetic interests of parents and their offspring are not identical, conflicts of reproductive interest characterize the parent–offspring dyad. This evolutionary view of conflict emphasizes that parents may not always care for their young in ways that are optimum from the viewpoint of the young, which could affect the importance to young of being able to distinguish among adults, including being able to recognize their parents.

Kin Signatures

If several independent and different kinds of discrimination tests failed to uncover differential responsiveness between parent and offspring, one might conclude that a recognition mechanism had not evolved to support the discrimination. It is just as likely, however, that discrimination did not occur because adequate kin signatures had not evolved (Beecher, 1982). To use a communication metaphor, a sender must provide cues (a kin signature) about its identity before a receiver can respond appropriately to the sender (Beecher, Loesche, Stoddard, & Medvin, in press). Phenotypic variation among individuals in traits such as plummage (Stoddard & Beecher, 1983), vocalizations (Gelfand & McCracken, 1986), location (Shugart, 1987), and odors (Beauchamp, Yamazaki, & Boyse, 1985) increases the potential for distinctive kin signatures and may decrease the probability of kin-recognition errors due to signature overlap.

For example, consider a newborn Mexican free-tailed bat left by its mother in the midst of a large group of over 10,000 young (a crèche). Adult females roost away from the crèche but visit it daily to search for their offspring, which they nurse selectively. (Davis, Herreid, and Short, 1962, suggested that female bats act as "one large dairy herd" and deliver milk indiscriminately to young. McCracken, 1984, however, demonstrated, by using allozyme genetic markers, that females nurse young selectively with respect to genotype.) Anything the pup can do to help its mother recognize it amid thousands of other hungry pups would increase the pup's survival prospects. At least three things enhance the distinctiveness of pups' signatures (Gustin & McCracken, 1987). First, pups have large sebaceous glands that may produce individually unique odor cues, which mothers could learn. Second, pups are marked by their mother with an odor from her muzzle gland, an odor that she can discriminate from those produced by other dams. Finally, pups emit individually distinct "isolation" calls when separated from their mothers (Gelfand & McCracken, 1986), and vocal exchanges occur between females and pups when females return to the crèche to search for their offspring. Thus several features contribute to the distinctiveness of pups' signatures (Gustin & McCracken, 1987), all of which increase the likelihood that pups will be recognized by and receive care from their mothers. A relatively unexplored area in kin-recognition research is whether the phenotypic variation used to discriminate among individual kin or groups of them evolved specifically in the context of kin recognition or evolved in other contexts and was subsequently coopted for recognizing kin.

Whether among-individual differences in kin signatures are of genetic or environmental origin or both (see the discussion in Gamboa, Reeve, Ferguson, & Wacker, 1986), close relatives are more likely than unrelated individuals to have similar signatures due to shared genes and/or environments (Bateson, Lotwick, & Scott, 1980; Carter-Saltzman & Scarr-Salapatek, 1975; Scarr & Grajeck, 1982). For parents that produce multiple young, offspring discrimination would be simplified if parents could learn a single shared kin signature rather than separate signatures for each offspring

(Jones, Falls, & Gaston, 1987). For instance, Doane and Porter (1978) in their study of spiny mice, found that dams discriminated between pups reared by dams fed on the same diet as themselves and pups reared by dams fed on different diets. Spiny mice dams apparently "label" their pups during nursing and, at least in laboratory tests, can use a single signature to distinguish their own offsrping from those of other females. That dams can discriminate between litters based on a single kin signature does not mean, however, that all littermates' signatures are identical (Porter, Matochik, & Makin, 1983). Indeed, there are theoretical reasons why individually unique signatures could be valuable. These reasons are often related to reducing the chances of classifying nonkin as kin (Waldman, 1987).

Parent–offspring recognition would also be simplified if kin signatures were stable over time so that new signatures did not have to be learned repeatedly. Within a few days of birth, human infants, for example, can distinguish between their own mother's breast odor and the breast odors of other lactating mothers (Cernoch & Porter, 1985; Macfarlane, 1975; Russell, 1976). This olfactory-discrimination ability led Macfarlane (1975) to suggest that encouraging nursing mothers to wash or apply odorous substances to their breasts might interfere with the infant's ability to recognize them by smell. Of course, the stability of kin signatures and the ability to remember them remain important only as long as parents and their young benefit by continued association and differential treatment.

"Errors" in Recognizing Kin

The distinction between proximate and ultimate explanations is particularly important when analyzing kin-recognition "errors." Consider parental recognition of offspring. A *proximate error* occurs when parents accept an alien young as their own, when parents reject their own offspring, or when both acceptance and rejection occur; an *ultimate error* occurs when parents differentiate between their own and alien young in ways that reduce the parents' reproductive success or, more exactly, their inclusive fitness (Hamilton, 1964). (Inclusive fitness is a measure of an individual's personal fitness [lifetime production of surviving offspring due to the individual's own reproductive efforts] *plus* the individual's effects on its relatives' personal fitnesses, taking into account the degree of relatedness between relatives [details are given in Dawkins, 1986].) Relations among proximate errors, ultimate errors, and inclusive fitness may be clarified by examining a study on northern elephant seals (Riedman & Le Boeuf, 1982).

Female northern elephant seals produce one pup annually amid densely packed harems of two to several hundred females. Females remain near and nurse their pups for about 35 days after birth, fasting throughout this period while they produce exceptionally fat-rich milk. Mother–pup separations occur frequently, especially during disruptive male–male fights or when pups wander away from their mothers. Mothers and their pups often reunite, but the fate of an orphaned pup is severe; 85% of nonadopted orphans die. Many orphans, however, are adopted by other females, and only 9% of adoptees die. In what sense, if any, do females that adopt aliens commit "offspring-recognition errors"? (In the following scenario, the focus is on females' ability to discriminate among pups rather than pups' discrimination abilities.)

Proximately, adoptive females may indeed be unable to discriminate between their own and alien young. This is especially likely if a female and her pup became separated before she learned her pup's vocal signature, which she could recognize in other circumstances (Petrinovich, 1974). In addition, recently parturient females are often behaviorally and hormonally primed to provide maternal care and may do so to any available young, with little attention to pups' signatures. Under normal circumstances, the most available pup would be a female's own offspring, but if an alien pup was more accessible or if the alien sought care more vigorously than the female's own pup, then she might care for the alien rather than her own, especially if her own offspring had disappeared.

From an ultimate perspective, caring for alien young *might* not represent an error, especially if a female had already lost her own offspring (Riedman & Le Boeuf, 1982). First, females could care for offspring of close relatives and gain an inclusive fitness benefit, although the breeding system and movement patterns of northern elephant seals make this unlikely. Second, because of their reproductive physiology, females may need to nurse to ensure that they will come into estrus and mate, which they typically do near the end of their 35-day lactation period. Finally, females may gain valuable parental experience by adopting an alien that could increase the survival probability of their own offspring in later years. Consistent with the last hypothesis, primiparous females are less successful in rearing young to weaning than multiparous females, and most adoptive mothers tend to be young and inexperienced (Riedman & Le Boeuf, 1982).

It would appear that female northern elephant seals have evolved to (1) remain near and protect their own pup, (2) learn to recognize their pup shortly after birth by becoming familiar with their pup's kin signature, and (3) rebuff alien pups that solicit care. Occasionally, a reproductive benefit may be gained by female northern elephant seals that adopt unrelated pups, but it seems unlikely that adoption is a primary reproductive tactic that evolved by natural selection in this species. Thus what seems to be an error in recognizing offspring at one level of analysis (proximate) does not reveal whether an error has been made at another level (ultimate).

Sex Differences in Parent–Offspring Recognition

In most species of mammals, males and females engage in different reproductive strategies such that mating opportunities are most critical to male reproductive success, whereas offspring production and care are most critical to female reproductive success (Trivers, 1972). This sex difference in basic reproductive strategies has at least two implications for parent–offspring recognition. First, mother–offspring recognition is more common taxonomically than father–offspring recognition. The latter has been described in laboratory mice (Huck, Soltis, & Coopersmith, 1982; Ostermeyer & Elwood, 1983), wild-strain house mice (Labov, 1980), collared lemmings (Mallory & Brooks, 1978), and spiny mice (Makin & Porter, 1984). However, recent reviews of kin recognition (e.g., Fletcher & Michener, 1987; Porter, 1987) reveal many more instances of maternal than paternal recognition of offspring. Moreover, studies demonstrating that males can discriminate between their own and alien young in certain test conditions do not reveal whether paternal recognition occurs in species-typical circumstances (discussion below).

Second, if parental ability to discriminate offspring has evolved in both sexes, females and males will often rely on different proximate mechanisms to make discriminations. Prior association between a female and her young controls maternal recognition in many mammalian species (see references throughout this chapter), and young reared apart from their biological mother are not usually recognized by her during tests (see Figure 21.1; see also Klopfer & Klopfer, 1968; Poindron & Le Neindre, 1980). In contrast, males in most mammalian species are unlikely to rely on the association mechanism in nature because, at the proximate level, they rarely interact directly with their young during early development or because they associate with young in environments that do not keep separate the offspring of different males (Elwood, 1983, and Dewsbury, 1985, on rodents; Busse, 1985, on primates; see chapter 20). At the ultimate level, a male cannot be certain that the young produced by the female he mated with are his own offspring because the female may have mated with more than one male during a single fertile period (Smith, 1984). More specifically, when strict monogamy or intense mate guarding (behaviors by a male that prevent other males from mating with his female partner) is practiced, a male's probability of paternity is increased—that is, the likelihood that he has sired a given offspring or litter. When males' probability of paternity is high, the association mechanism might reliably mediate father–offspring recognition if males had opportunities to learn the kin signatures of their mate's offspring. However, as males' probability of paternity declines, the accuracy of the association mechanism would diminish and selection would not likely have favored males that utilized association.

To suggest that prior association is unlikely to mediate father–offspring recognition in nature is not to say that association could not mediate recognition under any experimental conditions. Laboratory studies of father–offspring recognition often entail housing a male and a female together with their litter and, in test situations, observing that males treat familiar young (those they were housed with) differently from unfamiliar young. However, to interpret discriminative paternal care in an evolutionary context, one must know whether males routinely associate with their offspring in species-normal environments. If males do not so associate, one cannot legitimately explain the adaptive significance of paternal recognition by association, and, indeed, explanations for the adaptive significance of paternal care itself become problematic (Dewsbury, 1985; Hartung & Dewsbury, 1979). If, however, paternal care is an evolved trait characteristic of most males in a species, the discriminative basis and adaptive significance of paternal recognition warrant study (Gubernick & Alberts, 1987; see chapter 20).

In polygynous, polyandrous, or promiscuous mating relations, or when mate guarding by males is often unsuccessful, three recognition mechanisms other than prior association might facilitate father–offspring recognition. First, "mediated recognition" occurs "if two relatives previously unfamiliar with each other interact in the presence of a third conspecific . . . familiar to both (a 'go-between')" (Holmes & Sherman, 1983, p. 48). For example, Huck et al. (1982) used rates of infanticide to assess father–offspring relations in laboratory mice. Males exposed to their own offspring in their mate's presence were much less likely to kill them than males exposed to their offspring in the presence of a strange female (one never before encountered). The investigators wrote that "past association with the mother is the single most important factor mediating male discrimination of the young" (p. 1162; see also Labov, 1980). Mediated recognition could account for male–infant affiliations in some primate species if males recalled their prior mates and affiliated with prior mates' offspring (Busse, 1985), although there is as yet little experimental support for this hypothesis (Berenstain, Rodman, & Smith, 1981).

Second, if males' offspring were predictably and reliably distributed in space, differential treatment of males' own and alien young would occur if males varied their behavior relative to locational cues. Male Arctic ground squirrels *(Spermophilus parryii)* may respond to locational cues and, as a result, treat young they may have sired differently from unrelated young. McLean (1983) wrote, "I suggest that males behave paternally through defending territories because they increase the likelihood of survival of young that they may have sired" (p. 32). In other words, if females reared their litters on the territories of males they mated with and males protected young within males' territorial boundaries, then sires would be treating their own offspring preferentially as a result of location-specific behavior. Clearly, changes in territorial boundaries or ownership, or dispersal by females after mating, would compromise discriminative paternal care based on territoriality.

A third mechanism that males could employ to recognize offspring is phenotype matching (Holmes & Sherman, 1983; see also the discussion on "comparing phenotypes" in Alexander, 1979). Initially, males would acquire through learning a "kin template" by becoming familiar with their own phenotype or those of familiar kin. Assuming that phenotypes of close kin are more similar than those of nonkin (references above), males could subsequently compare the phenotypic attributes of unknown young with the learned kin template and distinguish between young whose phenotypic attribute(s) either matched or did not match this template (Holmes & Sherman, 1983; Waldman, 1987; see Lacy & Sherman, 1983, for a discussion of algorithms that males might use to compare phenotypes. Note that accepting matches or rejecting nonmatches could both produce phenotype matching).

Differential treatment of unfamiliar kin (e.g., paternal half-siblings) based on phenotype matching has been reported in various species (reviewed in Holmes, 1986), and studies of father–offspring recognition by phenotype matching seem warranted in species in which males behave paternally, but do not have opportunities to learn their own offsprings' signatures before encountering other males' offspring or if males' confidence of paternity is low. For example, Alexander (1979) suggests that for humans, "erratic confidence of paternity may have led to emphasis upon phenotypic attributes of putative offspring in determining whether or not to accept them as suitable objects of paternal care" (p. 159). Both mothers and untrained raters can match at a greater than chance level the full-face photographs of mothers and their 2-day-old babies (Porter, Cernoch, & Balogh, 1984), suggesting that facial resemblance can provide cues about relatedness. Whether father–offspring matching can be done by raters or fathers themselves remains to be determined, but there is intriguing evidence that father–offspring resemblance is important to humans. Daly and Wilson (1982) tested several hypotheses about whom newborn humans were said to resemble by mothers, fathers, and relatives of both parents. The authors found disproportionately high numbers of comments about paternal resemblance compared with resemblance to all other family members, including the newborns' mothers. It was not known whether phenotypic resemblances between fathers and their offspring were actually high, but Daly and Wilson's findings highlight the importance of putative paternal resemblance.

PARENT–OFFSPRING RECOGNITION IN PROXIMATE PERSPECTIVE

To clarify and expand the points made above on discriminative parental care, parent–offspring recognition in Belding's ground squirrels *(S. beldingi)* will be examined and compared with that in other mammals. Kin-correlated behaviors in which closely related individuals are favored over distantly related ones are common in several species of ground squirrels (reviewed in Michener, 1983; see papers in Murie & Michener, 1984) and are directly relevant here because such behaviors imply that kin-recognition abilities have evolved and mediate kin favoritism (reviewed in Schwagmeyer, 1988; Sherman & Holmes, 1985).

Behavior and Ecology of Belding's Ground Squirrels

S. beldingi is a 275- to 375-g, diurnally active rodent that lives in social groups in alpine and subalpine meadows of the western United States. The information below is based primarily on studies by Paul Sherman of Cornell University on a population near Tioga Pass, California, where adults ($\geq$ 1-year-olds) are active for about 5 months each year (early May to late September) and hibernate underground for about 7 months (Sherman, 1977, 1980, 1981a, 1981b, 1985; Sherman & Morton, 1984).

Female *S. beldingi* mate during their single annual estrus, shortly after emerging from hibernation. Because each female typically mates with three to four males, a litter usually has multiple sires (Hanken & Sherman, 1981). Below, littermates are referred to as "siblings," although litters routinely include full- and maternal half-siblings due to multiple mating by females. Each female rears her single litter of about five pups in an underground burrow (the natal burrow), from which pups first emerge aboveground at about 25 days of age as weaned juveniles (their natal emergence). Dams defend natal burrows from conspecific intruders during the lactational period, so siblings interact only with one another and their dam until natal emergence. For their first 2–3 days aboveground, juveniles (weaned young of the year) continue to interact only with siblings, but then they begin to encounter neighboring juveniles and adults as their social world expands.

Mechanisms of *S. beldingi* Parent–Offspring Recognition

In this section, the way in which the social experiences of fathers, mothers, and offspring affect parent–offspring recognition is considered. Three points suggest that male *S. beldingi* do not recognize their offspring. First, males do not behave parentally toward young, regardless of male–young relatedness. Second, males leave the meadow areas where they inseminated females shortly after mating (Sherman & Morton, 1984) and thus do not interact with the juveniles they sired. Finally, because most litters are of mixed paternity due to multiple mating by females (Hanken & Sherman, 1981), the identities of a male's former mates, even if he could recall them, would not correlate with male–young relatedness. Although the absence of paternal care implies that father–offspring recognition does not occur in *S. beldingi*, empirical tests must be conducted under ecological appropriate circumstances to determine whether such recognition *can* occur.

Mother–offspring recognition in *S. beldingi* is based on the association mechanism, as revealed by field (Holmes & Sherman, 1982) and laboratory (Holmes, 1984) tests (see also Michener & Sheppard, 1972, and Michener, 1974, on Richardson's ground squirrels). In the field, Sherman found that he could foster alien young by placing them near natal burrows and waiting for resident dams to retrieve them. When the litters later emerged, resident dams treated their biological and foster offspring alike and behaved aggressively toward young reared by other (unrelated) females. In the laboratory, Holmes fostered young reciprocally between females shortly after birth and later found that dams retrieved 22-day-old young they had reared (familiar young) faster than young they had not reared (unfamiliar young). Retrieval times did not vary with true relatedness (Holmes, 1988, Figure 3). Thus in field and laboratory tests, dams responded not to their biological offspring as such, but to familiar and unfamiliar young. Because dams in nature rear their litters isolated in underground burrows until the young are weaned, a "familiarity rule" (learn kin signatures of the young in your burrow) would result in discriminate treatment of a dam's own offspring. However, there is more to the story; the familiarity rule also has timing and contextual features.

In his field experiment (Holmes & Sherman, 1982), Sherman attempted to foster alien young of several ages, age matching aliens and resident young. He found that dams accepted aliens up to 25 days old (i.e., aliens were retrieved and remained with their adoptive dam for at least 1 week after the young emerged) but did not accept aliens thereafter. Prior to emergence of a dam's litter, which would occur when her offspring reached about 25 days of age, 75% of aliens were accepted, but only 21% were accepted thereafter (Holmes & Sherman, 1982, Figure 6). Similarly, in the laboratory, dams presented unfamiliar, unrelated young cared for them (carried, groomed, and/or nursed them) if (1) the alien young were up to 5 days older than the dams' own young and (2) the dams' own offspring were less than about 25 days old. However, dams whose young were more than 25 days old behaved aggressively toward aliens (Holmes, unpublished data; see also Holmes, 1984). Together, the field and laboratory results imply that a familiarity rule mediates dam–offspring recognition in S. beldingi if familiarity is established in or near natal burrows or about the time young would first come aboveground. (The role of offspring in dam–offspring recognition is considered below.)

It is important to distinguish between the time when dam–offspring recognition is first manifested from when it could be first manifested. In several bird species, parents first treat their own offspring discriminately about the time young become mobile and broods begin to mix (reviewed in Beecher, 1981; but see the discussion in Shugart, 1977). The behavior of S. beldingi females in the field (Holmes & Sherman, 1982) parallels the timing pattern in birds, but laboratory tests of pup retrieval revealed additional information about the onset of discrimination abilities. Retrieval times for familiar and unfamiliar 15-day-old young were similar, but familiar 22-day-old young were retrieved faster than unfamiliar 22-day-old young, suggesting that recognition occurred only when older pups were used (Holmes, 1984, Figure 2). However, when the same 15-day-old young were placed in dams' home cages, dams spent more time sniffing and engaging in direct contact with unfamiliar than familiar young (Holmes, unpublished data). The use of different behavioral measures and testing environments could explain why 15-day-old young were recognized in one laboratory test and appeared not to be in another. However, the point is that parent–

offspring recognition may be possible before it is first expressed, especially in species-typical environments where the appearance of discriminative care may be delayed until it is actually needed to prevent misdirected parental care. Learning techniques (e.g., habituation, classical or instrumental conditioning) may be required to specify the earliest age of young when parent–offspring *can first* occur.

The description above of *S. beldingi* mother–offspring recognition may imply that mothers were the only active participants in the discrimination process, but the results (Holmes, 1984; Holmes & Sherman, 1982) could also reflect discrimination by offspring. From the viewpoint of young confined to their natal burrow, if a dam prevented other ground squirrels from entering her burrow and if she cared only for the young in her own burrow, the young would gain little advantage from recognizing their dam because discriminative care was already being provided. However, circumstances change shortly after young come aboveground as weaned juveniles, because they begin encountering unrelated, aggressive adults (Sherman, 1981b). Thus after natal emergence, it would be valuable for juveniles to be able to discriminate related from unrelated animals. To study discrimination by juveniles, pups were cross-fostered reciprocally at birth (methods are discussed in Holmes, 1984), and preference tests were later conducted with 15-day-old (unweaned) and 22-day-old (nearly weaned) young.

Young *S. beldingi* were simultaneously presented two kinds of dams that varied in relatedness and familiarity to pups and that had been lactating for equal amounts of time. Pups were (1) either familiar with a dam (reared by her) or unfamiliar with her (reared by another dam) and (2) either related to a dam (the dam's offspring) or unrelated to her (another dam's offspring). Because dams were anesthetized and positioned under hardware cloth during the 5-min tests, it was assumed that pups' time in proximity to dams primarily reflected pups' discrimination abilities. Fifteen-day-old young spent about equal amounts of time near each of the various kinds of females, but the same young at 22 days of age spent about twice as much time near unfamiliar dams as familiar dams (Figure 21.2). Proximity times did not vary with pup–dam relatedness, regardless of pup–dam familiarity. Thus *S. beldingi* young (1) can distinguish between the dam that reared them and one that did not, (2) behave as though they cannot recognize their biological dam if reared apart from her, and (3) distinguish between a familiar and an unfamiliar dam at about the age young would first come aboveground in the field. It is not known why dams discriminate between familiar and unfamiliar 15-day-old young (above), whereas 15-day-old young do not discriminate between familiar and unfamiliar dams.

The picture emerging from laboratory tests of *S. beldingi* dam–offspring recognition must be interpreted carefully vis-à-vis recognition in nature. Laboratory data (Figure 21.2; Holmes, 1984) suggest that recognition does not occur between dams and *unweaned* young, yet in the field, dams care exclusively for their own unweaned young, probably by using cues associated with their natal burrow. In captivity, dams discriminate between familiar and unfamiliar 15-day-old young placed in dams' home cages (above), whereas dams do not discriminate between their own and alien young when 15-day-old aliens are introduced in field cross-fostering tests (Holmes & Sherman, 1982). An obvious explanation for these laboratory–field differences is that different cues are available to mediate discrimination in the two environments and that different recognition mechanisms are employed. The point is that to interpret

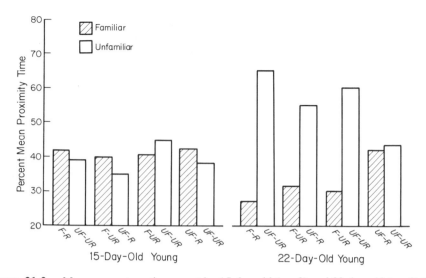

Figure 21.2.   Mean percentage time spent by 15-day-old ($n = 9$) and 22-day-old ($n = 9$) Belding's ground squirrels near two kinds of anesthetized dams during 5-min tests. Each pup was tested four times, and in each test two different kinds of dams were presented that varied in familiarity and relatedness to the pups. That is, pups had been reciprocally cross-fostered between dams, so they were either familiar (F) with a dam (reared by her) or unfamiliar (UF) with her (reared by another dam). They were also related (R) to a given dam (her biological offspring) or unrelated (UR) to her (another dam's biological offspring). The only discrimination that pups appeared to make was between familiar and unfamiliar dams when pups reached 22 days of age.

laboratory-derived discrimination abilities in an adaptive framework, one must be knowledgeable about the behavior and ecology of the organism in its natural environment. Indeed, this knowledge is critical in designing an ecologically relevant laboratory study in the first place.

Sensory Basis of *S. beldingi* Dam–Offspring Recognition

In many species, parent–offspring recognition can be mediated by more than one sensory system. In various bat species, for example, spatial, vocal, and olfactory cues can be important to females' selective nursing of their own young (Gustin & McCracken, 1987; Thompson, Fenton, & Barclay, 1984). Visual, olfactory, and auditory cues influence mother–offspring recognition in experimental tests with domestic horses (Wolski, Houpt, & Aronson, 1980) and domestic sheep (Poindron & Le Neindre, 1980). In the absence of any other information, human mothers can recognize auditory (Green & Gustafson, 1983); visual (Porter et al., 1984), and olfactory (Russell, Mendelson, & Peeke, 1983) cues produced by their own infants. Although discrimination tests can verify that cues in a particular sensory modality are sufficient for discrimination, the tests do not necessarily reveal the role (if any) of those cues in species-typical environments, nor do they necessarily identify the most crucial cues.

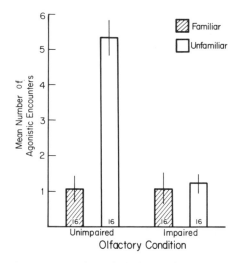

Figure 21.3.   Mean (±SE) number of agonistic interactions between Belding's ground squir-rel dams and young during paired-encounter tests. Olfactory discrimination abilities were im-paired (verified by independent tests) in dams and young given zinc sulfate or left unimpaired. Dams and young tested together as a pair were either familiar (F) or unfamiliar (UF) with each other because they had (familiar) or had not (unfamiliar) shared a rearing cage during the pups' development. That unfamiliar and familiar pairs differed in agonism only in the unim-paired condition suggests that the olfactory system is critical to dam–offspring recognition.

More than one sensory system often influences parent–offspring recognition in rodents, but the primary olfactory system is often crucial (reviewed in Porter, Bal-ogh, & Makin, 1988). To study the role of olfaction in *S. beldingi* dam–offspring recognition, intranasal injections of zinc sulfate, which produce olfactory impairment in several species, were used (Alberts, 1976). In control animals injected with water, familiar dam–juvenile pairs were about one-fifth as agonistic during paired-encounter tests as unfamiliar pairs (Figure 21.3; details in Holmes, 1984). After injections of zinc sulfate, however, familiar and unfamiliar pairs did not differ in agonism, veri-fying the importance of olfactory cues in these discrimination tests. The olfactory system is an appropriate one to mediate dam–offspring recognition for ground squir-rels that first interact in dark, underground burrows and for other species in which females remain closely associated with their offspring before encountering olfactory stimuli from other young (e.g., goats [Klopfer & Gamble, 1966] and wild rabbits [Mykytowycz & Dudzinski, 1972]). Although olfactory cues are not as salient to humans as to some other species of mammals, recent research implies that they may influence mother–infant relations shortly after birth, when infants can recognize their mothers and mothers can recognize their infants by olfactory cues alone (reviewed in Porter et al., 1988).

Kin Signatures in *S. beldingi* Dam–Offspring Recognition

Kin-signature differences among individuals that make parent–offspring recognition possible may result largely from genetic differences (e.g., Breed, 1981); environmental differences (e.g., Leon, 1978), including signatures acquired by learning (e.g., Beecher et al., 1986); or a combination of genetic and environmental factors that differ among individuals (Gamboa et al., 1986). Signature systems have been studied more intensively in insects (reviewed in Gamboa, Reeve, & Pfennig, 1986) and birds (Beecher et al., in press) than in mammals, but mammalian signature systems have received some attention, especially the influence of diet on signatures (e.g., Leon, 1978, on Wistar rats; Porter & Doane, 1976, 1977). Kin signatures in ground squirrels have yet to be studied.

CONCLUDING REMARKS

That mammalian parents give care to young may seem to be a trivial observation, and that parents care preferentially for their own rather than alien young may seem even more pedestrian. Thinking in an evolutionary framework, however, forces one to acknowledge that parental care did not always exist and that it must have evolved from a noncaregiving state, probably in response to several social and ecological factors (Barash, 1976; Klopfer, 1981; see chapter 3). It is also likely that as caregiving evolved, so did discriminative care.

This chapter has focused on a single aspect of discriminative care: the ability of parents to distinguish between their own young and those of other adults and, reciprocally, of young to discriminate their own parents. If parent–offspring interactions are to enhance the evolutionary (reproductive) success of parents and offspring, however, many variables other than kinship must be considered: the ages and health of adults and young; variation in the condition of individual young in a family; ecological factors like food, predators, and disease; and alternative ways to invest time and energy, to name a few. Thus the study of parental behavior requires an interdisciplinary approach and must proceed on several fronts to be successful. Modern Darwinian theory and the comparative method can be used to organize and help explain the diversity of parental behavior, including the proximate and ultimate factors that shape and control discriminitave parental care.

ACKNOWLEDGEMENTS

I would like to thank M. Beecher and R. Porter for providing me with copies of unpublished manuscripts, N. Krasnegor and B. Waldman for comments on earlier versions of this chapter, and P. Sherman for the many conversations we had during our collaborate research on Belding's ground squirrel kin recognition. My research was supported in part by funds from the University of Michigan and from the National Institute of Mental Health (MH 43861).

REFERENCES

Alberts, J. R. (1976). Olfactory contributions to behavioral development in rodents. In R. L. Doty (Ed.), *Mammalian olfaction, reproductive processes and behavior* (pp. 67–94). New York: Academic Press.

Alexander, R. D. (1979). *Darwinism and human affairs*. Seattle: University of Washington Press.

Barash, D. P. (1976). Some evolutionary aspects of parental behavior in animals and man. *American Journal of Psychology, 89*, 195–217.

Bateson, P., Lotwick, W., & Scott, D. K. (1980). Similarities between the faces of parents and offspring in Bewick's swan and the differences between mates. *Journal of Zoology, 191*, 61–74.

Beauchamp, G. K., Yamazaki, K., & Boyse, E. A. (1985). The chemosensory recognition of genetic individuality. *Scientific American, 253*, 86–92.

Beecher, M. D. (1981). Development of parent–offspring recognition in birds. In R. K. Aslin, J. R. Alberts, & M. R. Petersen (Eds.), *Development of perception* (Vol. 1, pp. 46–66). New York: Academic Press.

Beecher, M. D. (1982). Signature systems and kin recognition. *American Zoologist, 22*, 477–490.

Beecher, M. D., Beecher, I. M., & Hahn, S. (1981). Parent–offspring recognition in bank swallows *(Riparia riparia):* II. Development and acoustic basis. *Animal Behaviour, 29*, 95–101.

Beecher, M. D., Loesche, P., Stoddard, P. K., & Medvin, M. B. (in press). Individual recognition by voice in swallows: Signal or perceptual adaptation? In S. H. Hulse & R. J. Dooling (Eds.), *The comparative psychology of complex auditory perception*. Hillsdale, NJ: Erlbaum.

Beecher, M. D., Medvin, M. B., Stoddard, P. K., & Loesche, P. (1986). Acoustic adaptations for parent–offspring recognition in swallows. *Experimental Biology, 45*, 179–193.

Beecher, M. D., Stoddard, P. K., & Loesche, P. (1985). Recognition of parents' voices by young cliff swallows. *Auk, 102*, 600–605.

Bekoff, M. (1981). Mammalian sibling interactions: Genes, facilitative environments, and the coefficient of familiarity. In D. J. Gubernick & P. H. Klopfer (Eds.), *Parental care in mammals* (pp. 307–346). New York: Plenum Press.

Berenstain, L., Rodman, P. S., & Smith, D. G. (1981). Social relations between fathers and offspring in a captive group of rhesus monkeys *(Macaca mulatta). Animal Behaviour, 29*, 1057–1063.

Boyse, E. A., Beauchamp, G. K., & Yamazaki, K. (1983). The sensory perception of genotypic polymorphism of the major histocompatibility complex and other genes: Some physiological and phylogenetic implications. *Human Immunology, 6*, 177–183.

Breed, M. D. (1981). Individual recognition and learning of queen odors by worker honeybees. *Proceedings of the National Academy of Sciences. USA, 78*, 2635–2637.

Busse, C. D. (1985). Paternity recognition in multi-male primate groups. *American Zoologist, 25*, 873–881.

Byers, J. A., & Bekoff, M. (1986). What does "kin recognition" mean? *Ethology, 72*, 342–345.

Carter-Saltzman, L., & Scarr-Salapatek, S. (1975). Blood group, behavioral, and morphological differences among dizygotic twins. *Social Biology, 22*, 372–374.

Cernoch, J. M., & Porter, R. H. (1985). Recognition of maternal axillary odors by infants. *Child Development, 56*, 1593–1598.

Daly, M., & Wilson, M. I. (1982). Whom are newborn babies said to resemble? *Ethology and Sociobiology, 3*, 69–78.

Daly, M., & Wilson, M. (1983). *Sex, evolution, and behavior*. Boston: Willard Grant Press.

Davis, R. B., Herreid, C. F., II, & Short, H. L. (1962). Mexican free-tailed bats in Texas. *Ecological Monographs, 32*, 311–346.

Dawkins, M. S. (1986). *Unravelling animal behaviour*. Harlow, Eng.: Longman.

Dewsbury, D. A. (1985). Paternal behavior in rodents. *American Zoologist, 25*, 841–852.

Doane, H. M., & Porter, R. H. (1978). The role of diet in mother–infant reciprocity in the spiny mouse. *Developmental Psychobiology, 11*, 271–277.

Elwood, R. W. (Ed.). (1983). *Paternal care in rodents*. New York: Wiley.

Fletcher, D. J. C., & Michener, C. D. (Eds.). (1987). *Kin recognition in animals*. London: Wiley.

Gamboa, G. J., Reeve, H. K., Ferguson, D., & Wacker, T. L. (1986). Nestmate recognition in social wasps: The origin and acquisition of recognition odours. *Animal Behaviour, 34*, 685–695.

Gamboa, G. J., Reeve, H. K., & Pfenning, D. W. (1986). The evolution and ontogeny of nestmate recognition in social wasps. *Annual Review of Entomology, 31*, 431–454.

Gelfand, D. L., & McCracken, G. F. (1986). Individual variation in the isolation calls of Mexican free-tailed bat pups *(Tadarida brasiliensis mexicana)*. *Animal Behaviour, 34*, 1078–1086.

Gouzoules, S., & Gouzoules, H. (1987). Kinship. In B. B. Smuts, D. L. Cheney, R. M. Seyfarth, R. W. Wrangham, & T. T. Struhsaker (Eds.), *Primate societies* (pp. 299–305). Chicago: University of Chicago Press.

Green, J. A., & Gustafson, G. E. (1983). Individual recognition of human infants on the basis of cries alone. *Developmental Psychobiology, 16*, 485–493.

Gubernick, D. J., & Alberts, J. R. (1987). The biparental care system of the California mouse, *Peromyscus californicus*. *Journal of Comparative Psychology, 101*, 169–177.

Gubernick, D. J., & Klopfer, P. H. (1981). (Eds.). *Parental care in mammals*. New York: Plenum Press.

Gustin, M. K., & McCracken, G. F. (1987). Scent recognition between females and pups in the bat, *Tadarida brasiliensis mexicana*. *Animal Behaviour, 35*, 13–19.

Hamilton, W. D. (1964). The genetical evolution of social behaviour: I and II. *Journal of Theoretical Biology, 7*, 1–52.

Hanken, J., & Sherman, P. W. (1981). Multiple paternity in Belding's ground squirrel litters. *Science, 212*, 351–353.

Hartung, T. G., & Dewsbury, D. A. (1979). Paternal behavior in six species of muroid rodents. *Behavioral and Neural Biology, 26*, 466–478.

Holmes, W. G. (1984). Ontogeny of dam–young recognition in captive Belding's ground squirrels. *Journal of Comparative Psychology, 98*, 246–256.

Holmes, W. G. (1986). Kin recognition by phenotype matching in female Belding's ground squirrels. *Animal Behaviour, 34*, 38–47.

Holmes, W. G. (1988). Kinship and the development of social preferences. In E. Blass (Ed), *Developmental psychobiology and behavioral ecology* (pp. 389–413). New York: Plenum Press.

Holmes, W. G., & Sherman, P. W. (1982). The ontogeny of kin recognition in two species of ground squirrels. *American Zoologist, 22*, 491–517.

Holmes, W. G., & Sherman, P. W. (1983). Kin recognition in animals. *American Scientist, 71*, 46–55.

Hoogland, J. L. (1985). Infanticide in prairie dogs: Lactating females kill offspring of close kin. *Science, 230*, 1037–1040.

Hoogland, J. L. (1986). Nepotism in prairie dogs *(Cynomys ludovicianus)* varies with competition but not with kinship. *Animal Behaviour, 34*, 263–270.

Huck, W. U., Soltis, R. L., & Coopersmith, C. D. (1982). Infanticide in male laboratory mice: Effects of social status, prior sexual experience, and basis for discriminating between related and unrelated young. *Animal Behaviour, 30*, 1158–1165.

Jones, I. L., Falls, J. B., & Gaston, A. J. (1987). Vocal recognition between parents and young of ancient murrelets, *Synthliboramphus antiquus* (Aves: Alcidae). *Animal Behaviour, 35,* 1405–1415.

Klopfer, P. H. (1981). Origins of parental care. In D. J. Gubernick & P. H. Klopfer (Eds.), *Parental care in mammals* (pp. 1–12). New York: Plenum Press.

Klopfer, P. H., and Gamble, J. (1966). Maternal "imprinting" in goats: The role of chemical senses. *Zietschrift für Tierpsychologie, 23,* 588–592.

Kloper, P. H., & Klopfer, M. S. (1968). Maternal imprinting in goats: Fostering alien young. *Zietschrift für Tierpsychologie, 25,* 862–866.

Labov, J. B. (1980). Factors influencing infanticidal behavior in wild male house mice, *Mus musculus. Behavioral Ecology and Sociobiology, 6,* 297–303.

Lacy, R. W., & Sherman, P. W. (1983). Kin recognition by phenotype matching. *American Naturalist, 121,* 489–512.

Leon, M. L. (1978). Filial responsiveness to olfactory cues in the laboratory rat. In J. Rosenblatt, C. Beer, R. Hinde, & M.-C. Busnel (Eds.), *Advances in the study of behavior* (Vol. 3, pp. 117–153). New York: Academic Press.

Low, B. S. (1978). Environmental uncertainty and the parental strategies of marsupials and placentals. *American Naturalist, 112,* 197–213.

Macfarlane, A. (1975). Olfaction in the development of social preferences in the human neonate. In *Parent–infant interaction.* Ciba Foundation Symposium No. 33 (pp. 103–113). New York: Elsevier.

Makin, J. W., & Porter, R. H. (1984). Paternal behavior in the spiny mouse *(Acomys cahirinus). Behavioral and Neural Biology, 41,* 135–151.

Mallory, F. F., & Brooks, R. J. (1978). Infanticide and other reproductive strategies in the collared lemming, *Dicrostonyx groenlandicus. Nature, 273,* 144–146.

McCracken, G. (1984). Communal nursing in Mexican free-tailed bat maternity colonies. *Science, 223,* 1090–1091.

McLean, I. G. (1983). Paternal behaviour and killing of young in arctic ground squirrels. *Animal Behaviour, 31,* 32–44.

Michener, G. R. (1974). Development of adult–young identification in Richardson's ground squirrel. *Developmental Psychobiology, 7,* 375–384.

Michener, G. R. (1983). Kin identification, matriarchies, and the evolution of sociality in ground-dwelling sciurids. In J. E. Eisenberg & D. G. Kleiman (Eds.), *Advances in the study of mammalian behavior.* (Vol. 7, pp. 528–572). Special Publication of the American Society of Mammalogists.

Michener, G. R., & Sheppard, D. H. (1972). Social behavior between adult Richardson's ground squirrels, *Spermophilus richardsonii,* and their own and alien young. *Canadian Journal of Zoology, 50,* 1343–1349.

Murie, J. O., & Michener, G. R. (1984). *The biology of ground-dwelling squirrels.* Lincoln: University of Nebraska Press.

Mykytowycz, R. & Dudzinski, M. L. (1972). Aggressive and protective behaviour of adult rabbits, *Oryctolagus cuniculus* (L), toward juveniles. *Behaviour, 43,* 97–120.

Ostermeyer, M. C., & Elwood, R. W. (1983). Pup recognition in *Mus musculus:* Parental discrimination between their own and alien young. *Developmental Psychobiology, 16,* 75–82.

Petrinovich, L. (1974). Individual recognition of pup vocalisation by northern elephant seal mothers. *Zietschrift für Tierpsychologie, 34,* 308–312.

Poindron, P., & Le Neindre, P. (1980). Endocrine and sensory regulation of maternal behavior in the ewe. In J. S. Rosenblatt, R. A. Hinde, C. Beer, & M.-C. Busnel (Eds.), *Advances in the study of behavior* (Vol. 11, pp. 75–119). New York: Academic Press.

Porter, R. H. (1987). Kin recognition: Functions and mediating mechanisms. In C. Crawford, M. Smith, & D. Krebs (Eds.), *Sociobiology and psychology: Ideas, issues, and applications* (pp. 175–203). Hillsdale, NJ: Erlbaum.

Porter, R. H., Balogh, R. D., & Makin, J. W. (1988). Olfactory influences on mother–infant interactions. In C. Rovee-Collier & L. P. Lipsitt (Eds.), *Advances in infancy research* (Vol. 5, pp. 39–68). Norwood, NJ: Ablex.

Porter, R. H., Cernoch, J. M., & Balogh, R. D. (1984). Recognition of neonates by facial-visual characteristics. *Pediatrics, 74,* 501–504.

Porter, R. H., & Doane, H. M. (1976). Maternal pheromone in the spiny mouse *(Acomys cahirinus)*. *Physiology and Behavior, 16,* 75–78.

Porter, R. H., & Doane, H. M. (1977). Dietary-dependent cross-species similarities in maternal chemical cues. *Physiology and Behavior, 19,* 129–131.

Porter, R. H., Matochik, J. A., & Makin, J. W. (1983). Evidence for phenotype matching in spiny mice *(Acomys cahirinus)*. *Animal Behaviour, 31,* 978–984.

Porter, R. H., Tepper, V. J., & White, D. M. (1981). Experiential influences on the development of huddling preferences and "sibling" recognition in spiny mice. *Developmental Psychobiology, 14,* 375–382.

Porter, R. H., & Wyrick, M. (1979). Sibling recognition in spiny mice *(Acomys cahirinus)*: Influence of age and isolation. *Animal Behaviour, 27,* 761–766.

Porter, R. H., Wyrick, M., & Pankey, J. (1978). Sibling recognition in spiny mice *(Acomys cahirinus)*. *Behavioral Ecology and Sociobiology, 3,* 61–68.

Riedman, M. L., & Le Boeuf, B. J. (1982). Mother–pup separation and adoption in northern elephant seals. *Behavioral Ecology and Sociobiology, 11,* 203–215.

Rosenblum, L. A., & Moltz, H. (Eds.). (1983). *Symbiosis in parent–offspring interactions.* New York: Plenum Press.

Russell, M. J. (1976). Human olfactory communication. *Nature, 260,* 520–522.

Russell, M. J., Mendelson, T., & Peeke, H. V. S. (1983). Mothers' identification of their infant's odor. *Ethology and Sociobiology, 4,* 29–31.

Scarr, S., & Grajeck, S. (1982). Similarities and differences among siblings. In M. E. Lamb and B. Sutton-Smith (Eds.), *Sibling relationships* (pp. 357–381). Hillsdale, NJ: Erlbaum.

Schwagmeyer, P. L. (1988). Ground squirrel kin recognition abilities: Are there social and life-history correlates? *Behavior Genetics, 18,* 495–510.

Sherman, P. W. (1977). Nepotism and the evolution of alarm calls. *Science, 197,* 1246–1253.

Sherman, P. W. (1980). The limits of ground squirrel nepotism. In G. W. Barlow & J. Silverberg (Eds.), *Sociobiology: Beyond nature/nurture?* (pp. 505–544). Boulder, Co: Westview Press.

Sherman, P. W. (1981a). Kinship, demography, and Belding's ground squirrel nepotism. *Behavioral Ecology and Sociobiology, 8,* 251–259.

Sherman, P. W. (1981b). Reproductive competition and infanticide in Belding's ground squirrels and other organisms. In R. D. Alexander & D. W. Tindle (Eds.), *Natural selection and social behavior: Recent research and new theory* (pp. 311–331). New York: Chiron Press.

Sherman, P. W. (1985). Alarm calls of Belding's ground squirrels to aerial predators: Nepotism or self-preservation? *Behavioral Ecology and Sociobiology, 17,* 313–323.

Sherman, P. W. & Holmes, W. G. (1985). Kin recognition: Issues and evidence. In B. Holldobler & M. Lindauer (Eds.), *Experimental behavioral ecology and sociobiology* (pp. 437–460). Stuggart: Fischer.

Sherman, P. W., & Morton, M. M. (1984). Demography of Belding's ground squirrels. *Ecology, 65,* 1617–1628.

Shugart, G. W. (1977). The development of chick recognition by adult Caspian terns. *Proceedings of the Colonial Waterbird Group, 1,* 110–117.

Shugart, G. W. (1987). Individual clutch recognition by Caspian terns, *Sterna caspia. Animal Behaviour, 35,* 1563–1565.

Smith, R. L. (Ed.) (1984). *Sperm competition and the evolution of animal mating systems.* Orlando, FL: Academic Press.

Stoddard, P. K., & Beecher, M. D. (1983). Parental recognition of offspring in the cliff swallow. *Auk, 100,* 795–799.

Swartz, K. B., & Rosenblum, L. A. (1981). The social context of parental behavior: A perspective on primate socialization. In D. J. Gubernick & P. H. Klopfer (Eds.), *Parental care in mammals* (pp. 417–454). New York: Plenum Press.

Thompson, C. E., Fenton, M. B., & Barclay, R. M. R. (1985). The role of infant isolation calls in mother–infant reunions in the little brown bat, *Myotis lucifugus* (Chiroptera: Vespertilionidae). *Canadian Journal of Zoology, 63,* 1982–1988.

Tinbergen, N. (1963). On aims and methods of ethology. *Zeitschrift für Tierpsychologie, 20,* 410–433.

Trivers, R. L. (1972). Parental investment and sexual selection. In B. Campbell (Ed.), *Sexual selection and the descent of man, 1871–1971* (pp. 136–179). Chicago: Aldine.

Trivers, R. C. (1974). Parent–offspring conflict. *American Zoologist, 14,* 249–264.

Trivers, R. (1985). *Social evolution.* Menlo Park, CA: Benjamin/Cummings.

Waldman, B. (1987). Mechanisms of kin recognition. *Journal of Theoretical Biology, 128,* 159–185.

Waldman, B. (1988). The ecology of kin recognition. *Annual Review of Ecology and Systematics, 19,* 543–571.

Waldman, B., Frumhoff, P. C., & Sherman, P. W. (1988). Problems of kin recognition. *Trends in Ecology and Evolution, 3,* 8–13.

Williams, G. C. (1966). *Adaptation and natural selection.* Princeton, NJ: Princeton University Press.

Wolski, T. R., Houpt, K. A., & Aronson, R. (1980). The role of the senses in mare–foal recognition. *Applied Animal Ethology, 6,* 121–138.

22

Male Parental Behavior in Humans and Nonhuman Primates

MICHAEL W. YOGMAN

If sociocultural changes in Western society have led to a diversity of family structures and parenting styles, the impact on male parents has been to encourage even more diversity by lessening many of the stereotypes that have in the past constrained male behavior. Until the last decade, theories and empirical studies of infant social development largely ignored the father. The father's role with young infants was considered to be entirely indirect, supporting the mother, who was biologically adapted to be the infant's caregiver. This chapter reviews recent studies that suggest the father's role with young infants is far less biologically constrained than once thought and is more shaped by context. In what way biologically based sex differences constrain the social interactions of fathers with their infants must await future research, but such constraints now seem far more subtle than was once believed. Wide variability in the behavior and roles of fathers challenges many of the stereotypes of the father as incompetent or uninvolved with the infant and leaves ample opportunity for wide variations in the way fathers and their infants relate to each other.

The study of human behavior, and parental behavior in particular, has often confused two domains (what parents believe they do or should do vs. what parents actually do, as documented by independent observers) (Blurton-Jones, 1976). This chapter focuses on observational studies of male parental behavior in the hope of understanding the range of normal variation and the factors associated with adaptation across species, cultures, and gender. Ultimately, such knowledge may help clinicians to intervene with families at risk for problems such as child abuse in order to promote healthier parental adaptations.

Since many different variables have been shown to influence human development, any attempt to attribute levels of male parental care to a simple evolutionary trend based on phylogeny or anatomy is unlikely to be useful. However, comparative studies of humans and animals living under diverse ecological and social conditions with varying anatomy and genetic relatedness should help define the limits of variability and adaptation and the relative importance of ecological factors in regulating male parental involvement. Since paternity often cannot be specifically determined in field studies, the term "male parental care" or "investment" (see chapter 3) will be used rather than "fathering."

The evolution of parenting behavior has also been of great interest to sociobiologists as one example of altruistic behavior (Wilson, 1980). Trivers (1972) used the concept of "parental investment" (as opposed to simply "parental care") to describe the influence of natural selection on parent–infant behavior as an evolutionary mechanism that increases survival of the offspring. This review examines ecological factors associated with varying levels of male parental investment. Since male caregivers are part of a broader system of care, the impact of male involvement on the infant, the female caregiver, the adult male himself, and even the social group may regulate the level of male caregiving.

The chapter consists of four major sections. The first reviews cross-species and ethnographic data on male parental care in an attempt to illustrate the variability of and ecological influence on male involvement. The second section reviews the evidence for paternal competence in caring for and interacting with human infants. Similarities in the behavior and psychological experience of fathers and mothers are reviewed, and laboratory studies of parent–infant interaction are described. The third section examines the differences between maternal and paternal behaviors in an effort to gain an understanding of some unique aspects of fathering behavior, particularly play. The final section examines the antecedents and consequences of male involvement with infants (both humans and nonhuman primates) and concludes with speculation on the functional role of the father in the infant's life.

VARIABILITY IN MALE PARENTAL CARE:
CROSS-SPECIES AND ETHNOGRAPHIC EVIDENCE

While older theories of infant development suggest that parenting is predominantly instinctual, biologically determined, and exclusively maternal, such a view is not supported by either cross-species or anthropological empirical data. Although primary maternal caregiving is the predominant mode in most animal species, examples of primary paternal caregiving can be found among both nonprimates (wolf, ostrich, seahorse) (Rypma, 1976) and primates such as the marmoset (Hampton, Hampton, & Landwehr, 1966).

Phylogenetic evidence on the father's role shows no simple evolutionary trend (Redican, 1976), biological constraints seem less important than social and ecological influences, and variability best characterizes the data. Paternal involvement of nonhuman primates with their young varies not only within species, but also across settings for the same species and across conspecific males in the same social group. It can even vary from year to year in relation to male social dominance and age as social status in the group changes.

The range of male primate behavior is quite wide. The marmoset, a New World monkey, assists during birth, premasticates food during the first week, and carries the infant at all times, except during nursing in the first 3 months (Hampton et al., 1966). The Barbary macaque on Gibraltar begins in the newborn period to elicit social chatter and later encourages beginning locomotion, which functions to orient the infant toward interaction with other group members (Burton, 1972). However, male primates are by no means universally nurturant. Male tree shrews, bush babies, and langurs are known to be hostile to infants of their own species and are capable of infanticide, while feral male chimps show little interest in and have little contact with their infants (Redican & Taub, 1981).

Variability in male caretaking exists not only between species, but also among individuals within species. Hormonal variations are associated with the behavior of adult males toward their young. Androgen levels of Japanese male monkeys have a yearly cycle and are lowest during the birth season, when paternal behavior is highest (Alexander, 1970). Variations in prolactin levels have been related to levels of male caregiving in the marmoset (Dixson & George, 1982). Biological correlates of male parental behavior need further investigation.

In spite of this variability, male involvement in caretaking is typically higher in species in which adult males and females form monogamous relationships for prolonged periods and highest when such pair bonding lasts for life and when nonextended families (fewer than three generations) represent the modal social grouping (Parke & Suomi, 1980). Typical of such high-involvement males are marmosets, tamarins, gibbons, and siamangs (Chivers, 1972). Higher male involvement in marmosets and tamarins may be uniquely related to the mother's high reproductive burden: multiple births and closely spaced pregnancies (Redican & Taub, 1981). Squirrel monkeys which usually live in all-male groups, and orangutans, which live by themselves, are examples of species in which adult males almost never interact with infants.

While baboons, macaques, and chimps usually live in multimale, multifemale groups and father involvement is low, Redican (1976) has suggested that the degree of maternal restrictiveness regulates adult male involvement: when maternal restrictiveness is high, male involvement is low. In keeping with this hypothesis, Barbary and Japanese macaque males are more involved with infants than rhesus and pigtail macaques, playing with the infants for hours. Barbary macaque males encourage motor development (teach walking during the first week), orient infants to peers in their troop, and promote socialization (Figure 22.1) (Redican & Taub, 1981). Japanese macaque males are more involved with 1- and 2-year-olds, particularly when mothers are occupied with neonates (Itani, 1959). They have been observed to become primary caretakers of orphaned infants (Redican & Taub, 1981).

Infants reared in one-male/multifemale groups usually show intermediate levels of male involvement. Kinship ties have been invoked as a predictor of male involvement, since males are likely to provide care for their own offspring in their troop, while males that invade and displace a dominant male may kill the predecessor's dependent infants (e.g., langurs) (Redican & Taub, 1981). Dominance hierarchies within the troop may also influence the interactions of males with infants in species such as Hamadryas baboons. While controversy still abounds, some argue that males interact with infants as part of a triad as a means of allowing two adult males to interact with each other via an infant—that is, agonistic "buffering" to regulate interactions between a dominant and a submissive male (Kummer, 1967; Stein, 1984).

Male primate parenting behavior varies according to the context. Adoption of orphans by males has been observed among baboons, chimps, and macaques (Hrdy, 1976). Evan a species as aggressive as the rhesus male, when reared in a nuclear-family environment in a laboratory, interacts playfully with his infant (Suomi, 1977).

Anthropological data provide further insights into the way social and cultural variables influence paternal care of and involvement with infants. In cultures such as the Arapesh, fathers play an active and joint role with mothers during pregnancy as well as in caring for infants after birth (Howells, 1969). !Kung San (Bushmen) fa-

Figure 22.1. An adult male Barbary macaque in Morocco sitting with a young infant with a protective posture similar to that of a mother, with legs and arms around the infant. (Photograph by David M. Taub; reprinted by permission)

thers, representative of the earliest hunter-gatherer societies, were found to be affectionate and indulgent, often holding their infants, although they provided little routine care compared with mothers (West & Konner, 1981). Fathers from the Lesu village in Melanesia, who live in monogamous nuclear families and are gardeners, are reported to play with infants for hours every day (Powdermaker, 1933).

Analysis of social organization in different cultures suggests that males have a closer relationship with their infants when families are monogamous, when both parents live together in isolated nuclear families, when women contribute to subsistence by working, and when men are not required to be warriors (West & Konner, 1981; Whiting & Whiting, 1975).

In summary, phylogenetic and anthropological evidence underscores the diversity of the father–infant relationship across species and cultures. Fathers are involved with infants and play a competent role in many species and cultures. Moreover, the conditions of modern Western culture may help explain the currently increased involvement and interest of fathers in infant care.

SIMILARITIES OF MATERNAL AND PATERNAL BEHAVIOR

In contemporary Western society, the evidence that human fathers and infants can develop a direct relationship right from birth is impressive. Furthermore, the similarities between the psychological experiences of pregnancy and infant care for mothers and fathers are striking. Studies of the similarities can be grouped into four developmental periods: prenatal, perinatal, early infancy (1–6 months), and later infancy (6–24 months). A brief discussion of each will illustrate the similarity of maternal and paternal competencies with infants (for a more detailed review, see Yogman, 1982).

During the prenatal period, Bibring (1959) suggests, pregnancy represents a normative psychological crisis for women. Studies by Gurwitt (1976) and Ross (1975) have suggested that during pregnancy, men also rework significant relationships and events from early life. This turmoil has been called a "crisis of paternal identity" and involves conflicts of gender, generative, and generational identity (Ross, 1975) in which the men's roles as husband, son, and father become reintegrated. One father's drivenness at moving and settling into a new home had the symbolic quality of preparing a nest (Gurwitt, 1976).

The occurrence of physical complaints during pregnancy is probably one manifestation of this turmoil. Such symptoms are present in both men and women. Taboos and rituals such as the couvade both restrict and enhance the father's role in many cultures. In the traditional form of the couvade ritual, the father takes to bed during the mother's pregnancy, labor, and delivery as a means of sharing in the experience. The remnant of this ritual in modern cultures is evidenced by the couvade syndrome, in which men experience psychosomatic symptoms during their wives' pregnancies (Trethowan, 1972; Trethowan & Conlon, 1965).

A recent well-controlled epidemiological survey of patients seen by specialists in internal medicine suggests the clinical importance of couvade symptoms (Lipkin & Lamb, 1982). Almost one-quarter of all men whose wives were pregnant complained of nausea, vomiting, anorexia, abdominal pain, or bloating even though a diagnostic evaluation uncovered no objective explanation for these symptoms. These men made twice the number of visits to physicians and received twice the number of medications as their controls, in part because the health provider never asked if the man's wife was pregnant. In other studies, as many as 65% of men complained of physical symptoms, which also included backache and weight gain, and described dietary changes and smoking cessation (Liebenberg, 1973).

During the prenatal period, fathers are now almost routinely present during labor and delivery; this represents a dramatic change in father involvement. In 1972, only 27% of American hospitals allowed fathers to be present during birth. By 1980, the figure had risen to 80%, and it currently approaches 100%. In an interview study of fathers of healthy firstborns in London, fathers' descriptions of their feelings after having witnessed the birth were almost identical to those of mothers: extreme elation, relief that the baby was healthy, feelings of pride and increased self-esteem, and feelings of closeness when the baby opened its eyes (Greenberg & Morris, 1974; Robson & Moss, 1970). When fathers were given the opportunity to touch their babies, they did so in the same sequence as mothers: from fingertips to palms, first on the limbs and then on the trunk (Abbot, 1975), although the fathers took longer before they displayed this progression.

Studies by Parke and Sawin (1975, 1977) of father–newborn interaction in the postpartum period suggest that fathers and mothers are equally active and sensitive to newborn cues at this time. Parke, O'Leary, and West (1972) did similar studies with low-income as well as middle-income families, and in both dyadic (father–infant) and triadic (mother–father–infant) situations, so that the findings are quite generalizable. Fathers and mothers share not only the exhilaration of the perinatal period, but the lows or normative postpartum blues as well. In an interview study of men in the first few weeks postpartum, 62% reported feelings of sadness and disappointment (Zaslow et al., 1981).

In the ensuing weeks, fathers, just like mothers, adjusted their behavior to the slower processing times of their infants. They spoke more slowly, uttered shorter phrases, and used repetition (Phillips & Parke, 1981). When listening to infant cries or responding to infant smiles in a laboratory setting, mothers and fathers did not differ on psychophysiological responsiveness or reported moods, findings that suggest comparable sensitivity to infant cues (Frodi, Lamb, Leavitt, & Donovan, 1978).

Rhythms and Reciprocity: Laboratory Observations

During the first 6 months of life, infants become increasingly social as they begin to smile and vocalize. The social interaction of fathers with their infants of 2 weeks to 6 months of age has been studied during face-to-face play in the laboratory and will now be described as an example of observational research with human infants (Yogman, 1982). Laboratory observations allow the elicitation and study in detail of brief exchanges of expressive communication that may reflect the developing father–infant relationship. The technique has previously been used to characterize mother–infant interaction as a mutually regulated reciprocal process in which both partners rhythmically cycle to a peak of affective involvement and then withdraw (Brazelton, Tronick, Adamson, Als, & Wise, 1975).

Six families with healthy newborns were studied longitudinally from birth to 6 months at weekly intervals. All mothers were primary caretakers. Adults and infants were seen in the laboratory, with two video cameras recording a split-screen image of the dyad. Adults were instructed to "play without using toys." Each infant played for 2 min with the mother, the father and a stranger (order randomized), separated by 30 sec alone to ensure that she or he was alert and comfortable. The strangers were both male and female adults, and varied in their previous experience with infants.

The videotapes were analyzed by using a microbehavioral scoring system to describe the behaviors each participant displayed second by second. The scoring system included categories of infant and adult gaze patterns, facial expressions, and vocalizations, infant limb movements, and adult touching patterns and body position. Each category of behavior was scored separately, using mutually exclusive descriptive codes.

By 80 days of age, infants frowned significantly more of the time with strangers than with either fathers ($p < .05$) or mothers ($p < .05$). No differences existed in the amount of time infants displayed negative facial expressions with mothers compared with fathers or in the amount of time infants displayed positive facial expressions with the three adults (Yogman, 1982). Infants also differed in body movements, depending on the adult interactant. They remained still more often with fathers than with mothers ($p < .05$), and, while the finding was not significant, infants showed a tendency to remain still more often with fathers than with strangers ($p < .1$). Infants by 80 days of age and as young as 6 weeks of age interacted differently with their familiar parents than with unfamiliar strangers, as evidenced by differences in facial expression and limb movements (Dixon et al., 1981; Yogman, 1982).

Because it was hypothesized that the interactive messages and communicative meaning were carried in clusters of substitutable behaviors rather than discrete behaviors, in subsequent analyses the discrete behaviors were clustered into units of analysis called ''monadic phases.'' Monadic phases were derived by clustering discrete behaviors, using a priori decision rules derived in advance and guided by a theory of affect communicated in dyadic messages as either positive, neutral, or negative rather than factor analysis. The choice of phases and rules for clustering was guided by the expectation that each phase would convey an affective message from one partner to the other. For both the infant and the adult, these a priori decision rules were used to translate each second-by-second display into one of the following monadic phases: talk, play, set, elicit, monitor, avert, and protest/avoid. Each of the monadic phases is made up of a set of substitutable second-by-second displays and is not defined by a single behavior.

As displayed for 2 min of infant interaction with the mother and father in Figure 22.2, this analysis showed that infants and parents spent more than 90% of their interaction time in affectively positive phases and that transitions between phases were jointly regulated (indicated by lowercase letters and defined as both partners moving in the same direction within 1 to 2 sec of each other). In contrast to their interaction with parents, with strangers the infants seldom cycled above the neutral phase set, and simultaneous phase transitions did not occur. Infants by 3 months of age successfully interacted with both mothers and fathers with a similar, mutually regulated, reciprocal pattern, as evidenced by transitions between affective levels that occurred simultaneously for infant and parent (Yogman, 1977, 1982).

In a subsequent study, the rhythmicity in an infant's behavior and heart rate during social interaction with the mother, father, and a stranger were investigated to examine the microrhythms present in physiology as well as behavior. Infants and adult behavioral rhythms were synchronous with the father and mother but not with an unfamiliar stranger, but the 3-month-old infant's behavior and heart rate were synchronized during interaction with all three adults (Yogman, Lester, & Hoffman, 1983). These data suggested that the father's and mother's familiarity with their infant enabled them to synchronize their behavioral rhythms with those of the infant, while the stranger could not. In contrast, the relationship between infant measures

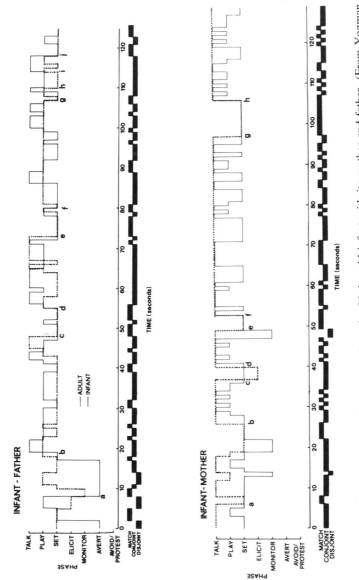

Figure 22.2. Monadic phases during a session of a 96-day-old infant with its mother and father. (From Yogman, 1982; reprinted with permission)

alone (behavior and physiology) remained synchronous with all three adults, perhaps reflecting the infant's intactness and organization. Finally, both mothers and fathers were equally able to involve the infant in games—that is, episodes of repeated adult behavior that engaged the infant's attention (Yogman, 1981).

The monadic phase analysis was also used to look in more detail at the ontogenesis of these patterns of father–infant communication. An example for a single father–infant dyad can be seen in Figure 22.3. The data suggested that as the infant's age increases, the range of the infant's affective displays with the father expands and the infant spends a greater proportion of time in more affectively positive phases. At 1 month of age, only brief peaks of play occur at the very end of the session. By 2 months, the play episodes occur earlier and are longer in duration. By 3 months, more modulated cycling with turn taking occurs, and simultaneous phase transitions are now present. Finally, by 5 months, having mastered social play, the infant now turns away to explore characteristics of the inanimate world (Yogman, 1982). In sum, studies of fathers and infants in the first 6 months of life supported the hypothesis that fathers are competent and capable of skilled and sensitive social interaction with young infants.

Later Attachment

Studies of the father–infant relationship with infants aged 6 to 24 months have focused primarily on the development of attachment as Bowlby (1969) and Ainsworth (1973) have defined it. These studies have asked questions such as, do infants greet, seek proximity with, and protest on separation from fathers as well as mothers? Such studies provide conclusive evidence that infants are attached to fathers as well as to mothers, although a preference for the mother becomes evident in high-stress situations. By 7–8 months of age, when the home environment tends to be relatively low in stress, infants are attached to both mothers and fathers and prefer either parent over a stranger (Lamb, 1977a). During the second year, most studies also show attachment to both mothers and fathers (Clarke-Stewart, 1980; Kotelchuck, 1976; Lamb, 1977b), although in a more stressful laboratory setting, some studies have shown that infants between 12 and 18 months prefer mothers (Cohen & Campos, 1974; Lamb, 1976).

Clinically important and yet not well studied are the influences on father–infant attachment. Pedersen and Robson (1969) report that the father's investment in caretaking, as well as the level of stimulation and play he provides, are positively correlated with the attachment greeting behaviors in 8- to 9-month-old babies. Infants whose fathers participate highly in caregiving show less separation anxiety and cry less with a stranger than infants whose fathers are less involved (Kotelchuck, 1975; Spelke, Zelazo, Kagan, & Kotelchuck, 1973). Qualitative aspects of infant attachment to mothers and fathers studied using Ainsworth's "strange situation" suggested that the relationships are independent and not redundant; that is, infants could develop a secure attachment to the father in spite of an insecure attachment to the mother (Lamb, 1978; Main & Weston, 1981).

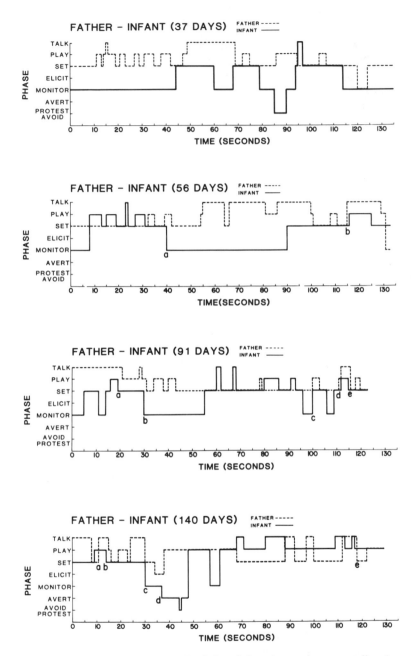

Figure 22.3. Developmental changes in father–infant interaction: monadic phases during sessions with infant aged 1, 2, 3, and 4 months. (From Yogman, 1982; reprinted with permission)

Intervention and Male Involvement

Several educational programs have shown that brief interventions with fathers in the perinatal period could influence their attitudes, caregiving skills, and knowledge of infant capabilities for as long as 3 months (Parke, Hymel, Power, & Tinsley, 1980). In Sweden, fathers who received simple instructions on bathing, changing, and feeding in the perinatal period were found to have higher degrees of infant caretaking activity, as recorded on a maternal questionnaire 6 weeks after discharge (Johannesson, 1969). Demonstrations of neonatal behavior to fathers in the newborn period have been shown to influence paternal involvement for as long as 6 months. In Australia, fathers of full-term and preterm newborns administered and scored a behavioral assessment of their baby and discussed a videotape of the baby. Compared with a control group, the fathers of term infants engaged in more face-to-face play at 6 months, and the fathers of both full-term and preterm infants provided more stimulation in a home observation (Dolby, English, & Warren, 1982). Simple interventions with fathers of older children (instructing them to play with their 12-month-old sons 50 min a day) were also effective. One month later, the infants in the intervention group showed greater degrees of proximity seeking to their fathers in a free play context than to fathers in the nonintervention group (Zelazo, Kotelchuck, Barber, & David, 1977).

In sum, studies of the father–infant relationship in each of these developmental epochs (the prenatal, perinatal, early, and later infancy periods) demonstrate the similarity of the father–infant and mother–infant relationships, the capacities of fathers and infants to interact successfully, and the plasticity of male parental behavior in response to fairly simple interventions.

DIFFERENCES BETWEEN FATHERS AND MOTHERS

While fathers can competently care for their infants in a manner similar to that of mothers, they seldom spend as much time with their infants as mothers do. Furthermore, they are more likely than mothers to play with their infants and to be less involved in feeding and other caretaking tasks.

In natural observations, the mother is clearly the predominant partner for human infants and assumes the primary responsibility for infant care (Clarke-Stewart, 1980). Reports of father involvement vary from a mean of 8 hr per week spent playing with babies and 26 hr per week at home with awake 9-month-old babies (Pedersen & Robson, 1969) to 3.2 hr per day (one-third the time of the mother) spent with the infant (Kotelchuck, 1976) or to 30 min per day spent alone with the infant (one-half the time of the mother) (Pedersen, Rubenstein, & Yarrow, 1979). When one looks at specific activities (feeding, cleaning, playing), fathers also spend much less time in these activities than mothers do. In contrast, a greater proportion of fathers' time with the baby is spent specifically in play (37.5% vs. 25.8%) (Kotelchuck, 1975). With toddlers (20 months of age), fathers spend an average of 3 hr per day in play, although the variability among fathers was extensive (Easterbrooks, 1982).

Play as a Context

In all species in which interactions between adult males and infants occur, play behavior accounts for a substantial portion of those interactions. For example, both in the feral rhesus (Taylor, Teas, Richie, & Shrestha, 1978) and in the laboratory-reared nuclear-family rhesus (Suomi, 1977), over 60% of all adult male interactions with infants involved social play. The proportion of interactions involving play increased with the age of the child.

Conversely, play is much less common among nonhuman primate mothers and their infants (Suomi, 1979). Infants are most likely to be in ventral contact with their mothers, often carried, while ventral contact of infants with adult males is exceedingly infrequent and, when observed, appears clumsy (Parke & Suomi, 1980).

Human Father–Infant Play

Regardless of the amount of time human fathers spend with their infants, they are more likely to be the infant's play partner than the mother. Moreover, fathers' play tends to be more stimulating, vigorous, arousing, and state disruptive for the infant.

In the longitudinal study of six families described earlier, episodes of play called "games" were studied. These were defined as a series of episodes of mutual attention in which the adult used a repeating set of behaviors with only minor variations during each episode of mutual attention (Stern, 1974). In order to describe the components of these games, the tapes were reviewed and episodes that met the definition of games were described in detailed narrative form. Descriptions of these games were then categorized as tactile, limb movement, visual, verbal, and combinations (Yogman, 1982). In studies of infant games during the first 6 months of life, fathers engaged their infants in tactile and limb-movement games with vigorous stimulation. Mothers more commonly played visual games in which they displayed distal motor movements that were observed by the infant (Yogman, 1981).

The visual games most often played by mothers may represent a more distal attention-maintaining form of interactive play than the more arousing proximal, idiosyncratic limb-movement games played more often by fathers.Studies of the games parents play with 8-month-old infants showed similar findings: mothers played more distal games, while fathers engaged in more physical games (Power & Parke, 1979). Stern (1974) has suggested that the goal of such games is to facilitate an optimal level of arousal in the infant in order to foster attention to social signals. The more proximal games of infants and fathers may serve to modulate the infant's attention and arousal in a more accentuated fashion than occurs during the distal games of infants and mothers. These findings are surprisingly robust in that they have been replicated with infants of different ages in different situations.

Fathers of newborns, while similar to mothers in most behaviors, tended to hold and rock their newborns more than mothers (Parke et al., 1972). In a separate interview study conducted after fathers witnessed the birth, fathers emphasized their desire to touch, pick up, move, hold, and play with their newborn and were particularly impressed by the liveliness, reflex activity, and movements of the baby. "When she starts moving I go and pick her up and she starts moving in your hands and your arms and you can feel her moving up against you. It's like a magnet" (Greenberg &

Morris, 1974, p. 525). Differences in interaction between mothers and fathers and their older infants all involved play: fathers engaged in more play than caretaking activities with 6-month-olds (Rendina & Dickersheid, 1976) and more often picked up their infants (8 months old) to play physical, idiosyncratic, rough-and-tumble games. In comparison, mothers were likely to hold infants, engage in caregiving tasks, and either play with toys or use conventional games such as peekaboo (Lamb, 1977a). When their children were 2.5 years old, fathers were better able than mothers to engage children in specific play activities. Father's play with his child was likely to be proximal (as was described for younger infants), physical, and arousing, and fathers reported that they enjoyed it more than mothers (Clarke-Stewart, 1978). Infants at 8 months responded more positively to play with fathers than with mothers (Lamb, 1977a), and by age 2.5 years not only preferred to play with fathers, but were judged to be more involved with and excited by them (Clarke-Stewart, 1978). It is fascinating to note that fathers' physical play with their infants correlates most highly with mothers' verbal stimulation (.89) and toy play (.96) (Clarke-Stewart, 1980). These are maternal behaviors found in other studies to be part of a pattern of "optimal maternal care" (Clarke-Stewart, 1973).

These differences in play and the quality of vigorous stimulation are quite robust and persist even in studies of primary caregiver fathers in the United States (Field, 1978; Yogman, 1982) and in studies of nontraditional fathers taking advantage of paternity leave in Sweden (Frodi, Lamb, Hwang, & Frodi, 1982; Lamb, Frodi, Hwang, Frodi, & Steinberg, 1982). It is interesting to speculate that these play differences may become less tied to gender as the socialization of young children changes. It is important to note that in contrast to these play differences, the performance of caregiving tasks seems more easily modifiable and closer in its relationship to role than to gender.

ANTECEDENTS AND CONSEQUENCES OF MALE INVOLVEMENT

Antecedents of Paternal Involvement

The father's involvement with his infant in influenced by a number of interpersonal, intrapersonal, social, cultural, and economic variables. One of the major influences on the father's involvement is the mother's preference or restrictiveness, as suggested by Redican (1976) regarding male involvement. The mother seems to function much as a gatekeeper, regulating the father's involvement with the infant. In families in which the mother is at work all day, she spends more time with the infant in the evening; this pattern is associated with lower levels of father–infant interaction, at least for the first several months (Pedersen, Zaslow, Suwalsky, & Cain, 1982). The mother's feelings about the father's involvement become one determinant of this gatekeeper function, and they are influenced by her relationship with her own father. Surprisingly, mothers who perceive their own fathers as having had a minimal role tend to be married to men who are highly involved in child care (Radin, 1981). The mother continues to influence the father's relationship with his infant even when he is not home, since the mother conveys a representation of the father in his absence (Atkins, 1981) that influences the father–infant relationship after the father's return.

The marital relationship can also exert a strong influence on the parent–infant relationship. Strong paternal feelings of competition with the spouse have been shown to occur after the birth of an infant, particularly if the father has a close relationship with the baby and the mother is nursing. Husbands of nursing mothers described feelings of inadequacy, envy, and exclusion, and the competition may actually undermine the mother's attempts to breast-feed unless these feelings are addressed (Lerner, 1979; Waletzky, 1979).

The father's positive relationship with his own father was also associated with his high involvement in infant care (Cordell, Parke, & Sawin, 1980); conversely, the father's negative self-esteem was associated with less involvement (Gamble & Belsky, 1982). The most satisfied fathers were those who were more strongly identified with the parenting role than the average person (Dickie, Van Gent, Hoogerwerf, Martinez, & Dieterman, 1981).

Cultural practices, economic constraints, and employment policies such as paternity leave and flexible work schedules have major influences on paternal involvement. Stresses such as job loss may be associated with paternal depression, and although unemployment may result in additional free time for these men, they are less likely to assume infant-care responsibilities (Cordell et al., 1980).

Consequences of Paternal Involvement

The influence of the father–infant relationship on the later cognitive, social, and emotional development of the infant has not been well studied. One report suggests that at least for males, increased involvement of the father at home is associated with greater social responsiveness of the infant at 5 months of age during a Bayley test (Pedersen et al., 1979), which is used to assess psychological and motor development from 3 months to about 2 years of age. Concurrent predictions of infant Bayley scores at 16 and 22 months were related to the father's positive perceptions of the child and to his ability to engage the child in play and to anticipate independence on the part of the child. Predictions of concurrent social competence were related to the father's verbal and playful behavior and, for girls, his expectation of independence (Clarke-Stewart, 1978, 1980). Boys, in particular, have been found to be more autonomous when both parents are warm and affectionate (Baumrind & Black, 1967). The use of cross-lagged correlations led Clarke-Stewart (1980) to suggest that the mother's warmth, verbal stimulation, and play with toys with infants at 15 months of age were related to higher infant Bayley scores at 30 months, which, in turn, influenced fathers to engage in play more often, to expect more independence, and to perceive their children more positively. All these studies were done with healthy, full-term infants.

In a longitudinal study of fathers' relationships with 20 preterm infants from birth to 18 months, father–child play (games) at 5 months was significantly correlated with Bayley Mental Development Index (MDI) scores at 18 months (Table 22.1) (Yogman, 1984). These correlations were as high as or higher than those for games with the mother alone. Furthermore, considering the impact of arousing and nonarousing games separately, it is the frequency of arousing games that better predicts the Bayley MDI outcome at 18 months.

While the consequences of male adult involvement for nonhuman primate infants have seldom been studied, the infants in troops of Barbary macaques where

Table 22.1   Father–infant games and premature infant outcome

	Bayley 9		Bayley 18	
	MDI	PDI	MDI	PDI
Number of all games	.61	.43	.77*	.59
With father	.45	.16	.79*	.15
With mother	.59	.45	.68	.63
Number of arousing games	.50	.15	.80*	.28
Number of nonarousing games	.55	.43	.44	.62
Duration of all games	.70*	.61	.85*	.72*
Duration of arousing games	.45	.03	.77*	.15
Duration of nonarousing games	.67*	.62	.40	.80*

*$p \leq .05$.

Source: From Yogman (1984). Reprinted with permission.

male adults teach walking during the first week showed an advantage in motor skills by 1.5 months of age compared with troops where male teaching was low (Burton, 1972; Redican & Taub, 1981).

Fathers also seem to exert a prominent influence on early sex typing. Right after birth, fathers rate their male newborns, as evidenced by interviews on day 1, as firmer, more alert, stronger, and hardier, while they rate their female newborns as softer, finer-featured, and more delicate (Rubin, Provenzano, & Luria, 1974). Fathers also vocalize, touch, and stimulate their newborn sons more than their daughters; this is especially true for firstborn sons (Parke & Sawin, 1975). They spend more time watching their infant sons (Rendina & Dickersheid, 1976), and play about 30 min longer daily with firstborn sons than firstborn daughters (Kotelchuck, 1976). Fathers seem to play a more important role in the sex typing of their infants; fathers more than mothers interact differently with their male and female infants.

Nonhuman primate males also seem to be more interested in and involved with male infants. Male rhesus and stumptail macaques have been observed, both in the wild and in the laboratory, to provide more frequent care to and play more with male infants (Redican & Taub, 1981). While male sex differentiation and a preference for male infants are more common, social structure and ecological conditions may influence the pattern, since adult male Hamadryas baboons show a preference for female infants (Kummer, 1967), in contrast to Chacma, Olive, and other baboons (Rowell, Din, & Omar, 1968).

CONCLUSION: A UNIQUE ROLE FOR FATHERS?

These studies of fathers and infants are beginning to form a basis for theorizing about the unique nature of paternal behavior and the father–infant relationship.

The fact that both parents can have a direct, sensitive, and responsive relationship with their infant suggests that highly invested human adults are capable of forming a consistent, loving relationship sensitive to an infant's emotional state and immature attentional capacities. The fact that either parent can learn to perform certain caretaking tasks may be a reflection of both the infant's eliciting capabilities and the human adult's capability to learn. As Rossi (1977) points out, however, while what men and women do is not entirely genetically determined, the biological contributions shape what is learned, and there are differences in the ease with which the two sexes learn certain things.

Therefore, it seems unlikely that the relationships of fathers and mothers with their infants are identical and redundant. The biological fact of pregnancy, the 9-month period of physical symbiosis, sets the stage for an intimacy between mother and baby from birth and a facilitated process of entrainment that a father can develop only over a period of time. While there is little evidence that human systems are unmodifiable, the father is likely to require more learning in order to achieve a level of intimacy with his baby comparable with that of the mother. In the studies described in this chapter, fathers were able to achieve mutually regulated, reciprocal interactions with their infants by 3 months of age, but the father offered the infant a qualitatively different experience than the mother.

Most research has applied to the understanding of fathering paradigms useful for the understanding of mothering. It has used constructs such as attachment and measures such as separation anxiety to assess whether fathers can be good enough mothers. The unique aspects of father–infant play and the relationship to infant arousal have not yet been adequately accounted for in theoretical constructs and empirical measures of fathering. While proximity to the infant may be most salient for mothering, fathers may add flexibility to the infant's social experience by approaching (moving in) and withdrawing (leaving). Sigel (1980) suggests that such relative distancing may enhance symbolization by requiring that the infant construct a label for the absent figure. Bischof (1975) speculates that the infant has an appetance not only for security, which can be satisfied by a secure attachment relationship, but also for arousal. The infant may be predisposed to seek simultaneously two very different kinds of stimulation: soothing, nonarousing, integrative stimuli as well as exciting and arousing stimuli. It is certainly intriguing that these opposite stimulus qualities are related to but not necessarily identical with the differences in behavior that have been described for male and female parents.

It may be useful to conceptualize the consequences of the father–infant relationship along two dimensions: (1) relative distance from the infant and (2) the quality of direct proximal interactions. Some have suggested that the father–infant relationship be characterized as an affinity system that emphasizes protection and provisioning of the young rather than proximity, as in an attachment system (Bailey, 1982). Perhaps the father's relative distancing and approach–withdrawal behavior, together with protection and provisioning, are complementary to the mother's continual availability for proximity and contact. In a parallel way, the direct proximal interactions of fathers and mothers with their infants may also be complementary: arousing versus soothing. The paucity of available data makes it abundantly clear that the study of the unique dimensions of male parental behavior is still in its scientific infancy.

REFERENCES

Abbot, L. (1975). *Paternal touch of the newborn: Its role in paternal attachment.* Unpublished thesis, Boston University School of Nursing, Boston.

Ainsworth, M. D. S. (1973). The development of infant–mother attachment. In B. Caldwell & H. Ricciuti (Eds.), *Review of child development research* (Vol. 3, pp. 1–94). Chicago: University of Chicago Press.

Alexander, B. K. (1970). Paternal behavior of adult male Japanese monkeys. *Behavior, 36,* 270–285.

Atkins, R. (1981). *Discovering daddy: The mother's contributions to father-representations.* Paper presented at the meeting of the American Psychiatric Association, New Orleans.

Bailey, W. T. (1982). *Affinity: An ethological theory of the infant–father relationship.* Paper presented to the International Conference on Infant Studies, Austin, TX.

Baumrind, D., & Black, A. E. (1967). Socialization practices associated with dimensions of competence in preschool boys and girls. *Child Development, 38,* 291–327.

Bibring, G. (1959). Some considerations of the psychological processes in pregnancy. *Psychoanalytic Study of the Child, 14,* 113–121.

Bischof, N. (1975). A systems approach toward the functional connections of fear and attachment. *Child Development, 46,* 801–817.

Blurton-Jones, N. G. (1976). Growing points in human ethology: Another link between ethology and the social sciences? In P. G. Bateson & R. A. Hinde (eds.), *Growing points in ethology* (pp. 427–450). Cambridge: Cambridge University Press.

Bowlby, J. (1969). *Attachment and loss* (Vol. 1). New York: Basic Books.

Brazelton, T. B., Tronick, E., Adamson L., Als, H., & Wise, S. (195). Early mother–infant reciprocity. In R. Hinde (Ed.), *Parent–infant interaction.* Ciba Foundation symposium No. 33. Amsterdam: Elsevier.

Burton, F. D. (1972). The integration of biology and behavior in the socialization of *Macaca sylvana* of Gibraltar. In F. Poirier (Ed.), *Primate socialization* (pp. 29–62). New York: Random House.

Chivers, D. J. (1972). The Siamang and the gibbon in the Malay peninsula. *Gibbon and Siamang, 1,* 103–135.

Clarke-Stewart, K. A. (1973). Interactions between mothers and their young children. *Monographs of the Society for Research in Child Development, 38,* 1–109.

Clarke-Stewart, K. A. (1978). And daddy makes three: The father's impact on mother and child. *Child Development, 49,* 466–478.

Clarke-Stewart, K. A. (1980). The father's contribution to children's cognitive and social development in early childhood. In F. A. Pedersen (Ed.), *The father–infant relationship: Observational studies in a family setting* (pp. 111–146). New York: Holt, Rinehart and Winston.

Cohen, L. J., Campos, J. J. (1974). Father, mother, and stranger as elicitors of attachment behaviors in infancy. *Developmental Psychology, 10,* 146–154.

Cordell, A. S., Parke, R. D., & Sawin, D. B. (1980). Father's views on fatherhood with special referency to infancy. *Family Relations, 29,* 331–338.

Dickie, J., Van Gent, R., Hoogerwerf, E., Martinez, I., & Dieterman, B. (1981). *Mother–father–infant triad: Who affects whose satisfaction?* Paper presented at the biennial meeting of the Society for Research in Child Development, Boston.

Dixon, S., Yogman, M. W., Tronick, E., Als, H., Adamson, L., & Brazelton, T. B. (1981). Early social interaction of infants with parents and strangers. *Journal of the American Academy of Child Psychiatry, 20,* 21–52.

Dixson, A. F., & George, L. (1982). Prolactin and parental behaviour in a male New World primate. *Nature, 299,* 551–553.

Dolby, R., English, B., & Warren, B. (1982). *Brazelton demonstrations for mothers and fathers.* Paper presented to the International Conference on Infant Studies, Austin, TX.

Easterbrooks, M. A. (1982). *Father involvement, parenting characteristics and toddler development*. Paper presented to the International Conference on Infant Studies, Austin, TX.

Field, T. M. (1978). Interaction behaviors of primary versus secondary caretaker fathers. *Developmental Psychology, 14*, 183–184.

Frodi, A. M., Lamb, M., Hwang, C., & Frodi, M. (1982). *Increased paternal involvement and family relationships*. Paper presented to the International Conference on Infant Studies, Austin, TX.

Frodi, A. M., Lamb, M., Levitt, L., & Donovan, W. L. (1978). Fathers' and mothers' responses to infant smiles and cries. *Infant Behavior and Development, 1*, 187–198.

Gamble, W. C., & Belsky, J. (1982). *The determinants of parenting within a family context: A preliminary analysis*. Paper presented to the International Conference on Infant Studies, Austin, TX.

Greenberg, M., & Morris, N. (1974). Engrossment: The newborn's impact upon the father. *American Journal of Orthopsychiatry, 44*, 520–531.

Gurwitt, A. R. (1976). Aspects of prospective fatherhood. *Psychoanalytic Study of the Child, 31*, 237–271.

Hampton, J. K., Hampton, S. H., & Landwehr, B. T. (1966). Observations on a successful breeding colony of the marmoset, *Oedipomidas Oedipus*. *Folia Primatologica, 4*, 265–287.

Howells, J. G. (1969). Fathering. In J. G. Howells (Ed.), *Modern perspectives in international child psychiatry* (pp. 125–156). Edinburgh: Oliver & Boyd.

Hrdy, S. B. L. (1976). Care and exploitation of nonhuman primate infants by conspecifics other than mothers. In J. R. Rosenblatt, R. A. Hinde, E. Shaw, & C. Beer (Eds.), *Advances in the study of Behavior* (Vol. 6, pp. 101–158). New York: Academic Press.

Itani, J. (1959). Paternal care in the wild Japanese monkey, *Macaca fuscata*. *Primates, 2*, 61–93 (Reprinted in English in C. Southwick (Ed.), *Primate social behavior*. Princeton, NJ: Van Nostrand, 1963).

Johannesson, P. (1969). *Instruction in child care for fathers*. Unpublished dissertation, University of Stockholm.

Kotelchuck, M. (1975). *Father-caretaking characteristics and their influence in infant–father interaction*. Paper presented at the meeting of the American Psychological Association, Chicago.

Kotelchuck, M. (1976). The infant's relationship to the father: Experimental evidence. In M. Lamb (Ed.), *The role of the father in child development* (pp. 329–344). New York: Wiley.

Kummer, H. (1967). Tripartite relations in hamadryas baboons. In S. A. Altmann (Ed.), *Social communication among primates* (pp. 63–71). Chicago: University of Chicago Press.

Lamb, M. E. (1976). Twelve-month-olds and their parents: Interaction in a laboratory playroom. *Developmental Psychology, 12*, 237–244.

Lamb, M. E. (1977a). Father–infant and mother–infant interaction in the first year of life. *Child Development, 48*, 167–181.

Lamb, M. E. (1977b). The development of mother–infant and father–infant attachments in the second year of life. *Developmental Psychology, 13*, 637–648.

Lamb, M.E. (1978). Qualitative aspects of mother– and father–infant attachments. *Infant Behavior and Development, 1*, 265–275.

Lamb, M., Frodi, A., Hwang, C., Frodi, M., & Steinberg, J. (1982). Mother– and father–infant interaction involving play and holding in traditional and nontraditional Swedish families. *Developmental Psychology, 18*, 215–221.

Lerner, H. (1979). Effects of the nursing mother–infant dyad on the family. *American Journal of Orthopsychiatry, 49,* 339–348.

Liebenberg, B. (1973). Expectant fathers. In P. Shereshfsky & L. Yarrow (Eds.), *Psychological aspects of a first pregnancy and early postnatal adaptation* (pp. 103–114). New York: Raven Press.

Lipkin, M., & Lamb, G. S. (1982). The couvade syndrome: An epidemiologic study. *Annals of Internal Medicine, 96,* 509–511.

Main, M., & Weston, D. R. (1981). The quality of the toddler's relationship to mother and to father. *Child Development, 52,* 932–940.

Parke, R. D., Hymel, S., Power, T., & Tinsley, B. (1980). Fathers and risk: A hospital based model of intervention. In D. B. Sawin, R. C. Hawkes, L. O. Walker, & J. H. Penticuff (Eds.), *Psychosocial risks in infant environment transactions* (Vol. 4). New York: Bruner Mazel.

Parke, R. D., O'Leary, S. E., & West, S. (1972). Mother–infant–newborn interaction: Effects of maternal medication, labor, and sex of infant. *Proceedings of the American Psychological Association,* 85–86.

Parke, R. D., & Sawin, D. (1975). *Infant characteristics and behavior as elicitors of maternal and paternal responsivity in the newborn period.* Paper presented to the Society for Research in Child Development, Denver.

Parke, R. D., & Sawin, D. (1977). *The family in early infancy: Social interactional and attitudinal analyses.* Paper presented to the Society for Research in Child Development, New Orleans.

Parke, R. D., & Suomi, S. (1980). Adult male–infant relationships: Human and nonhuman primate evidence. In K. Immelmann, G. Barlow, M. Main, & L. Petrinovich (Eds.), *Behavioral development: The Bielefeld Interdisciplinary Project* (pp. 1–35). Cambridge: Cambridge University Press.

Pedersen, F. A., & Robson, K. S. (1969). Father participation in infancy. *American Journal of Orthopsychiatry, 39,* 466–472.

Pedersen, F. A., Rubenstein, J., & Yarrow, L. J. (1979). Infant development in father-absent families. *Journal of Genetic Psychology, 135,* 51–61.

Pedersen, F., Zaslow, M., Suwalsky, J., & Cain, R. (1982). *Infant experience in traditional and dual wage-earner families.* Paper presented to the International Conference on Infant Studies, Austin, TX.

Phillips, D. M., & Parke, R. D. (1981). *Parental speech to prelinguistic infants.* Unpublished manuscript, University of Illinois.

Powdermaker, H. (1933). *Life in Lesu.* New York: Norton.

Power, T.G., & Parke, R. D. (1979). *Toward a taxonomy of father–infant and mother–infant play patterns.* Paper presented to the Society for Research in Child Development, San Francisco.

Radin, N. (1981). Child rearing in intact families: I. *Merrill Palmer Quarterly, 27,* 489–514.

Redican, W. K. (1976). Adult male–infant interactions in nonhuman primates. In M. E. Lamb (Ed.), *The role of the father in child development* (pp. 345–385). New York: Wiley.

Redican, W. K., & Taub, D. M. (1981). Male parental care in monkeys and apes. In M. E. Lamb (Ed.), *The role of the father in child development* (pp. 203–258). New York: Wiley.

Rendina, I., & Dickerscheid, J. D. (1976). Father involvement with first-born infants. *Family Coordinator, 25,* 373–379.

Robson, K., & Moss, H. (1970). Patterns and determinants of maternal attachment. *Journal of Pediatrics, 77,* 976–985.

Ross, J. M. (1975). The development of paternal identity: A critical review of the literature on nurturance and generativity in boys and men. *Journal of the American Psycho-analytic Association, 23*, 783–817.

Rossi, A. (1977). A biosocial perspective on parenting. *Daedalus, 106*, 1–32.

Rowell, T. E., Din, N. A., & Omar, A. (1968). The social development of baboons in their first three months. *Journal of Zoology, 155*, 461–483.

Rubin, J. Z., Provenzano, F. J., & Luria, Z. (1974). The eye of the beholder: Parent's views on sex of newborns. *American Journal of Orthopsychiatry, 44*, 512–519.

Rypma, C. B. (1976). The biological bases of the paternal responses. *Family Coordinator, 25*, 335–341.

Sigel, I. (1980). The distancing hypothesis: A causal hypothesis for the acquisition of repre-sentation thought. In M. Jones (Ed.), *Miami symposium on the prediction of behav-ior. Effects of early experiences*. Coral Gables, FL: University of Miami Press.

Spelke, E., Zelazo, P., Kagan, J., & Kotelchuck, M. (1973). Father interaction and separation protest. *Developmental Psychology, 9*, 83–90.

Stein, D. M. (1984). *The sociobiology of infant and adult male baboons*. Norwood, NJ: Ablex.

Stern, D. (1974). The goal and structure of mother–infant play. *Journal of the American Academy of Child Psychiatry, 13*, 402–421.

Suomi, S. J. (1977). Adult male–infant interactions among monkeys living in nuclear families. *Child Development, 48*, 1215–1270.

Suomi, S. J. (1979). Differential development of various social relationships by rhesus mon-key infants. In M. Lewis & L.A. Rosenblum (Eds.), *The child and its family* (pp. 219–244). New York: Plenum Press.

Taylor, H. J., Teas, T., Richie, C. S., & Shrestha, R. (1978). Social interactions between adult male and infant rhesus monkeys in Nepal. *Primates, 19*, 343–351.

Trethowan, W. H. (1972). The couvade syndrome. In J. Howells (Ed.), *Modern perspectives in psycho-obstetrics* (pp. 67–93). Edinburgh: Oliver & Boyd.

Trethowan, W., & Conlon, M. F. (1965). The couvade syndrome. *Journal of British Psychia-try, 111*, 57–66.

Trivers, R. L. (1972). Parental investment and sexual selection. In B. Campbell (Ed.), *Sexual selection and the descent of man, 1871–1971* (pp. 136–179). Chicago: Aldine.

Waletzky, L. (1979). Husband's problems with breast-feeding. *American Journal of Ortho-psychiatry, 49*, 349–352.

West, M. M., & Konner, M. J. (1981). The role of the father: An anthropological perspective. In M. E. Lamb (Ed.), *The role of the father in child development* (pp. 185–216). New York: Wiley.

Whiting, B., & Whiting, J. (1975). *Children of six cultures*. Cambridge: Cambridge Univer-sity Press.

Wilson, E. O. (1980). *Sociobiology: The abridged edition*. Cambridge: Belknap Press.

Yogman, M. W. (1977). *The goals and structure of face-to-face interaction between infants and fathers*. Paper presented to the Society for Research in Child Development, New Orleans.

Yogman, M. W. (1981). Games fathers and mothers play with their infants. *Infant Mental Health Journal, 2*, 241–248.

Yogman, M. W. (1982). Development of the father–infant relationship. In H. Fitzgerald, B. Lester, & M. W. Yogman (Eds.), *Theory and research in behavioral pediatrics* (Vol. I, pp. 221–279). New York: Plenum Press.

Yogman, M.W. (1984). The father's role with preterm and full-term infants. In J. Call, E. Galenson, & R. Tyson (Eds.), *Frontiers in infant psychiatry* (Vol. 2, pp. 363–374). New York: Basic Books.

Yogman, M., Lester, B., & Hoffman, J. (1983). Behavioral and cardiac rhythmicity during mother–father–stranger–infant social interaction. *Pediatric Research, 17*, 872–876.

Zaslow, M., Pedersen, R., Kramer, E., Cain, R., Suwalsky, J., & Fivel, M. (1981). *Depressed mood in new fathers*. Paper presented to the Biennial Meeting of the Society for Research in Child Development, Boston.
Zelazo, P. R., Kotelchuck, M., Barber, L., & David, J. (1977). *Fathers and sons: An experimental facilitation of attachment behaviors*. Paper presented to the Society for Research in Child Development, New Orleans.

V
FUTURE DIRECTIONS

23

Future Directions in Research on Mammalian Parenting

NORMAN A. KRASNEGOR AND ROBERT S. BRIDGES

The preceding chapters attest to the substantial knowledge that has been acquired on biological and behavioral mechanisms underlying parenting behaviors in mammals. The common themes highlight both the commonalities in parenting among mammals and the opportunities for guiding future research. Among the topics that appear to be fruitful are functional studies of brain sexual dimorphisms, studies of sensory cues in parenting, studies of the role of experience in parenting, studies of hormonal mechanisms, studies of mechanisms that maintain parenting, and studies of human parenting.

BRAIN SEXUAL DIMORPHISMS

Roger Gorski's chapter provides ample evidence that certain structures (sexually dimorphic nucleus of the preoptic area) of the rat brain are different in males and females. Might it also be possible that functional differences between male and female parenting behaviors relate in part to different brain structures in the sexes? The work of Alberts and Gubernick on biparental care in the California mouse *(Peromyscus californicus)* provides exciting evidence of not only differences, but also changes in the male's brain when he becomes a parent. Their findings raise the possibility of brain plasticity as a result of male parenting in species in which both parents care for young. The evidence, while preliminary, suggests a whole series of CNS/behavior studies that these and other investigators will no doubt pursue. Researchers should also investigate mammals other than the California mouse in which triadic systems of care are the norm to determine whether the generality of the observations made about brain plasticity holds or whether this is a unique adaptation of the California mouse.

SENSORY CUES

An important theme in many of these chapters is that sensory factors, particularly olfaction, appear to be essential for the onset of parenting in rodents and sheep. Odors act as triggers to initiate caregiving and apparently help to stimulate the secretion of hormones involved in aspects of mammalian parenting. The olfactory system is, from the evolutionary point of view, part of the "old brain." Some neuroscientists have even argued that the olfactory system is the anlage for the cortex in mammals. It has also been shown that newborn mammals are heavily dependent on olfactory cues for adapting to their postnatal environment and becoming attached to their caregiver. Other evidence strongly suggests that the sense of smell may change in females as the time for parturition nears. Such a change may make the odor of the newborn more salient and perhaps less aversive. More research is needed to elucidate how such putative changes occur, as well as the neurochemical, neurophysiological, and behavioral (learning) mechanisms that underpin this aspect of parenting.

EXPERIENCE

Reproductive and parental experiences have been identified as potent determinants regulating the expression of caregiving in general and maternal behavior in particular. In rodents, sheep, nonhuman primates, and possibly humans, multiparous females have been shown to become maternal more readily as a result of biological and behavioral manipulations or to exhibit greater responsivity to young. Little is known about the biological bases for these changes in mammals, even though most parous females are multiparous rather than primiparous. The roles of biological, experiential, and environmental factors, each of which can modulate the expression of parenting, have not been studied in depth. An important question for future research is whether and how prior parity produces long-term changes in the neurochemical and neuroanatomical substrates of parenting behavior. The work of Modney and Hatton (see chapter 15) elegantly demonstrates how parturition modifies neuroanatomical substrates associated with lactation. Similar studies of the neuroplasticity of brain regions directly regulating maternal behavior should provide information relevant to the study of both parenting and the neurosciences. Additional research should also focus on the developmental aspects of parenting, as outlined by Brunelli and Hofer in this book. They have suggested that parenting behavior may begin around the time of puberty and may relate to play behaviors observed in rats of that age. This aspect of social development deserves further study to delineate the antecedent behaviors that may be necessary for the expression of parenting during adulthood. The work of Fleming presented here also points to a significant new direction for studies of the role of experience. Her innovative work on parenting is helping to link data on human caregiving with the animal literature. Research of this type should be vigorously pursued.

HORMONAL MECHANISMS

Remarkable progress has been made during the past two decades in elucidating the role of hormones during pregnancy, at parturition, and during lactation in the initiation and regulation of parenting in mammals. Research on the gonadal hormones, oxytocin, and prolactin has been particularly productive. Studies of these hormones have demonstrated their roles in triggering the complex pattern of maternal behavior in both sheep and rats. Further work on the hormonal bases of maternal behavior should examine how hormones interact with an array of neuroactive substances, such as the neuropeptides and catecholamines (see chapters 4 and 6). These substances may mediate the effects of hormones or help to regulate their action. Research on the endogenous opioids should therefore be of high priority. Bridges (see chapter 6) has speculated that opioids might alter the olfactory system or have an effect on the thermoregulatory system (see chapter 19). The sensory effects could have a general impact on parenting, since selective changes in pup odor could alter the threshold for pup acceptance by the mother. If opioids regulate thermoregulatory mechanisms, they could influence a particular aspect of maternal behavior in rats—the number and duration of contact bouts between mother and litter during lactation. Comparative studies should also be carried out to determine which hormones and neurochemicals are commonly involved in the regulation of parenting across the mammalian species studied, and the similarities and differences in their mechanisms of action at the neural and behavioral levels.

MECHANISMS THAT MAINTAIN PARENTING

The focus of this book is primarily on the biological and behavioral mechanisms involved in the initiation of caregiving during pregnancy, at birth, and during lactation. Little has been said about factors that underpin the *maintenance* of parenting over the life span of parent(s) and offspring. As was pointed out earlier, one evolutionary strategy for survival is the parental trade-off of reducing the metabolic cost of producing fewer young against the investment of time and energy in sustaining offspring. The altricial mammals, in general, are good exemplars of this evolutionary mechanism. Among the mammals, nonhuman primates and humans probably invest the most time and energy in the long-term care and sustenance of their relatively few offspring. What are the biological and behavioral factors that maintain caregiving beyond the initiation phase? Research on this aspect of parenting is essential to gain an understanding of the development of such patterns of support across an organism's life span and within its natural ecology.

HUMAN PARENTING

The study of the biological bases of human parental behavior is in its infancy. Animal studies now provide us with an understanding of those factors that regulate parental behavior in other mammals. To what extent do these or similar biological systems influence parenting capacity and performance in women and men over the course of their development? Asking questions such as this may allow us to address some of the serious health concerns related to parenting that confront our society. For example, what are the factors that underlie poor parenting in drug-addicted mothers? What are the relative contributions of social, biological, and economic factors in parental care displayed by teenage mothers? The major challenge facing researchers in the field of mammalian parenting during the next decade is to determine the extent of involvement of biological factors in the regulation of parenting in humans and the way these factors interact during development with experiential and genetic events to shape the responsiveness of mothers and fathers toward their children. The primary objective of studies that examine the biological regulation of human parenting would be to apply new research knowledge toward improving the health, stability, and well-being of the parent–child relationship.

In summary, research on parenting is a richly diverse field of scientific inquiry. We hope that this book exhibits the gains in knowledge made thus far and the vital questions that have yet to be answered.

Index